Staphylococcus

Molecular Genetics

Edited by Jodi A. Lindsay

Caister Academic Press

Caister Academic Press
Norfolk, UK

www.caister.com

British Library Cataloguing-in-Publication Data
A catalogue record for this book is available from the British Library

ISBN: 978-1-904455-29-5

Printed and bound in Great Britain

Cover images courtesy of Jodi Lindsay, Matt Holden and Michael Otto

Contents

Contributors

Mark C. Enright
Department of Infectious Disease Epidemiology
St. Mary's Hospital Campus
Imperial College London
London
UK

m.c.enright@imperial.ac.uk

J. Ross Fitzgerald
Centre for Infectious Diseases
University of Edinburgh
Edinburgh
UK

Ross.Fitzgerald@ed.ac.uk

Bénédicte Fournier
Department of Pathology
Emory School of Medicine
Atlanta, GA
USA

bfourni@emory.edu

Patrice Francois
Genomic Research Laboratory
Geneva University Hospitals
Geneva
Switzerland

patrice.francois@genomic.ch

Matthew T. G. Holden
The Wellcome Trust Sanger Institute
Cambridge
UK

mh3@sanger.ac.uk

Malcolm J. Horsburgh
Bacterial Pathogenesis Group
Division of Integrative Biology
School of Biological Sciences
University of Liverpool
Liverpool
UK

M.J.Horsburgh@liverpool.ac.uk

Jodi A. Lindsay
Centre for Infection
Department of Cellular and Molecular Medicine
St George's, University of London
London
UK

jlindsay@sgul.ac.uk

Peter J. McNamara
Department of Medical Microbiology & Immunology
University of Wisconsin
School of Medicine and Public Health
Madison, WI
USA

pjmcnamara@wisc.edu

Michael Otto
Rocky Mountain Laboratories
Laboratory of Human Bacterial Pathogenesis
National Institute of Allergy and Infectious Diseases
The National Institutes of Health
Hamilton, MT
USA

MOtto@niaid.nih.gov

José R. Penadés
Centro de Investigación y Tecnología Animal
Instituto Valenciano de Investigaciones Agrarias (CITA-IVIA)
Castellón
Spain

jpenades@ivia.es

Mariana G. Pinho
Laboratory of Bacterial Cell Biology
Instituto de Tecnologia Química e Biológica (ITQB)
Oeiras
Portugal

mgpinho@itqb.unl.pt

Shu Yeong Queck
Rocky Mountain Laboratories
Laboratory of Human Bacterial Pathogenesis
National Institute of Allergy and Infectious Diseases
The National Institutes of Health
Hamilton, MT
USA

quecks@niaid.nih.gov

Jacques Schrenzel
Genomic Research Laboratory
and Clinical Microbiology Laboratory
Geneva University Hospitals
Geneva
Switzerland

jacques.schrenzel@genomic.ch

Preface

Introduction

The aim of this book is to provide an overview of the major areas of staphylococcal genetics research in recent years, and to indicate the likely future research areas. We have endeavoured to include information to interest the both the specialist reader trying to cover a large field, as well as those new to bacterial genetics. In particular, we have explained many technologies and techniques and defined many specialist terms.

The staphylococci are important human and animal pathogens. In the past, the staphylococi were treated as 'model' pathogens and their biochemistry, ecology, serology, virulence and antibiotic susceptibility was widely studied (Elek, 1959). However, with the emergence of genetic technologies, it was quickly discovered that staphylococci were difficult to genetically manipulate. In a subsequent era of bacterial research dominanted by genetic technology, other organisms such as *Eschericia coli* and its pathogenic relatives became the dominant 'model' organisms. But things are changing. This is due to improved genetic technologies, as well as an increased urgency to understand and prevent infections caused by a new generation of resistant and pathogenic staphylococci.

Although this is a book about genetics, much genetic research on staphylococci is performed because of the clinical importance of this organism. Therefore, it is important to understand the complex relationships the staphylococci have with their mammalian hosts. More than thirty species of staphylococci have been described, with about ten colonizing the skin and mucous membranes of humans. By far the major human pathogen is *S. aureus*, which normally resides in the nose, but also the throat, armpit, groin and intestinal tract. Approximately 25% of healthy humans carry *S. aureus* in their nose all of the time, while another 50% carry *S. aureus* intermittently. *S. aureus* is a common cause of minor skin infections, particularly when introduced into a wound or skin incision. The typical immune response is pus formation (pyogenesis), with some inflammation and 'reddening'. Most infections do not require treatment and will resolve on their own after a few days, and this is due to a successful immune response mediated by polymorphonuclear lymphocytes. *S. aureus* can also cause toxin-mediated disease, such as toxic shock sydrome associated with surgery or tampon use, acute-onset food poisoning and scalded skin syndrome in babies.

S. aureus is the most common cause of hospital acquired infection. *S. aureus* enter at surgical incision sites or at vascular access sites such as catheters and injection sites. Patients who are immunocompromised and elderly are at most risk, probably as they have greater trouble mounting a routine immune response to *S. aureus*. However, young and healthy patients can also be affected. The bacteria often cause localized infection at the skin entry site, but can also develop into systemic infections affecting multiple tissues such as blood (bacteraemia), heart (endocarditis), lungs (pneumonia), muscle (abscess), bone (osteomyelitis), eye (conjunctivitis), and joints (bacterial arthritis). The infection may develop rapidly (acute) or may be long and drawn out (chronic); most will be minor but others will be severe, even resulting in death. Some infections do not require treatment, others respond well to treatment, and some respond poorly to treatment. In the UK, it is estimated that around 2% of surgical incisions become infected with *S. aureus*, and nearly 20,000 cases of bacteraemia occur every year.

The situation in hospitals has changed quite dramatically in the last 15 years due to the evolution of *S. aureus* that are resistant to the most useful antibiotics, the beta-lactmase resistant penicillins such as flucloxacillin, oxacillin, and methicillin. These methicillin-resistant *S. aureus* (MRSA) account for 40–80% of all hospital *S. aureus* in some countries. There is some evidence that these strains haven't replaced other *S. aureus* but represent addition infection numbers. Prophylaxis (prevention of infection using antibiotics prior to surgery) is more likely to fail, and so is treatment. The only remaining class of antibiotics that *S. aureus* is resistant to are the glycopeptides such as vancomycin. These antibiotics are more expensive, must be administered intravenously requiring patient hospitalization, have more side-effects and are less effective. However the first fully resistant VRSA are now being described. New antibiotics have recently been licensed, but they are significantly more expensive and resistance is already reported.

Outside of hospitals, MRSA have also evolved. These CA-MRSA have been described since the late 1990s, but their incidence seems to be increasing dramatically. CA-MRSA cause infections in healthy patients, particularly those in close communities such as athletic teams, the armed forces, prisons, men who have sex with men, and schools. The most common infections are severe skin and soft tissue infections, but rare and fatal necrotizing pneumonia in children is also described. The infections do not respond to GP-prescribed antibiotics and have resulted in dramatic increases in accident and emergency admissions, particularly in the USA.

S. aureus is an important cause of bovine mastitis, leading to economic losses in the dairy industry. MRSA incidence is domestic pets (dogs, cats, horses) is on the increase leading to changes in veterinary practice and potential public health issues for human contacts. MRSA in pigs and pig farmers has also recently been described.

S. epidermidis is the most common staphylococcus found on human skin, and is a common cause of infection associated with prosthetic devices. Other species of staphylococci such as *S. haemolyticus*, *S. hominis*, *S. capitis* can also cause infections. *S. saprophyticus* is the second most common cause of cystitis (urinary tract infection) in women of child-bearing age. *S. intermedius* is a common cause of infection in dogs.

So what is staphylococcal genetics teaching us? To begin with, whole genome technologies such as sequencing projects and microarrays have shown there is an enormous variety of *S. aureus* strains. They each contain different combinations of surface proteins that interact with host as well as different toxins. Teasing apart these differences and relating them to carriage and pathogenic behaviour is becoming one of the major staphylococcal research areas (Chapter 1).

The development of robust typing methods has enabled the spread of different lineages of *S. aureus* to be tracked. This has lead to surprising discoveries about how *S. aureus* are distributed and spread, and in the future may lead to better predictions and control of outbreak strains (Chapter 2).

The evolution of more virulent and resistant strains has clearly lead to new clinical problems. A greater understanding of how the bacteria evolve, predominantly due to the acquisition of mobile genetic elements encoding resistance and virulence genes is now possible. Better understanding of these processes is helping us to rapidly identify new outbreak strains and even prevent their emergence (Chapter 3).

Diagnostic microbiology laboratories and reference laboratories are key for identifying outbreaks and new strains of *S. aureus*. The development of reliable, simple and inexpensive techniques useful in these laboratories is progressing rapidly (Chapter 4).

In the research laboratory, understanding *S. aureus* biology is dependent on genetic manipulation of bacteria. The construction of knock-outs, reporter fusions, mutant libraries, etc has always been problematic in the staphylococci, but is essential to really understand key processes. The latest developments in these techniques is described (Chapter 5).

Staphylococci use genes encoding toxins, host receptors and immune evasion mechanisms to cause disease. But having the genes is not enough, they need to be expressed at the right time. Global regulators that control expression of virulence genes have been well characterized and are prime candidates for new interventionist therapies (Chapter 6).

Staphylococcal gene expression is also highly dependent on environmental triggers such as temperature, low iron concentration and stress. Understanding these triggers and their receptors helps us to understand pathogenesis better, and may also lead to novel therapeutic options (Chapter 7).

S. aureus resistance to antibiotics, predominantly those that act on the cell wall, has impacted on healthcare enormously. Genetic techniques have recently described how the bacteria acquire resistance, and how resistant mechanisms work. Potentially this will lead to better therapies (Chapter 8).

Other species of staphylococci that cause human disease are widespread and cause significant morbidity and mortality. Rapid progress is being made in understanding key virulence mechanisms and how these bacteria are evolving (Chapter 9).

Animal staphylococcal disease is an economic and social burden, but also developing into a human public health threat. Understanding these bacteria is leading to greater understanding about host specificity, spread, evolution and pathogenesis of all species (Chapter 10)

In summary, a bacterium cannot do anything it doesn't have the genes for. Understanding the nature, distribution, variability, expression and function of these genes in different strains with different pathogenic and transmission potential will lead to better strategies to prevent and control staphylococcal infections in the future. This book aims to bring all readers up to speed with these developments, enlarging the staphylococcal genetics community, and enabling continued high-quality research on these important pathogens.

Jodi A. Lindsay
University of London
UK

Reference

Elek, S.D. (1959). *Staphylococcus pyogenes* and its relation to disease. (E. & S. Livingstone Ltd, Edinburgh and London).

Other Books of Interest

Whole Genomes: Sequence, Microarray and Systems Biology

1

Matthew T. G. Holden and Jodi A. Lindsay

Abstract

Whole genome technologies facilitate the investigation of the genetic processes that underpin bacterial cells. Whole genome sequences for twelve diverse *Staphylococcus aureus* isolates are available and their annotation provides enormous insight into *S. aureus* physiology, capabilities and virulence. Whole genome microarrays, built using the sequences, have enabled whole genome regulatory responses to environmental conditions or global regulators to be investigated. Comparative genomics by sequencing and by multi-strain microarrays have identified the *S. aureus* population structure and how genomes vary, as well as suggesting that invasive isolates do not carry more virulence genes than carriage isolates. Whole genomes provide the framework for other systems biology approaches such as RNomics, proteomics and metabolomics. In a rapidly changing field, this chapter summarizes the major achievements so far, and what is likely to be achieved in the near future.

Introduction

The field of staphylococcal genetics has been revolutionized in the last few years because of new technologies such as whole genome sequencing and microarrays. In this chapter, we explain how these technologies have been applied to *S. aureus* and indicate key findings. For non-experts we describe how these technologies work, variations and limitations, as this helps to interpret the data generated. We also discuss how technologies are likely to advance in the near future and how systems biology will encourage integration of a range of data types. The expansive nature of whole genome data means that we can not review all the findings, so we also highlight specific tools to enable readers to investigate their own genes and areas of interest.

The advantage of these new techniques is they allow a snapshot of the whole cell to be examined at once. This is much more useful than investigating one gene at a time. In particular, it suggests functional relationships between genes that are homologous or are regulated similarly. It also broadens our understanding of what a cell can do when we identify unexpected genes or pathways. Since a cell can only do what it has the genes to allow it to do, the limitations of the cell also become clearer. But even with these technologies, it is important to remember that many features of the genome are still unknown, such as the large number of coding regions for proteins of unknown function. So there is still much to learn and this is an exciting time to be studying the staphylococci.

Genome sequencing

If you were placed in the situation of having to negotiate your way round an unknown city, surrounded with unfamiliar views, but recognizable features, the first thing you would turn to is a map. A map allows you to find your location from the visible landmarks, and once pinpointed, you can navigate your way beyond your local knowledge, and explore parts of the city that you had never been to before. The connections between neighbouring districts become immediately apparent, as do the routes that link the distant parts of the city together. You may not necessarily find what you are looking for immediately, as the destination may not be obvious, or there may not be enough detail on the map, but it provides a useful tool to navigate your way in the search. By analogy, bacterial genome sequencing can be considered genetic cartography for bacteria. The complete genome of an organism is the template for the components of the cell that enable it to thrive and survive. It provides new insights into the makeup of the constituent parts of the cell, and relationships of those elements. New horizons in research have opened up with the advent of genomics.

The first *S. aureus* genomes to be completed where those of N315 and Mu50 in 2001 (Kuroda *et al.*, 2001). At the time of writing (July 2007) a further 10 complete *S. aureus* genomes have been submitted to the public databases, which made *S. aureus* one of the most extensively sequenced bacteria at the time. The use of genomic data and their accompanying annotation is now widespread, and provides a valuable resource for researchers working with *S. aureus*. It is therefore pertinent to focus on how these genomes have been annotated to better understand what information is captured, and perhaps more importantly what limitations there are. We will then discuss the key findings from the *S. aureus* genome sequencing projects.

How are genomes annotated?

In the case of *S. aureus* the start point is normally a file containing a string As, Cs, Gs and Ts that represent the ~2.8 million nucleotides that comprise the genome. The goal of the process is to generate an accurate catalogue of the genes and features contained within the genome, which can then deposited in the public sequence databases (GenBank, EMBL Nucleotide Sequence Database, and DDBJ), and used by the scientific community.

The *ab initio* annotation of genome sequences is a multistage process that often relies on the application of several different programs or algorithms. The results of these analyses are merged and used to generate a prediction of features contained within a genome and their role or function. There is a wealth of bioinformatic tools available that can be used for genome annotation analysis, and these can roughly be divided up into those tools that predict features associated with DNA, and those that predict features of proteins. Whilst genome annotations deposited in the public sequence databases (GenBank, EMBL and DDBJ) will contains standard types of features with a minimal level of annotation (as required for submission), there is no standardized methodology for genome annotation. The strategies employed by the major sequencing centres to annotate their sequences vary in the tools used, how they are applied, and the level of human intervention. Some centres have developed automated annotation pipelines in which individual tools have been merged into a single program that will output annotation from a DNA sequence input. Using a computer pipeline such as this it is possible to reduce costs, increase the throughput of analysis, and produce bioinformatically controlled output. Conversely, some sequencing centres, such as the Sanger Institute, utilized computer tools, but employ biologists to curate the data, thus incorporating biologically relevant annotation and generating high-quality annotation. Fig. 1.1 shows a schematic diagram of a typical annotation process used for bacterial sequencing projects, highlighting the features predicted and the tools used to generate the annotation.

Gene prediction

The starting point for the annotation process of complete genomes is the primary DNA sequence that will have been finished to an acceptable standard, i.e. contiguous and with a predicted error rate below an arbitrary quality standard (standards vary from centre to centre). The first task in producing the annotation is to predict the genes, both protein coding, and non-coding; this later category includes tRNAs, rRNAs, ncRNAs, and pseudogenes. In addition, non-coding features such as repeats can also be predicted (Fig. 1.1). The prediction of protein coding sequences, or CDSs, is a crucial stage in the annotation process, as it provides the genome protein set that finds its way into public protein databases that are regularly queried by researchers, via web tools such as the NCBI BlastP resource. Mispredictions can therefore lead to confusion; false positive results populate protein databases with

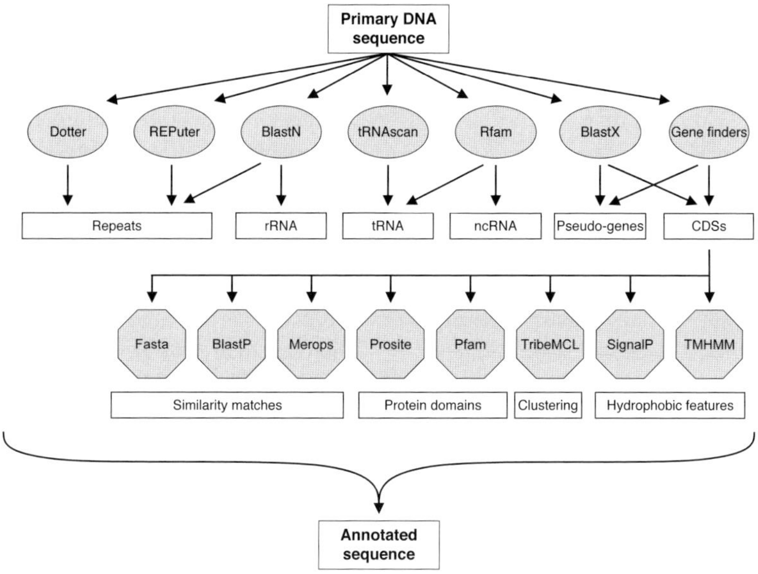

Figure 1.1 Schematic diagram of genome annotation. The figure illustrates the steps, and some of the tools used, in identify and annotating the feature contained within of bacterial genome sequences. Bioinformatic tools and processes are shaded in grey.

incorrect sequences, conversely, missed CDSs result in incorrect distributions and absences from predicted proteomes.

As with the annotation process in general, the prediction of genes in a genome is a process that relies on bringing several lines of evidence together, which is combined and reviewed, to generate an overall prediction. For the prediction of CDSs there are several automated gene finder programs that have been written, for example Glimmer3 (Delcher *et al.*, 2007), Orpheus (Frishman *et al.*, 1998), EasyGene (Larsen and Krogh, 2003). These typically work by training the program to recognize the typical features of protein coding DNA in a genome, using a small training set of sequences that are predicted to be 'real' from that sequence. These training set sequences are often the translated products of long open reading frames (ORFs), or BlastX matches against a protein database.

The performance of these programs varies considerably and is often affected by the DNA properties of the starting sequence; bacterial genomes vary markedly in their %G+C composition, some organisms like *S. aureus* are AT rich, whilst others, such as streptomycetes, are GC rich. Due to the performance differences of the programs it is often useful to run more than one, and then combine the outputs and curate the prediction. At this stage it is also possible to utilize other analysis, such as codon usage and GC frame plot, to refine the prediction. These analyses use features of the coding properties of DNA and can be displayed as plots overlaying sequence in a genome viewing software such as Artemis (Rutherford *et al.*, 2000). An additional line of evidence that can be used in gene prediction is the results of BlastX (Altschul *et al.*, 1990) searches against a public protein database such as UniProt (Apweiler *et al.*, 2004). Although useful, the results can sometimes be misleading where matches are to false CDSs that now litter the databases.

Non-protein coding RNAs

The prediction of non-protein coding genes in genomes is less straightforward due to the diverse nature of these features. tRNAs and rRNAs (5S, 16S and 23S rRNA) are the easiest to identify; specific prediction software such as tRNA-scan (Lowe and Eddy, 1997) can be used for tRNAs, and BlastN searches and alignment to bacterial rRNA databases can be used to identify rRNA sequences. Non-coding RNA (ncRNA) including small RNAs (sRNA) and regulatory RNAs are difficult to identify by BlastN, as their function is often associated with their secondary structure and therefore there is often covariance of nucleotides in a linear sequence, thus leading to higher levels of diversity than seen for protein coding genes. The Rfam database (www.sanger.ac.uk/Software/Rfam/) is a searchable database of functionally, and experimentally described ncRNA families (Griffiths-Jones et al., 2003). These include regulatory RNAs such as 6S RNA, tmRNA, RNAIII; regulatory elements such as riboswitches, T-boxes and ribosomal protein leaders; and ribozymes, like Ribonuclease P. The number of characterized families limits the prediction of ncRNA using Rfam. Recent global analyses, both bioinformatic and experimental, have identified many new ncRNAs in bacterial genomes (for a review see Livny and Waldor, 2007), highlighting the paucity of information about these non-coding genes. With more of these sequences identified, the prediction and annotation of these important, but often-overlooked, features in bacterial genomes will become easier, and therefore more widespread.

Pseudogenes

Both pseudogenes and partial genes are effectively non-coding as they are not translated into functional products, although they often will have the properties of protein-coding DNA. The identification of these sequences can therefore be made using the same tools that are used for CDSs. The challenge with annotating pseudogenes and partial genes is to firstly identify that they are non-functional, and secondly to identify the mutation responsible. Using comparative analysis it is often possible to identify mutations that will disrupt the translation of a gene (e.g. nonsense and frameshift mutations), as well as mutations that disrupt the function of the protein (e.g. larger scale insertions and deletions that alter the number of amino acids in a protein). Typically, the proportion of pseudogenes identified in bacterial genomes ranges from 1% to 8% (Lerat and Ochman, 2005), although there are exceptions, such as *Mycobacterium leprae*, where the pseudogenes make up 27% of the genome (Cole *et al.*, 2001). Where several strains of the same species have been sequenced, the identification of pseudogenes can be very important, as it identifies genotypic differences that may phenotypically differentiate strains, as well as provide clues as to the selective pressures experienced by the strains. It is worth noting that the number of pseudogenes identified in a genome annotation is probably an underestimation, as it is difficult to identify point mutations that effect protein function and gene expression by bioinformatics methods alone.

Repeats

There are many bioinformatic programs that can be used to identify DNA repeats, from basic dot matrix display programs such as Dotter (Sonnhammer and Durbin, 1995), through to more sophisticated programs such as REPuter (Kurtz *et al.*, 2001), which can identify different types of repeats, for example direct, palindromic, inverted repeats, as well as degenerate repeat families. The functions of repeats in genomes are diverse and often poorly understood. A growing interest is now being paid to these features as more and more genomes become available. Programs and databases that focus on specific types of repeats have appeared which provide detailed descriptions and annotation. For example, the CRISPRdb database (Grissa *et al.*, 2007) catalogues the repeats and spacers in CRISPRs elements (*c*lustered *r*egularly *i*nterspaced *s*hort *p*alindromic *r*epeats) that are believed to participate in the bacterium's defence against bacteriophage (Barrangou *et al.*, 2007).

Predicting function of protein coding sequences (CDSs)

The annotation of CDSs often brings together the results of several analyses to predict function and role. Similarity searches to proteins in the UniProt database is a commonly used approach to ascribe a product description for a CDS. The most frequently used alignment search tool is BlastP (Altschul *et al.*, 1990), which produces local alignments of protein sequences. Another commonly used tool is Fasta (Pearson and Lipman, 1988), this differs from BlastP as it produces a global alignment of two sequences, i.e. extended comparison from N- to C-termini, rather than an alignment to a portion of the sequence. The choice of which tool to use can be important where functions are inferred from similarity data; BlastP results may only report matches to a domain rather than the whole protein, therefore for multi domain proteins the overall function may not be the same. As a general rule of thumb Fasta is a more useful tool for investigating similarity at the level of the whole protein, whereas BlastP is useful for identifying domain matches. The identification of similarity to an experimentally characterized protein in the public database is often used to infer function. However the identification of those proteins that have been experimentally characterized is difficult due to the volume of proteins in the database, and the lack of tags that identify entries that contain experimental evidence. Many of the proteins that populate the protein database originate from genome sequencing projects, and have product descriptions that are the result of transitory annotation based on similarity, with little reference to the original description of function. A 'gold standard' set of proteins whose functions have been curated based on experimental evidence are SwissProt proteins. This dataset forms part of the UniProt database and is curated based on the literature or an automated annotation procedure (HAMAP) (Gattiker *et al.*, 2003). In addition to this curated version of the public protein database, there are a growing number of specialist databases that provide high-quality annotation of a limited, but related number of proteins. For example, the MEROPS database is a collection of peptidases that have been grouped into families and annotated (Rawlings *et al.*, 2006). The scope and level of detail in these databases often surpasses what is available elsewhere due to expert curation, and therefore is a useful addition to the annotation process.

Searching protein databases for similar proteins is a useful method for predicting function, however deciding what alignment statistical values and criteria to apply as cut-offs for the transfer of annotation is problematic. At what level of protein identity do you draw the line to infer conservation of function? What if the similarity only extends to part of the protein rather than the whole sequence? In the latter case, searches against protein domain databases may prove useful in identifying functional units, which can then be used to infer function. There are several protein domain databases that are available for searching, for example: Prosite, Pfam, Prints, SMART, Superfamily. These databases differ in the methods used to identify and build domain families as well as the degree of curation, therefore searching with a combination of databases increases the levels of coverage and detail obtained. Many of these domain databases have been amalgamated into a single collection, InterPro (Mulder *et al.*, 2007), which can be searched as a single entity (http://www.ebi.ac.uk/InterProScan).

A large proportion of CDSs identified in a genome sequence will be similar to proteins with no known function. These proteins are often categorized as conserved hypothetical proteins. Additionally, CDSs that do not have any matches in the public databases are often designated hypothetical proteins. CDSs of unknown function comprise approximately 38% of the *S. aureus* genome. Protein domain information can be informative for assigning functions to this group of protein, as the algorithms used to identify the domain matches are often have a greater degree of sensitivity than standard search tools such as BlastP.

Where protein similarity data have not provided clear evidence for a function, protein domain information often gives an insight, albeit often a less precise one, for example, assigning proteins to families rather than specific functions. This can be particularly useful for expanded groups of proteins found in genomes such as transporters and transcriptional regulatory proteins. Programs such as TribeMCL (Enright *et al.*, 2003) can be used to cluster the proteins of a genome based on an all-against-all BlastP comparison. Using the clustering information it is possible to identify groups of related proteins and annotate them accordingly.

Additional functional information can also be gleaned from structural domains of the predicted protein, for example, hydrophobic domains such as N-terminal signal sequences and transmembrane domains are important determinants of the location of the protein. Programs such as SignalP (Bendtsen *et al.*, 2004) and TMHMM (Krogh *et al.*, 2001) can be used to generate predictions for the cellular location of the protein, which in turn can be used to annotate a protein prediction in the absence of any other evidence, for example membrane protein or exported protein.

Beyond the annotation – genome databases

As a result of the *S. aureus* genome sequencing activity, the annotation of future sequences will be a far simpler proposition, as much of the genome is shared between strains, and therefore can be transferred directly from previous annotations, rather than being predicted from scratch. *Ab initio* annotation need only be applied to the unique regions, which encompass strain specific sequences. The question then arises as to where to transfer annotation from? The public sequence databases such as Genbank are archival repositories, and only the original submitters can update entries. A consequence of this is that the annotation accompanying a genome can become out of date rapidly if it is not regularly updated. Although there are third party annotation, and database-generated re-annotated versions (Genome Reviews produced by the EBI, and RefSeq produced by the NCBI), the annotation available in the public database lags someway behind the latest publications, even in curated databases such as SwissProt. There are various web databases that host *S. aureus* genomes, which can be used for searching, browsing and downloading data, for example: GeneDB (http://www.genedb.org/), Comprehensive Microbial Resource CMR (http://cmr.tigr.org), DOGAN (http://www.bio.nite.go.jp/dogan/Top), AureoList (http://genolist.pasteur.fr/AureoList/index.html). These databases vary in the coverage of *S. aureus* strains included, the tools and features that are available, as well as the amount and frequency of curation; for any one gene found in all sequenced *S. aureus* strains there are numerous annotations available at a multitude of locations. With more genome sequences on the way, and technologies that speed up and cut the cost of sequencing round the corner, one of the challenges of the coming years will be how to integrate and present these data in a user-friendly way, that is both accurate and informative.

In addition to web-databases for presenting genome annotation there is a growing selection of databases that can be used for interpreting genomic data, for example sites for comparative genomics and metabolic mapping. These databases can add value to genome data, presenting additional information that is not found with the primary annotation, and placing new contextual data on them. For example, the *S. aureus* genome annotations contain EC numbers describing the enzymatic reactions catalysed by the components of the cell. Whilst accurately categorizing the enzymatic function of protein encoded in the genome, the numerical format of the EC system is not particularly intuitive, and therefore can be time consuming to interpret. The KEGG database (Kyoto Encyclopaedia of Genes and Genomes) (Kanehisa *et al.*, 2006) is a 'biological systems' database integrating both molecular building block information and higher-level systemic information. Some of the systemic information that is stored is metabolic pathway data from *in silico* analysis of the genomes. It is possible to view the predicted metabolic reactions overlaid on reference maps, and therefore explore the predicted metabolome for an organism.

Features of the *S. aureus* genomes

The *S. aureus* genome varies between isolates, but we will first concentrate on universal features. The genomes are circular chromosomes of approximately 2.8 million base pairs, and may include one or more plasmids. The chromosome encodes approximately 2700 CDSs as well as structural and regulatory RNAs, and shares features found in other bacterial chromosomes.

The chromosome has an origin of replication from which bidirectional replication takes place. Typically genome sequences deposited in public databases start from the origin of replication, with the terminus of replication being found approximately in the middle of the sequence. Bacterial circular chromosomes often have features that make it possible to estimate the position of the origin and terminus of replication. The base composition of the leading and lagging replication strands exhibit a skew in GC content (Lobry, 1996) which can be seen clearly using a GC skew (G+C/G-C) plot (Fig. 1.2). In addition, the density of CDSs on the leading replication strand is often greater than the lagging strand (Fig. 1.2). Like GC skew, the precise origins of this property of bacterial chromosomes is not clear, although it is speculated that orientating CDSs such that the direction

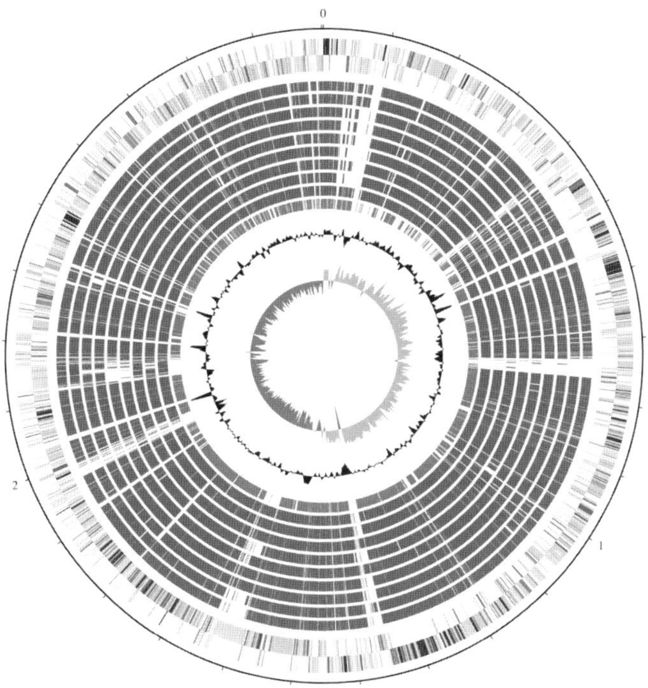

Figure 1.2 Schematic circular diagram of the *S. aureus* MRSA252 genome. Key for the circular diagram: scale (in Mb); annotated CDSs coloured according to predicted function represented on a pair of concentric circles, representing both coding strands; orthologue matches shared with other *S. aureus* strains, N315, Mu50, MW2, MSSA476, COL, RF122, USA300_FPR3757, NCTC8325, JH9, and *S. epidermidis* RP62a; G + C% content plot; GC skew plot. A colour version of this figure is located in the plate section at the back of the book.

of transcription and replication are in the same direction may prevent interference between these two machineries. In addition to identifying features associate with replication, GC deviation plots can also indicate chromosomal rearrangements and horizontal gene transfer in bacterial genomes: inversion within replicors reverses GC skew within that region; and the DNA of mobile genetic elements (MGEs) may not have the same base composition as the surrounding DNA. Plots such as percentage G + C content (Fig. 1.2) can also highlight the anomalous DNA properties sometimes associated with MGEs, and can be used to identify these regions in genomes.

Most of CDSs in the *S. aureus* genome have a function assigned to them. The majority of predicted functions are based on significant homology with genes in other species with a known function, and only a small proportion CDSs have been characterized in *S. aureus*. Functions fall into a number of classes that describe the processes and metabolisms that are required for the growth, replication and survival of cells as well as interactions with the host: cell division and replication (0.5% of total CDSs), chaperones (0.3%), adaptation and protection responses (1.7%), transport/binding proteins (9.2%), macromolecule degradation (1.5%), macromolecule biosynthesis and modification (4.7%), amino acid biosynthesis (2.0%), cofactor and carriers biosynthesis (2.3%), central and intermediary metabolism (2.2%), small molecules degradation (3.2%), energy metabolism (1.4%), fatty acid biosynthesis (0.6%), nucleotide biosynthesis (0.9%), cell envelope constituents (20.3%), ribosome components (2.4%), MGEs (7.2%), regulators (4.7%), toxins (0.8%) and pseudogenes (2.6%).

Without doubt, the sequencing projects identified many genes that were previously unknown. These included genes involved in cell division, cell wall synthesis, transport, osmoprotection, pigmentation, iron uptake and storage, capsule synthesis, adhesins, surface proteins, exoenzymes, toxins, regulators, two-component signal transduction systems, and more. Some of the functions newly identified in the genome could be assigned to phenotypic characteristics that have long been associated with *S. aureus*. For example, *S. aureus* is able to utilize D-galactose and lactose as a carbon sources (Bissett and Anderson, 1974); the genome revealed a cluster of eight genes (*lacRABCDFEG*) encoding components for the transport and utilization of these two sugars (Kuroda *et al.*, 2001).

The functional information garnered from genome sequencing has made it possible to perform metabolic reconstruction on *S. aureus*. There are various resources available now for investigating theses maps, for example, KEGG pathways and BioCyc (http://biocyc.org/). Tools such as these allow visualization of typical metabolic pathways in *S. aureus* and other organisms, so that it is clear which functions are present and absent in *S. aureus*. The complete mapping permits researchers a more holistic view of *S. aureus*, and provides a framework for the integrated analysis of cellular processes. *In silico* reconstructions have also been used to identify vaccine candidates. For example, sortase-processed cell wall-anchored proteins (LPxTG family proteins) have been shown to play important role in the binding of *S. aureus* to a wide range of host molecules (Mazmanian *et al.*, 2001). Using the genome sequences it has been possible to define the full complement of these molecules in *S. aureus*, as well as predict those proteins that are secreted (Sibbald *et al.*, 2006). Using a reverse vaccinology approach, Stranger-Jones and colleagues tested LTxTG proteins for their ability to generate protective immune responses against invasive *S. aureus* disease in mice (Stranger-Jones *et al.*, 2006). By combining antigens with the highest level of protective immunity, they generated a vaccine that protected mice against strains isolated from human clinical infections.

Many *S. aureus* features that have been studied for many years have benefited enormously from the sequencing projects. For example, virulence factor regulation (Chapters 6 and 7) is understood in much greater detail now that all the predicted regulators and two-component signal transduction systems have been identified. Specifically, newly identified homologues of important regulators, such as *sar* and *fur*, are now known to be functional and interact in complex networks. The 12 signal transduction systems have all been knocked out to determine which are essential (Clausen *et al.*, 2003). The full range of surface proteins is predicted, and increasingly their variation between strains has become appreciated (Kuhn *et al.*, 2006; Lindsay *et al.*, 2006). Identification of components in very complex pathways such as cell wall biosynthesis is now possible (Chapter 8). The benefits will continue to be felt.

There is still large scope for discovery in the *S. aureus* genome; ~38% of CDSs have no assigned function (for example hypothetical proteins, membrane proteins, exported proteins etc.), and a further ~10% only having a tentative description (family assignment based on domain matches). One of the most interesting groups of proteins contained within these FUN (*fun*ction *un*known) proteins (Hinton, 1997) are the *S. aureus*-specific CDSs, or *S. aureus* orfans. Contained within this group there are likely to be CDSs that encode traits that are uniquely expressed in this species.

In silico comparative genomics

There are now 12 whole genome sequences that are completed and available in the public domain (Table 1.1). Much of the chromosome (~75%) is conserved in all strains and encodes proteins associated with central metabolism and other housekeeping functions essential for the growth and replication of *S. aureus* (Fig. 1.2). A large proportion of the variable component of *S. aureus* genomes is composed of MGEs that have been horizontally acquired. Aside from the functions associated with mobility, these elements often carried genes associated with virulence and drug resistance. The variable distribution of these elements and their cargos accounts for some of the important differences observed in the clinical behaviours of *S. aureus* strains (Chapter 3 and 4). Comparative genomics is therefore a very useful tool for investigating the genetic basis of strain differences.

Alignment programs such as BlastN and tBlastX make it possible to compare genome sequences and identify regions of similarity and divergence. The results of these analyses

Table 1.1 *S. aureus* genome sequencing projects

Strain	Description and source	Lineage CC	Key toxins	Reference	Accession number
N315	HA-MRSA, pharynx, Japan 1982	5,	*tst, sec, sak, chp*	Kuroda et al., 2001	BA000018
Mu50	HA-MRSA, VISA, wound, Japan, 1997	5	*tst, sec, sak, sea*	Kuroda et al., 2001	BA000017
MW2	CA-MRSA, paediatric fatal septicemia, USA, 1998	1	PV-*luk*, *sec, sak, sea*	Baba et al., 2002	BA000033
MRSA252 (EMRSA-16)	HA-MRSA, fatal bacteraemia, UK, 1997	30	*sak, chp, sea*	Holden et al., 2004	BX571856
MSSA476	CA-MSSA, osteomyelitis, UK, 1998	1	*sak, sea*	Holden et al., 2004	BX571857
COL	MRSA, UK, 1961	8	*seb*	Gill et al., 2005	CP000046
NCTC8325	MSSA, UK, <=1949	8	*sak, chp*	Gillaspy et al., 2006	CP000253
FPR3757 (USA300)	CA-MRSA, abscess, USA, >2000	8	PV-*luk*, *sak, chp*	Diep et al., 2006	CP000255
RF122	MSSA, bovine mastitis, Ireland	151	*tst*	www.pathogenomics.umn.edu	AJ938182
JH1	MRSA, endocarditis, USA 2000	5	*chp, sak*	Mwangi *et al.*, 2007	CP000736
JH9	*in vivo* VISA derivative of JH1	5	*chp, sak*	Mwangi *et al.*, 2007	CP000703
Newman	MSSA, clumping factor overproducer UK, <1941	8	*sak, chp, sea*	www.ebi.ac.uk/genomes/bacteria	AP009351

VISA = intermediate level vancomycin resistant.

can be visualized using tools such as Artemis Comparison Tool (ACT) (Carver *et al.*, 2005) to provide an interactive display for comparative genomics. *In silico* comparisons using tools such as ACT is helping us to exploring the mechanisms of evolution of *S. aureus* genomes, and has identified horizontally transferred regions carrying virulence and drug resistance determinants. A web resource that can be used to visualize genome comparison is WebACT (www.webact.org) This online database contains sequence comparisons between all publicly available prokaryotic genome sequences, allowing the online visualization of comparisons using webstart version of ACT (Abbott *et al.*, 2005).

Using ACT, it is relatively easy to identify large variations between related genomes, such as the insertion or deletion of MGEs between N315 and Mu50, or MW2 and MSSA476. It is also possible to identify regions of variation within genes, such as between capsule types, or between variant regions in the accessory gene regulator (*agr*) locus or coagulase gene or restriction modification specificity subunits (*hsdS*). Closer inspection reveals minor changes such as small insertions or point mutations in relatively conserved genes, such as the housekeeping genes used for multi-locus sequence typing (MLST). It has subsequently been shown that a large proportion of this variation is associated with independently evolving lineages (see Microarrays section below and Chapter 3).

Pan genomes – can we identify all *S. aureus* genes?

With the achievement of complete genome sequence behind us, attention has now turned to the pan genome of a species. This expanded genome includes the core genes shared between all members of the species, as well as the accessory genes that are variably distributed. An insight into challenges of surveying a species' genome has been provided by Tettelin *et al.* (2005), who investigated the *Streptococcus agalactiae* pan genome. Using comparative analysis of eight *S. agalactiae* genomes Tettelin and colleagues were able predict that unique genes will continue to be identified even after sequencing hundreds of genomes, therefore making the pan genome effectively open ended. However, the size of the pan genome appears to be organism specific, with some organism populations being more diverse than others, and exhibiting greater rates of horizontal gene transfer. For a highly clonal organism such as *Bacillus anthracis* the 'pan' genome was estimated to be small; approximately four genome sequences are predicted to be required to capture the diversity of the whole population (Tettelin *et al.*, 2005).

So how big is the *S. aureus* pan genome? So far 12 genome sequences have been sequenced, and considering the diversity of the *S. aureus* population as predicted by the MLST, they are far from representative. With each new *S. aureus* genome, the variable component of the *S. aureus* genome grows; the recently published genome, USA300, yielded new MGEs, for example SaPI5 and ACME, not previously seen in the other sequenced strains (Diep *et al.*, 2006). Genome comparison of the available *S. aureus* genomes shows that approximately 75% is conserved in all the isolates sequenced, with the remainder being variably distributed. Many of these accessory regions are composed of MGEs that can be horizontally transferred between *S. aureus* isolates, and possibly beyond. It is therefore apparent that the pan genome for *S. aureus* will be very large, and the ability to capture it is probably beyond the immediate capability (both in capacity and cost) of current sequencing. However there are developments in sequencing technology that will make the rapid sequencing of hundreds of isolates a reality.

Genomics has delivered a wealth of data that provide an unprecedented glimpse of the genetics of *S. aureus*. Many of the advances that have delivered complete whole genomes sequences have relied on technological developments that have increased the throughput of sequence production, with the underlying chemistry of sequencing remaining fundamentally unchanged. New sequencing approaches have recently emerged that promise to propel the genomic revolution forward in the coming years.

The future of sequencing technologies

The first genome to be completed was that of a 5386 bp bacteriophage phiX174 (Sanger *et al.*, 1977). At that time, it was estimated that by using the di-deoxy termination and gel electrophoresis methodology devised by Sanger and colleagues that it would take ~ 3000 years to complete a genome of the size of *S. aureus* (~2.8 Mbp). Over the next couple of decades technological refinements in sequencing made the genome sequencing of bacteria a reality. The first bacterial genome to be published was that of *Haemophilus influenza* in 1995 (Fleischmann *et al.*, 1995). This 1.8 Mb genome was completed in little over a year using a random shotgun sequencing strategy. Since then, in excess of 450 bacterial genomes (Liolios *et al.*, 2006) have been deposited in the public databases, utilizing similar sequencing strategies and methodologies, i.e. Sanger chemistry and capillary electrophoresis. The cost of sequencing has fallen dramatically since the first projects, to a level where the cost per base is one tenth of what it would have been 10 years ago (for a finished genome rather than draft). The time taken to sequence a genome has also fallen as the capacity of modern sequencing machines has increased and automation has been applied to sample preparation and handling. This rate of fall has not been as dramatic for the economic costs, as the underlying technology of sequencing has remained the same, and the finishing of sequences, where remaining gaps from the shotgun are closed, is labour intensive and therefore expensive.

In the last couple of years several new sequencing technologies have appeared which have dramatically increased sequencing capacities that promises to revolutionize the way in which genomic sequences are produced. The increase in performance has also seen a decrease in costs. One such new technology that is now commercially available is that provided by 454 Life Sciences Corp (Margulies *et al.*, 2005). The method is called pyrosequencing, and utilizes PCR to amplify DNA on beads in an aqueous-oil emulsion, to generate starting material for sequencing; a single random fragment of DNA is captured on a bead and then amplified by PCR to generate a uniform/clonal population of DNA which provides the template for sequencing. Beads are then separated on a plate containing as many as 1.6 million nano-wells (a single bead fits into a well) that act as sequencing reactors. Sequencing reagents and nucleotides are sequentially supplied to the plate, where the incorporation of nucleotides releases pyrophosphate (PPi), which in turn is a substrate for enzymes that generate light (ATP sulfurylase, generates ATP from the PPi, which in turn is utilized by luciferase to produce light). By measuring the light production in the nano wells and correlating it with the supplied nucleotides, it is possible to generate sequence data for each of the templates isolated in the wells.

Another technology that promises to dramatically speed up sequencing in the near future is Solexa sequencing. This technique utilizes the clonal amplification of DNA molecules on a surface to generate template molecules for sequencing (Johnson *et al.*, 2007; Robertson *et al.*, 2007). Single template molecules are covalently attached to a surface and then amplified *in situ* to generate a localized clonal population of template. The sequencing reaction is performed using DNA polymerase and a mixture of the four fluorescently labelled reversible chain terminators. Fluorescent signals can be detected across the surface, and therefore the sequence for each template can be recorded for a sequencing cycle. Reversible blocks are then removed and the sequencing reaction is repeated, thus building the sequence for the templates.

The increase in performance of these new technologies in comparison to the Sanger sequencing and capillary electrophoresis technologies is impressive: in a single run it is possible to generate ~1,000 Mb and ~100 Mb of sequence for the Solexa and 454 technologies respectively, in comparison to ~0.25 Mb for a standard capillary electrophoresis sequencer. Not only do these technologies promise more rapid sequencing, but they also have the potential to cut costs as well, and open up new exciting new possibilities in microbiology. In field of human genomics there is a race towards 'the $1,000 genome'. This goal is seen as an important landmark in genomic science, as it represents commercially viable personalized genomics, and would herald in a new age of human medicine. By the same measure, it would also herald the $1 bacterial genome, making personalized microbiological diagnostics a possibility, as well as revolutionizing microbial genetic research.

For all the potential that these technologies have for speeding up genomics, there are also some drawbacks and hurdles that mean that they are not yet ready to overtake Sanger capillary electrophoresis as the method of choice for complete genome sequencing. For both of the new sequencing technologies the length of sequence they are able to generate for a single template remains low in comparison to capillary electrophoresis (~30 bp for Solexa, and ~100 bp for 454, ~1,000 bp for capillary electrophoresis). Read length can become an important criterion in the assembly process if a sequence contains repeats. Using 454 it is also difficult to resolve nucleotide repeats greater than 5 bp. Whilst some of these drawbacks mean that these technologies are not suitable for *de novo* sequencing by themselves, they can be successfully teamed up with capillary electrophoresis for whole genome sequencing. In addition, 454 and Solexa are useful tools for comparative sequencing, where the sequence generated can be compared to a related scaffold sequence for SNP detection, or investigating pan genomes.

So what do these developments mean for *S. aureus* genomics? Where in the past genomics has been exclusive in the strains it has focused upon, we will be able to sample the population more rigorously and gain a greater knowledge of *S. aureus* diversity. The ability to use genomics to rapidly genotype a strain will provide unprecedented levels of diagnostics information that could be used to tailor clinical treatment regimes, depending on the drug resistance and virulence genes complements.

Microarrays

One of the most useful ways to exploit genome sequence data is to construct whole genome microarrays. A microarray is typically a glass slide printed with thousands of tiny spots of DNA. Each spot has a known sequence and usually represents a gene or gene fragment. Microarrays are used to identify the composition of complex mixtures of nucleic acid. They use the concept of hybridization – the ability of the DNA double helix to unwind exposing its nucleotide bases, and allowing it to preferentially bind to labelled DNA with the correctly matching base pairs.

Microarrays are used for two types of experiments:

1 Determining the presence or absence of DNA in a complex mixture. The most common type of experiment is identifying which genes are found in a bacterial isolate, and this is often referred to as comparative genomics hybridization (CGH). When many isolates are compared in large scale experiments it is possible to determine the population structures of some species, and genes associated with pathogenic behaviour.
2 Determining gene expression patterns. In this case the complex nucleic acid mixture is mRNA, which is reverse transcribed to cDNA for hybridization to the microarray. The most common types of experiment are to identify how a bacterial isolate responds to environmental triggers, or how gene expression is altered when a particular gene regulator is mutated.

Microarray data are currently revolutionizing certain fields of staphylococcal study. For non-experts, an understanding of the technology is necessary to interpret the literature, especially issues about microarray design, controls, data analysis, replicates, data deposition and the use of nucleic acid stabilizing agents. Large-scale comparative genomics studies have identified the clonal nature of *S. aureus* populations and how they are evolving, and provided information about the ability of strains to cause disease. For transcriptional studies, it is now possible to identify the transcriptional response of every gene in the genome to a particular regulator in a single experiment at moderate cost, and this supersedes technology such as Northern blotting or RT-PCR which only investigate one gene at a time. However, it is still early days for this technology, and it is expected that many more studies will be published in the next few years, with long-term consequences for the field.

The technology

The development of technology to design, print and hybridize microarrays is the focus of many specialist research teams, including those who supply microarrays at low cost and train collaborators in their use such as the Bacterial Microarray Group at St George's (BμG@S) and

The Institute for Genomic Resources (TIGR). Commercial suppliers of microarrays can custom design, supply microarrays and dedicated equipment and software for their use, such as Affymetrix, Oxford Gene Technology, Agilent and Illumina. Many universities, companies and other research institutions now have in-house facilities for hybridizing arrays, as the major equipment required (scanners, software) is now affordable. Each approach uses slightly different technology, but the major factors to be considered are discussed here.

Design and printing

Microarrays can only detect the presence or absence of genes printed on them. Thus the design of a microarray is crucial to its usefulness. For transcriptional studies of a given bacterial isolate, a whole genome microarray containing probes for every gene from the isolate to be tested is desirable, and ensures that the response of the whole genome is considered and the most important transcriptional changes are identified. For comparative genomics, a microarray designed with a single genome is often insufficient if there is substantial population variation, such as within the *S. aureus* species. Therefore, multi-strain whole genome microarrays have been developed. Multi-strain microarrays can also be used for transcriptomics and have the advantage that they can be used for a range of bacterial isolates. Other microarrays may include only a selection of genes, and these can have specialist applications such as diagnostics or strain typing.

Microarray spots or probes can vary from oligonucleotides of between 25 and 70 bases (the widespread terminology is 70-mer) through to PCR products of hundreds of bases. This has implications for the performance of the microarray, as each probe will have its own unique hybridizing efficiency based on length, nucleotide composition and ability to hybridize to slightly variant homologous DNA. Oligoarrays may be designed to have only one probe per gene, or to improve the reliability of data, may have dozens of probes per gene. Some oligoarrays include internal 'control spots' that have a single base pair variant, and these can help determine the levels of background hybridization. All arrays have positive and negative controls to measure array performance and to assist with data normalization (see below).

Most microarrays are glass slides coated with agents to encourage even binding of DNA, such as poly-L-lysine or amino-silane, and printed using specialized robots under vacuum, such as the MicroGrid system from BioRobotics. Affymetrix synthesize oligos directly onto the GeneChip slides using photolithography. Slides can also carry multiple arrays, and be separated by spacers during the hybridization process, allowing multiple samples to use the same slide, thus reducing costs. Microarray slides are generally not re-usable.

Labelling and hybridization

For CGH, high-quality *S. aureus* genomic DNA is suitable for labelling, such as that from Qiagen columns, caesium chloride purification (Lindsay *et al.*, 1998), or using the kit from Edge Biosystems (Cockfield *et al.*, 2007). mRNA is much more unstable, and many transcripts have a half-life of only seconds. Therefore, the use of nucleic acid stabilization agents such as RNAprotect (Qiagen) or Trizol is highly recommended, otherwise the profiles detected will likely be strongly influenced by cold shock or centrifugation rather than the conditions intended (Mangan *et al.*, 1997). Once stabilized, mRNA is extracted using a FastPrep ribolyser or lysostaphin and chemical extraction. mRNA then needs to be reverse transcribed to cDNA prior to labelling.

There are two main ways of labelling DNA prior to hybridization, single colour or two colour. Single colour labelling means that a sample will be labelled with a single type of marker, such as biotin. This can later be fluorescently labelled using specific antibodies. Biotin-labelled DNA is then hybridized to the slide. During data analysis, the ratio of hybridization to 'real spots' compared to 'control spots' (such as those with single base variations on an oligoarray) is used to identify if a spot has preferentially bound to labelled DNA.

Two colour labelling means that two samples will be hybridized to the same slide, and one of those samples will usually act as a 'control'. The labelling compounds are fluorescent dyes, Cy3 and Cy5. A typical example would be cDNA from a wild-type isolate is labelled with Cy5 and cDNA from a mutant isolate is labelled with Cy3. After hybridization the ratio of each signal to the spot allows the differences between the isolates to be easily compared. Another approach is to use a common 'control' for a series of experiments, such as genomic DNA labelled with one dye. In this case a series of gene expression studies in various media can be performed, with each sample hybridized with the same control, and this allows all the samples to be compared during subsequent data analysis. In general, two colour experiments are more tightly controlled and allow for variations in hybridizing ability of individual DNA spots, thus making data analysis much more reliable.

Once the labelled DNA is hybridized to the microarray, it is generally left overnight, although more rapid binding is achieved with very low volume systems. The slides are then washed to remove non-specific binding of probes. The stringency of these washes determines the percentage of DNA identity required for a labelled DNA fragment to bind to its corresponding spot. In the author's laboratory, and using comparative genomics data compared to the sequencing projects, this cut-off is in the region of 90–92% identity using our typical conditions (Witney *et al.*, 2005).

Microarray scanners detect both fluorescent dyes directly from slides, or can detect a fluorescently labelled antibody bound to biotin after the appropriate binding and washing steps. Scanners are now relatively inexpensive and are easily purchased, such as the Affymetrix 428 scanner. Software to visualize the fluorescent signals typically is included with the scanner. Turning these pictures into numbers assigned to specific DNA spots for subsequent data analysis may require additional software such as ImaGene (BioDiscovery, CA) or BlueFuse (BlueGnome, Cambridge, UK).

Replicates of microarray experiments must be considered. For comparative genomics, a single microarray per strain is suitable for large-scale populations studies, but two or more should be considered for making detailed analysis of specific isolates (Witney *et al.*, 2005). For gene expression studies, we recommend three individual experiments with each experiment microarrayed twice, and this is due to the instability of mRNA and typical experimental variation (Wernisch *et al.*, 2003). Some laboratories also suggest dye-swap experiments, although this is becoming less common.

Data analysis and deposition

There are several types of software available that are necessary for analysing microarray data, the most comprehensive currently available is GeneSpring (Silicon Genetics). The first step is to normalize the data, which is a complex mathematical procedure to enable slides from independent experiments to be compared, and this is achieved using the signals from control spots. The software can then plot data in a variety of ways making data analysis of thousands of spots practical.

A common analysis in gene expression studies is to compare replicates of two conditions or strains, and to generate lists of genes that show a statistically significant 2-fold up- or downregulation. Some groups regard a twofold cut-off as arbitrary, and certainly it can lead to very long lists of genes. Two-fold is approximately the reliable detection limit of most microarray experiments. Higher cut-offs are appropriate if a major conclusion is to be made from the microarray data. All important data should be confirmed with another technology such as RT-PCR or a relevant biological test (Fox *et al.*, 2007).

GeneSpring (and some other software) can also display data in gene order with a colour system to identify which genes are up or down regulated, identifying operons. Or it can select out the most upregulated spots and plot them over a series of experiments, identifying regulons. Statistical analysis of replicate experiments using appropriate methods that can filter out error due to thousands of replicates (such as Bonferroni type methods) are possible. Spearman clustering of data between experiments can indicate relationships between different experiments, such as similar stress responses to a range of conditions. So there are many ways the data can be investigated or interpreted.

While most software is designed for gene expression analysis, it is being adapted to comparative genomics studies, and hopefully will improve in the future. In the meantime, Spearman clustering in Genespring (Lindsay *et al.*, 2006) or GACK conversion of ratios to binary data and Bayesian clustering (Champion *et al.*, 2005) has been used to identify relationships between isolates. Multi-strain microarrays are much more complex than single genome microarrays to analyse, and may require the construction of 'virtual genomes' to visualize the relationships of different genes to each other in different genomes (Witney *et al.*, 2005). It should not be underestimated how complex and time-consuming data analysis is.

Increasingly journals are requiring deposition of data into public databases prior to publication of microarray studies. This is to encourage confidence in the data, as well as to allow others the opportunity to re-analyse data or analyse them with newly developing technologies, such as systems biology. Microarray data are required to be MIAME compliant (Brazma *et al.*, 2001), and this refers to a range of stringent requirements covering the design of the microarray, detailed protocols for every step of the laboratory experiments and data analysis, including dates, persons and controls. Deposition of microarray data that incorporate all of the MIAME guidelines is therefore a complex process. Public databases include ArrayExpress and BuG@SBase. Note that the design of some commercial microarrays is not publicly available, making data deposition incomplete, and some journals have yet to make judgements on this issue.

Comparative genomics of *S. aureus*

Whole genome comparative genomics – comparing strain genomes to identify differences correlated with phenotype – was begun using the first sequencing projects. However, only a limited number of genomes will be sequenced. To investigate gene variation in larger populations of bacteria, microarray technology can be used. Its main drawback is that only known genes can be included on microarrays, and unknown/unsequenced genes are missing. Nevertheless, *S. aureus* microarray comparative genomics studies have told us a lot about *S. aureus* population structure, evolution of the species, host specificity, and allowed us to make conclusions about virulence and host-pathogen interactions.

Virulence

The first major *S. aureus* microarray CGH study involved a microarray constructed with 90% of the predicted genes from the COL genome, and compared 36 strains, predominantly MRSA or associated with toxic shock syndrome. This study identified large deletions in many of the clinical isolates, and these were not just in MGE (Fitzgerald *et al.*, 2001). At this point it became clear that there was a large amount of variation between *S. aureus* genomes, even with those causing similar diseases and that were unable to be discriminated by other typing methods.

Since microarrays can only detect genes that are printed onto them, there was a need to develop complex multi-strain *S. aureus* microarrays to investigate variation in detail. To date the most comprehensive *S. aureus* microarray is the seven-strain whole genome *S. aureus* designed by BμG@S (Witney *et al.*, 2005), containing DNA spots for every predicted gene from the first seven genome sequence projects. This microarray was validated and showed strong correlation between known presence of genes from sequencing projects compared to signals on the microarray (Witney *et al.*, 2005). The first and most comprehensive comparative genomics study using this array compared 61 community acquired *S. aureus* isolates causing invasive disease to 100 nasal carriage isolates from healthy blood donors in the same Oxford community (Lindsay *et al.*, 2006). The aim was to identify differences between the populations, and thus genes that might be preferentially associated with invasive disease that could be targeted by interventional therapies. Other aims were to identify which regions of the genome varied, and if there was a population structure. A large amount of variation between isolates was identified, particularly in MGEs encoding toxin and resistance genes, but surprisingly none of the variation correlated with the invasive isolates. In other words, the invasive isolates were not clearly different to the carriage isolates. A major conclusion of this study was that in the community setting, *S. aureus* invasive disease is likely to be more dependent on host factors than previously thought. [It should be noted that the incidence of Panton–Valentine leukocidin (PV-*luk*) carried on a bacteriophage in this population was low. The incidence of PV-*luk* in *S. aureus* from the community is increasing in some parts of the world and has been implicated in community acquired skin infections (but not in invasive disease) (Dufour *et al.*, 2002; Gillet *et al.*, 2002; Herold *et al.*, 1998).] An alternative explanation is that whatever factors make *S. aureus* a successful colonizer of the human nose are the same factors necessary for invasive disease (Lindsay *et al.*, 2006).

Lineages

Another unexpected finding from the Lindsay *et al.* (2006) study was determined from further analysis of the regions of the *S. aureus* genome that varied. If all the genes on the microarray were used to Spearman cluster the isolates, there was no obvious population structure. However, many MGEs had been identified by the sequencing projects and were known to be able to move between isolates at high frequency. On the other hand, a large proportion of genes were found in >95% of isolates, and these are referred to as core genes. If a list of genes was generated that excluded both the MGE and core genes (corresponding to 723 microarray spots), and the Spearman clustering was repeated, a population structure for *S. aureus* became very clear. It consisted of approximately ten dominant lineages and several minor lineages. The population structure matched almost perfectly with the clonal clusters (CC) derived using multi-locus sequence typing (MLST) and *spa* typing (Chapter 2). The 723 genes were called the core variable (CV) genes. Each lineage did not have unique CV genes, but unique combinations of multiple CV genes.

What does this mean for *S. aureus* population structure? It is likely that a common *S. aureus* ancestor diverged and recombined many times to form multiple types of *S. aureus*, and only some of these were successful. These successful clones became the ancestors for each of the dominant lineages. The lineages have evolved relatively independently of each other, and this may be due to restriction modification system differences between lineages that control exchange of DNA (Waldron and Lindsay, 2006) (see Chapter 3 for a fuller description of the evolution of lineages). Because each lineage is independent, there are many differences between lineages, including large differences (insertions/deletions of multiple genes) and small differences (single nucleotide polymorphisms (SNPs), such as those utilized for MLST typing).

What are the CV genes? Of those with known or predicted function, many encode proteins or structures displayed on the bacterial cell surface, or regulate the expression of these genes (Table 1.2). They include variants of fibronectin-binding proteins (FnbA, FnbB), coagulase (Coa), the large host binding protein Ebh, a haemagglutinin-like protein (SasA), collagen binding protein (Cna) and a protein implicated in biofilm production (SasG). The regulators include the major variants of the accessory gene regulators (Agr), staphylococcal accessory regulator (SarT), and target of RNAIII activating protein (TRAP). Some CV genes are next to each other on a long segment of DNA that is present or absent in some strains. Others represent a single gene that is present or absent. Some CV genes have variant regions internal to the gene, often encoding key functional or binding domains.

Apart from the conserved core genes, and the lineage defining CV genes, the third major component of *S. aureus* genomes are MGEs. The biology of these elements will be discussed in more detail in Chapter 3. Microarray studies have shown that some mobile genetic elements are associated with particular lineages, while others have spread between lineages. It has also emphasized the large amount of recombination and variation in MGE between strains, including those within the same lineage.

MRSA

Further comparative genomics studies have investigated the evolution of hospital epidemic MRSA. In the UK, 'epidemic' MRSA types EMRSA-1 to -14 were associated with low incidence and some spread between hospitals prior to the mid 1990s. EMRSA-15 (CC22) and EMRSA-16 (CC30) have displaced the earlier types in UK hospitals and caused a massive increase in the incidence of MRSA infection, from 1–2% of all *S. aureus* infections to >40%. The CC22 and CC30 do not appear very different to MSSA from these lineages, except that they contain an SCC*mec* element encoding methicillin resistance (Cockfield and Lindsay, unpublished). Thus it does not seem at this stage that epidemic behaviour is caused by the acquisition of a known gene. A new type of epidemic MRSA, TW (ST239), has been described that causes increased bacteraemia associated with vascular

Table 1.2 Selected putative virulence factor and resistance genes and their distribution in *S. aureus* genomes. Genes identified by protein name, abbreviated gene name and representative annotated gene name(s). R is from MRSA252, N from N315, V from Mu50, M from MW2, C from COL, S from MSSA476

Core genes highly conserved in all strains	Core variable (CV) genes present/absent or highly variable between lineages	Mobile genetic elements (MGEs) carried on bacteriophage (B), SaPI (SP), plasmid (P), transposon (T) or SCC (SC)
Siderophore uptake (sir) R0116–8	Protein A (spa) R0114	Bleomycin (ble) R0032 SC
Siderophore synthesis (sbn) R0119–27)	Capsule (cap) R0151–66	Kanamycin (kan) R0033 SC
Superoxide dismutase (sodM) R0135	Coagulase (coa) R0222	Methicillin (mecA) R0039 SC
Lipase (geh) R0317	Bacillus-like toxin (bctE) R0284	Erythromycin (ermA) R0050 T
Multidrug efflux (mep) R0330–2	Enterotoxin family (set)* R0422–31	Streptomycin (spc) R0051 T
Staphylococcal accessory regulator (sarA) R0625	Lipoprotein (lpl)* R0438–45	Arsenic (ars) R0690–2 P
Teichoic acid biosynthesis (tag) R646–9	Serine-aspartate repeat surface protein (sdrD, sdrE) V0562, R0567	Cadmium (cad) R0723–4 P/T
Siderophore uptake (fhu) R0657–9	Large surface protein (ebh) R1447	Beta-lactamase (bla) R1829–31 T
Staphylococcal exoprotein regulator (sae) R0758–61	Hyaluronate lyase (hysA)* R1892	Chemotaxis-inhibitory protein (chp) R2036 B
Clumping factor (clfA) R0842	Proteases (spl)* R1900–1,06	Complement inhibitor (scin) R2039 B
Nuclease (nuc) R0847	Superantigen family (se)* R1916–21	Staphylokinase (sak) R2039 B
Two-component regulator (arl) R1426–7	Mammalian protein binding (mapW)R2030	Enterotoxin A (sea) R2043 B
Sigma B (sigB) R2152	Accessory gene regulator (agr) R2123–6	Enterotoxin C (sec) N1817 SP
GAMMA-haemolysin (hlg) R2509–11	Hyaluronate lyase (hysA2) R2292	Toxic shock syndrome toxin1 (tst) N1819 SP
Clumping factor (clfB) R2709	Aureolysin protease (aur) R2716	Tetracycline (tetM) V0398 T
Intracellular adhesion (ica) R2746–50	Immunodominant antigen (isaB) R2717	Siderophore transport (fhuD) V0803 SP
	Haemagglutinin-like protein (haem) N2447	Preventing adhesion (pls) C0050 SC
	Collagen-binding protein (cna) R2774	Enterotoxin K (sek) C0886 SP
	Leukocidin (lukDE)* N1637–8	Enterotoxin Q (seq) C0887 SP
	Accumulation-associated proteins (aap) N2284–5	Enterotoxin B (seb) COL0909 SP
	Staphylococcal accessory regulators (sarH2, sarH3) N2286-7 Fibrinogen binding protein (fnbB) N2290	Tetracycline (tet) CP0002 P
	Fibrinogen binding protein (fnbA) R2580	Enterotoxin H (seh) M0051 T
	Lantibiotic synthesis (epi) C1877–78	Panton–Valentine leukocidin (PVluk) M1379–80 B
		Fusidic acid (far1) S0043 SC
		Aminoglycosides (aap/aph) VP026 P
		Quarternary ammonium chlorides (qacA) VP032
		Exfoliative toxin A (eta) NP_510960 B
		Exfoliative toxin B (etb) AAA26628 P
		Exfoliative toxin D (etd) BAC22944 T
		Enterotoxin D (sed) AAB06195 P
		Biofilm-associated protein (bap) AAK38834 SP

*Carried on a genomic island (GIa or GIb), highly variable.

access devices (Edgeworth *et al.*, 2007). TW appears to have evolved from earlier EMRSA types and has accumulated all of their identifiable MGEs (Edgeworth *et al.*, 2007). While a specific gene necessary for the TW phenotype has not been identified, future sequencing of a TW isolate, construction of an expanded microarray, and comparison of epidemic strains may identify such a gene.

Animal S. aureus

Animal isolates of *S. aureus* have been compared to human isolates to determine how they are related, if they could exchange DNA, and if there is evidence for host-specific receptors. *S. aureus* is a common cause of bovine mastitis that adversely affects the dairy industry. *S. aureus* can also cause infections in other animals such as horses, pigs, goats and domestic pets (Chapter 10). Bovine mastitis isolates in the UK were found to carry the vast majority of human isolate core genes, but to cluster into their own multiple bovine lineages (Sung and Lindsay, unpublished). This suggests that animal *S. aureus* are not distinct from human *S. aureus*, but just have unique surface gene combinations. Investigation of CV or surface proteins found in human isolates but not animal isolates failed to find candidates for host-specific binding interactions. However, it could be that there are animal specific markers not on the microarray that are important. Other animals such as horses can be infected with *S. aureus* isolates from either cows or humans, and this is likely because horses are transient carriers rather than niches that *S. aureus* has adapted specifically to. There was some similarity in MGEs between animal and human isolates, suggesting gene exchange was possible.

Future of CGH

In the future, we will see four major developments. Firstly, larger populations of *S. aureus* will be investigated. Of particular value will be those studies comparing isolates with complex phenotypes that are difficult to study in the laboratory such as epidemic vs. non-epidemic, hospital vs. community, or those causing specific disease types (skin vs. pneumonia vs. bacteraemia, etc.). The second development will be the building of more complex microarrays, containing additional DNA spots, as a direct result of the proliferation of *S. aureus* sequencing projects. In particular, isolates of the remaining dominant lineages, and those representing new epidemic MRSA clones will be the most valuable. This will ensure that the widest number of genes is interrogated in these population studies, and that no important genes are missed. The third development will be an improvement in data analysis tools so that associations between gene markers and phenotype are easier to identify with confidence.

The final development will be the reduction in costs and complexity of this technology, as commercial suppliers of microarrays support the basic research community. In addition diagnostic *S. aureus* arrays will be promoted once markers of *S. aureus* clinical importance are identified. As more microarray data are collected and deposited, it will be possible to identify worldwide epidemiological patterns in dissemination and evolution of *S. aureus*, and specific interactions between host and pathogen, potentially identifying virulent strains, susceptible hosts, predicting infection outcomes, and identifying targets for intervention strategies.

Gene expression of *S. aureus*

The use of microarrays for gene expression studies allows a rapid snapshot of all mRNAs present in an isolate at a particular time under particular conditions. Provided all of the genes from that isolate are on the microarray, it is possible to ensure that every gene is considered when deciding how an isolate has responded. This is a technological leap from previous technologies such as RT-PCR and Northern blot that only allow a single gene to be investigated at a time.

Experiments

Consequently, dozens of labs have used *S. aureus* microarrays to investigate a range of isolates, and their response to growth conditions or mutation. Some of these studies are summarized in Table 1.3, and a useful website for comparing datasets from published papers is at www.bioinformatics.org/sammd/. It is difficult to summarize such a large number of data, and the reader is urged to look at individual papers and datasets to search for their favourite genes or pathways. But in general, environmental conditions such as stress cause a wide range of gene expression changes in specific pathways, as well as general metabolic pathways, regulators, surface proteins and toxins. These profiles often vary significantly after prolonged exposure to the condition, so it seems likely the bacteria adapt in a complex manner and in multiple stages. Similarly, mutants in the global regulatory genes investigated so far usually show complex gene expression alterations, often in both virulence and metabolic functions. It is expected that many regulators in *S. aureus* will

Table 1.3 Selected studies of *S. aureus* gene expression using microarray

Condition/ mutant	Strain	Array	Major findings	Reference
Δ*agr* Δ*sarA*	RN27*	Affymetrix (partial genome)	Virulence and metabolic genes affected	Dunman *et al.*, 2001
Cell wall antibiotics	RN450*	Affymetrix	Up: cell wall synthesis, metabolic Down: range	Utaida *et al.*, 2003
ΔSigma B	COL GP267 Newman*	Affymetrix	Up: adhesins Down: toxins	Bischoff *et al.*, 2004
Δ*vraSR*, vancomycin resistance	N315	Scienion N315	Down: *proP*, *tcaAB*, *tag*, *fmt*, *recU*, cell wall synthesis	Kuroda *et al.*, 2003
Δ*arlR*, toxin regulator	WCUH29*	Affymetrix	Up: toxins, agr, adhesins, autolysins Down: capsule, metabolism, other toxins.	Liang *et al.*, 2006
Glycopeptide resistance	Mu50 N315*	N315	Up: *gra*, *mgrA*, *smr*, *lysC*, *malR*, *sigB*	Cui *et al.*, 2005
Biofilm vs. planktonic bacteria	8325 variant	Scienion N315	Up: *ica*, surface proteins	Resch *et al.*, 2005
Δ*clpP*, protease	8325	Scienion N315	Up: heat shock, stress, SOS, autolysis Down: *agr*, anaerobic/*fnr*	Michel *et al.*, 2006
Δ*saeRS*, toxin regulator	WCUH29*	Affymetrix	Up: *agr* Down: adhesion	Liang *et al.*, 2006
Nitric oxide stress	COL*	Affymetrix	Up: fermentation, iron uptake, heat shock, *clpBL*, *ctsR*	Richardson *et al.*, 2006
Δ*cidR*, acetoin metabolism	UAMS-1*	Affymetrix	Down: *cidABC*, *lrgAB*, *alsSD*, acetoin production	Yang *et al.*, 2006
Δ*mgrA*	Newman*	Affymetrix	Up: urease, adhesins, autolysins, nor Down: capsule, toxins	Luong *et al.*, 2006
Shock (cold, heat, stringent, SOS)	UAMS-1*	Affymetrix	Stabilization of mRNA	Anderson *et al.*, 2006
Hydrogen peroxide	8325*	Affymetrix	Up: anaerobic, DNA repair	Chang *et al.*, 2006
Vancomycin stress	Newman	Affymetrix	Many differences	McCallum *et al.*, 2006
Glycopeptide resistance Autolysis	MRGR3, 14-4*	Agilent	Up: amino acid and ion transporters, regulators Autolysis not due to changes in expression of autolysin genes	Scherl *et al.*, 2006; Renzoni *et al.*, 2006
Δ*murF*, cell wall synthesis	COL	TIGR	Slow growth, thin cell walls, up: surface and toxins, anaerobic Down: iron transport, *ica*, phage	Sobral *et al.*, 2007
Ciprofloxacin stress	8325	TIGR	Up: SOS, prophage induction, TCA, mutation	Cirz *et al.*, 2007
Salicylate	SH1000	TIGR	Minor differences only	Riordan *et al.*, 2007

*No reference to a mRNA stabilization agent used.
Notes about strains: RN450 and RN27 are derived from 8325. SH1000 is 8325–4 with a complemented *rsbU* gene. WCUH29 is a sequenced strain, but the sequence is not publicly available. UAMS-1 is a clinical isolate from osteomyelitis.

only control expression of a small number of genes, but understandably microarrays have not been used to investigate these regulators as yet.

The experiments have raised several issues with regard to interpretation of data. Data interpretation is complex, and to make sound conclusions the experiments have to be performed to the highest standards, usually indicated by deposition in a public database. This also allows others to challenge the data when appropriate. Secondly, *S. aureus* do not respond to environment or regulators in a simple one step process. The cell will adapt, then adapt to the new cellular environment, then adapt again, etc. In addition, it is known that gene expression is highly dependent on growth phase (exponential vs. stationary), and in response to different nutrients, shaking/aeration, bacterial numbers, etc. Therefore, if we wish to draw conclusions about gene expression and virulence, we should think seriously about the growth conditions of the bacterial cells and how well they mimic *in vivo* conditions. Thirdly, it is becoming increasingly clear that not all strains use regulators in the same way or react to environmental conditions equally. All of these issues confound data interpretation and are leading to a rethink about our understanding of *S. aureus* gene expression.

Future

In the future, it is expected that this technology will become relatively inexpensive and simpler to use due to enhanced commercial supply of arrays and technology. It will become less acceptable to use strains that do not have all of their genes on the array used, so complex arrays will be more widely employed. mRNA stabilization agents will become standard (RNAProtect, RNALater, Trizol). All studies will deposit their raw data in public databases (ArrayExpress, BuG@SBase), as this will be a condition for manuscript acceptance by journals. This will enable researchers interested in specific genes or loci to be able to access unprecedented amounts of information about their expression, enabling more focused gene expression studies in the future. It will also provide raw data for systems biologists to produce models of gene regulatory networks.

Amplification technologies will likely improve in the future. At present, approximately 1 $\times 10^9$ (1 billion) bacteria are necessary to extract enough mRNA for one microarray. It can be difficult to collect this many bacteria from *in vivo* models, and this likely explains why there are currently no microarray datasets for *S. aureus* during colonization or infection. In eukaryotic cells, this problem is overcome by amplifying the mRNA using a primer matching the conserved polyA tail of eukaryotic mRNA and production of multiple copies using T7 polymerase. Because bacterial mRNA does not have polyA tails, an alternative approach is to add them (Francois *et al.*, 2007; Motley *et al.*, 2004), although this is relatively difficult with small amounts of mRNA. Unfortunately other polymerases and primer sets cannot be used as they amplify mRNA in a non-linear way, and this can skew any microarray signals generated.

There will also be developments in the technology and new ways to use it. For example, ChIP-chip technology is used to identify exactly which regions of DNA specific proteins bind to (Wade *et al.*, 2007). A protein of interest is mixed with fragmented genomic DNA, and then a specific monoclonal antibody is used to collect the protein using chromatin immunoprecipitation. DNA recovered from the protein is labelled and hybridized to the microarray. Microarrays can also be used to identify the location of genes, such as where a transposon has inserted into a chromosome. A primer facing outward from the gene of interest is used to generate labelled cDNA, which is then hybridized to the microarray (Martinez-Vaz *et al.*, 2005).

Systems biology and other whole genome technologies

The definition of systems biology as applied to bacteria is still evolving. In general it involves the use of engineering and mathematical modelling methods to interpret complex networks, and/or the integration of whole genome data combined with microarrays, proteomics, metabolomics and bioinformatics to address important biological questions. This systems approach should explain cellular functioning in a more meaningful way than by investigation of individual genes or pathways. It is worth noting that a whole genome approach was also taken by several investigators prior to the sequencing projects, and some of these studies will also be mentioned in this section as they are now ready for reinterpretation. Much is expected of systems biology in the future, including the ability to test hypotheses without the need for 'wet' experiments (Stelling, 2004), and this could be especially useful for studying host-pathogen interactions. This will require co-operation from all researchers in depositing high-quality data in the public domain.

Metabolic modelling and essential genes

Whole genome sequencing and annotation makes it possible to construct mathematical models of metabolic pathways. This can help predict which growth factors and chemical reactions are essential, and how the organism will respond to changes in nutrients. Some bacteria such as *E. coli* have become the 'workhorses' of systems biology. However, a few studies have begun to exploit the large number of *S. aureus* sequenced genomes and increasing amounts of gene expression data.

Two groups constructed metabolic models for *S. aureus* using the N315 genome at the same time. Both Becker and Palsson (2005) and Heinemann *et al.* (2005) identified a number of enzymes that were necessary for growth in rich media that could be targeted by known inhibitors. These then become antibiotic candidates. They were also able to show how oxygen is essential for maximum growth mass, as were particular carbon and amino acid sources. Heinemann *et al.* also reconstructed the pathways for cell envelope components (a major target of antimicrobials), respiratory chain, anaerobic growth and small colony variants.

Like all mathematical models, their worth is dependent on how well they correlate with real-life experiments (Kell *et al.*, 2005). These kinds of studies are likely to be increased in the future, especially as their accuracy will improve as more information about the regulation of gene expression (in public databases such as ArrayExpress and BuG@SBase), thermodynamics of reactions, and evolutionary mutation or species variation becomes incorporated (Reed *et al.*, 2006).

Genes essential for growth of *S. aureus* have been identified prior to the sequencing projects by constructing inhibitory RNA (iRNA) libraries (Ji *et al.*, 2002). Random fragments of *S. aureus* were cloned into an inducible expression system and used to generate libraries of mutants able to express iRNA on demand. Those bacteria that survived without induction but not with induction were identified. Many of the essential genes were involved in transcription, translations and metabolism.

Similarly, genes necessary for survival *in vivo* during infection of mouse models were identified (Benton *et al.*, 2004; Coulter *et al.*, 1998; Mei *et al.*, 1997). Signature tagged mutagenesis involves libraries of mutants with multiple transposon insertions being injected into mice, and screening for those that could survive in rich laboratory medium but could not survive during infection. Other models such as nematodes have also been used (Bae *et al.*, 2004). A large number of the mutants identified were also in basic metabolic processes.

The RNome

Genome sequences reveal a range of RNA molecules including rRNA and tRNA for translation. However, ncRNA molecules can also be present which fold into complex structures and bind to DNA, mRNA or proteins altering gene expression, translation or cell physiology (Pichon and Felden, 2005). The well-known accessory gene regulation (*agr*) system uses RNAIII, a ncRNA, as an effector which regulates gene expression via an unknown intermediate, but also controls translation of alpha haemolysin mRNA by binding directly to the transcript (Morfeldt *et al.*, 1995). ncRNA can also control plasmid replication in *S. aureus* (Kwong *et al.*, 2004).

ncRNA in the *S. aureus* genome have been identified by scanning genomes and constructing microarrays to identify ncRNAs that are expressed (O'Neill *et al.*, 2001). Twelve ncRNAs have been proposed in the N315 genome, and more than half are found on MGEs (Pichon and Felden, 2005). Many are also found in multiple copies on the genome. One, *sprA*, has been shown to bind to mRNA of an ABC transporter gene, presumably affecting its translation. Another, 6S RNA, may play a role in binding and sequestering sigma factors and therefore alter gene expression. Surprisingly, the expression of ncRNA varies enormously between strains and may contribute to the complex regulatory networks in clinical isolates of *S. aureus*.

Proteomics

Proteomics aims to identify a range of proteins expressed during particular growth conditions at a particular time and are a useful adjunct or even replacement for microarrays. Proteins are extracted, separated on gels in two dimensions (by pH gradient and size/charge), stained, and compared using image analysis software such as Proteomweaver. Individual spots can be cut out of gels, digested with trypsin, and the size of peptides determined using mass spectrometry. These sizes are then compared to data from the genome sequencing projects to identify the protein using Mascot software. The advantage of proteomics over microarrays is that proteins are considered to be a better predictor of cell function than mRNA, which can be affected by post-transcriptional modification, although in practice the differences are relatively small (Scherl *et al.*, 2006). Another advantage is that proteomics is not constrained by the choice of proteins to investigate, in the way that microarrays are. The disadvantages are that only a fraction of proteins can be separated on 2D gels, and transmembrane proteins, small or highly charged proteins can easily be lost. This is a major drawback if we are interested in whole cell responses. In addition, experiments in complex media such as *in vivo* are difficult due to contamination with non-bacterial proteins.

Proteomics has been successfully used to identify differences between isolates that are susceptible or resistant to vancomycin, identifying important peptidoglycan synthesis components (Gatlin *et al.*, 2006; Pieper *et al.*, 2006; Scherl *et al.*, 2006) (see Chapter 8). Resch *et al.* (2006) identified proteins associated with biofilm production including those involved in cell attachment, while some proteases and toxins were downregulated. Plikat *et al.* (2007) used proteomics to investigate formylation of *S. aureus* proteins due to *fmt*, as this is a target for antibiotics. Friedman *et al.* (2006) have investigated responses to iron and haem identifying a novel transport system.

Host genetics

If *S. aureus* is an opportunistic pathogen, then host must play a major role in disease. Many mutations in immune response genes in humans have been identified, and theoretically they could play a vital role in susceptibility to infection. Early studies have associated some SNPs with host susceptibility to carriage and/or disease, although the frequency of the mutations is rare (Claassen *et al.*, 2005; Emonts *et al.*, 2007; Moore *et al.*, 2004; van den Akker *et al.*, 2006). The identification of large numbers of human genome SNPs is frenetic with the human sequencing projects and SNP detection programmes such as HapMap (www.hapmap.org). The development of commercial SNP arrays allowing up to 1 million SNPs to be interrogated per array are expected to revolutionize the efficiency of experiments to compare susceptible versus non-susceptible hosts (www.affymetrix.com; www.illumina.com).

Conclusions

Technologies that exploit whole genomes or whole cell responses have only become widespread in bacteriology in the last ten or so years, and the first *S. aureus* whole genome was deposited in 2001. The progress made in only six years is incredible, particularly in sequencing, microarrays, metabolic modelling, the RNome and proteomics. Already we have made major advances in our understanding of *S. aureus* metabolism, virulence, gene expression and regulation, evolution, host–pathogen interactions, diagnostics and more. Because the technologies, and the tools to analyse the enormous volumes of data, are still being developed, we expect the study of *S. aureus* to continue at breakneck speed.

Future

The future of whole genome technologies is very exciting. In general, we have discussed the likely future developments for each technology in the relevant section of text. Overall, we expect to see developments in rapid sequencing, microarray design strategies, inexpensive microarray platforms and detection technologies, amplification of small mRNA levels for microarrays, simpler microarray data deposition, improved proteomic separation and spot detection, and commercial human SNP arrays with >1 million targets.

Another major step in the future will be improved methods of data analysis. In the future, we expect advances in microarray comparative genomics data analysis, proteomics data analysis, metabolic modelling, and host specificity. Just as important will be the ability to compare data from multiple technologies in a format that is easy to use, publicly available and inexpensive. This will be even more important when the volume of data from whole genome technologies increases dramatically, as these types of studies become more accepted, inexpensive and practical.

There are also emerging technologies that haven't yet been exploited by *S. aureus* biologists. These include Chip-Chip microarray analysis, the use of microarrays to determine the location of genes, and metabolomics. There is also a great need to develop better models of infection, so that these technologies can be used in a more biologically relevant way. It is likely that we have only reached the tip of the iceberg, and the future of whole cell technologies for studying *S. aureus* is very bright indeed.

Acknowledgements

JAL is indebted to the Bacterial Microarray Group at St Georges (BμG@S), Philip Butcher, Jason Hinds, Adam Witney, Sally Husain, Kate Gould, Richard Stabler, Gemma Marsden and Lucy Brooks, for excellent technical and collaborative support for many years. BμG@S are funded by The Wellcome Trust. She also wishes to thank her many collaborators who have worked on microarray projects. Work in JAL's laboratory is funded by BBSRC, St George's Charitable Foundation and DEFRA. MTGH is funded by The Wellcome Trust.

Web resources

Bacterial Microarray Group at St George's
www.bugs.sgul.ac.uk
Designs, prints, supplies microarrays to collaborators and fosters education and training in using microarray technology.

ArrayExpress
www.ebi.ac.uk/arrayexpress
Microarray data deposition

Genbank
http://www.ncbi.nlm.nih.gov/Entrez/
Entrez sequence search site at the NCBI

EMBL
http://srs.ebi.ac.uk/
SRS (sequence retrieval system) sequence search for EMBL databases

DDBJ
http://srs.ddbj.nig.ac.jp/
SRS (sequence retrieval system) sequence search for DNA Data Bank of Japan database

EBI
http://www.ebi.ac.uk/genomes/bacteria.html
List of complete bacterial genomes in the public database with links for downloading

Database for *S. aureus* genomes:
GeneDB
www.genedb.org/
Contains MRSA252 and MSSA476

Comprehensive Microbial Resource – CMR
cmr.tigr.org
Contains COL and the other sequenced *S. aureus* genomes

DOGAN
www.bio.nite.go.jp/dogan/Top
Contains N315 and MW2

AureoList
genolist.pasteur.fr/AureoList/index.html
Contains Mu50 and N315

BioCyc
biocyc.org/
Metabolic mapping of *S. aureus*

Kyoto Encyclopedia of Genes and Genomes – KEGG
www.genome.jp/kegg/
Metabolic mapping of *S. aureus*

Webact
http://www.webact.org/WebACT/home
Comparative genomic resource for generating, or download pre-computed, ACT comparison files

BugsBase
http://bugs.sgul.ac.uk/bugsbase/
Microarray data deposition

SAMMD – *S. aureus* microarray meta-database
bioinformatics.org/sammd/
Allows interrogation of the microarray data published in the literature but not in microarray databases

BLAST
http://www.ncbi.nlm.nih.gov/BLAST/
NCBI BLAST server

FASTA
http://www.ebi.ac.uk/fasta33/index.html
EBI FASTA server for searching UniProt

References

Abbott, J.C., Aanensen, D.M., Rutherford, K., Butcher, S., and Spratt, B.G. (2005). WebACT – an online companion for the Artemis Comparison Tool. Bioinformatics *21*, 3665–3666.

Altschul, S.F., Gish, W., Miller, W., Myers, E.W., and Lipman, D.J. (1990). Basic Local Alignment Search Tool. J. Molec. Biol. *215*, 403–410.

Anderson, K.L., Roberts, C., Disz, T., Vonstein, V., Hwang, K., Overbeek, R., Olson, P.D., Projan, S.J., and Dunman, P.M. (2006). Characterization of the *Staphylococcus aureus* heat shock, cold shock, stringent, and SOS responses and their effects on log-phase mRNA turnover. J. Bacteriol. *188*, 6739–6756.

Apweiler, R., Bairoch, A., Wu, C.H., Barker, W.C., Boeckmann, B., Ferro, S., Gasteiger, E., Huang, H., Lopez, R., Magrane, M., *et al.* (2004). UniProt: the Universal Protein knowledgebase. Nucleic Acids Res. *32*, D115–119.

Baba, T., Takeuchi, F., Kuroda, M., Yuzawa, H., Aoki, K., Oguchi, A., Nagai, Y., Iwama, N., Asano, K., Naimi, T., *et al.* (2002). Genome and virulence determinants of high virulence community- acquired MRSA. Lancet *359*, 1819–1827.

Bae, T., Banger, A.K., Wallace, A., Glass, E.M., Aslund, F., Schneewind, O., and Missiakas, D.M. (2004). *Staphylococcus aureus* virulence genes identified by bursa aurealis mutagenesis and nematode killing. Proc. Natl. Acad. Sci. USA *101*, 12312–12317.

Barrangou, R., Fremaux, C., Deveau, H., Richards, M., Boyaval, P., Moineau, S., Romero, D.A., and Horvath, P. (2007). CRISPR provides acquired resistance against viruses in prokaryotes. Science *315*, 1709–1712.

Becker, S.A., and Palsson, B.O. (2005). Genome-scale reconstruction of the metabolic network in *Staphylococcus aureus* N315: an initial draft to the two-dimensional annotation. BMC Microbiol. *5*, 8.

Bendtsen, J.D., Nielsen, H., von Heijne, G., and Brunak, S. (2004). Improved prediction of signal peptides: SignalP 3.0. J. Mol. Biol. *340*, 783–795.

Benton, B.M., Zhang, J.P., Bond, S., Pope, C., Christian, T., Lee, L., Winterberg, K.M., Schmid, M.B., and Buysse, J.M. (2004). Large-scale identification of genes required for full virulence of *Staphylococcus aureus*. J. Bacteriol. *186*, 8478–8489.

Bischoff, M., Dunman, P., Kormanec, J., Macapagal, D., Murphy, E., Mounts, W., Berger-Bachi, B., and Projan, S. (2004). Microarray-based analysis of the *Staphylococcus aureus* sigmaB regulon. J. Bacteriol. *186*, 4085–4099.

Bissett, D.L., and Anderson, R.L. (1974). Genetic evidence for the physiological significance of the D-tagatose 6-phosphate pathway of lactose and D-galactose degradation in *Staphylococcus aureus*. J. Bacteriol. *119*, 698–704.

Brazma, A., Hingamp, P., Quackenbush, J., Sherlock, G., Spellman, P., Stoeckert, C., Aach, J., Ansorge, W., Ball, C.A., Causton, H.C., *et al.* (2001). Minimum information about a microarray experiment (MIAME)-toward standards for microarray data. Nat. Genet. *29*, 365–371.

Carver, T.J., Rutherford, K., Berriman, M., Rajandream, M.A., Barrell, B., and Parkhill, J. (2005). ACT: the Artemis comparison tool. Bioinformatics *21*, 3422–3423.

Champion, O.L., Gaunt, M.W., Gundogdu, O., Elmi, A., Witney, A.A., Hinds, J., Dorrell, N., and Wren, B.W. (2005). Comparative phylogenomics of the food-borne pathogen *Campylobacter jejuni* reveals genetic markers predictive of infection source. Proc. Natl. Acad. Sci. USA *102*, 16043–16048.

Chang, W., Small, D.A., Toghrol, F., and Bentley, W.E. (2006). Global transcriptome analysis of *Staphylococcus aureus* response to hydrogen peroxide. J. Bacteriol. *188*, 1648–1659.

Cirz, R.T., Jones, M.B., Gingles, N.A., Minogue, T.D., Jarrahi, B., Peterson, S.N., and Romesberg, F.E. (2007). Complete and SOS-mediated response of *Staphylococcus aureus* to the antibiotic ciprofloxacin. J. Bacteriol. *189*, 531–539.

Claassen, M., Nouwen, J., Fang, Y., Ott, A., Verbrugh, H., Hofman, A., van Belkum, A., and Uitterlinden, A. (2005). *Staphylococcus aureus* nasal carriage is not associated with known polymorphism in the Vitamin D receptor gene. FEMS Immunol. Med. Microbiol. *43*, 173–176.

Clausen, V.A., Bae, W., Throup, J., Burnham, M.K., Rosenberg, M., and Wallis, N.G. (2003). Biochemical characterization of the first essential two-component signal transduction system from *Staphylococcus aureus* and *Streptococcus pneumoniae*. J. Mol. Microbiol. Biotechnol. *5*, 252–260.

Cockfield, J.D., Pathak, S., Edgeworth, J.D., and Lindsay, J.A. (2007). Rapid determination of hospital-acquired meticillin-resistant *Staphylococcus aureus* lineages. J. Med. Microbiol. *56*, 614–619.

Cole, S.T., Eiglmeier, K., Parkhill, J., James, K.D., Thomson, N.R., Wheeler, P.R., Honore, N., Garnier, T., Churcher, C., Harris, D., *et al.* (2001). Massive gene decay in the leprosy bacillus. Nature *409*, 1007–1011.

Coulter, S.N., Schwan, W.R., Ng, E.Y., Langhorne, M.H., Ritchie, H.D., Westbrock-Wadman, S., Hufnagle, W.O., Folger, K.R., Bayer, A.S., and Stover, C.K. (1998). *Staphylococcus aureus* genetic loci impacting growth and survival in multiple infection environments. Mol. Microbiol. *30*, 393–404.

Cui, L., Lian, J.Q., Neoh, H.M., Reyes, E., and Hiramatsu, K. (2005). DNA microarray-based identification of genes associated with glycopeptide resistance in *Staphylococcus aureus*. Antimicrob. Agents Chemother. *49*, 3404–3413.

Delcher, A.L., Bratke, K.A., Powers, E.C., and Salzberg, S.L. (2007). Identifying bacterial genes and endosymbiont DNA with Glimmer. Bioinformatics *23*, 673–679.

Diep, B.A., Gill, S.R., Chang, R.F., Phan, T.H., Chen, J.H., Davidson, M.G., Lin, F., Lin, J., Carleton, H.A., Mongodin, E.F., *et al.* (2006). Complete genome sequence of USA300, an epidemic clone of community-acquired meticillin-resistant *Staphylococcus aureus*. Lancet *367*, 731–739.

Dufour, P., Gillet, Y., Bes, M., Lina, G., Vandenesch, F., Floret, D., Etienne, J., and Richet, H. (2002). Community-acquired methicillin-resistant *Staphylococcus aureus* infections in France: Emergence of a single clone that produces Panton–Valentine leukocidin. Clin. Infect. Dis. *35*, 819–824.

Dunman, P.M., Murphy, E., Haney, S., Palacios, D., Tucker-Kellogg, G., Wu, S., Brown, E.L., Zagursky, R.J., Shlaes, D., and Projan, S.J. (2001). Transcription profiling-based identification of *Staphylococcus aureus* genes regulated by the *agr* and/or *sarA* loci. J. Bacteriol. *183*, 7341–7353.

Edgeworth, J.D., Yadegarfar, G., Pathak, S., Batra, R., Cockfield, J.D., Wyncoll, D., Beale, R., and Lindsay, J.A. (2007). An outbreak in an intensive care unit of a strain of methicillin-resistant *Staphylococcus aureus* sequence type 239 associated with an increased rate of vascular access device-related bacteremia. Clin. Infect. Dis. *44*, 493–501.

Emonts, M., de Jongh, C.E., Houwing-Duistermaat, J.J., van Leeuwen, W.B., de Groot, R., Verbrugh, H.A., Hermans, P.W., and van Belkum, A. (2007). Association between nasal carriage of *Staphylococcus aureus* and the human complement cascade activator serine protease C1 inhibitor (C1INH) valine vs. methionine polymorphism at amino acid position 480. FEMS Immunol. Med. Microbiol. *50*, 330–332.

Enright, A.J., Kunin, V., and Ouzounis, C.A. (2003). Protein families and TRIBES in genome sequence space. Nucleic Acids Res. *31*, 4632–4638.

Fitzgerald, J.R., Sturdevant, D.E., Mackie, S.M., Gill, S.R., and Musser, J.M. (2001). Evolutionary genomics of *Staphylococcus aureus*: Insights into the origin of methicillin-resistant strains and the toxic shock syndrome epidemic. Proc. Natl. Acad. Sci. USA *98*, 8821–8826.

Fleischmann, R.D., Adams, M.D., White, O., Clayton, R.A., Kirkness, E.F., Kerlavage, A.R., Bult, C.J., Tomb, J.F., Dougherty, B.A., Merrick, J.M., and *et al.* (1995). Whole-genome random sequencing and assembly of *Haemophilus influenzae* Rd. Science *269*, 496–512.

Fox, P.M., Climo, M.W., and Archer, G.L. (2007). Lack of relationship between purine biosynthesis and vancomycin resistance in *Staphylococcus aureus*: a cautionary tale for microarray interpretation. Antimicrob. Agents Chemother. *51*, 1274–1280.

Francois, P., Garzoni, C., Bento, M., and Schrenzel, J. (2007). Comparison of amplification methods for transcriptomic analyses of low abundance prokaryotic RNA sources. J. Microbiol. Methods *68*, 385–391.

Friedman, D.B., Stauff, D.L., Pishchany, G., Whitwell, C.W., Torres, V.J., and Skaar, E.P. (2006). *Staphylococcus aureus* redirects central metabolism to increase iron availability. PLoS Pathog. *2*, e87.

Frishman, D., Mironov, A., Mewes, H.W., and Gelfand, M. (1998). Combining diverse evidence for gene recognition in completely sequenced bacterial genomes. Nucleic Acids Res. *26*, 2941–2947.

Gatlin, C.L., Pieper, R., Huang, S.T., Mongodin, E., Gebregeorgis, E., Parmar, P.P., Clark, D.J., Alami, H., Papazisi, L., Fleischmann, R.D., *et al.* (2006). Proteomic profiling of cell envelope-associated proteins from *Staphylococcus aureus*. Proteomics *6*, 1530–1549.

Gattiker, A., Michoud, K., Rivoire, C., Auchincloss, A.H., Coudert, E., Lima, T., Kersey, P., Pagni, M., Sigrist, C.J., Lachaize, C., *et al.* (2003). Automated annotation of microbial proteomes in SWISS-PROT. Comput. Biol. Chem. *27*, 49–58.

Gill, S.R., Fouts, D.E., Archer, G.L., Mongodin, E.F., DeBoy, R.T., Ravel, J., Paulsen, I.T., Kolonay, J.F., Brinkac, L., Beanan, M., *et al.* (2005). Insights on evolution of virulence and resistance from the complete genome analysis of an early methicillin-resistant *Staphylococcus aureus* strain and a biofilm-producing methicillin-resistant *Staphylococcus epidermidis* strain. J. Bacteriol. *187*, 2426–2438.

Gillaspy, A.F., Worrell, V., Orvis, J., Roe, B.A., Dyer, D.W., and Iandolo, J.J. (2006). The *Staphylococcus aureus* NCTC 8325 genome, In Gram positive pathogens, V. Fischetti, R. Novick, J. Ferretti, D. Portnoy, and J. Rood, eds. (ASM Press), pp. 381–412.

Gillet, Y., Issartel, B., Vanhems, P., Fournet, J.C., Lina, G., Bes, M., Vandenesch, F., Piemont, Y., Brousse, N., Floret, D., and Etienne, J. (2002). Association between *Staphylococcus aureus* strains carrying gene for Panton–Valentine leukocidin and highly lethal necrotising pneumonia in young immunocompetent patients. Lancet *359*, 753–759.

Griffiths-Jones, S., Bateman, A., Marshall, M., Khanna, A., and Eddy, S.R. (2003). Rfam: an RNA family database. Nucleic Acids Res. *31*, 439–441.

Grissa, I., Vergnaud, G., and Pourcel, C. (2007). The CRISPRdb database and tools to display CRISPRs and to generate dictionaries of spacers and repeats. BMC Bioinformatics *8*, 172.

Heinemann, M., Kummel, A., Ruinatscha, R., and Panke, S. (2005). *In silico* genome-scale reconstruction and validation of the *Staphylococcus aureus* metabolic network. Biotechnol. Bioeng. *92*, 850–864.

Herold, B.C., Immergluck, L.C., Maranan, M.C., Lauderdale, D.S., Gaskin, R.E., Boyle-Vavra, S., Leitch, C.D., and Daum, R.S. (1998). Community-acquired methicillin-resistant *Staphylococcus aureus* in children with no identified predisposing risk. JAMA *279*, 593–598.

Hinton, J.C. (1997). The *Escherichia coli* genome sequence: the end of an era or the start of the FUN? Mol. Microbiol. *26*, 417–422.

Holden, M.T.G., Feil, E.J., Lindsay, J.A., Peacock, S.J., Day, N.P.J., Enright, M.C., Foster, T.J., Moore, C.E., Hurst, L., Atkin, R., *et al.* (2004). Complete genomes of two clinical *Staphylococcus aureus* strains: Evidence for the rapid evolution of virulence and drug resistance. Proc. Natl. Acad. Sci. USA *101*, 9786–9791.

Ji, Y., Woodnutt, G., Rosenberg, M., and Burnham, M.K. (2002). Identification of essential genes in *Staphylococcus aureus* using inducible antisense RNA. Methods Enzymol. *358*, 123–128.

Johnson, D.S., Mortazavi, A., Myers, R.M., and Wold, B. (2007). Genome-wide mapping of *in vivo* protein-DNA interactions. Science *316*, 1497–1502.

Kanehisa, M., Goto, S., Hattori, M., Aoki-Kinoshita, K.F., Itoh, M., Kawashima, S., Katayama, T., Araki, M., and Hirakawa, M. (2006). From genomics to chemical genomics: new developments in KEGG. Nucleic Acids Res. *34*, D354-D357.

Kell, D.B., Brown, M., Davey, H.M., Dunn, W.B., Spasic, I., and Oliver, S.G. (2005). Metabolic footprinting and systems biology: the medium is the message. Nat. Rev. Microbiol. *3*, 557–565.

Krogh, A., Larsson, B., von Heijne, G., and Sonnhammer, E.L.L. (2001). Predicting transmembrane protein topology with a hidden Markov model: Application to complete genomes. J. Molec. Biol. *305*, 567–580.

Kuhn, G., Francioli, P., and Blanc, D.S. (2006). Evidence for clonal evolution among highly polymorphic genes in methicillin-resistant *Staphylococcus aureus*. J. Bacteriol. *188*, 169–178.

Kuroda, M., Kuroda, H., Oshima, T., Takeuchi, F., Mori, H., and Hiramatsu, K. (2003). Two-component system VraSR positively modulates the regulation of cell-wall biosynthesis pathway in *Staphylococcus aureus*. Mol. Microbiol. *49*, 807–821.

Kuroda, M., Ohta, T., Uchiyama, I., Baba, T., Yuzawa, H., Kobayashi, I., Cui, L.Z., Oguchi, A., Aoki, K., Nagai, Y., *et al.* (2001). Whole genome sequencing of meticillin-resistant *Staphylococcus aureus*. Lancet *357*, 1225–1240.

Kurtz, S., Choudhuri, J.V., Ohlebusch, E., Schleiermacher, C., Stoye, J., and Giegerich, R. (2001). REPuter: the manifold applications of repeat analysis on a genomic scale. Nucleic Acids Res. *29*, 4633–4642.

Kwong, S.M., Skurray, R.A., and Firth, N. (2004). *Staphylococcus aureus* multiresistance plasmid pSK41: analysis of the replication region, initiator protein binding and antisense RNA regulation. Mol. Microbiol. *51*, 497–509.

Larsen, T.S., and Krogh, A. (2003). EasyGene – a prokaryotic gene finder that ranks ORFs by statistical significance. BMC Bioinformatics *4*, 21.

Lerat, E., and Ochman, H. (2005). Recognizing the pseudogenes in bacterial genomes. Nucleic Acids Res. *33*, 3125–3132.

Liang, X., Yu, C., Sun, J., Liu, H., Landwehr, C., Holmes, D., and Ji, Y. (2006). Inactivation of a two-component signal transduction system, SaeRS, eliminates adherence and attenuates virulence of *Staphylococcus aureus*. Infect. Immun. *74*, 4655–4665.

Lindsay, J.A., Moore, C.E., Day, N.P., Peacock, S.J., Witney, A.A., Stabler, R.A., Husain, S.E., Butcher, P.D., and Hinds, J. (2006). Microarrays reveal that each of the ten dominant lineages of *Staphylococcus aureus* has a unique combination of surface-associated and regulatory genes. J. Bacteriol. *188*, 669–676.

Lindsay, J.A., Ruzin, A., Ross, H.F., Kurepina, N., and Novick, R.P. (1998). The gene for toxic shock toxin is carried by a family of mobile pathogen.icity islands in *Staphylococcus aureus*. Mol. Microbiol. *29*, 527–543.

Liolios, K., Tavernarakis, N., Hugenholtz, P., and Kyrpides, N.C. (2006). The Genomes On Line Database (GOLD) v.2: a monitor of genome projects worldwide. Nucleic Acids Res. *34*, D332–334.

Livny, J., and Waldor, M.K. (2007). Identification of small RNAs in diverse bacterial species. Curr. Opin. Microbiol. *10*, 96–101.

Lobry, J.R. (1996). Asymmetric substitution patterns in the two DNA strands of bacteria. Molec. Biol. Evol. *13*, 660–665.

Lowe, T.M., and Eddy, S.R. (1997). tRNAscan-SE: A program for improved detection of transfer RNA genes in genomic sequence. Nucleic Acids Res. *25*, 955–964.

Luong, T.T., Dunman, P.M., Murphy, E., Projan, S.J., and Lee, C.Y. (2006). Transcription Profiling of the *mgrA* Regulon in *Staphylococcus aureus*. J. Bacteriol. *188*, 1899–1910.

Mangan, J.A., Sole, K.M., Mitchison, D.A., and Butcher, P.D. (1997). An effective method of RNA extraction from bacteria refractory to disruption, including mycobacteria. Nucleic Acids Res. *25*, 675–676.

Margulies, M., Egholm, M., Altman, W.E., Attiya, S., Bader, J.S., Bemben, L.A., Berka, J., Braverman, M.S., Chen, Y.J., Chen, Z., *et al.* (2005). Genome sequencing in microfabricated high-density picolitre reactors. Nature *437*, 376–380.

Martinez-Vaz, B.M., Xie, Y., Pan, W., and Khodursky, A.B. (2005). Genome-wide localization of mobile elements: experimental, statistical and biological considerations. BMC Genomics *6*, 81.

McCallum, N., Spehar, G., Bischoff, M., and Berger-Bachi, B. (2006). Strain dependence of the cell wall-damage induced stimulon in *Staphylococcus aureus*. Biochim. Biophys. Acta. *1760*, 1475–1481.

Mei, J.M., Nourbakhsh, F., Ford, C.W., and Holden, D.W. (1997). Identification of *Staphylococcus aureus* virulence genes in a murine model of bacteraemia using signature-tagged mutagenesis. Mol. Microbiol. *26*, 399–407.

Michel, A., Agerer, F., Hauck, C.R., Herrmann, M., Ullrich, J., Hacker, J., and Ohlsen, K. (2006). Global regulatory impact of ClpP protease of *Staphylococcus aureus* on regulons involved in virulence, oxidative stress response, autolysis, and DNA repair. J. Bacteriol. *188*, 5783–5796.

Moore, C.E., Segal, S., Berendt, A.R., Hill, A.V., and Day, N.P. (2004). Lack of association between Toll-like receptor 2 polymorphisms and susceptibility to severe disease caused by *Staphylococcus aureus*. Clin. Diagn. Lab. Immunol. *11*, 1194–1197.

Morfeldt, E., Taylor, D., von Gabain, A., and Arvidson, S. (1995). Activation of alpha-toxin translation in

Staphylococcus aureus by the trans-encoded antisense RNA, RNAIII. EMBO J *14*, 4569–4577.

Motley, S.T., Morrow, B.J., Liu, X., Dodge, I.L., Vitiello, A., Ward, C.K., and Shaw, K.J. (2004). Simultaneous analysis of host and pathogen interactions during an *in vivo* infection reveals local induction of host acute phase response proteins, a novel bacterial stress response, and evidence of a host-imposed metal ion limited environment. Cell. Microbiol. *6*, 849–865.

Mulder, N.J., Apweiler, R., Attwood, T.K., Bairoch, A., Bateman, A., Binns, D., Bork, P., Buillard, V., Cerutti, L., Copley, R., *et al.* (2007). New developments in the InterPro database. Nucleic Acids Res. *35*, D224–228.

Mwangi, M.M., Wu, S.W., Zhou, Y., Sieradzki, K., de Lencastre, H., Richardson, P., Bruce, D., Rubin, E., Myers, E., Siggia, E.D., and Tomasz, A. (2007). Tracking the *in vivo* evolution of multidrug resistance in *Staphylococcus aureus* by whole-genome sequencing. Proc. Natl. Acad. Sci. USA *104*, 9451–9456.

O'Neill, G.L., Murchan, S., Gil-Setas, A., and Aucken, H.M. (2001). Identification and characterization of phage variants of a strain of epidemic methicillin-resistant *Staphylococcus aureus* (EMRSA-15). J. Clin. Microbiol. *39*, 1540–1548.

Pearson, W.R., and Lipman, D.J. (1988). Improved Tools for Biological Sequence Comparison. Proc. Natl. Acad. Sci. USA *85*, 2444–2448.

Pichon, C., and Felden, B. (2005). Small RNA genes expressed from *Staphylococcus aureus* genomic and pathogenicity islands with specific expression among pathogenic strains. Proc. Natl. Acad. Sci. USA *102*, 14249–14254.

Pieper, R., Gatlin-Bunai, C.L., Mongodin, E.F., Parmar, P.P., Huang, S.T., Clark, D.J., Fleischmann, R.D., Gill, S.R., and Peterson, S.N. (2006). Comparative proteomic analysis of *Staphylococcus aureus* strains with differences in resistance to the cell wall-targeting antibiotic vancomycin. Proteomics *6*, 4246–4258.

Plikat, U., Voshol, H., Dangendorf, Y., Wiedmann, B., Devay, P., Muller, D., Wirth, U., Szustakowski, J., Chirn, G.W., Inverardi, B., *et al.* (2007). From proteomics to systems biology of bacterial pathogens: approaches, tools, and applications. Proteomics *7*, 992–1003.

Rawlings, N.D., Morton, F.R., and Barrett, A.J. (2006). MEROPS: the peptidase database. Nucleic Acids Res. *34*, D270–272.

Reed, J.L., Famili, I., Thiele, I., and Palsson, B.O. (2006). Towards multidimensional genome annotation. Nat. Rev. Genet. *7*, 130–141.

Renzoni, A., Barras, C., Francois, P., Charbonnier, Y., Huggler, E., Garzoni, C., Kelley, W.L., Majcherczyk, P., Schrenzel, J., Lew, D.P., and Vaudaux, P. (2006). Transcriptomic and functional analysis of an autolysis-deficient, teicoplanin-resistant derivative of methicillin-resistant *Staphylococcus aureus*. Antimicrob. Agents Chemother. *50*, 3048–3061.

Resch, A., Leicht, S., Saric, M., Pasztor, L., Jakob, A., Gotz, F., and Nordheim, A. (2006). Comparative proteome analysis of *Staphylococcus aureus* biofilm and planktonic cells and correlation with transcriptome profiling. Proteomics *6*, 1867–1877.

Resch, A., Rosenstein, R., Nerz, C., and Gotz, F. (2005). Differential gene expression profiling of *Staphylococcus aureus* cultivated under biofilm and planktonic conditions. Appl. Environ. Microbiol. *71*, 2663–2676.

Richardson, A.R., Dunman, P.M., and Fang, F.C. (2006). The nitrosative stress response of *Staphylococcus aureus* is required for resistance to innate immunity. Mol. Microbiol. *61*, 927–939.

Riordan, J.T., Muthaiyan, A., Van Voorhies, W., Price, C.T., Graham, J.E., Wilkinson, B.J., and Gustafson, J.E. (2007). Response of *Staphylococcus aureus* to salicylate challenge. J. Bacteriol. *189*, 220–227.

Robertson, G., Hirst, M., Bainbridge, M., Bilenky, M., Zhao, Y., Zeng, T., Euskirchen, G., Bernier, B., Varhol, R., Delaney, A., *et al.* (2007). Genome-wide profiles of STAT1 DNA association using chromatin immunoprecipitation and massively parallel sequencing. Nat. Methods *4*, 651–7.

Rutherford, K., Parkhill, J., Crook, J., Horsnell, T., Rice, P., Rajandream, M.A., and Barrell, B. (2000). Artemis: sequence visualization and annotation. Bioinformatics *16*, 944–945.

Sanger, F., Air, G.M., Barrell, B.G., Brown, N.L., Coulson, A.R., Fiddes, C.A., Hutchison, C.A., Slocombe, P.M., and Smith, M. (1977). Nucleotide sequence of bacteriophage phi X174 DNA. Nature *265*, 687–695.

Scherl, A., Francois, P., Charbonnier, Y., Deshusses, J.M., Koessler, T., Huyghe, A., Bento, M., Stahl-Zeng, J., Fischer, A., Masselot, A., *et al.* (2006). Exploring glycopeptide-resistance in *Staphylococcus aureus*: a combined proteomics and transcriptomics approach for the identification of resistance-related markers. BMC Genomics *7*, 296.

Sibbald, M.J., Ziebandt, A.K., Engelmann, S., Hecker, M., de Jong, A., Harmsen, H.J., Raangs, G.C., Stokroos, I., Arends, J.P., Dubois, J.Y., and van Dijl, J.M. (2006). Mapping the pathways to staphylococcal pathogenesis by comparative secretomics. Microbiol. Mol. Biol. Rev. *70*, 755–788.

Sobral, R.G., Jones, A.E., Des Etages, S.G., Dougherty, T.J., Peitzsch, R.M., Gaasterland, T., Ludovice, A.M., de Lencastre, H., and Tomasz, A. (2007). Extensive and genome-wide changes in the transcription profile of *Staphylococcus aureus* induced by modulating the transcription of the cell wall synthesis gene *murF*. J. Bacteriol. *189*, 2376–2391.

Sonnhammer, E.L., and Durbin, R. (1995). A dot-matrix program with dynamic threshold control suited for genomic DNA and protein sequence analysis. Gene *167*, GC1–10.

Stelling, J. (2004). Mathematical models in microbial systems biology. Curr Opin Microbiol *7*, 513–518.

Stranger-Jones, Y.K., Bae, T., and Schneewind, O. (2006). Vaccine assembly from surface proteins of *Staphylococcus aureus*. Proc. Natl. Acad. Sci. USA *103*, 16942–16947.

Tettelin, H., Masignani, V., Cieslewicz, M.J., Donati, C., Medini, D., Ward, N.L., Angiuoli, S.V., Crabtree, J., Jones, A.L., Durkin, A.S., *et al.* (2005). Genome analysis of multiple pathogenic isolates of *Streptococcus agalactiae*: implications for the microbial 'pan-genome'. Proc. Natl. Acad. Sci. USA *102*, 13950–13955.

Utaida, S., Dunman, P.M., Macapagal, D., Murphy, E., Projan, S.J., Singh, V.K., Jayaswal, R.K., and Wilkinson, B.J. (2003). Genome-wide transcriptional profiling of the response of *Staphylococcus aureus* to cell-wall-active antibiotics reveals a cell-wall-stress stimulon. Microbiol. *149*, 2719–2732.

van den Akker, E.L., Nouwen, J.L., Melles, D.C., van Rossum, E.F., Koper, J.W., Uitterlinden, A.G., Hofman, A., Verbrugh, H.A., Pols, H.A., Lamberts, S.W., and van Belkum, A. (2006). *Staphylococcus aureus* nasal carriage is associated with glucocorticoid receptor gene polymorphisms. J. Infect. Dis. *194*, 814–818.

Wade, J.T., Struhl, K., Busby, S.J., and Grainger, D.C. (2007). Genomic analysis of protein-DNA interactions in bacteria: insights into transcription and chromosome organization. Mol. Microbiol. 65, 21–26.

Waldron, D.E., and Lindsay, J.A. (2006). *Sau*1: a novel lineage-specific type I restriction-modification system that blocks horizontal gene transfer into *Staphylococcus aureus* and between *S. aureus* isolates of different lineages. J. Bacteriol. *188*, 5578–5585.

Wernisch, L., Kendall, S.L., Soneji, S., Wietzorrek, A., Parish, T., Hinds, J., Butcher, P.D., and Stoker, N.G. (2003). Analysis of whole-genome microarray replicates using mixed models. Bioinformatics *19*, 53–61.

Witney, A.A., Marsden, M., Holden, T.G., Stabler, R.A., Husain, S.E., Vass, J.K., Butcher, P.D., Hinds, J., and Lindsay, J.A. (2005). Design, validation, and application of a seven-strain *Staphylococcus aureus* PCR product microarray for comparative genomics. Appl. Environ. Microbiol. *71*, 7504–7514.

Yang, S.J., Dunman, P.M., Projan, S.J., and Bayles, K.W. (2006). Characterization of the *Staphylococcus aureus* CidR regulon: elucidation of a novel role for acetoin metabolism in cell death and lysis. Mol. Microbiol. *60*, 458–468.

The Population Structure of *Staphylococcus aureus*

2

Mark C. Enright

Abstract

Staphylococcus aureus is a major human pathogen causing a wide spectrum of diseases from minor ailments to severe life-threatening conditions. Methicillin-resistant *S. aureus* (MRSA) are endemic in most hospitals in many industrialized countries and they are considered the most serious hospital-acquired pathogen as they can cause large outbreaks that are frequently difficult to treat using antibiotics. A large amount of genetic information is available on several examples of this species and this, together with multilocus sequence typing (MLST) data on >1400 isolates from many countries has provided unique insights into the biology of the species and in particular, its ability to exploit a wide variety of niches. These studies show that although the great majority of *S. aureus* genes share a high degree of homology, virulence and antibiotic resistance genes carried on mobile genetic elements can drastically alter strain characteristics in the short-term giving the species a high degree of adaptability. These allow it to survive in many different human and animal tissues and provide the adaptability necessary to evolve resistance to new antibiotics.

Introduction

Knowledge of the population structure and epidemiology of bacterial populations has proven to be crucial when examining and modelling how pathogenic species spread, when following the evolution and movement of antibiotic resistance and virulence genes within and between species and in estimating the potential efficacy of vaccination programmes. Less obviously, population studies on the bioterrorism agent *Bacillus anthracis* (Pearson *et al.*, 2004) have provided information on its likely origins and such approaches may well prove to be useful for other potential agents in future. The increasingly rich and diverse sources of genomic sequences of bacteria that have recently become publicly available has provided easy access to the genetic information required to examine the genetic diversity within many pathogenic bacterial species. This has spurred many studies of the epidemiology and evolution of major human bacterial pathogens that are key in outbreak analysis and will be invaluable in the future to help us formulate novel, effective therapies.

Reservoirs

Staphylococcus aureus is overwhelmingly a colonizer of animals and is especially common in mammals. In humans the organism primarily resides in the anterior nares (vestibulum nasi) and is often shed onto healthy skin and can frequently be sampled from the axilla and perineum (Williams, 1963). In healthy humans *S. aureus* can also occasionally be found as part of the flora of the digestive and vaginal tracts (Smith *et al.*, 1982). Carriage rates vary between different human populations and between different studies; however, it is widely accepted that regarding carriage, three types of individuals exist (Williams, 1963; VandenBergh *et al.*, 1999). Humans are either non-carriers (approximately 20% of the population); persistent carriers (20–25%) or intermittent carriers (55–60%). The carrier state is extremely important because *S. aureus* carriage has been shown to be an important risk factor for invasive disease (Peacock *et al.*, 2001) with most community-acquired infections due to the colonizing strain (Yu *et al.*, 1986; Luzar *et al.*, 1990).

Disease

Community-acquired

All, or nearly all individuals will suffer from *S. aureus* infections in their lifetime. Pimples, boils, styes, conjunctivitis and uncomplicated wound infections are very common, and minor ailments frequently caused by this species. Invasive disease, the infection of normally sterile body sites, is comparatively speaking very rare. In 1999 in England and Wales the prevalence of bacteraemia was 194 cases per 100 000 of the population (Anonymous, 2000). However *S. aureus* is a major cause of community-acquired pneumonia and can also cause sepsis, toxic-shock syndrome, osteomyelitis, meningitis and other fulminant illnesses in the previously healthy individual.

Hospital-acquired

Patients in hospitals are at risk from many species of microorganism but *S. aureus* has always been considered one of the most important. In the 1940s and before antibiotics, one Boston study found 82% mortality in patients with staphylococcal bacteraemia (Skinner and Keefer, 1941) and in that period simple surgical procedures were often life-threatening. Patients in the modern hospital environment are still at risk from *S. aureus* infections due to large numbers of surgical wounds, venous catheter entry sites, stents, ventilators and intravenous lines that are present in modern units – all areas readily colonized with the organism. The widespread use of broad-spectrum antibiotics and immunosuppressive agents as well as increased support of major organs and more adventurous surgery in an ageing population have all been proposed as factors that have led to a dramatic increase in the number of serious *S. aureus* infections, especially those due to MRSA (methicillin-resistant *S. aureus*).

MRSA

First isolated in 1961 in the UK (Jevons, 1961), penicillin-resistant *S. aureus* with decreased susceptibility to semi-synthetic β-lactam antibiotics such as methicillin, oxacillin and flucloxacillin, soon spread globally to many countries in the 1960s and 1970s. Such infections were still relatively rare and were usually readily treatable with other agents. In the late 1980s and early 1990s MRSA prevalence increased in many countries. Fig. 2.1 shows how MRSA (as a percentage of total *S. aureus* bacteraemia) increased from 1–2% in the England and Wales in 1990 to 46.1% in 2001 (Johnson *et al.*, 2005). Such dramatic increases were also mirrored in countries such as the United States and Japan as well as countries in mainland Europe, elsewhere in Asia, Oce-

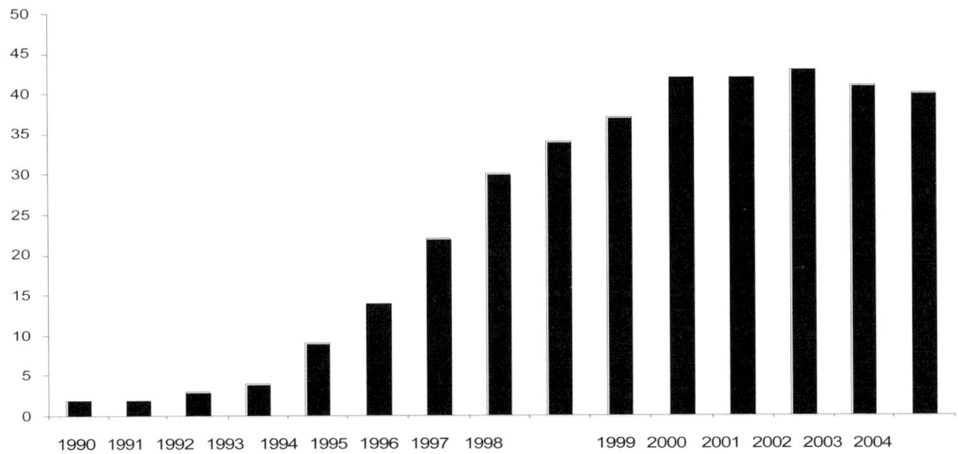

Figure 2.1 Data from voluntary reporting of the proportion of isolates (%) of *Staphylococcus aureus* from blood culture that are methicillin resistant, England and Wales. Data from Health Protection Agency website (www.hpa.org.uk).

ania and South America (Santos Sanches *et al.*, 2000; Anonymous, 2002a; Bell and Turnidge, 2002). Modern MRSA is normally treated with intravenous vancomycin in most countries but resistance to this drug has been found in isolates that were already resistant to multiple antibiotics (Anonymous, 2002b) and the threat of untreatable MRSA infections in the future is a realistic prospect.

S. aureus population structure

In the 1950s the population structure of *S. aureus* was first hinted at in studies by Rountree (Rountree and Freeman, 1955; Rountree and Beard, 1958) using bacteriophage typing. This method relies on analyzing patterns of susceptibility of isolates to a battery of phages. In the 1950s a particular type of penicillin-resistant *S. aureus* was causing severe disease both in communities and in hospitals. The organism, first found in Australia and Canada was found to lyse using phages 80 and 81 (Rountree and Beard, 1958) and it therefore became known as the 80/81 strain. 80/81 spread globally throughout the 1950s becoming increasingly resistant to antibiotics and it was notable for its enhanced virulence due to its production of an unusual leukocidin (Donahue and Baldwin, 1966) nowadays known as Panton–Valentine leukocidin (PVL) (Robinson *et al.*, 2005a). The rapid spread of 80/81 was particularly notable in the UK where it was responsible for more than 60% of staphylococcal infections in England and Wales (Anonymous, 1959) in 1957 compared to just 13% three years earlier (data from staphylococcal reference laboratory, Colindale, UK). The clonal nature of 80/81 isolates, the fact that indistinguishable isolates of the same species were found years apart and in different continents, has since been borne out by a recent study using more exacting technologies (Robinson *et al.*, 2005a) and this is one of the main features of *S. aureus* populations.

Short-term/local studies of population structure

The largest burden of *S. aureus* mortality is in hospitalized patients and nosocomial spread of this organism is extremely common. In the UK *S. aureus* is by far the most common nosocomial pathogen and it is second only to *Escherichia coli* as the most frequent cause of bacteraemia in hospitals there (Anonymous, 1997a). The analysis of the spread of *S. aureus* in healthcare settings has therefore been the subject of hundreds of studies using a sometimes bewildering variety of typing methodologies. These range from the use of antibiotic resistance data and other phenotypic tests to DNA-based 'fingerprinting techniques' including rapid PCR based typing systems and pulsed-field gel electrophoresis (PFGE) to DNA sequence-based technologies.

Pulse field gel electrophoresis (PFGE)

By far the most commonly used genotypic typing technique in analyzing the spread of *S. aureus* is PFGE. The technique relies on separating large restriction fragments of chromosomal DNA in a switching electric field to produce a restriction pattern or 'fingerprint' on agarose gels (Fig. 2.2). This restriction-fragment length polymorphism (RFLP) – based method was first applied to the study of *S. aureus* in an intensive care setting (Prevost *et al.*, 1991) to track the spread of MRSA strains within a unit. PFGE has proven immensely popular as an epidemiological tool as it is highly discriminatory; however, difficulties arise in comparing the complex fingerprints generated, especially when these are on different agarose gels run in different laboratories.

PFGE is ideal for studying small numbers of isolates in outbreak settings but the method is difficult to apply to very large sample sizes or in long-term epidemiological studies as small amounts of genetic change can radically change PFGE banding patterns. Attempts have been made to standardize the criteria used to decide whether isolates are outbreak associated or non-outbreak associated by limiting the number of restriction fragment mismatches allowed when comparing patterns (Tenover *et al.*, 1995). However such criteria are arbitrary and are therefore scientifically unsatisfactory. Attempts to standardize PFGE methodologies between laboratories in Europe (Murchan *et al.*, 2004; Cookson *et al.*, 2007) have been partially successful whereas in the United States a national database using PFGE as its primary method has been in use for several years (McDougal *et al.*, 2003).

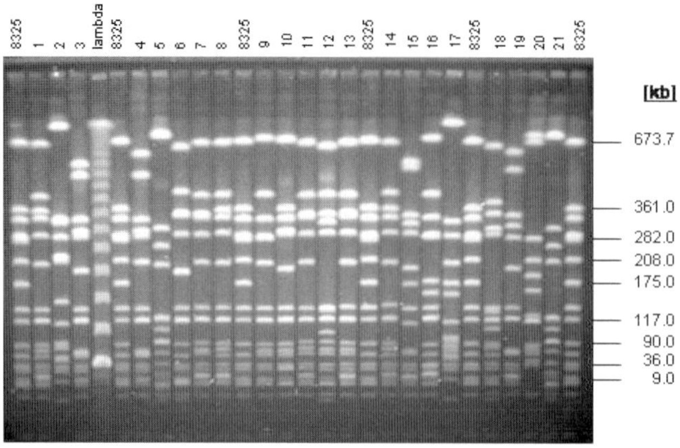

Figure 2.2 A pulsed-field electrophoresis agarose gel of DNA from 21 European MRSA isolates with size standards (bacteriophage lamda DNA and strain 8325). Thanks to Barry Cookson of the Health Protection Agency, London for supplying the gel image.

spa typing

DNA sequencing of the polymorphic X region of the *S. aureus* protein A gene (*spa*) has been widely accepted as a useful tool in investigating the organism's epidemiology. The X-region of *spa* consists of a variable number of 24 nucleotide repeats which are highly diverse due to deletion and duplication of the units themselves as well as by point mutation (Brigido Mde *et al.*, 1991; Shopsin *et al.*, 1999). *spa* typing has slightly inferior discriminatory value compared to PFGE (Strommenger *et al.*, 2006; Cookson *et al.*, 2007) but it is very well suited to multicentre studies as DNA sequence data quality can be controlled and checked and the results can be stored digitally in online databases that can be queried remotely (Harmsen *et al.*, 2003) (spa.ridom.de).

Local spread of clones

Most studies of outbreaks of *S. aureus* and particularly those caused by MRSA, report broadly similar findings. In most hospital studies a small number of genetically similar isolates tend to predominate and these clones may become endemic in such settings, often excluding other genotypes. Examples of such events are the emergence and national dominance of epidemic MRSA clones (EMRSA) -15 and -16 in the UK which overtook EMRSA-3 as the most common cause of bacteraemia (Anonymous, 1997b) and have now spread to other continents including Australia (Pearman *et al.*, 2001), mainland Europe (Aires de Sousa *et al.*, 2005), Asia (Hsu *et al.*, 2007) and the United States (McDougal *et al.*, 2003). These two clones were first recognized in the early 1990s but are now currently responsible for >95% of UK cases of MRSA bacteraemia and are endemic in the majority of UK hospitals. The emergence in Portugal of the so-called 'Iberian' clone (Sanches *et al.*, 1996) of MRSA shows a similar pattern of emergence, dominance and national then global spread (de Sousa *et al.*, 1998; Mato *et al.*, 1998). At the local level it seems that most geographical locations tend to only have one or two dominant hospital MRSA (Aires de Sousa and de Lencastre, 2003; Cockfield *et al.*, 2007).

Global population studies

The study of the emergence and spread of successful genotypes of *S. aureus* is of great interest from clinical, epidemiological and hospital infection perspectives but it provides little information regarding the population structure of the species as a whole. Methods that are used to investigate outbreaks and even track the spread of a small number of genotypes are only required to be highly discriminatory and to have a limited degree of reproducibility – so that outbreak-associated isolates should always differ from non-outbreak isolates. Limited information is available from such studies regarding the relationship between MRSA and MSSA (Crisostomo *et al.*, 2001) and between disease causing isolates and those from carriage in hospital and community settings. The methodologies of RFLP methods limit their applicability to such studies as isolates are either grouped as identical or similar. Where genetic similarity is enumerated (Struelens *et al.*, 1992) the results are simple numeric coefficients of RFLP band similarities (Dice, 1945) and deeper associations between genotypes cannot reliably be inferred.

Multilocus enzyme electrophoresis (MLEE)

Until recently the technique that had provided most information regarding bacterial population structure was MLEE. This technique allowed the genetic similarity between bacterial isolates to be measured based on shared allelic similarity at a number of individual loci. In MLEE protein extracts of bacterial isolates are separated by electrophoresis on starch gels. These gels are typically sliced thinly and each section is treated with reagents that stain for individual metabolic (housekeeping) enzymes. Allelic variants are detected by comparing their electrophoretic mobility with standard isolates containing previously characterized alleles. For any given enzyme only a few allelic variants are usually resolved as electrophoretic mobility is dependent on the net charge of the enzyme and as housekeeping genes are thought to be selectively neutral or close to neutral major charge variations are uncommon. MLEE, however, does give high levels of strain to strain differentiation as typically 15–25 enzymes are studied.

MLEE was first used to analyse genetic diversity in eukaryotes such as *Drosophila melanogaster* (Laurie-Ahlberg and Weir, 1979) and was first applied to the study of *Escherichia coli* in seminal studies by R.K. Selander and colleagues in the 1980s (Selander and Levin, 1980; Caugant *et al.*, 1981; Caugant *et al.*, 1984; Caugant *et al.*, 1985). Studies on other organisms such as *Neisseria meningitidis* (Caugant *et al.*, 1986a; Caugant *et al.*, 1986b), *Salmonella* spp. (Beltran *et al.*, 1988; Reeves *et al.*, 1989) and *Haemophilus influenzae* (Musser *et al.*, 1985) followed soon afterwards. In these studies the isolation of strains with identical alleles from diverse geographical sources separated in time was taken as strong evidence for a clonal population structure in which strains derived from a common ancestor diversify predominantly by mutation rather than homologous recombination within lineages. Recombination, if a common mechanism of bacterial diversification would, it was thought, have a strong 'mixing' effect on the bacterial chromosome so that there would be little or no association between individual alleles at different loci. This 'clonal paradigm' of bacterial evolution was largely accepted until a later study by Maynard Smith (Smith *et al.*, 1993) demonstrated a larger role for homologous recombination in shaping population structure. In this work population structures were described as clonal (e.g. *E. coli*, *Salmonella* spp.), panmictic (free-mixing or random mating, e.g. *N. gonorrhoeae*) or epidemic (e.g. *N. meningitidis*) depending on the degree of association found between alleles. An epidemic population structure is one in which there is a large degree of recombination where particularly well-adapted genotypes 'explode', becoming widely distributed in a short period of time.

MLEE studies of the population structure of *S. aureus* (Musser *et al.*, 1990; Musser and Selander, 1990; Musser and Kapur, 1992) demonstrate an overwhelmingly clonal population structure with frequent recovery of the same electrophoretic type in temporally and spatially distinct isolates from disease and carriage. These studies include a sample of >2,000 isolates from human and animal sources (Musser and Selander, 1990) and represent one of the largest *S. aureus* studies using any method. Musser and colleagues also used MLEE to characterize isolates of *S. aureus* from menstruation associated toxic-shock syndrome (Musser *et al.*, 1990). Isolates producing the superantigen toxin TSST-1 (toxic shock syndrome toxin 1) from female patients with a urogenital focus were predominantly (in 88% of cases) of one MLEE electrotype (ET) which was also found in a large proportion (53%) of non-urogenital TSS cases indicating limited spread of the *tst-1* gene between lineages. The same group also reported on bovine mastitis isolates (Kapur *et al.*, 1995) – an economically important disease in the dairy industry. They found 39 ETs within 357 isolates with seven ETs shared between humans and cows although lineages were mainly either exclusively human or bovine. An early study of MRSA isolates, predominantly from the United States and Canada, by Musser and Kapur (Musser and Kapur, 1992) showed an association of the methicillin resistance gene *mecA* with six different MRSA lineages (separated by a genetic distance > 0.2). This was taken as evidence that *S. aureus* may have acquired the gene on multiple occasions in contrast to a study published the following year that argues for a single acquisition of the gene by the species (Kreiswirth *et al.*, 1993) that has since been shown to be incorrect (Fitzgerald *et al.*, 2001; Enright *et al.*, 2002).

Multilocus sequence typing (MLST)

The usefulness of MLEE has been limited by the technical impracticalities of the technique in terms of the level of expertise required, the large number of reagents and the requirement of having to run allelic standards on each gel that necessitates the frequent exchange of isolates between laboratories. Despite being an excellent tool that has given many new insights into bacterial (and non-bacterial) population biology MLEE was largely confined in its use to a relatively small number of laboratories. The development of a DNA based method in 1998 (Maiden *et al.*, 1998) that relied on the same genetic principles underlying MLEE encouraged many more new studies of the population biology of many different bacterial pathogens including *S. aureus* (Enright *et al.*, 2000). Multilocus sequence typing differentiates alleles directly, at the DNA level, rather than by the electrophoretic mobility of proteins. Most MLST schemes nowadays (www.mlst.net) examine allelic diversity at seven housekeeping gene loci by DNA sequencing of approximately 500bp internal fragments. Sequences on both strands are compared to known alleles at each locus to generate an allelic profile consisting of seven integers, for example sequence type (ST) 1 isolates have the allelic profile 1–1-1–1-1–1-1 in the *S. aureus* scheme.

MLST is an improvement in several ways over MLEE as it requires only DNA sequencing technology that is currently available in many research laboratories. DNA sequences are unambiguous and quality control can therefore readily be employed to ensure nucleotide base-calls are accurate. Individual gene sequence traces, allelic data and organism information are curated over the internet ensuring databases contain only accurate information. Typically laboratories request the assignment of novel allele numbers or STs for their isolates via email. The two main sites hosting MLST databases are at Imperial College London (www.mlst.net) and Oxford University (www.pubmlst.org) hosting between them 48 different species databases including bacteria and fungi. Users around the world can characterize their local isolates in their own laboratory before querying the MLST servers via the internet to determine whether isolates similar to their own have been seen elsewhere before.

S. aureus MLST

The first *S. aureus* MLST study examined 155 consecutive isolates from invasive *S. aureus* disease in the Oxford region of the UK between 1997 and 1998 (Enright *et al.*, 2000). From 14 loci that were partially sequenced (~500 bp internal fragments) in ten isolates the seven most polymorphic were chosen and these regions were sequenced in all isolates. Within the 61 community-acquired infection isolates and 94 hospital-acquired isolates there were 29 MRSA (19%) isolates, all except one of which were hospital-acquired. 53 allelic profiles or sequence types (STs) were found with the most common MRSA ST being ST36 which corresponds to the UK epidemic MRSA clone (EMRSA-) 16 clone (Cox *et al.*, 1995). The remainder of isolates were ST22 which corresponds to UK EMRSA-15 (Richardson and Reith, 1993) with the exception of two isolates found to differ at one of the MLST loci to ST36. These were assumed to be recent descendants of this clone. The only other MRSA found in the study was an isolate of ST12 that does not correspond to any other MRSA isolate in the *S. aureus* MLST database. The remainder of isolates were MSSA of which four shared identical profiles with EMRSA-15 isolates (ST22). In this study parsimony analysis was used to propose an ancestor for EMRSA-16 isolates from a community- and hospital-acquired MSSA genotype – ST30. No significant differences were detected in the STs of isolates from community- and hospital-acquired disease in this study with the exception of MRSA isolates.

The MLST website currently (September 2007) contains genotypic and other information on 1,803 isolates (829 of which are MRSA) from more than thirty countries. 106 individuals have deposited information on their isolates and the popularity of the technique is further seen in the scientific literature where MLST has become the accepted method for describing strain genetic background, particularly for MRSA strains that are identified by ST and staphylococcal chromosomal cassette (SCC*mec*) type. In the United States however, isolates are primarily named based on PFGE pattern through the Pulsenet initiative (http://www.cdc.gov/pulsenet/) however these do correspond to MLST/SCC*mec* designations and most publications refer to both PFGE and MLST/SCC*mec* types together.

The usage of different molecular typing methods to characterize isolates of *S. aureus* presents an impediment to the exchange of information regarding isolates that are important in terms of disease or antibiotic resistance. The lack of access to sequencing facilities due to a lack of laboratory infrastructure or cost means that MLST and SCC*mec* typing cannot be used to examine every isolate, indeed from a practical point of view simpler and less expensive techniques can be used with one example of each genotype subjected to MLST. In studies *spa* typing and PFGE have been used to group isolates and MLST and SCC*mec* typing has then been performed to relate selected isolates to the published literature or the information held at the MLST website. Comparative studies of *spa* typing, PFGE and MLST typing show that they provide approximately comparable levels of resolution and clonal groupings are largely congruent (Cookson *et al.*, 2007)

S. aureus epidemiology

Hospital-acquired MRSA

The first MRSA emerged in the UK in 1961 (Jevons, 1961) and spread to many countries throughout the 1960s and early 1970s. This clone of MRSA was identical using MLST to 1950s isolates of MSSA that were common in Denmark in the 1950s and were also probably widely distributed throughout Europe at that time (Crisostomo *et al.*, 2001). The main difference between the MSSA and MRSA strain was the presence of *mecA* and the SCC*mec* on which it resides (Katayama *et al.*, 2000). SCC*mec* is the mobile genetic element that carries the *mecA* genes and its adjacent regulatory genes as well as recombinases necessary for the element's integration and excision from the chromosome. In certain SCC*mec*s resistance genes for antibiotics and heavy metals may also be present in the cassette on insertion sequences, plasmids and transposons. There are five main classes of SCC*mec* (Ito *et al.*, 2001; Okuma *et al.*, 2002) that differ in the *mec* region and the number and/or type of recombinase genes and these are usually detected by PCR (Milheirico *et al.*, 2007; Oliveira and de Lencastre, 2002).

A 2002 study of 912 globally distributed isolates of *S. aureus* that included 359 MRSA, used MLST in conjunction with PCR analysis of the SCC*mec* region to investigate the evolutionary history of MRSA (Enright *et al.*, 2002). Using an algorithm BURST (based upon related sequence types) to group together similar allelic profiles (in this case those sharing ≥ 5 alleles in common with at least one other isolate) five main MRSA containing lineages, or clonal complexes (CCs) were found.

The BURST algorithm, later implemented online as eBURST (Feil *et al.*, 2004), also predicts an ancestral ST for each CC. This is the genotype that has the highest number of single-locus variants (SLVs) or in the case of a tie, double locus variants (DLVs). In this way the ancestries of all five nosocomial MRSA CCs (5, 8, 22, 30 and 45) were investigated. The origins of CC8 were of particular interest as ST8-MSSA, the ancestral genotype of this CC, was found to be ancestral to ST250-MSSA, which upon acquisition of a class I SCC*mec* element gave rise to the first MRSA clone (Jevons, 1961). This genotype, ST250-MRSA-I (using the nomenclature introduced in this work), gave rise to other successful MRSA clones all with SCC*mec* I elements including the Iberian clone of MRSA (Sanches *et al.*, 1995) (ST247-MRSA-I) which is also known as UK EMRSA-5 and SCCmec (Aucken *et al.*, 2002). From this work it was found that the acquisition of SCC*mec*s by genetically similar MSSA to form novel MRSA clones has occurred on multiple occasions, for example ST8-MSSA has on individual occasions evolved into four different MRSA clones following integration of four different SCC*mec*s (types I–IV). In a later study using sequence from 15 gene fragments (seven MLST genes, seven surface-associated genes and *spa*) and SCC*mec* typing, evolutionary scenarios for the emergence of MRSA from all five CCs were proposed. This study also demonstrated the large number of novel MRSA formed by the acquisition of SCC*mec* IV by MSSA. SCC*mec* IV is smaller than SCC*mec*s I-III which are commonly found in hospital-acquired MRSA whereas SCC*mec* IV is common in community-acquired and hospital clones (see below – Community-Acquired MRSA).

Community-acquired MRSA

MRSA clones that are endemic within hospitals were, until recently, rare in healthy individuals with no exposure to healthcare settings. However in the late 1990s MRSA were found at higher prevalence than expected in Australia and in the United States (Okuma *et al.*, 2002). Four paediatric cases of fatal lung disease caused by community-acquired MRSA isolates (Anonymous, 1999) attracted much attention from scientists and physicians. These community MRSA isolates were all peculiar in their relatively susceptible phenotype – most were only resistant to β-lactam antibiotics. This was due to their carriage of SCC*mec* IV which carries no ancillary resistance genes. These isolates also contained the genes (*lukS-PV* and *lukF-PV*) for the Panton–Valentine leukocidin (PVL) toxin a potent bi-component, pore-forming toxin that in purified form is sufficient to cause pneumonia in a mouse model (Labandeira-Rey *et al.*, 2007).

The origins of community MRSA isolates were obscure until a 2003 study (Vandenesch *et al.*, 2003) showed that PVL producing community-acquired MRSA carrying SCC*mec* IV were of six main types and these were in most cases genetically distinct from nosocomial MRSA clones (Fig. 2.3) and more closely resembled community-acquired strains. The five major hospital genotypes of HA-MRSA are STs 5, 8, 22, 36 and 45 (2002, Enright *et al.*) but CA-MRSA have emerged from STs 1, 30 (the ancestral genotype of ST36), 59, 80, 93 as well as ST8. Studies have now reported PVL+ CA-MRSA in most countries (Vandenesch *et al.*, 2003; Tristan *et al.*, 2007). Such isolates tend to be associated with disease in younger, healthier individuals with none of the risk factors associated with nosocomial MRSA disease for review see (Chambers, 2001). CA-MRSA is mostly found in skin and soft-tissue infection although in some cases it can cause necrotizing pneumonia

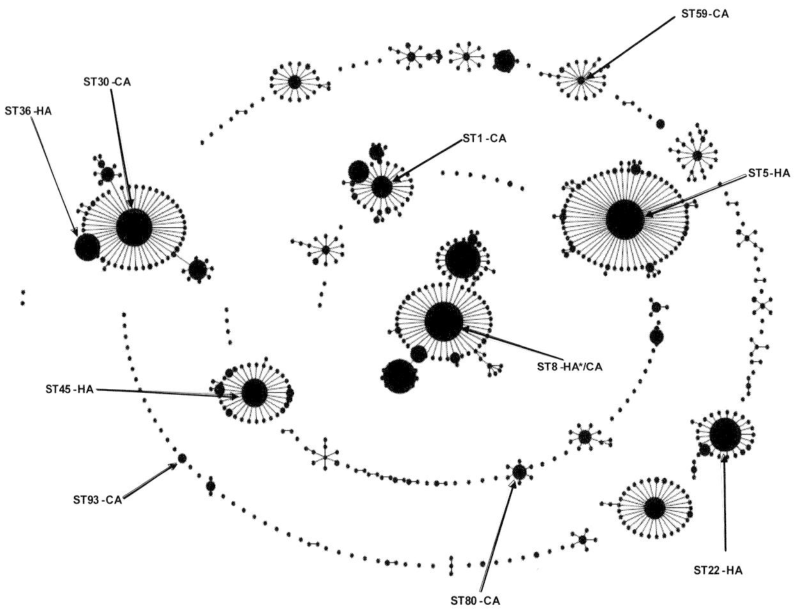

Figure 2.3 eBURST analysis of all isolates in the *S. aureus* MLST databases. The sequence types most commonly found in hospital and community-acquired cases of serious MRSA disease are shown. Each circle represents a genotype (MLST sequence type – ST) whose abundance in the MLST database is proportional to it's area. The major hospital-acquired (HA) and community-acquired (CA) STs are indicated by arrows.

which has a high mortality rate in young patients (Lina *et al.*, 1999; Gillet *et al.*, 2002).

In common with HA-MRSA disease there are risk factors for CA-MRSA and most early reports of serious infections, particularly in the United States, were in populations of intravenous drug users, men who have sex with men, prison inmates and members of contact sports teams although increasingly infections are occurring in the general population (Moran *et al*, 2006).

Community carriage and disease isolates

The population structure of *S. aureus* from carriage and invasive disease has until recently been unclear. The associations between isolates with certain prominent virulence factors and specific diseases have been established including toxic-shock syndrome toxin-1 and enterotoxins with superantigen activity with toxic shock syndrome (Dinges and Schlievert, 2001); exfoliative toxins A and B and staphylococcal scalded skin syndrome (Lee *et al.*, 1987); enterotoxin A and staphylococcal food poisoning (Dinges *et al.*, 2000); and as stated before, necrotizing pneumonia and the PVL genes (Gillet *et al.*, 2002). It is known that many toxins reside on mobile genetic elements including prophages and pathogenicity islands (Novick, 2003) which are limited in their distribution between *S. aureus* lineages (Chapter 4). A study by Peacock *et al* (Peacock *et al.*, 2002) looked at a collection of *S. aureus* isolates from community-acquired disease and carriage isolates in the same population and found an association between seven individual virulence factors and invasive disease so that the more of these factors were present, the more likely the strain was to have come from a disease rather than a carried isolate. An MLST analysis of the same sample of 334 isolates (Feil *et al.*, 2003) however found that disease and carried isolates had similar population structures so that unlike, species such as *N. meningitis* (Caugant *et al.*, 1986b) and *S. pneumoniae* (Enright and Spratt, 1998) there are no hyper-virulent lineages. A subset of the same strains were examined by microarray for the presence or absence of all of the putative virulence genes found in the first seven genome sequencing projects, and did not find any marker or lineage associated with disease (Lindsay *et al.*, 2006). Taken together these studies indicate that the potential of any ST of *S. aureus* to cause invasive disease depends on its frequency in the human population, and the role of its repertoire of virulence factors is unclear. Host susceptibility must also play a major role in invasive disease.

Mechanisms of clonal divergence – recombination and mutation

The availability of large databases of sequence information from thousands of isolates of *S. aureus* from carriage and disease has allowed the study of the main mechanisms of evolutionary divergence. Using the program eBURST (Feil *et al.*, 2004) ancestral isolates can be compared to their SLV descendants. The divergent alleles can be ascribed to having occurred by point mutation (single unique polymorphism amongst all alleles) or recombination (shared single nucleotide polymorphism amongst alleles or multiple polymorphisms). The ratio of recombination to mutation (R:M) is a useful measure of the clonality of a species and this has been extremely useful in ordering species between the highly clonal to the panmictic (Feil *et al.*, 2000; Spratt *et al.*, 2001). *S. aureus* genotypes diversify from shared ancestors mainly by point mutation with alleles and individual nucleotide sites 15-times more likely to change by mutation than recombination (Feil *et al.*, 2003).

Although point mutation is much more common than recombination in the recent divergence events seen in the evolution from ancestral genotypes to single-locus variants, the importance of recombination in the longer term evolution of this species is largely unknown. The extremely successful genotype ST239-MRSA-III is a major, globally distributed multidrug resistant clone whose recent evolution is the result of a major recombination event (Robinson and Enright, 2004). The alleles of MLST genes and other markers of ST239 are either from ST30 or ST8 and further genetic analysis of genes from around the chromosome provided evidence that demonstrated that this MRSA clone was the result of the transfer of at least 555kb (approximately 20% of the chromosome) of DNA from an ST30 isolate into an ST8 genetic background. This event appears to be relatively recent as all genes studied were alleles of either ST8 or ST30. The success of this MRSA clone may be due to its multidrug resistance phenotype masking the presumably significant fitness disadvantage from having a large part of the chromosome which does not share a recent evolutionary history with the genetic background of the ST8 lineage. Such significant homologous recombination events are rare (Robinson and Enright, 2004) but could significantly influence the species evolution. One other example was cited in this study where the 'donor' was an ST30 isolate but the size of the DNA transfer involved was less than half as large.

The richness of DNA sequence data from genomic sequencing projects and housekeeping gene sequences from the thousands of isolates from disease and carriage present in the MLST databases are invaluable tools in reconstructing the population structure of microbial species. Using concatenated MLST gene sequences and other genes Feil *et al* (Feil *et al.*, 2003) (Cooper and Feil, 2006) and other groups (Robinson *et al.*, 2005b) have constructed robust phylogenies of the species. From these studies it can be seen that *S. aureus* can be divided into two main groups as reported by other workers using multilocus sequencing approaches (Holden *et al.*, 2004). A phylogeny based on housekeeping gene sequences kindly provided by E. J. Feil from University of Bath shows these as groups 1 and 3 (Fig. 2.4). Group 2 is a subgroup closely related to group 1. In these studies the isolates corresponding to each group do not seem to differ in the range of clinical disease caused or antibiotic resistance pattern. However, any phylogeny is dependent on the genes used to construct it (Feil *et al.*, 2003), and phylogenies become less representative when recombination between 'unrelated' clones occurs.

Conclusion

S. aureus is an extremely adaptable organism causing an enormous range of clinical disease yet it resides asymptomatically in a large proportion of the human population. It is very tolerant of environmental change and is particularly noted for its acquisition of antibiotic resistance genes causing real challenges in providing adequate chemotherapy. The adaptability of the species is seen in the remarkable flexibility in gene content seen which is the more remarkable as the organism is on the whole very conserved with a strongly clonal population structure. The acquisition of mobile genetic elements enables the rapid adaptation to new challenges and although exchange is usually between genetically

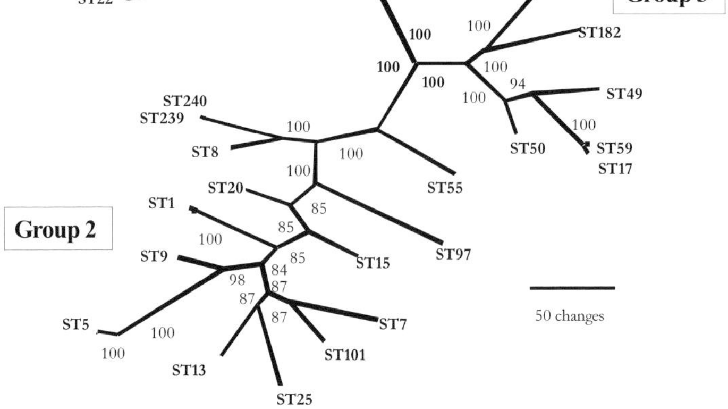

Figure 2.4 Bayesian reconstruction of the phylogeny of 26 sequence types of *Staphylococcus aureus* based on 17.8 kb of housekeeping gene sequences. Numbers indicate bootstrap support from 100 trials. Thanks to E.J. Feil, University of Bath for providing this figure.

similar isolates there is strong evidence of recent exchanges with other species, the best example of which is the movement of SCC*mec* from outside the species although most probably from *Staphylococcus epidermidis*. The concentration of genomic sequencing efforts on MRSA points to the large amount of genetic variability within the species that is perhaps better pictured in the 'shallower' but 'wider' view of the species pictured using MLST which has been applied to large numbers of MRSA and MSSA from humans, animals, carriage and disease. The mobile genetic elements, in particular in MSSA isolates may yet yield more surprises regarding this exceptional organism. Despite recent efforts to understand the population biology of *S. aureus* we still cannot predict how the organism will change in the future although it is certain that novel antibiotic resistances and virulence characteristics will continue to evolve to present ever more challenges to human health.

Web resources

MLST databases at Imperial College London
www.mlst.net
Hosts MLST schemes for 23 species of bacteria and yeast including data on > 1800 isolates of *S. aureus*.

MLST database at Oxford
www.pubmlst.org
Hosts MLST schemes for 26 species of bacteria

Ridom SpaServer spa type database
spa.ridom.de
Hosts spa types and compares them to MLST profiles

References

Aires de Sousa, M., Conceicao, T., Simas, C., and de Lencastre, H. (2005). Comparison of genetic backgrounds of methicillin-resistant and -susceptible *Staphylococcus aureus* isolates from Portuguese hospitals and the community. J. Clin. Microbiol. *43*, 5150–5157.

Anonymous (1959). Staphylococcal disease in hospitals. Lancet *1*, 185–186.

Anonymous (1997a). Epidemic methicillin resistant *Staphylococcus aureus*. Commun. Dis. Rep. Wkly *7*, 1.

Anonymous (1997b). Epidemic methicillin resistant *Staphylococcus aureus*. Commun. Dis. Rep. Wkly *7*, 191.

Anonymous (1999). From the Centers for Disease Control and Prevention. Four pediatric deaths from community-acquired methicillin-resistant *Staphylococcus aureus* – Minnesota and North Dakota, 1997–1999. JAMA *282*, 1123–1125.

Anonymous (2000). *Staphylococcus aureus* bacteraemia: England and Wales, January to December 1999. Commun. Dis. Rep. Wkly *7*, 23–24.

Anonymous (2002a). National Nosocomial Infections Surveillance (NNIS) System Report, data summary from January 1992 to June 2002, issued August 2002. Am. J. Infect. Control *30*, 458–475.

Anonymous (2002b). *Staphylococcus aureus* resistant to vancomycin – United States, 2002. MMWR Morb. Mortal. Wkly Rep. *51*, 565–567.

Aucken, H. M., Ganner, M., Murchan, S., Cookson, B. D., and Johnson, A. P. (2002). A new UK strain of epidemic methicillin-resistant *Staphylococcus aureus* (EMRSA-17) resistant to multiple antibiotics. J. Antimicrob. Chemother. *50*, 171–175.

Baba, T., Takeuchi, F., Kuroda, M., Yuzawa, H., Aoki, K., Oguchi, A., Nagai, Y., Iwama, N., Asano, K., Naimi, T., *et al.* (2002). Genome and virulence determinants of high virulence community-acquired MRSA. Lancet *359*, 1819–1827.

Bell, J. M., and Turnidge, J. D. (2002). High prevalence of oxacillin-resistant *Staphylococcus aureus* isolates from hospitalized patients in Asia-Pacific and South Africa: results from SENTRY antimicrobial surveillance program, 1998–1999. Antimicrob. Agents Chemother. *46*, 879–881.

Beltran, P., Musser, J. M., Helmuth, R., Farmer, J. J., 3rd, Frerichs, W. M., Wachsmuth, I. K., Ferris, K., McWhorter, A. C., Wells, J. G., Cravioto, A., and *et al.* (1988). Toward a population genetic analysis of Salmonella: genetic diversity and relationships among strains of serotypes S. choleraesuis, S. derby, S. dublin, S. enteritidis, S. heidelberg, S. infantis, S. newport, and S. typhimurium. Proc. Natl Acad. Sci. USA *85*, 7753–7757.

Brigido Mde, M., Barardi, C. R., Bonjardin, C. A., Santos, C. L., Junqueira, M. L., and Brentani, R. R. (1991). Nucleotide sequence of a variant protein A of *Staphylococcus aureus* suggests molecular heterogeneity among strains. J. Basic Microbiol. *31*, 337–345.

Caugant, D. A., Bovre, K., Gaustad, P., Bryn, K., Holten, E., Hoiby, E. A., and Froholm, L. O. (1986a). Multilocus genotypes determined by enzyme electrophoresis of Neisseria meningitidis isolated from patients with systemic disease and from healthy carriers. J. Gen. Microbiol. *132*, 641–652.

Caugant, D. A., Froholm, L. O., Bovre, K., Holten, E., Frasch, C. E., Mocca, L. F., Zollinger, W. D., and Selander, R. K. (1986b). Intercontinental spread of a genetically distinctive complex of clones of Neisseria meningitidis causing epidemic disease. Proc. Natl Acad. Sci. USA *83*, 4927–4931.

Caugant, D. A., Levin, B. R., Orskov, I., Orskov, F., Svanborg Eden, C., and Selander, R. K. (1985). Genetic diversity in relation to serotype in Escherichia coli. Infect Immun *49*, 407–413.

Caugant, D. A., Levin, B. R., and Selander, R. K. (1981). Genetic diversity and temporal variation in the E. coli population of a human host. Genetics *98*, 467–490.

Caugant, D. A., Levin, B. R., and Selander, R. K. (1984). Distribution of multilocus genotypes of Escherichia coli within and between host families. J. Hyg. (Lond) 92, 377–384.

Chambers, H. F. (2001). The changing epidemiology of *Staphylococcus aureus*? Emerg. Infect. Dis. *7*, 178–182.

Cookson, B. D., Robinson, D. A., Monk, A. B., Murchan, S., Deplano, A., de Ryck, R., Struelens, M. J., Scheel, C., Fussing, V., Salmenlinna, S., *et al.* (2007). Evaluation of molecular typing methods in characterizing a European collection of epidemic methicillin-resistant *Staphylococcus aureus* strains: the HARMONY collection. J. Clin. Microbiol. *45*, 1830–1837.

Cooper, J. E., and Feil, E. J. (2006). The phylogeny of *Staphylococcus aureus* – which genes make the best intra-species markers? Microbiology *152*, 1297–1305.

Cox, R. A., Mallaghan, C., Conquest, C., and King, J. (1995). Epidemic methicillin-resistant *Staphylococcus aureus*: controlling the spread outside hospital. J. Hosp. Infect. *29*, 107–119.

Crisostomo, M. I., Westh, H., Tomasz, A., Chung, M., Oliveira, D. C., and de Lencastre, H. (2001). The evolution of methicillin resistance in *Staphylococcus aureus*: similarity of genetic backgrounds in historically early methicillin-susceptible and -resistant isolates and contemporary epidemic clones. Proc. Natl Acad. Sci. USA *98*, 9865–9870.

de Sousa, M. A., Sanches, I. S., Ferro, M. L., Vaz, M. J., Saraiva, Z., Tendeiro, T., Serra, J., and de Lencastre, H. (1998). Intercontinental spread of a multidrug-resistant methicillin-resistant *Staphylococcus aureus* clone. J. Clin. Microbiol. *36*, 2590–2596.

Dice, L. R. (1945). Measures of the amount of ecologic association between species. Ecology *26*, 297–302.

Dinges, M. M., Orwin, P. M., and Schlievert, P. M. (2000). Exotoxins of *Staphylococcus aureus*. Clin. Microbiol. Rev. *13*, 16–34.

Dinges, M. M., and Schlievert, P. M. (2001). Role of T cells and gamma interferon during induction of hypersensitivity to lipopolysaccharide by toxic shock syndrome toxin 1 in mice. Infect. Immun. *69*, 1256–1264.

Donahue, J. A., and Baldwin, J. N. (1966). Hemolysin and leukocidin production by 80/81 strains of *Staphylococcus aureus*. J. Infect. Dis. *116*, 324–328.

Enright, M. C., Day, N. P., Davies, C. E., Peacock, S. J., and Spratt, B. G. (2000). Multilocus sequence typing for characterization of methicillin-resistant and methicillin-susceptible clones of *Staphylococcus aureus*. J. Clin. Microbiol. *38*, 1008–1015.

Enright, M. C., Robinson, D. A., Randle, G., Feil, E. J., Grundmann, H., and Spratt, B. G. (2002). The evolutionary history of methicillin-resistant *Staphylococcus aureus* (MRSA). Proc. Natl Acad. Sci. USA *99*, 7687–7692.

Enright, M. C., and Spratt, B. G. (1998). A multilocus sequence typing scheme for Streptococcus pneumoniae: identification of clones associated with serious invasive disease. Microbiol. *144 (Pt 11)*, 3049–3060.

Feil, E. J., Cooper, J. E., Grundmann, H., Robinson, D. A., Enright, M. C., Berendt, T., Peacock, S. J., Smith, J. M., Murphy, M., Spratt, B. G., *et al.* (2003). How clonal is *Staphylococcus aureus*? J. Bacteriol. *185*, 3307–3316.

Feil, E. J., Enright, M. C., and Spratt, B. G. (2000). Estimating the relative contributions of mutation and recombination to clonal diversification: a comparison between Neisseria meningitidis and Streptococcus pneumoniae. Res. Microbiol. *151*, 465–469.

Feil, E. J., Li, B. C., Aanensen, D. M., Hanage, W. P., and Spratt, B. G. (2004). eBURST: inferring patterns of evolutionary descent among clusters of related bacterial genotypes from multilocus sequence typing data. J. Bacteriol. *186*, 1518–1530.

Fitzgerald, J. R., Sturdevant, D. E., Mackie, S. M., Gill, S. R., and Musser, J. M. (2001). Evolutionary genomics of *Staphylococcus aureus*: insights into the origin of methicillin-resistant strains and the toxic shock syndrome epidemic. Proc. Natl Acad. Sci. USA *98*, 8821–8826.

Gillet, Y., Issartel, B., Vanhems, P., Fournet, J. C., Lina, G., Bes, M., Vandenesch, F., Piemont, Y., Brousse, N., Floret, D., and Etienne, J. (2002). Association between *Staphylococcus aureus* strains carrying gene for Panton–Valentine leukocidin and highly lethal necrotising pneumonia in young immunocompetent patients. Lancet *359*, 753–759.

Harmsen, D., Claus, H., Witte, W., Rothganger, J., Turnwald, D., and Vogel, U. (2003). Typing of methicillin-resistant *Staphylococcus aureus* in a university hospital setting by using novel software for spa repeat determination and database management. J. Clin. Microbiol. *41*, 5442–5448.

Hiramatsu, K., Hanaki, H., Ino, T., Yabuta, K., Oguri, T., and Tenover, F. C. (1997). Methicillin-resistant *Staphylococcus aureus* clinical strain with reduced vancomycin susceptibility. J. Antimicrob. Chemother. *40*, 135–136.

Holden, M. T., Feil, E. J., Lindsay, J. A., Peacock, S. J., Day, N. P., Enright, M. C., Foster, T. J., Moore, C. E., Hurst, L., Atkin, R., *et al.* (2004). Complete genomes of two clinical *Staphylococcus aureus* strains: evidence for the rapid evolution of virulence and drug resistance. Proc. Natl Acad. Sci. USA *101*, 9786–9791.

Hsu, L. Y., Loomba-Chlebicka, N., Koh, Y. L., Tan, T. Y., Krishnan, P., Lin, R. T., Tee, N. W., Fisher, D. A., and Koh, T. H. (2007). Evolving EMRSA-15 epidemic in Singapore hospitals. J. Med. Microbiol. 56, 376–379.

Ito, T., Katayama, Y., Asada, K., Mori, N., Tsutsumimoto, K., Tiensasitorn, C., and Hiramatsu, K. (2001). Structural comparison of three types of staphylococcal cassette chromosome mec integrated in the chromosome in methicillin-resistant *Staphylococcus aureus*. Antimicrob. Agents Chemother. *45*, 1323–1336.

Jevons, M. P. (1961). Celbenin-resistant staphylococci. Br. Med. J. *1*, 124–125.

Johnson, A. P., Pearson, A., and Duckworth, G. (2005). Surveillance and epidemiology of MRSA bacteraemia in the UK. J. Antimicrob. Chemother. *56*, 455–462.

Kapur, V., Sischo, W. M., Greer, R. S., Whittam, T. S., and Musser, J. M. (1995). Molecular population genetic analysis of *Staphylococcus aureus* recovered from cows. J. Clin. Microbiol. *33*, 376–380.

Katayama, Y., Ito, T., and Hiramatsu, K. (2000). A new class of genetic element, staphylococcus cassette chromosome mec, encodes methicillin resistance in *Staphylococcus aureus*. Antimicrob. Agents Chemother. *44*, 1549–1555.

Kreiswirth, B., Kornblum, J., Arbeit, R. D., Eisner, W., Maslow, J. N., McGeer, A., Low, D. E., and Novick, R. P. (1993). Evidence for a clonal origin of methicillin resistance in *Staphylococcus aureus*. Science *259*, 227–230.

Kuroda, M., Ohta, T., Uchiyama, I., Baba, T., Yuzawa, H., Kobayashi, I., Cui, L., Oguchi, A., Aoki, K., Nagai, Y., *et al.* (2001). Whole genome sequencing of meticillin-resistant *Staphylococcus aureus*. Lancet *357*, 1225–1240.

Labandeira-Rey, M., Couzon, F., Boisset, S., Brown, E. L., Bes, M., Benito, Y., Barbu, E. M., Vazquez, V., Hook, M., Etienne, J., *et al.* (2007). *Staphylococcus aureus* Panton–Valentine leukocidin causes necrotizing pneumonia. Science *315*, 1130–1133.

Laurie-Ahlberg, C. C., and Weir, B. S. (1979). Allozymic Variation and Linkage Disequilibrium in Some Laboratory Populations of DROSOPHILA MELANOGASTER. Genetics *92*, 1295–1314.

Lee, C. Y., Schmidt, J. J., Johnson-Winegar, A. D., Spero, L., and Iandolo, J. J. (1987). Sequence determination and comparison of the exfoliative toxin A and toxin B genes from *Staphylococcus aureus*. J. Bacteriol. *169*, 3904–3909.

Lina, G., Piemont, Y., Godail-Gamot, F., Bes, M., Peter, M. O., Gauduchon, V., Vandenesch, F., and Etienne, J. (1999). Involvement of Panton–Valentine leukocidin-producing *Staphylococcus aureus* in primary skin infections and pneumonia. Clin. Infect. Dis. *29*, 1128–1132.

Lindsay, J. A., Moore, C. E., Day, N. P., Peacock, S. J., Witney, A. A., Stabler, R. A., Husain, S. E., Butcher, P. D., and Hinds, J. (2006). Microarrays reveal that each of the ten dominant lineages of *Staphylococcus aureus* has a unique combination of surface-associated and regulatory genes. J. Bacteriol. *188*, 669–676.

Luzar, M. A., Coles, G. A., Faller, B., Slingeneyer, A., Dah, G. D., Briat, C., Wone, C., Knefati, Y., Kessler, M., and Peluso, F. (1990). *Staphylococcus aureus* nasal carriage and infection in patients on continuous ambulatory peritoneal dialysis. N. Engl. J. Med. *322*, 505–509.

Maiden, M. C., Bygraves, J. A., Feil, E., Morelli, G., Russell, J. E., Urwin, R., Zhang, Q., Zhou, J., Zurth, K., Caugant, D. A., *et al.* (1998). Multilocus sequence typing: a portable approach to the identification of clones within populations of pathogenic microorganisms. Proc. Natl Acad. Sci. USA *95*, 3140–3145.

Mato, R., Santos Sanches, I., Venditti, M., Platt, D. J., Brown, A., Chung, M., and de Lencastre, H. (1998). Spread of the multiresistant Iberian clone of methicillin-resistant *Staphylococcus aureus* (MRSA) to Italy and Scotland. Microb. Drug Resist. *4*, 107–112.

McDougal, L. K., Steward, C. D., Killgore, G. E., Chaitram, J. M., McAllister, S. K., and Tenover, F. C. (2003). Pulsed-field gel electrophoresis typing of oxacillin-resistant *Staphylococcus aureus* isolates from the United States: establishing a national database. J. Clin. Microbiol. *41*, 5113–5120.

Milheirico, C., Oliveira, D. C., and de Lencastre, H. (2007). Update to the multiplex PCR strategy for the assignment of mec element types in *Staphylococcus aureus*. Antimicrob. Agents Chemother. *51*, 3374–7.

Moran, G. J., Krishnadasan, A., Gorwitz, R. J., Fosheim, G. E., McDougal, L. K., Carey, R. B., and Talan, D. A. (2006). Methicillin-resistant *S. aureus* infections among patients in the emergency department. N. Engl. J. Med. *355*, 666–674.

Murchan, S., Aucken, H. M., O'Neill G, L., Ganner, M., and Cookson, B. D. (2004). Emergence, spread, and characterization of phage variants of epidemic methicillin-resistant *Staphylococcus aureus* 16 in England and Wales. J. Clin. Microbiol. *42*, 5154–5160.

Musser, J. M., Granoff, D. M., Pattison, P. E., and Selander, R. K. (1985). A population genetic framework for the study of invasive diseases caused by serotype b strains of Haemophilus influenzae. Proc. Natl Acad. Sci. USA *82*, 5078–5082.

Musser, J. M., and Kapur, V. (1992). Clonal analysis of methicillin-resistant *Staphylococcus aureus* strains from intercontinental sources: association of the mec gene with divergent phylogenetic lineages implies dissemination by horizontal transfer and recombination. J. Clin. Microbiol. *30*, 2058–2063.

Musser, J. M., Schlievert, P. M., Chow, A. W., Ewan, P., Kreiswirth, B. N., Rosdahl, V. T., Naidu, A. S., Witte, W., and Selander, R. K. (1990). A single clone of *Staphylococcus aureus* causes the majority of cases of toxic shock syndrome. Proc. Natl Acad. Sci. USA *87*, 225–229.

Musser, J. M., and Selander, R. K. (1990). Genetic analysis of natural populations of *Staphylococcus aureus*. In Molecular biology of the staphylococci, R. Novick, and R. A. Skurray, eds. (New York, VCH Publishers), pp. 59–67.

Novick, R. P. (2003). Mobile genetic elements and bacterial toxinoses: the superantigen-encoding pathogenicity islands of *Staphylococcus aureus*. Plasmid *49*, 93–105.

Okuma, K., Iwakawa, K., Turnidge, J. D., Grubb, W. B., Bell, J. M., O'Brien, F. G., Coombs, G. W., Pearman, J. W., Tenover, F. C., Kapi, M., *et al.* (2002). Dissemination of new methicillin-resistant *Staphylococcus aureus* clones in the community. J. Clin. Microbiol. *40*, 4289–4294.

Oliveira, D. C., and de Lencastre, H. (2002). Multiplex PCR strategy for rapid identification of structural types and variants of the mec element in methicillin-resistant *Staphylococcus aureus*. Antimicrob. Agents Chemother. *46*, 2155–2161.

Peacock, S. J., de Silva, I., and Lowy, F. D. (2001). What determines nasal carriage of *Staphylococcus aureus*? Trends Microbiol. *9*, 605–610.

Peacock, S. J., Moore, C. E., Justice, A., Kantzanou, M., Story, L., Mackie, K., O'Neill, G., and Day, N. P. (2002). Virulent combinations of adhesin and toxin genes in natural populations of *Staphylococcus aureus*. Infect. Immun. *70*, 4987–4996.

Pearman, J. W., Coombs, G. W., Grubb, W. B., and O'Brien, F. (2001). A British epidemic strain of methicillin-resistant *Staphylococcus aureus* (UK EMRSA-15) in Western Australia. Med. J. Aust. *174*, 662.

Pearson, T., Busch, J. D., Ravel, J., Read, T. D., Rhoton, S. D., U'Ren, J. M., Simonson, T. S., Kachur, S. M., Leadem, R. R., Cardon, M. L., *et al.* (2004). Phylogenetic discovery bias in Bacillus anthracis using single-nucleotide polymorphisms from whole-genome sequencing. Proc. Natl Acad. Sci. USA *101*, 13536–13541.

Prevost, G., Pottecher, B., Dahlet, M., Bientz, M., Mantz, J. M., and Piemont, Y. (1991). Pulsed field gel electrophoresis as a new epidemiological tool for monitoring methicillin-resistant *Staphylococcus aureus* in an intensive care unit. J. Hosp. Infect. *17*, 255–269.

Reeves, M. W., Evins, G. M., Heiba, A. A., Plikaytis, B. D., and Farmer, J. J., 3rd (1989). Clonal nature of Salmonella typhi and its genetic relatedness to other salmonellae as shown by multilocus enzyme electrophoresis, and proposal of Salmonella bongori comb. nov. J. Clin. Microbiol. *27*, 313–320.

Richardson, J. F., and Reith, S. (1993). Characterization of a strain of methicillin-resistant *Staphylococcus aureus* (EMRSA-15) by conventional and molecular methods. J. Hosp. Infect. *25*, 45–52.

Robinson, D. A., and Enright, M. C. (2004). Evolution of *Staphylococcus aureus* by large chromosomal replacements. J. Bacteriol. *186*, 1060–1064.

Robinson, D. A., Kearns, A. M., Holmes, A., Morrison, D., Grundmann, H., Edwards, G., O'Brien, F. G., Tenover, F. C., McDougal, L. K., Monk, A. B., and Enright, M. C. (2005a). Re-emergence of early pandemic *Staphylococcus aureus* as a community-acquired meticillin-resistant clone. Lancet *365*, 1256–1258.

Robinson, D. A., Monk, A. B., Cooper, J. E., Feil, E. J., and Enright, M. C. (2005b). Evolutionary genetics of the accessory gene regulator (agr) locus in *Staphylococcus aureus*. J. Bacteriol. *187*, 8312–8321.

Rountree, P. M., and Beard, M. A. (1958). Further Observations on Infection With Phage Type 80 Staphylococci in Australia. Med. J. Aust. *2*, 789–795.

Rountree, P. M., and Freeman, B. M. (1955). Infections Caused by a Particular Phage Type of *Staphylococcus aureus*. Med. J. Aust. *2*, 157.

Sanches, I. S., Aires de Sousa, M., Cleto, L., de Campos, M. B., and de Lencastre, H. (1996). Tracing the origin of an outbreak of methicillin-resistant *Staphylococcus aureus* infections in a Portuguese hospital by molecular fingerprinting methods. Microb. Drug Resist. *2*, 319–329.

Sanches, I. S., Ramirez, M., Troni, H., Abecassis, M., Padua, M., Tomasz, A., and de Lencastre, H. (1995). Evidence for the geographic spread of a methicillin-resistant *Staphylococcus aureus* clone between Portugal and Spain. J. Clin. Microbiol. *33*, 1243–1246.

Santos Sanches, I., Mato, R., de Lencastre, H., and Tomasz, A. (2000). Patterns of multidrug resistance among methicillin-resistant hospital isolates of coagulase-positive and coagulase-negative staphylococci collected in the international multicenter study RESIST in 1997 and 1998. Microb. Drug Resist. *6*, 199–211.

Selander, R. K., and Levin, B. R. (1980). Genetic diversity and structure in Escherichia coli populations. Science *210*, 545–547.

Shopsin, B., Gomez, M., Montgomery, S. O., Smith, D. H., Waddington, M., Dodge, D. E., Bost, D. A., Riehman, M., Naidich, S., and Kreiswirth, B. N. (1999). Evaluation of protein A gene polymorphic region DNA sequencing for typing of *Staphylococcus aureus* strains. J. Clin. Microbiol. *37*, 3556–3563.

Skinner, D., and Keefer, C. S. (1941). Significance of bacteraemia caused by *Staphylococcus aureus*. Arch. Intern. Med. *68*, 851–875.

Smith, C. B., Noble, V., Bensch, R., Ahlin, P. A., Jacobson, J. A., and Latham, R. H. (1982). Bacterial flora of the vagina during the menstrual cycle: findings in users of tampons, napkins, and sea sponges. Ann. Intern. Med. *96*, 948–951.

Smith, J. M., Smith, N. H., O'Rourke, M., and Spratt, B. G. (1993). How clonal are bacteria? Proc. Natl Acad. Sci. USA *90*, 4384–4388.

Spratt, B. G., Hanage, W. P., and Feil, E. J. (2001). The relative contributions of recombination and point mutation to the diversification of bacterial clones. Curr. Opin. Microbiol. *4*, 602–606.

Strommenger, B., Kettlitz, C., Weniger, T., Harmsen, D., Friedrich, A. W., and Witte, W. (2006). Assignment of Staphylococcus isolates to groups by spa typing, SmaI macrorestriction analysis, and multilocus sequence typing. J. Clin. Microbiol. *44*, 2533–2540.

Struelens, M. J., Deplano, A., Godard, C., Maes, N., and Serruys, E. (1992). Epidemiologic typing and delineation of genetic relatedness of methicillin-resistant *Staphylococcus aureus* by macrorestriction analysis of genomic DNA by using pulsed-field gel electrophoresis. J. Clin. Microbiol. *30*, 2599–2605.

Tenover, F. C., Arbeit, R. D., Goering, R. V., Mickelsen, P. A., Murray, B. E., Persing, D. H., and Swaminathan, B. (1995). Interpreting chromosomal DNA restriction patterns produced by pulsed-field gel electrophoresis: criteria for bacterial strain typing. J. Clin. Microbiol. *33*, 2233–2239.

Tristan, A., Bes, M., Meugnier, H., Lina, G., Bozdogan, B., Courvalin, P., Reverdy, M. E., Enright, M. C., Vandenesch, F., and Etienne, J. (2007). Global distribution of Panton–Valentine leukocidin – positive methicillin-resistant *Staphylococcus aureus*, 2006. Emerg. Infect. Dis. *13*, 594–600.

VandenBergh, M. F., Yzerman, E. P., van Belkum, A., Boelens, H. A., Sijmons, M., and Verbrugh, H. A. (1999). Follow-up of *Staphylococcus aureus* nasal carriage after 8 years: redefining the persistent carrier state. J. Clin. Microbiol. 37, 3133–3140.

Vandenesch, F., Naimi, T., Enright, M. C., Lina, G., Nimmo, G. R., Heffernan, H., Liassine, N., Bes, M., Greenland, T., Reverdy, M. E., and Etienne, J. (2003). Community-acquired methicillin-resistant *Staphylococcus aureus* carrying Panton–Valentine leukocidin genes: worldwide emergence. Emerg. Infect. Dis. *9*, 978–984.

Williams, R. E. (1963). Healthy carriage of *Staphylococcus aureus*: its prevalence and importance. Bacteriol. Rev. *27*, 56–71.

Wisplinghoff, H., Rosato, A. E., Enright, M. C., Noto, M., Craig, W., and Archer, G. L. (2003). Related clones containing SCCmec type IV predominate among clinically significant Staphylococcus epidermidis isolates. Antimicrob. Agents Chemother. *47*, 3574–3579.

Yu, V. L., Goetz, A., Wagener, M., Smith, P. B., Rihs, J. D., Hanchett, J., and Zuravleff, J. J. (1986). *Staphylococcus aureus* nasal carriage and infection in patients on hemodialysis. Efficacy of antibiotic prophylaxis. N. Engl. J. Med. *315*, 91–96.

S. aureus Evolution: Lineages and Mobile Genetic Elements (MGEs) 3

Jodi A. Lindsay

Abstract

There is enormous variation between strains of *S. aureus*. Evolution occurs when genomes vary and the fittest bacteria are selected. Variation occurs in three major ways : single nucleotide polymorphisms (SNP) and other minor changes in conserved core genes; variation in hundreds of genes (particularly those encoding proteins that interact with host) that are associated with lineages of *S. aureus*; and acquisition and loss of mobile genetic elements (MGE) which often encode virulence and resistance genes. Barriers that block horizontal transfer of DNA, such as restriction modification, influence lineages and MGE. *S. aureus* MGE have their own complex life cycles that control their spread and survival. Selection of the fittest bacteria is likely being driven by mammalian host factors and antibiotic use, and new strains of *S. aureus* are emerging that are increasingly virulent and resistant to antibiotics, causing novel healthcare issues.

Introduction

Evolution requires genetic change then selection of the fittest. With new genetic tools of sequencing, MLST and whole genome microarrays, we have learned there is a large amount of genetic variation in this species. Not only is there a range of variation, but there is evidence that variation occurs frequently. The *S. aureus* genome consists of three components, the core genome which is relatively conserved between all strains but may have single nucleotide polymorphisms (SNPs), core variable (CV) genes which vary according to lineage, and mobile genetic elements (MGE) that move into and out of strains and encode virulence and resistance genes.

What selection pressures are relevant for *S. aureus*? They survive and grow in a wide variety of locations, such as in the human nose, throat, armpit, groin and mucous membranes. *S. aureus* can survive attack by the immune system during infection (predominantly phagocytosis by polymorphonuclear leukocytes), and starvation and/or desiccation on inanimate objects for months. Survival and growth on other mammalian species is also common, as is survival and toxin expression in food. In ideal conditions, *S. aureus* can replicate every 20 min. A mutation that may be beneficial in some locations may be detrimental in others. This can also lead to divergent evolution as strains adapt to unique environments. However, *S. aureus* appears to be a very adaptable organism, and there seem few examples of isolates that are unable to cause disease, or that cannot survive in a range of environments.

In recent years, the widespread use of antibiotics for preventing and treating infections has undoubtedly altered the evolution of *S. aureus*, and led to the emergence of strains that have systematically acquired multiple resistance genes. However, there is still a long way to go before truly multidrug-resistant isolates are widespread and this is likely because *S. aureus* are still evolving and adapting to the antibiotic challenge. However, it is important to note that the supply of new antibiotics is diminishing sharply due to costs of development and licensing issues (Projan, 2003) and it is likely that few new antibiotics will become available in the short to medium term. This means we can expect to see widespread multidrug resistant *S. aureus* before there is impetus for new drugs to enter the marketplace.

On several occasions, new strains of *S. aureus* have appeared that cause new types of disease, epidemiological spread and/or response to interventions, impacting on healthcare. Each new strain prompts the search for the 'genes responsible', to increase our understanding of pathogenesis, diagnostics or targets for therapy. Examples include the first methicillin-resistant *S. aureus* (MRSA) which had acquired a SCC*mec* element (section 4.5 SCC), and the strain COL is an example (Gill *et al.*, 2005). Epidemic hospital-acquired MRSA (HA-MRSA) are another example, and their incidence has increased dramatically in the last 15 years or so in many countries of the world (Chapter 2), although why they are successful is not known. Community-acquired MRSA (CA-MRSA) have evolved independently of hospitals, and are associated with severe skin and soft tissue infections and rare fatal haemolytic pneumonia in healthy adults and children (Lina *et al.*, 1999; Kluytmans-Vandenbergh and Kluytmans, 2006). There is a strong association between these strains and carriage of the toxin Panton–Valentine leukocidin, although there is currently debate over whether this is the key toxin (see Bacteriophage section, below). VRSA, which are fully vancomycin resistant HA-MRSA, have acquired resistance genes horizontally from enterococci (see Transpsons section, below); only six cases have been described so far (Weigel *et al.*, 2003). A new HA-MRSA called TW is more invasive than other HA-MRSA (Edgeworth *et al.*, 2007). Our increasing knowledge about how *S. aureus* evolve and what makes them successful shows there is potential for *S. aureus* to accumulate all of these abilities and to continue to become even more resistant and virulent.

Point mutation versus horizontal transfer

There are two major ways that bacterial genomes evolve, and that is the slow process of point mutation and occasional selection of bacteria with advantageous mutations, and the rapid process of importing foreign DNA with beneficial genes.

Point mutation

Point mutations, frame shift mutations and similar are generally caused by environmental agents that mutate DNA (UV light, mutagenic chemicals, etc.), or errors in genetic replication or gene repair mechanisms. Many point mutations have little impact on the *S. aureus* genome because the nucleotide change does not result in a coding amino acid change (synonymous), or the resultant amino acid change is in a non-important region of the protein. However, some single mutations are proposed to have major effects on bacterial pathogenicity or survival. The laboratory strain *S. aureus* Newman has stop mutations in fibrinogen binding proteins (*fnbA* and *fnbB*) which reduce its binding to fibrinogen and other host proteins and ability to invade mammalian cells (Grundmeier *et al.*, 2004). Resistance to antibiotics can also be due to the accumulation of point mutations, such as fluoroquinolone resistance due to DNA gyrase (*gyrA* and *gyrB*) and topoisomerase IV (*parE*) mutations (Fujimoto-Nakamura *et al.*, 2005; Deplano *et al.*, 1997), mupirocin resistance due to isoleucyl-tRNA synthetase (*ileS*) mutation (Antonio *et al.*, 2002) or fusidic acid resistance due to mutations in elongation factor (*fusA*) (Besier *et al.*, 2003). Point mutations in promoter regions of genes can alter gene expression (Ince and Hooper, 2003). An important study has recently illustrated how *S. aureus* accumulate point mutations during the course of disease in infected patients, particularly under antibiotic pressure (Mwangi *et al.*, 2007). Strain JH1 was isolated from a patient with endocarditis who was subsequently treated with vancomycin. During the course of infection, a series of isolates were collected, including JH9, an intermediate-level vancomycin resistant isolate that had also acquired resistance to daptomycin, rifampicin and β-lactams and had altered virulence gene expression patterns. Both strains were sequenced and only 35 point mutations differed between them (Mwangi *et al.*, 2007). Accumulation of beneficial mutations over time is presumably how many *S. aureus* genes have evolved.

Horizontal transfer

Horizontal transfer of DNA into most bacteria is via transformation, conjugation or transduction. Transformation involves uptake of freely available DNA by bacteria that have gone into a competence phase and are expressing pilus- or pore-like structures (Chen and Dubnau, 2004). However, the necessary genes for competence are not identified in the *S. aureus* genome, and early reports of *S. aureus* transformation were due to contamination with phage tail fragments (Birmingham and Pattee, 1981). Conjugation involves the mobile DNA carrying a series of *tra* genes which encode pilus or pore formation between the donor and recipient bacteria followed by DNA transfer. Conjugation occurs in *S. aureus* and there are several conjugative plasmids and transposons described. The 12 published multiple whole genome *S. aureus* sequencing projects have identified only two strains with an MGE encoding conjugation genes (Mu50 and FPR3757).

Transduction involves the packaging of DNA into bacteriophage (phage) particles during the normal phage lifestyle and delivery to a recipient host. Most *S. aureus* have between one and four intact bacteriophage genomes incorporated into their genome, and most bacterial cultures contain phage particles. When phage undergo normal induction (usually in response to a stress such as UV light or other mutagen), they will switch the host bacterial transcription and translation systems to produce large quantities of bacteriophage-encoded head and tail proteins and then package copies of phage genome inside these phage particles. Eventually the host cell is lysed, releasing virulent bacteriophage particles that can bind to new *S. aureus* cells and inject their DNA. These recipient cells may undergo a lytic pathway where new phage is synthesized and released, or may undergo a lysogenic pathway where the bacteriophage inserts into a specific site in the bacterial DNA and shuts down. At this point the phage genome behaves like any other part of the chromosome, and is replicated into daughter cells. Under stress it can be induced again.

Some bacteriophage, in particular ϕ11 (also known as phi11) and $\phi 80\alpha$ in the laboratory, are known to be generalized transducing phage. Generalized transduction is a variation in the normal phage lifestyle, and involves the packaging of bacterial DNA into phage particles. It is not clear if this packaging is accidental. Once the phage particles are released from the donor cell they bind to and inject the donor DNA into recipient bacteria. The newly injected DNA can be a mobile genetic element (MGE) or chromosomal DNA. MGE encode genes that allow themselves to replicate independently in the new host (plasmid) or to integrate specifically into the host chromosome (SaPI, transposons), and their transfer occurs at a relatively high frequency of 10^{-5} to 10^{-6} per donor phage in the laboratory. Chromosomal DNA cannot survive in the new recipient bacteria unless it can recombine with homologous chromosomal DNA via the *recA* mediated pathway (Bayles *et al.*, 1994). This is relatively inefficient in *S. aureus*, and thus transfer of chromosomal markers occurs at 100–1,000 times less efficiently than most MGE markers (Wyman *et al.*, 1974).

Frequency of horizontal transfer of DNA into and out of strains in nature is probably high. Bacteriophage lysis is very efficient in the laboratory, generating high titres of phage. Exchange of markers or MGE between strains has been well documented *in vivo* on skin (Noble *et al.*, 1994), and *in vivo* in patients with chronic infection (Moore *et al.*, 2001; Kahl *et al.*, 2003; Goerke *et al.*, 2004; Goerke *et al.*, 2006a). 'Clonal' lineages of MRSA also vary substantially in their carriage of MGE (Moore *et al.*, 2002; Lindsay, unpublished).

Horizontal transfer into *S. aureus* is predominantly from other *S. aureus*, and it is suggested this is because phage are species specific, but could also be due to barriers that block transfer between species such as restriction enzymes. A few exceptions where foreign DNA has moved into *S. aureus* from other species will be described in the sections on SaPI, SCC and large insertions, below.

Lineages

S. aureus populations consist of dominant lineages and minor lineages. Most human nasal carriage isolates belong to about ten dominant lineages (Feil *et al.*, 2003). In general, human lineages are thought to have been relatively stable for the last 40 years (Gomes *et al.*, 2006). Most hospital *S. aureus* also belong to a few of these same lineages, CC5, CC8, CC22, CC30 and

CC45. Surprisingly, in each geographical location, only one or two of these MRSA lineages predominate in hospitals (Cockfield *et al.*, 2007). However, some of the newer MRSA belong to rarer lineages such as ST239 (hospital) and ST80 and ST59 (community) and perhaps selection of drug resistance is driving the success of these lineages. Bovine isolates of *S. aureus* that cause dairy cow mastitis have their own unique lineages (Smith *et al.*, 2005; Sung and Lindsay, 2007). A more thorough description of the dominant lineages (CCs) and how they are defined by microarray is in Chapter 1. Here is described in detail how they vary and evolve.

How do lineages vary?

Each lineage is remarkably distinct. The sequencing projects and comparative genomics by microarray show that hundreds of genes vary between lineages, but are relatively stable within lineages (Lindsay *et al.*, 2006). Variations can be point mutations within conserved genes (such as those detected by MLST), substantial variant regions within genes (such as putative binding regions in surface proteins), presence or absence of single genes (such as collagen binding protein), substantial variation in sequence of several genes in an operon (such as the *agr* regulatory operon and capsule operon), presence or absence of a series of genes (such as the putative *bctE* genomic islet), and substantial gene variation and/or deletion in dozens of genes (such as the genomic islands α and β).

Isolates from the same lineage but from diverse geographical locations or time can be remarkably similar (except for their MGEs which will be discussed below). This was first noted by comparing MW2 and MSSA476, both CC1 and from Dakota USA and Oxford UK respectively, and is also true of 8325 and FPR3757, both CC8 and from pre-1940s UK and recent USA CA-MRSA outbreaks respectively (Holden *et al.*, 2004; Diep *et al.*, 2006).

The role of mammalian host in evolution

Why has *S. aureus* evolved into a series of independent but stable lineages? A clue to those selective pressures is the fact that a large proportion of the lineage defining CV genes encode known or predicted surface proteins or structures (Lindsay *et al.*, 2006). Many are known to bind or interact with human proteins (Patti *et al.*, 1994; Clarke *et al.*, 2002; Roche *et al.*, 2003). They include capsule, fibrinogen binding proteins A and B, SasG that binds to nasal epithelial cells, collagen binding protein, immunodominant gene *ebh*, haemagglutinin-like protein SasA, and multiple staphylococcal superantigen like (*ssl*) gene variants implicated in control of the host immune response. Regulators that control the expression of these genes, such as *agr* and *sar* homologues, also vary according to lineage.

It therefore seems likely that human host interaction may play an important part in selecting for successful lineages. The normal habitat for *S. aureus* is the human nose and mucous membranes, so the key receptors are likely to be found there. A further support for this hypothesis is that most human carriers seem to carry only one 'type' of *S. aureus*, and when cleared of this *S. aureus* and rechallenged with a mixture of *S. aureus* lineages, will preferentially become recolonized with their own type (Nouwen *et al.*, 2004). Furthermore, animal *S. aureus* belong to different lineages than human strains (Sung and Lindsay, 2007)

Does this mean that different hosts carry specific receptors that are recognized by specific lineages? This seems simplified but could be true. Surprisingly it doesn't seem that each lineage has unique surface proteins, but unique combinations of surface proteins found in other lineages. So the interactions between pathogen and host are likely to be complex.

The genetic controllers of *S. aureus* lineage evolution are becoming clearer. Each lineage has a unique combination of CV genes, but that there has been some recombination between lineages. If each of the lineages were diverging continually from each other, it would be expected that unique surface proteins would be found in each lineage, but this is not the case. On the other hand, if all isolates could exchange DNA at an equal rate, we would see no lineages at all. The variation actually seen predicts a system of control that allows DNA exchange within lineages at a higher frequency than between lineages. Such a system has now been identified, the *Sau*1 (or *Sau*1I) restriction modification system.

Restriction modification and the evolution of lineages

Restriction enzymes are the workhorse of molecular biology laboratories as they target specific sequences of DNA and cut them, allowing unrelated DNA fragments to be artificially ligated together. This is the key step in cloning and other genetic manipulations. Most bacteria investigated have the genes for, and produce, restriction enzymes, and each enzyme targets a specific sequence of typically 4 to 8 base pairs. It is thought restriction enzymes protect bacteria from 'invasion' by foreign DNA, such as bacteriophage which cause bacterial lysis or from DNA encoding features that are a burden.

How does a restriction enzyme identify and digest foreign DNA and not its own? Type I restriction enzymes are part of a system that includes a modification enzyme which recognizes the same specific sequence, modifies it, and therefore protects it from digestion by the restriction enzyme (Murray, 2000). Both the restriction and modification enzymes recognize the same specific DNA sequence by a common subunit of the enzyme called the specificity subunit.

Investigation of the *S. aureus* genome sequencing projects identified a putative type I RM system found in all strains called Sau1. It consists of a restriction enzyme subunit (*sau1hsdR*) and two copies each of a modification enzyme subunit (*sau1hsdM1* and *sau1hsdM2*) and specificity subunit (*sau1hsdS1* and *sau1hsdS2*). Each *sau1hsdM* and *sau1hsdS* gene appear to be co-located, and sit in the middle of the genomic islands GIα and GIβ. Waldron and Lindsay (2006) have proven that *sau1hsdR* encodes a functional restriction enzyme that is responsible for digesting foreign DNA delivered by electroporation from *E. coli* and by conjugation from *Enterococcus faecalis*. The *sau1hsdR* gene is mutated in the strain RN4220, a chemical mutagen of 8325–4, which itself has had three prophage removed from the early clinical isolate 8325 (also known as PS47 or RN1). RN4220 is able to accept foreign DNA and is virtually the only reliable *S. aureus* strain that can be genetically manipulated in the laboratory. Therefore, Sau1 is responsible for the difficulties associated with genetically manipulating wild *S. aureus* isolates.

Further investigation showed that each of the *sau1hsdS* genes from the different whole genome sequencing projects were highly divergent. Each *sau1hsdS* gene has two putative DNA binding regions and these are the variant regions. Microarray studies have shown that variation in *Sau1hsdS* correlates strongly with lineage (Waldron and Lindsay, 2006), and indeed now forms the basis of a rapid method to discriminate between lineages (Cockfield *et al.*, 2007). This suggests that each lineage will have the same Sau1 RM system but recognize, digest and modify unique sequences. Waldron and Lindsay (2006) also proved that *sau1hsdR* could recognize and digest DNA from a foreign *S. aureus* lineage.

This model suggests a common *S. aureus* ancestor that has evolved into a number of lineages by variation and recombination, with selection of a few successful lineages. The Sau1 RM system has developed lineage specificity, and is beautifully designed for allowing gene exchange between isolates of the same lineage at greater frequency than exchange between isolates of different lineages. This is likely the reason why the lineages have remained independent yet still exchange DNA at low frequency. It also suggests that alteration of an *sau1hsdS* gene may be the first step in the evolution of a new lineage.

Sung and Lindsay (2007) have identified a naturally occurring lineage of bovine *S. aureus*, ST151, that has mutations in both *sau1hsdS* genes. These isolates are able to accept DNA from *E. coli* and *E. faecalis* and *E. faecium* at high frequency. Since *E. faecium* carry a transposon encoding full vancomycin resistance (*vanA*) and these strains are found in the agricultural setting, there is potential for transfer. The phenotype of ST151 is expected to be unable to digest or modify foreign DNA or its own, and therefore it may be an efficient acceptor of foreign DNA, but it should also be a poor donor of DNA to other *S. aureus* with intact RM systems.

Detection of lineages

Because each lineage is so distinct, there are a number of methods that can be used to assign strains to their appropriate lineage. The original method was MLST, which involves sequencing and comparing seven housekeeping genes, and assigning a sequence type (ST) number (Enright *et al.*, 2000). Related STs (typically having five or more identical genes) are grouped as clonal clusters (CC) or lineages. MLST is useful because thousands of isolates have been typed and their profiles are available on public databases (www.mlst.org), allowing worldwide epidemiological trends to be followed. For non-specialist laboratories, the 14 sequencing reactions (both directions for each gene) per strain required can be relatively expensive, and a single base pair error can result in assignment to the wrong ST.

spa typing involves sequencing part of the protein A gene, and comparing the patterns of repeats to a public database that is easily matched to MLST CCs (Harmson *et al.*, 2003). Some variation within CCs is detectable but does not necessarily correlate with MLST ST variation within CCs. The advantage of spa typing is that only two sequencing reactions are required.

The most comprehensive method for characterizing a *S. aureus* isolate is by hybridization to a multi-strain *S. aureus* microarray (Lindsay *et al.*, 2006). This has the advantage of assigning a lineage based on the presence or absence of signals to hundreds of core variable genes, but also can detect presence of MGE. However, the method is very expensive and technically demanding. Smaller microarrays designed to identify lineage and the presence of particular virulence genes are feasible, and could be made available on a cheap and flexible hybridization platform in the future.

We have recently exploited variation in the *sau1hsdS* gene sequence to design a rapid, simple and inexpensive method for assigning MRSA to their lineages (Cockfield *et al.*, 2007). The advantage is that it uses PCR and requires no specialist equipment or skills. Data can be available the same day. The RM test has potential to be exploited using a mini array platform or by light-generating PCR in a format already available off-the-shelf for use in diagnostic laboratories. Because *sau1hsdS* gene variation is thought to define the lineages, it is the most useful gene to target for a rapid typing method, and any variation should indicate the evolution of a new lineage. At present the RM test does not cover all the major *S. aureus* lineages, but the test is currently being expanded.

Other methods for identifying lineages have been proposed that exploit the fact that each lineage is very distinct. These include multi-locus variation analysis (MLVA) (Sabat *et al.*, 2003) and variable number tandem repeats (VNTR) (Hardy *et al.*, 2006). Some markers vary according to lineage but are shared by several lineages, thus they are not useful on their own but may have benefit when combined with other markers (capsule typing, *agr* typing, *egc* typing). No lineage detection method is suitable for investigating local MRSA outbreaks as too few lineages are found in any geographical location (Cockfield *et al.*, 2007). Variation within clonal MRSA lineages is due to MGE variation.

Mobile genetic elements (MGEs)

MGE are defined segments of DNA that can replicate on their own, or have specific mechanisms to insert into and out of chromosomes or plasmids that replicate. MGE account for approximately 10–20% of a *S. aureus* chromosome, with multiple distinct elements inserted at specific sites. *S. aureus* MGE typically have a G+C content that is equivalent to chromosomal DNA, suggesting that horizontal transfer predominantly occurs between isolates of *S. aureus* or other low G+C content bacteria. This is in contrast to many Gram negative bacteria such as *E. coli* where MGE have altered G+C contents (Lawrence and Ochman, 1997) and probably integrate infrequently.

Many MGE encode virulence or antibiotic resistance genes (Table 3.1). Therefore, the movement of these elements into and out of isolates can contribute to the ability of *S. aureus* to cause disease or respond to treatment. In particular, Panton–Valentine leukocidin (*PV-luk*), enterotoxin A (*sea*), chemotaxis inhibitory protein (*chip*), staphylokinase (*sak*), staphylococcus complement inhibitor (*scin*) and exfoliative toxin (*eta*) are found on bacteriophage, toxic shock syndrome toxin 1 (*tst*), enterotoxins B (*seb*) and C (*sec*) are found on *S. aureus* pathogenicity islands (SaPI), exfoliative toxins are also found on plasmids, and variant capsules found on staphylococcal chromosome cassettes (SCC). Resistance to β-lactams, tetracyclines, erythromycin, aminoglycosides, fusidic acid, mupirocin, trimethoprim, vancomycin, quarternary ammonium chlorides and heavy metals are found on plasmids and/or transposons, while resistance to β-lactamase resistant penicillins (such as methicillin) are found on SCC.

Horizontal transfer of MGE occurs by generalized transduction as described above, or by MGE encoded mechanisms which will be described individually below. Transfer frequency of most MGE is probably high in nature. Transfer frequency can be controlled by the Sau1 RM system which blocks transfer of MGE between strains of different lineages, regardless of their transfer mechanism (Waldron and Lindsay, 2006). Some MGE encode their own functional RM systems, such as the Sau42I system found on ϕ42 (Dempsey *et al.*, 2005). Other putative RM systems have been identified from the sequencing projects. RM on MGE serve the element rather than the host bacterium, perhaps by preventing infection of the cell by other MGE that may displace them. However, these systems likely impact on the ability of the host to evolve, and contribute to the difficulties still associated with genetically manipulating wild type *S. aureus* isolates in the laboratory. Despite the wide range of MGE found in the sequenced strains, only a few suggest they carry RM. Other MGE specific systems may control horizontal transfer and will be described individually below. There are likely to be further systems controlling genetic exchange that have yet to be described.

MGEs may provide beneficial resistance genes or virulence factors that aid in survival. However, they are often large pieces of DNA encoding many genes that may be a burden to the cell. This has been shown for SCCmec type I (Ender *et al.*, 2004).

The distribution of MGE amongst *S. aureus* strains is becoming known because of the sequencing projects and large scale comparative genomics projects using microarrays (Table 3.1). There is enormous variation between strains, even those of the same lineage. Variation occurs in two ways. Firstly, the distribution of an MGE can be widespread suggesting stability or even selection, while others are rare, and still others show signs of association with particular lineages. Secondly, there is substantial variation within MGE, such that each element appears to be composed of a number of mosaic fragments of other elements. It is not known how this occurs, since the sequences vary too much between elements for it to be due to homologous recombination.

In the next sections, each of the major MGE types will be described, emphasizing their structure, horizontal transfer and control of transfer, variation and distribution, and their effect on the *S. aureus* cell which includes the carriage of virulence and resistance genes.

Bacteriophage

Horizontal transfer

Most *S. aureus* bacteriophage are approximately 45 kb and lysogenic (temperate). This means they insert into the chromosome at a specific site and behave like any other piece of DNA, but under stress conditions can be induced to excise, replicate their phage genome and produce hundreds of phage progeny, which are released causing lysis of the host cell. The released phage can then bind to other recipient bacteria, and inject their DNA. Once injected into a recipient bacteria, this phage genome is likely to induce its own replication, producing more phage particles and lyses of the cell (lytic pathway). A small proportion of phage genomes will integrate into a specific location on the *S. aureus* chromosome, where they will reside in a lysogenic and dormant state (prophage). These prophage will be replicated as part of the chromosome and thus spread to daughter cells, but can be induced into the lytic pathway by stress at any point.

A number of prophage have been sequenced as part of the whole genome sequencing projects, and a range of other phage have also been sequenced (Iandolo *et al.*, 2002; Vybiral *et al.*, 2003; Kwan *et al.*, 2005). Lysogenic phage that are widespread in *S. aureus* are Siphoviridae, typically 40–45 kb and have a relatively conserved gene order. There are also non-lysogenic phage found in *S. aureus* such as the large Twort phage that was first used to show the potential of phage therapy, and podoviridae (phi29-like) that have much smaller genomes. The following discussion will focus on lysogenic phage.

Many of the phage genes necessary for this complex lifecycle have been determined. At each terminal end of the phage are short inverted

Table 3.1 Summary of major MGE in sequenced *S. aureus* strains

Strain	MRSA252	N315	Mu50	COL	8325	MW2	MSSA476	FPR3757	JH1/JH9	Newman	RF122
GenBank	BX571856	BA000018	BA000017	CP000046	CP000253	BA000033	BX571857	CP000255	CP000703	AP009351	AJ938182
Lineage/CC	30	5	5	8	8	1	1	8	5	8	151
Bacteriophage											
ϕSa1			NI								lukFM (trunc)
ϕSa2	NI				NI (ϕ12)	PV-luk		PV-luk	NI		
ϕSa3	chip scin sak sea	chip scin sak sep	scin sak sea		chip scin sak	scin sak sea seg sek	scin sak sea seg sek	chip scin sak	chip scin sak	chip scin sak sea	
ϕSa4							NI		NI		
ϕSa5					NI (ϕ11)					NI	
ϕSa6				NI (L54a)					NI	NI	NI (trunc)
ϕSa7										NI	
ϕSa8											NI
SaPIs											
SaPI1				seb ear seq sek				ear seq sek			
SaPI2		sel sec tst	sel sec tst								mdr
SaPI3			fhuD			ear sel sec					
SaPI4	NI						NI				
SaPIbov											tst sel

Plasmids									
I	ble kn*	ble kn*	ble kn*	tet				tetK; NI	
II	cadAC arsBC*	cadDX arsBC				blaZ cadD	blaZ cadD		blaZ arsR cadD aac/ aph
III			aac/aph qacA					erm ileS	
Transposons									
Tn*552*	blaZ								
Tn*554*	erm spc	erm spc	erm spc						
Tn*5801*			tetM						
Tn*916*-like	NI			NI					
SCC									
mec I				mecA					
mec II	mecA	mecA	mecA						mecA
mec III									
mec IV						mecA		mecA arc opp	
non-*mec*							far1		

Phage and SaPI families are based on homology of integrase genes and insertion site. NI, element present but no known virulence factor or resistance gene; * integrated plasmid; FPR3757 SCC*mec* is fused to an ACME element; 'SaPI5' (Diep *et al*., 2006) belongs to the SaPI1 family based on integrase and insertion site. RF122 has two phage fragments. Abbreviations: aac/aph, aminoglycoside resistance; arc, arginine catabolism; arsBC, arsenic resistance genes; blaZ, penicillin resistance; bsa, bacteriocin biosynthesis genes; cadACDX, cadmium resistance genes; chip, chemotaxis inhibitory protein; ear, putative b – lactamase type protein; erm, erythromycin resistance; far1, fusidic acid resistance; fhuD, siderophore transporter; ileS, mupiricin resistance; lukFM, leukocidin; mdr, multidrug resistance; opp, oligopeptide uptake; mecA, penicillin-binding protein 2a conferring resistance to methicillin; qacA, quarternary ammonium compound (antiseptic) resistance; PV-luk, Panton–Valentine leukocidin; sak, staphylokinase; SaPI, *Staphylococcus aureus* pathogenicity island; SCC, staphylococcal cassette chromosome; scin, staphylococcal chemotaxis inhibitory protein; sea to sep, enterotoxin A to enterotoxin P; spc, spectinomycin resistance; tst, toxic shock syndrome toxin – 1; tet and tetM, tetracycline resistance. Updated from Lindsay and Holden (2004).

repeats (attL and attR) and these match a short sequence at the specific insertion site (attP) (Carroll *et al.*, 1995). At one end of the phage is integrase (*int*), a gene necessary for site-specific integration of the phage genome into the attP site (Lee and Iandolo, 1986). Activation of ϕ11 *int* is enhanced by *rinA* and *rinB* (Ye and Lee, 1993), although these genes are not found in most other phage. Some phage also have genes for excision (*xis*) (Ye and Lee, 1989) and regulators of the lytic-lysogenic switch (*cI*, *cro*, other repressors and anti-repressors) similar to those found in the well-studied *E. coli* lambda phage (Carroll *et al.*, 1995). It is likely the well conserved divergent promoter region upstream of *int* is bound by cI, Cro and other regulatory proteins that alternatively control expression of lytic (excision and replication) and lysogenic (integration) pathways. Regulators such as cI are probably cleaved by RecA during stress, alleviating repression and causing phage induction. However, amongst multiple sequenced phage there is sufficient variation to suggest the story will be more complex (Kwan *et al.*, 2005). The model does suggest an important mechanisms for controlling the transfer of phage. It is well known that a lysogenic strain cannot be infected with a related phage (Lowbury and Hood, 1953), and this may be due to the prophage encoded regulatory proteins binding to incoming phage and preventing expression of essential genes.

Bacteriophage replication proteins are described on staphylococcal phage, but most of the replication machinery is supplied by the host cell. Some phage encoded polypeptides can bind to the cell's DNA polymerase and reduce host genome replication during phage infection (Belley *et al.*, 2006). Several phage genes have been identified that encode structural phage head, phage tail and tape-measure proteins and these are necessary for packaging of the phage DNA (Tallent *et al.*, 2007). Once phage particles are made, holins are used to punch a hole in the cell membrane to enable delivery of endolysins (murine hydrolases/amidases) to the cell wall that digest peptidoglycan and allow lysis of the bacterial cell and release of phage particles. These endolysins may also play a role in penetrating the peptidoglycan layer when the phage delivers DNA to a recipient cell (Navarre *et al.*, 1999; Loessner *et al.*, 1999; Takac *et al.*, 2005; Sass and Bierbaum, 2007).

Induction of prophage into the lytic state is common and most normal bacterial cultures will contain some free phage. Other stresses such as antibiotics (Goerke *et al.*, 2006b), growth in biofilms (Resch *et al.*, 2005) or infection with another phage also induce the phage lytic pathway. In the laboratory, mutagens such as mitomycin C, or UV light stress are used to induce lysis of whole cultures within a few hours. The released phage probably infect recipient *S. aureus* with high frequency in nature. Studies of *S. aureus* strains during the course of normal or chronic infection have suggested phage acquisition and loss *in vivo* is a typical feature (Moore and Lindsay, 2001; Kahl *et al.*, 2003; Goerke *et al.*, 2006a).

The *S. aureus* cell surface receptor for phage binding is speculated to be teichoic acid (Chatterjee, 1969). *S. aureus* phage are considered to be species specific. Attachment seems to require calcium ions, and can be inhibited by chelators such as citrate (Rountree, 1955).

Generalized transduction is a variation of the normal lysogenic pathway, and involves host bacterial DNA being packaged and delivered. Only one naturally occurring phage has been studied that can achieve this, ϕ11. ϕ11 is found integrated into the genome of 8325 (also known as RN1 or PS47) (Iandolo *et al.*, 2002). A variant phage, 80α, that can also cause generalized transduction is probably the result of mixing ϕ11 with ϕ80. The distribution of other generalized transducing phage in *S. aureus* is unknown. The mechanism for generalized transduction could be a packaging error for a normal lysogenic phage, or could be a specifically evolved pathway for distributing *S. aureus* DNA between strains, and the mechanism is currently unknown.

Control of bacteriophage spread is likely to occur by a phage immunity mechanism as described earlier. Another mechanism for controlling the spread of phage is the Sau1 RM system (Waldron and Lindsay, 2006). In addition, a few phage carry their own RM systems (Dempsey *et al.*, 2005).

Variation and distribution

Bacteriophage are widespread in *S. aureus*, with most strains carrying between one to four prophage genomes integrated into the chromosome. Each phage in a genome will carry a different integrase gene that specifies its insertion site in the *S. aureus* chromosome. Several integrase/insertion site combinations are currently known, and it has been proposed that this should be used as the basis of a classification system of phage families (Lindsay and Holden, 2004). This concept is supported by the fact that putative phage immunity is also encoded in the same region, and thus a strain lysogenic for a phage of one family is likely to be resistant to infection by phage of the same family.

Extensive sequencing, and microarray studies have allowed the distribution of hundreds of phage genes in hundreds of strains to be compared (Kwan *et al.*, 2005; Lindsay *et al.*, 2006). Phage distribution varies in two major ways. Firstly, the presence or absence of phage varies. Some phage are very common such as of the ϕ3 family, and this suggests that they may not move very much. Others are common in certain *S. aureus* lineages but not others, suggesting their spread is limited by the Sau1 system. Others are rare, and this could indicate they are newly evolving or that they are dying out. Even those that are common in some lineages will be missing in some strains, and this suggests frequent acquisition and loss.

The second way that phage vary is in their mosaic structure. Each phage appears to be a combination of modular units from other phages (Kwan *et al.*, 2005). The amount of rearrangement is very high. How this rearrangement occurs is unknown, but if cells typically have two or more phage perhaps rearrangement occurs in the cell and likely during induction. Presumably, only those phage that have all the necessary components for replication and spread will survive.

Effect on the cell

The phage lytic cycle has serious consequences for the cell, particularly under stressful conditions when the cell is in danger of lysis and death. However, phage genomes are so widespread in *S. aureus*, they must have some selective benefit. One advantage they may provide is the carriage of virulence and toxin genes that can benefit *S. aureus* during infection. These are usually carried at one end of the phage, typically the opposite end to the integrase, and it is speculated that they were initially acquired by accident. When phage are induced to replicate, the number of copies of any toxin gene encoded on them increases dramatically, and this can increase toxin production (Sumby and Waldor, 2003).

A number of important *S. aureus* toxins are encoded on bacteriophage. They include enterotoxin A (*sea*) a major cause of food poisoning, exfoliative toxin A (*eta*) a cause of scalded skin syndrome, and Panton–Valentine leukocidin (*PV-luk*) (Betley and Mekalanos, 1985; Kaneko *et al.*, 1998; Yamaguchi *et al.*, 2000). *PV-luk* is necessary for haemolytic pneumonia (Labandeira-Rey *et al.*, 2007). There is also strong epidemiological association with *PV-luk* found in many CA-MRSA strains but its role in CA-MRSA disease such as severe skin infections is debated by some (Voyich *et al.*, 2006). There is no doubt that CA-MRSA cause severe skin and soft tissue infections in healthy populations, such as athletes, military personnel, children and men who have sex with men. These strains have been spreading rapidly in the USA in particular, leading to enormous demands on accident and emergency departments, deaths in young children, and now are an increasing cause of hospital acquired infection. If *PV-luk* is not the contributing factor, it is not clear what has made these strains so virulent.

Phage of the family ϕ3 have recently been shown to carry genes that help *S. aureus* evade the host immune response. A region that has been referred to as the 'innate immune evasion cluster' (van Wamel *et al.*, 2006) is located at one end of the phage, and may carry any of the following genes : chemotaxis inhibitory protein (*chip*) which binds to host chemokine receptors and prevents neutrophil chemotaxis (de Haas *et al.*, 2004), staphylococcal complement inhibitor (*scin*) which blocks the formation of complement products that bind to *S. aureus* and thus prevents phagocytosis (Rooijakkers *et al.*, 2005a), staphylokinase (*sak*) which modulates the effect of defensins and is antiopsonic (Jin *et al.*, 2004; Rooijakkers *et al.*, 2005b), and the enterotoxins A and P (*sea, sep*). There is substantial gene rearrangement in this region, and it is not clear if strains with different combinations have differing abilities to affect host immune responses.

Small RNA (sRNA) are also known to be encoded on phage, and again the ϕ3 family is implicated (Pichon and Felden, 2005). sRNA can specifically bind to mRNA controlling translation of toxin or other key proteins (Morfeldt *et al.*, 1995; Pichon and Felden, 2005). sRNA can also bind to promoter regions controlling gene transcription, or can bind to proteins altering their function (Pichon and Felden, 2007).

There are also many genes of unknown function in the 'virulence factor zone' of many phage, and many have signal sequences suggesting they are secreted. Recently, Bae *et al.*, (2006) has shown that the ϕ3 phage in strain Newman is necessary for virulence. Interestingly, the other three phage carried by this strain were, in combination, also necessary for virulence despite no known toxins being encoded on them.

Some phage integrate into virulence genes such as β-haemolysin (*hlb*) or lipase (*geh*), thus preventing production of these toxins (Lee and Iandolo, 1986; Coleman *et al.*, 1989).

Considering that generalized transduction may be the major mechanism for the spread of genetic material between *S. aureus*, and recent studies show a lot of variation between *S. aureus* strains, phage may play an important role in enabling *S. aureus* to mutate and adapt to new environments. This may be in a flexible manner, allowing frequent acquisition and loss of elements, and contribute to the extraordinary range of diseases, hosts and habitats it can adapt to.

Exploitation of bacteriophage

S. aureus phage have been exploited for many years. Firstly, they were used to develop a very successful typing scheme for investigating *S. aureus* outbreaks in the 1940s (Williams and Rippon, 1952; Blair, 1956; Elek, 1959), which has only been superseded by genetic methods in the last ten or so years. Sets of phage were distributed to laboratories all over the world, and strains were classified according to their susceptibility on agar plates with multiple phage spots. Notably, the phage sets had to be updated as the *S. aureus* strains systematically became resistant to them, and there was significant geographical variation in *S. aureus* strains. We can now suggest that the phage typing system was successful because susceptibility was partly related to the Sau1 system that defines lineages, although there is no doubt that the patterns were also complicated by the presence of prophage.

In the laboratory, generalized transduction is the most widespread method used for transferring DNA between *S. aureus* strains and the construction of genetically modified organisms for studying *S. aureus* virulence and biology (Chapter 5). More recently, phage have been proposed as delivery mechanisms for targeting antibiotics to *S. aureus* (Embleton *et al.*, 2005; Yakoby *et al.*, 2007). Phage endolysins have shown promise as bactericidal agents (Donovan *et al.*, 2006a; Sass and Bierbaum, 2007). Genetically modified cows that secrete *S. aureus* phage endolysins to protect them from infection have also been proposed (Donovan *et al.*, 2006b).

Bacteriophage therapy for treating infections has been widespread in Eastern Europe for decades. However there are few rigorous clinical trails. In animal models of disease, bacteriophage are highly effective treatments (Matsuzaki *et al.*, 2003; Matsuzaki *et al.*, 2005; Capparelli *et al.*, 2007). Systemic phage are cleared from the host very quickly by the reticuloendothelial system, but longer lasting phage can be selected (Merril *et al.*, 1996). The advantages of phage therapy are that it is highly species (and even lineage) specific meaning that only the infecting organism will be targeted and not protective commensal bacteria. However, routine diagnosis of infecting organisms might have to improve before this therapy is feasible. The disadvantages include the potential risk of resistance evolving very quickly. In practical terms, companies who invest in bringing phage therapy to the market may have difficulties in patenting their technology, getting licensing for a living organism, and adapting phage as *S. aureus* quickly evolve, and these are issues that need to be tackled by government regulating bodies before we will see any further progress.

Staphylococcus aureus pathogenicity islands (SaPIs)

Horizontal transfer

SaPIs are discrete genetic sequences of approximately 15 kb. They are dependent on a specific interaction with a helper bacteriophage to transfer between *S. aureus* cells at very high frequency (Lindsay *et al.*, 1998). SaPIs integrate into the *S. aureus* genome at specific sites via an integrase and terminal inverted repeats. Under stress they can be excised from the chromosome, replicate their genomes and are packaged into miniature phage heads before lysis of the host bacterium and release. SaPI particles can bind to recipient bacteria and inject their genome, which can be integrated into the recipient chromosome. They are much more efficient than bacteriophage transfer because recipient cells are not in danger of lysis by a lytic phage. Amongst Gram positive bacteria, they were the first mobile pathogenicity islands to be described (Lindsay *et al.*, 1998).

Excision of the SaPI from the chromosome can be induced by stress such as UV light or mitomycin (Lindsay *et al.*, 1998) or antibiotics (Ubeda *et al.*, 2005; Maiques *et al.*, 2006), and occurs via the host encoded SOS induction pathway involving *recA* and *lexA* (Ubeda *et al.*, 2007a). The SaPI integrase gene and the two terminal repeat regions are sufficient for excision, circularization of the element and integration (Ubeda *et al.*, 2003). For replication, an origin of replication has been identified that is bound by a replication initiation protein (*rep*). In combination with another protein (*pri*), this is sufficient for replication of the circular intermediate (Ubeda *et al.*, 2007b) and this mechanism is similar to those found in some plasmids. Maintenance of the circular state is unstable, probably due to incorrect segregation into daughter cells during cell replication, and this is an essential function found in successful plasmids.

SaPIs do not encode the structural genes necessary for phage head, tail, tape measure or scaffolding, and thus are dependent on the helper phage for these functions (Lindsay *et al.*, 1998; Tallent *et al.*, 2007). Because phage packaging occurs using a scaffolding mechanism, where the element's genome is used to attach proteins sequentially, this is likely to explain why SaPI particles are approximately one third of the size of phage particles (Ruzin *et al.*, 2001). Some SaPI proteins are also suggested to play a role in the synthesis of the small phage particles (Ubeda *et al.*, 2007a), although they are not found in purified SaPI particles (Tallent *et al.*, 2007). The specific interaction between a helper phage and a SaPI has not been fully elucidated, but there appears to be a special relationship between each SaPI and particular phage. For example, the prototype element SaPI1 is transferred by generalized transducing phage $\phi 11$ at a frequency of 10^{-7} per plaque forming unit, but by generalized transducing phage 80α at a frequency of 10^{-1}. This may be due to a phage factor that encourages SaPI replication (Lindsay *et al.*, 1998).

It is likely that SaPIs are capable of moving into and out of *S. aureus* at high frequency during infection by the same mechanisms as *S. aureus* bacteriophage. SaPI transfer between *S. aureus* is probably blocked by the Sau1 RM system. In addition, it is likely that transfer can also be blocked by a mechanism similar to phage immunity, described above. The integrase promoter region of SaPIs shows high homology to that of *S. aureus* phage (Lindsay *et al.*, 1998) and suggests that phage and SaPI may interact to control the induction and immunity to each other. Proof is currently lacking.

Maiques *et al.*, (2007) have reported that a SaPI found in a bovine strain of *S. aureus* was able to be transferred via a helper phage to other species of staphylococci, such as *S. epidermidis*. *S. aureus* bacteriophage were thought to be species specific, but this result suggests the host range of both phage and SaPI elements may be wider than previously thought.

Variation and distribution

The whole genome sequencing projects and microarray comparative genomics experiments suggest that most *S. aureus* carry between zero and two SaPIs. Like phage, each insertion site appears to correlate with a site specific integrase, and these have been proposed as a method for classification (Lindsay and Holden, 2004) (Table 3.1).

Like phage, the SaPI vary substantially between strains in two ways (Lindsay *et al.*, 2006). Firstly, the distribution of SaPI varies in that some are common and others less so. Some are associated with specific lineages, while others are more widespread. Some are so common within lineages that they may be relatively stable and not move much. Secondly, there is a very strong mosaic structure, such that each SaPI is composed of modular units of other SaPI. Interestingly, there is little evidence that these units have derived directly from phage, suggesting rearrangements may be restricted to interaction with other SaPI elements only.

Effect on the cell

SaPI are called pathogenicity islands because a large proportion carry known virulence genes. In particular, superantigens that are capable of non-specific activation of T cell populations leading to shock are carried on SaPI (Lindsay *et al.*, 1998, Fitzgerald *et al.*, 2001; Yarwood *et al.*, 2002). They include toxic shock syndrome toxin 1 (*tst*), enterotoxin B (*seb*) and enterotoxin C (*sec*) all able to cause toxic shock syndrome, while the enterotoxins can also cause food poisoning. The sequenced strain Mu50 has a putative siderophore iron uptake protein, *fhuD*, encoded on a SaPI (Kuroda *et al.*, 2001). In animal strains of *S. aureus*, biofilm associated protein (*bap*) can also be carried on SaPIs (Cucarella *et al.*, 2001). It is likely that induction of SaPI, like phage, leads to increased copy number of toxin genes and therefore release of toxin (Sumby and Waldor, 2003).

SaPI are widespread enough in *S. aureus* that they likely provide some sort of selective benefit, but it is currently not clear what that may be. The toxins encoded on SaPI can cause disease independent of the host bacterium, so it is difficult to imagine how they benefit the cell. Some of the toxins are suggested to act as regulators of *S. aureus* gene expression (Vojtov *et al.*, 2002). Many SaPI encode open reading frames with no known function but have signals for secretion and they may provide some advantage. Similarly, small RNAs may also be present. SaPIs may also contribute to *S. aureus* genome flexibility.

Related elements

Recently, elements that look very much like SaPI but carry resistance genes at the non-integrase end have been described. O'Neill *et al.* (2007) identified an element similar to SaPI1 and other SaPI from *S. haemolyticus* and *S. aureus* N315 that also included the fusidic acid resistance gene (*fusB*). *fusB* and its leader peptide CDS appear to have integrated into the end of the element from plasmid pUB101. The element has been named a *Staphylococcus aureus* resistance island, $SaRI_{fusB}$. It was found in a successful epidemic *S. aureus* clone that causes impetigo in Europe. Since fusidic acid is prescribed for impetigo, the emergence and spread of this clone is concerning.

S. aureus strain RF122 from bovine mastitis (GenBank AJ938182) also appears to carry a SaRI. A pathogenicity island-like element encodes a gene SAB1892c which shows homology to multidrug resistance proteins, although its function has not yet been proven.

Plasmids

Horizontal transfer

Plasmids are circular pieces of DNA that are found in bacterial cells and can replicate independently of the chromosome. They vary in size substantially, from approximately 3 to 150 kb in *S. aureus*. Some plasmids encode their own conjugative horizontal transfer mechanism, and these genes are called *tra*. Conjugation involves the production of pores or pili that bridge between a donor and recipient cell allowing direct transfer of DNA. Other plasmids do not encode their own transfer mechanism, and are typically transferred by generalized transduction (Dyer *et al.*, 1985). Plasmids are also capable of integrating into other transferable elements, such as a larger conjugative plasmid (Berg *et al.*, 1998).

The unique feature of autonomous replication occurs by either the rolling circle or theta mechanism. Plasmids that replicate by an unusual rolling circle mechanism are typically less than 5 kb (Khan, 2005). Replication involves a plasmid encoded replication (*rep*) protein nicking the plasmid DNA at a specific site, the unwinding of the double-stranded DNA, and the sequential release of one strand and synthesis of a new double strand on the remaining template. The free single stranded plasmid is then used as a template for synthesis of another new double strand. Most rolling circle plasmids occur at high copy number.

Most other plasmids replicate by the theta mechanism, where a section of DNA around the origin is unwound, and new DNA is synthesized in both directions until the two resulting plasmids are nicked to separate them (Kwong *et al.*, 2004). These plasmids can range in size up to hundreds of kilobases, and may contain integrated smaller plasmids, transposons or insertion sequences (Berg *et al.*, 1998).

Regulation of replication is often controlled by an antisense RNA molecule that controls rep protein and thus controls copy number (Kwong *et al.*, 2006). Partition (*par*) genes are often involved in stable maintenance and the correct segregation of plasmids into daughter cells during bacterial replication. Plasmids that have similar replication mechanisms can be 'incompatible' in the same cell, as the regulation of copy number will ensure that less copies of each plasmid are produced, and once the cells split into daughter cells, plasmids are likely to be unevenly distributed and therefore quickly lost (Novick, 1987). This is an important mechanism for controlling the horizontal transfer and stability of plasmids. Sau1 RM is also important in blocking plasmid transfer (Waldron and Lindsay, 2006).

Conjugative plasmids from enterococci can be used to deliver enterococcal transposons to *S. aureus*. This is the mechanism of transfer of the *vanA* carrying transposon to *S. aureus* in at least one naturally occurring VRSA isolate from Michigan (Flannagan *et al.*, 2003; Weigel *et al.*, 2003). At present, only six cases of VRSA have been documented (Anonymous, 2004) and some have evolved independently. The spread of VRSA in our hospitals would be a catastrophe as vancomycin and related antibiotics are widely depended on for preventing and treating MRSA infections, and no other antibiotics are reliably useful.

Other conjugative plasmids in enterococci are pheromone sensitive, meaning they only transfer to recipient cells when they have been activated by a small molecule pheromone (Clewell *et al.*, 1985). Surprisingly, of many species tested, only *S. aureus* produce appropriate pheromone triggers for enterococcal plasmid transfer. The enterococcal plasmid replication system does not function in *S. aureus* and the plasmids are unable to replicate and are therefore quickly lost. However, any hitch-hiking mobile elements such as transposons have the opportunity to jump into the *S. aureus* chromosome (Clewell *et al.*, 1985). This is an extremely efficient transfer system in the laboratory, and can be blocked by the Sau1 RM system (Waldron and Lindsay, 2006). Some animal *S. aureus* that are deficient in *sau1hsdS* genes are hyper-susceptible to this transfer (Sung and Lindsay, 2007), which is concerning as vancomycin resistant enterococci are often found in animals.

Variation and distribution

Plasmids are often classified according to size and mechanism of replication. The small rolling circle plasmids are often referred to as group I plasmids. Larger plasmids are often called group II, and larger plasmids with tra genes for conjugation group III. Incompatibility groups defined by the replication mechanism similarity and ability of related plasmids to survive in the same cell can be used to classify plasmids, and many groups exist (Novick, 1987).

There is enormous variation in the distribution of *S. aureus* plasmids. About half of the sequenced strains have no free plasmids, while others have up to three. Several strains have plasmids integrated into their chromosomes and/or into the SCC*mec* element. Many of the newly sequenced plasmids are unique, suggesting a wide variety of plasmids and plasmid genes in *S. aureus*. Microarray studies have indicated only a few strains have plasmid genes (Lindsay *et al.*, 2006), but this is likely to be due to the large amount of plasmid variation and therefore the relatively small proportion of plasmid genes found on multi-strain microarrays.

Effect on the cell

Plasmids often carry antibiotic resistance genes. Thus, it is expected that they provide an advantage to the host bacterium under conditions of antibiotic exposure. Resistances to the following compounds have been described on plasmids: penicillin, tetracycline, erythromycin, kanamycin, trimethoprim, fusidic acid, aminoglycosides, vancomycin, chloramphenicol, mupirocin, heavy metals, and quarternary ammonium chlorides (disinfectants). Some plasmids are so large that they must be a physical burden on the cell, and presumably they can be lost or consolidated when antibiotic pressure of the environment is relaxed.

Virulence genes are also found on some plasmids, such as the exfoliative toxin B (*etb*) responsible for scalded skin syndrome (Yamaguchi *et al.*, 2001). It is not clear what selective advantage this provides for the bacterium, and notably the plasmid sequenced that contained these genes also contained resistance to heavy metals and lantibiotic synthesis. Lantibiotics (also known as bacteriocins) are small molecules produced by *S. aureus* that kill other *S. aureus* or species; lantibiotic immunity genes are an important component of this operon and protect the producer cell from lethal action (Navaratna *et al.*, 1999). Their presence may be a selective advantage for *S. aureus* in mixed populations, particularly if nutrients are scarce or other bacteria are producing lethal toxins. Many plasmids are cryptic, and it is unknown what they do or the function of the genes carried on them.

Since many plasmids are large, or found in high copy number, maintaining them must be a burden to the cell. Presumably even cryptic plasmids must provide some selective advantage to the host cell or they will be lost.

Transposons

Horizontal transfer

Transposons are discrete pieces of DNA that encode their own transposase. Transposases enable replication of transposon DNA and integration into other DNA. Many transposases are not site specific, so multiple copies of the same transposon can appear in the bacterial cell, on chromosomes and/or plasmids. Transposons in *S. aureus* are typically 3–60 kb in size. Horizontal transfer of transposons between *S. aureus* typically involves integration into another mobile element and hitch-hiking across via conjugation or generalized transduction. Some transposons carry *tra* genes and can be transferred by conjugation. The mechanisms for blocking transposon transfer likely include those that block the transfer of any element they are carried on, including Sau1 RM.

Transposons that integrate randomly into the chromosome have been exploited by *S. aureus* geneticists for constructing transposon libraries, such as Tn*917*, Tn*551* and *bursa aurealis* elements. A transposon library is generated by introducing a transposon into a cell, and then selecting for its random integration into the chromosome, usually by temperature selection. If thousands of mutants are isolated, then theoretically, every non-essential gene will be deleted in at least one mutant. Libraries can be screened for particular phenotypic changes and the insertion site of the transposon identified to identify the gene responsible for the phenotype (Chapter 5).

Variation and distribution

Most sequenced strains of *S. aureus* carry between one and two transposons. A number of transposons have been characterized in detail, including Tn*551*, Tn*552* and Tn*554*. (Khan and Novick, 1980; Murphy *et al.*, 1985; Rowland and Dyke, 1989; Wu *et al.*, 1999). However, several of the *S. aureus* transposons have been identified only by sequencing projects and it is not clear how mobile they are. Composite transposons are those that consist of two transposons with a random piece of DNA in between that is accidentally replicated and transferred, such as *S. aureus* Tn*4001* (Mahairas *et al.*, 1989).

Transposons from other Gram positive species have successfully transferred into *S. aureus*, and include the *vanA* positive transposon from enterococci (Flannagan *et al.*, 2003; Clark *et al.*, 2005), and the conjugative transposon Tn*918* from streptococci (Clewell *et al.*, 1985).

Effect on the cell

Many *S. aureus* transposons encode antibiotic resistance genes, and this is presumably a very important selective advantage they provide to the host cell. Resistance to penicillin, erythromycin, tetracycline, aminoglycosides and vancomycin has been found on transposons. Presumably large transposons can be a burden to the host cell when antibiotic levels are low, but it is not known whether they are unstable. Some transposons are found in multiple copies in the cell, such as Tn554 which occurs five times in N315 and twice in Mu50 (Kuroda, *et al.*, 2001). Theoretically transposons can insert into important genes or their promoters. The Tn552 transposon encodes β-lactamase resistance and the regulatory system of these genes can also regulate methicillin resistance genes (Lewis and Dyke, 2000)

Insertion sequences (IS)

Insertion sequences are transposons that only encode the transposase. They are very common in some *S. aureus* but not others. Their distribution appears to correlate with lineage, suggesting they are rarely transferred between strains (Lindsay *et al.*, 2006). Multiple copies of the same insertion sequence in the cell are common. In *S. epidermidis*, it has been shown that insertion sequences can insert into genes and/or promoter regions affecting transcription and translation of important cell proteins (Ziebuhr *et al.*, 1999). This is also likely to occur in *S. aureus* (Valle *et al.*, 2007). IS can also act as composite transposons, when two IS copies flank an unrelated piece of DNA and move it to another location. Once moved, IS elements can be deleted, and many *S. aureus* whole genome sequences show the scars of this sort of transfer. Recently, it has been suggested that such a mechanism is responsible for large genome rearrangements in *S. haemolyticus* (Watanabe *et al.*, 2007). It is not clear what the selective advantage of insertion sequences are, but being so small, it is possible they act as some sort of parasite that can be quite difficult for the bacterial cell to eradicate. Alternatively, they may contribute to *S. aureus* adaptability, enabling genes to be quickly mutated and/or restored.

Staphylococcal cassette chromosomes (SCC)

Horizontal transfer

SCC elements have defined left and right junctions and are always found inserted into one site, the *orfX* gene (Ito *et al.*, 1999). Thus *S. aureus* strains have either zero or one copy, ranging in size from about 3 to 60 kb (Ito *et al.*, 2001). SCC elements encode *ccr* genes that catalyse excision and integration of the element (Katayama *et al.*, 2000). The mechanism of horizontal transfer between staphylococci is unknown. It is speculated that elements less than 45 kb, such as the SCC*mec* type IV elements, are transferred by generalized transduction. SCC appear to transfer at very low frequency, and are relatively stable compared to other MGE (Lina *et al.*, 2006). The first SCC elements encoding methicillin resistance (SCC*mec*) probably came from coagulase negative staphylococci (Katayama *et al.*, 2003).

Variation and distribution

SCC*mec* elements have been grouped into types I to IV (Ito *et al.*, 2001, Ma *et al.*, 2002), but recently there appears to be a burst of new variants described and this number is expanding rapidly (Ito *et al.*, 2004; Shore *et al.*, 2005; Qi *et al.*, 2005; Jansen *et al.*, 2006; Milheirico *et al.*, 2007; Heusser *et al.*, 2007). The first four types were classified by the types of *ccr* genes they carried, and by variations in the *mec* genes themselves. Each SCC*mec* often carried plasmids and transposons integrated into them (Ito *et al.*, 1999), and thus were considered to be integration hot spots. Now that the SCC*mec* elements appear to be getting smaller and spreading more quickly, likely via transduction, the amount of variation also appears to be increasing.

Some CA-MRSA such as FPR3757 have an element fused to the end of their SCCmec type IV called ACME (Diep *et al.*, 2006). This element encodes an arginine deaminase pathway and oligopeptide uptake genes but no known mobilization genes. ACME is found in other species of staphylococci, but its role in virulence is unknown.

Some SCC elements do not encode methicillin resistance. They include the SCC element in MSSA476 which encodes a trimethoprim resistance gene (Holden *et al.*, 2004), and another SCC that encodes a mucoid capsule type (Luong *et al.*, 2002).

Effect on the cell

SCC*mec* elements carry resistance to methicillin, which no other MGE appears able to do. Early SCC*mec* elements also carried a number of other resistance genes on plasmids and transposons including erythromycin, kanamycin, bleomycin (Ito *et al.*, 2001) The widespread use of methicillin type antibiotics in hospitals for preventing and treating infections likely contributed to the selection of MRSA, which are now endemic in hospitals in most countries of the world (HA-MRSA). In addition, community-acquired MRSA (CA-MRSA) are becoming prevalent in the USA in particular, and these strains cause severe skin infections in healthy people. It is not clear if SCC*mec* provides a selective advantage in CA-MRSA spread or ability to cause disease, but since it is resistant to first line antibiotics prescribed in the community, it seems likely that this has contributed to spread.

SCC*mec* type I element was a burden on the bacterial cell during growth in rich media without antibiotic pressure (Ender *et al.*, 2004,

Lee *et al.*, 2007). This may explain why multiple antibiotic resistances were necessary on SCC*mec* type I in order for it to survive in a sufficient number of environments. In contrast, SCC*mec* type IV is not a burden to the cell, and this may contribute to its rapid spread amongst *S. aureus*. Furthermore, it has been shown that SCC*mec* type I inhibits the ability of *S. aureus* to invade cells (Werbick *et al.*, 2007).

Large-scale genome rearrangements

Genomic islands (GI)

Every sequenced *S. aureus* isolate contains two genomic islands, GIα and GIβ, each of about 20–30 kb. They vary substantially between strains, but variation is strongly correlated with lineage (Lindsay *et al.*, 2006). There is no evidence that GIs are mobile, and they carry no known mobility or replication genes. Presumably they can be transferred between *S. aureus* by generalized transduction, but at low frequency like all other non-mobile regions of the chromosome.

Each GI encodes one copy of the *sau1hsdM* and *sau1hsdS* genes in the centre of the element (Waldron and Lindsay, 2006). The *sau1hsdS* genes vary according to lineage. Since they are one of the key controllers of lineage evolution, it is proposed that any variation in the *sau1hsdS* gene represents the first step in establishing a new lineage. Thus successful horizontal transfer of a GI between *S. aureus* would represent a particularly significant event.

A number of known or putative virulence factors are encoded on GIs. GIα typically carry a series of related staphylococcal enterotoxin-like genes called '*set*', and between 7 and 11 copies are common. Recent studies have shown that at least some of these proteins don't function as enterotoxins but have alternative effects on the host immune system and are thus being renamed staphylococcal superantigen-like.. *Ssl5* inhibits neutrophil rolling and thus neutrophil recruitment to the site of infection (Bestebroer *et al.*, 2007). *Ssl7* bind to IgA antibodies and C5 complement, and inhibit phagocyte binding and killing (Langley *et al.*, 2005, Wines *et al.*, 2007). GIα also carry multiple copies of related lipoprotein genes, although their function is unknown.

The GIβ elements carry multiple copies of related genes for serine proteases (*spl*). Some are known to be functional and are activated by signal peptidases (Popowicz *et al.*, 2006). Multiple copies of variant superantigen toxin genes are also found, some of which have been named as *seg, sei, sel, sem,* or *seo* (Jarraud *et al.*, 2001). Each GIβ can also carry any of the following: a second homologue of the chromosomally encoded hyaluronate lyase gene (*hysA*) (Makris *et al.*, 2004), a two-component leukocidin toxin (*lukDE*) related to Panton–Valentine leukocidin (Gravet *et al.*, 1998) or lantibiotic biosynthesis (*bsa*) genes for production of bacteriocins to kill other bacteria (Navaratna, *et al.*, 1999).

The ability of GIs to carry multiple but variant copies of toxin genes (including pseudogenes) is a unique feature. It has been speculated that the multiple copies occur by recombinations and repeats of the genes leading to a seemingly unlimited number of variants (Jarraud *et al.*, 2001). Why this should be associated with GIs and not occur elsewhere in the *S. aureus* chromosome is unknown. Although GIs are classified as core variable (CV) regions as they are strongly associated with lineage, they are slightly more variable that most CV genes and deletions can occasionally be detected when comparing isolates of the same lineage in large scale studies (Lindsay JA, unpublished).

Large insertions

The *S. aureus* chromosome of isolates of the ST239 lineage are actually composed of 75% of the chromosome of CC8 isolates and 25% of CC30 isolates (Robinson and Enright, 2004). The likely mechanism is that the CC30 DNA has homologously recombined with and replaced a section of DNA in a CC8 isolate. The DNA segment corresponds to the origin of replication and the first quarter of the genome. Significantly, this includes the SCCmec type I element found in some CC30 isolates. Since no ST239s that are methicillin sensitive have been reported, this may have been a crude method for a CC8 isolate to acquire methicillin resistance. ST239 isolates have been very successful MRSA, spreading to hospitals all over the world. The mechanism of DNA transfer is unknown, but since such a large piece of DNA is involved, it may have required fusion of two *S. aureus* strains. Fusions have been demonstrated in the laboratory (Stahl and Pattee, 1983).

The first whole genome sequence of an ST239 isolate is currently underway. The isolate is a TW strain, associated with enhanced invasion in an intensive care setting (Edgeworth *et al.*, 2007). Surprisingly, another large piece of DNA (>130 kb) has been identified in the genome, and appears to have come directly from *S. epidermidis*. It includes antibiotic resistance genes which may have driven evolution (Holden, Pathak, Edgeworth, Lindsay, unpublished). The mechanism of transfer is currently unknown.

Large inversions

The *S. aureus* core genomes are relatively conserved, with each gene in the same order and orientation. In contrast, two *S. epidermidis* isolates have been sequenced, and only one of them has a massive inversion around the origin of replication (Zhang *et al.*, 2003; Gill *et al.*, 2005). Even more surprisingly, the *S. saprophyticus* and *S. haemolyticus* sequenced isolates had a similar inversion, and it seems as if one of the *S. epidermidis* isolates had the inversion, but inverted again at new breakpoints via IS elements (Lindsay and Holden, 2007). A similar proposed mechanism for large inversions and deletions via IS elements in *S. haemolyticus* has also been suggested (Watanabe *et al.*, 2007). It remains to be seen whether any of these will be found in a *S. aureus* genome. The advantages of inverting the genome are unclear, but it could influence the order that genes are expressed and their relative concentrations in the bacterial cell, and thus have quite large effects on cellular function. Otherwise, inversions, even small ones, are uncommon in staphylococcal genomes.

Conclusions

The last few years has seen an explosion in our knowledge about *S. aureus* genomes because of sequencing and microarray projects, and improved molecular typing methods. We can now see that *S. aureus* are evolving via specific pathways, and responding to selective pressures such as antibiotic use. Further understanding of these pathways will help us to predict how the *S. aureus* have evolved so far, how they may continue to evolve, and identify what we can do to prevent the evolution of increasingly pathogenic and virulent strains. At present we have strains that are methicillin-resistant and endemic in hospitals (HA-MRSA) with rare vancomycin resistant strains (VRSA). We also have community adapted strains that can infect healthy patients. In the future, we must be vigilant about the evolution of strains that are resistant to multiple antibiotics as well as capable of infecting healthy patients and persisting in the hospital environment. These strains will be a very heavy burden on our healthcare systems, potentially altering them irreparably.

Future

There is undoubtedly a lot we still don't know about *S. aureus* genomes variation and selection. The importance of evolving strains to human health is imperative and will dictate how infection is prevented and treated in the future.

Likely areas of study will be on the selective pressures driving evolution in normal *S. aureus* habitats such as the human nose. Identification of the key pressures and interactions in the nose may identify new therapeutic targets, for drugs, vaccines or public health campaigns, and may also allow better predictions about how strains will spread. It will also be beneficial to understand how drug resistance spreads and stabilizes, allowing prudent use of a diminishing supply of antibiotics, and rapid identification and isolation of strains exhibiting key steps in the evolution to successful multidrug resistant strains. The spread of toxins proven to be associated with unique disease profiles, such as *PV-luk*, will also be important to understand. The use of bacteriophage as a potential therapy may be developed, but it will be important to understand the limitations and risks, and how new phage can be altered to extend their life. Finally, associating particular lineages, toxins, resistance patterns or other features of strains with their ability to cause disease, spread in certain populations or survive extreme conditions will enable us to prevent, diagnose and treat *S. aureus* infections more effectively.

References

Anonymous. (2002). Vancomycin resistant *Staphylococcus aureus* – New York, 2004. Morb. Mort. Wkly. Rep. 53, 322–323.

Antonio, M., McFerran, N., and Pallen, M.J. (2002). Mutations affecting the Rossman fold of isoleucyl-tRNA synthetase are correlated with low-level mupirocin resistance in *Staphylococcus aureus*. Antimicrob. Agents Chemother. *46*, 438–442.

Bae, T., Baba, T., Hiramatsu, K., and Schneewind, O. (2006). Prophages of *Staphylococcus aureus* Newman and their contribution to virulence. Mol. Microbiol. *62*, 1035–1047.

Bayles, K.W., Brunskill, E.W., Iandolo, J.J., Hruska, L.L., Huang, S., Pattee, P.A., Smiley, B.K., and Yasbin, R.E. (1994). A genetic and molecular characterization of the recA gene from *Staphylococcus aureus*. Gene *147*, 13–20.

Belley, A., Callejo, M., Arhin, F., Dehbi, M., Fadhil, I., Liu, J., McKay, G., Srikumar, R., Bauda, P., Bergeron, D., Ha, N., Dubow, M., Gros, P., Pelletier, J., and Moeck, G. (2006). Competition of bacteriophage polypeptides with native replicase proteins for binding to the DNA sliding clamp reveals a novel mechanism for DNA replication arrest in *Staphylococcus aureus*. Mol. Microbiol. *62*, 1132–1143.

Berg, T., Firth, N., Apisiridej, S., Hettiaratchi, A., Leelaporn, A., Skurray, R.A. (1998). Complete nucleotide sequence of pSK41: evolution of staphylococcal conjugative multiresistance plasmids. J. Bacteriol. *180*, 4350–4359.

Besier, S., Ludwig, A., Brade, V., Wichelhaus, T.A. (2003). Molecular analysis of fusidic acid resistance in *Staphylococcus aureus*. Mol. Microbiol. *47*, 463–469.

Bestebroer, J., Poppelier, M.J., Ulfman, L.H., Lenting, P.J., Denis, C.V., van Kessel, K.P., van Strijp, J.A., and de Haas, C.J. (2007). Staphylococcal superantigen-like 5 binds PSGL-1 and inhibits P-selectin-mediated neutrophil rolling. Blood. *109*, 2936–2943.

Betley, M.J., and Mekalanos, J.J. (1985). Staphylococcal enterotoxin A is encoded by phage. Science. *229*, 185–7.

Birmingham, V.A., and Pattee, P.A. (1981). Genetic transformation in *Staphylococcus aureus*: isolation and characterisation of a competence-conferring factor from bacteriophage 80α lysates. J. Bact. *148*, 301–307.

Blair, J.E. (1956). Epidemiological implications of staphyloocccal phage typing. Ann. NY Acad. Sci. *65*, 152–159.

Capparelli, R., Parlato, M., Borriello, G., Salvatore, P., and Iannelli, D. (2007). Experimental phage therapy against *Staphylococcus aureus* in mice. Antimicrob. Agents Chemother. *51*, 2765–2773.

Carroll, D., Kehoe, M.A., Cavanagh, D., and Coleman, D.C. (1995). Novel organization of the site-specific integration and excision recombination functions of the *Staphylococcus aureus* serotype F virulence-converting phages phi 13 and phi 42. Mol. Microbiol. *16*, 877–893.

Chatterjee, A.N. (1969). Use of bacteriophage-resistant mutants to study the nature of the bacteriophage receptor site of *Staphylococcus aureus*. J. Bacteriol. *98*, 519–527.

Chen, I., and Dubnau, D. (2004). DNA uptake during bacterial transformation. Nat. Rev. Microbiol. *2*, 241–249.

Clark, N.C., Weigel, L.M., Patel, J.B., and Tenover, F.C. (2005). Comparison of Tn1546-like elements in vancomycin-resistant *Staphylococcus aureus* isolates from Michigan and Pennsylvania. Antimicrob. Agents Chemother. *49*, 470–472.

Clarke, S.R., Harris, L.G., Richards, R.G., and Foster, S.J. Analysis of Ebh, a 1.1-megadalton cell wall-associated fibronectin-binding protein of *Staphylococcus aureus*. Infect. Immun. *70*, 6680–6687.

Clewell, D.B., An, F.Y., White, B.A., and Gawron-Burke, C. (1985). *Streptococcus faecalis* sex pheromone (cAM373) also produced by *Staphylococcus aureus* and identification of a conjugative transposon (Tn918). J. Bacteriol. *162*, 1212–1220.

Cockfield, J.D., Pathak, S., Edgeworth, J.D., and Lindsay, J.A. (2007). Rapid determination of hospital-acquired meticillin-resistant *Staphylococcus aureus* lineages. J. Med. Microbiol. *56*, 614–619.

Coleman, D.C., Sullivan, D.J., Russell, R.J., Arbuthnott, J.P., Carey, B.F., and Pomeroy, H.M. (1989). *Staphylococcus aureus* bacteriophages mediating the simultaneous lysogenic conversion of beta-lysin, staphylokinase and enterotoxin A: molecular mechanism of triple conversion. J. Gen. Microbiol. *135*, 1679–1697.

Cucarella, C., Solano, C., Valle, J., Amorena, B., Lasa, I., and Penades, J.R. (2001). Bap, a *Staphylococcus aureus* surface protein involved in biofilm formation. J. Bacteriol. *183*, 2888–2896.

de Haas, C.J., Veldkamp, K.E., Peschel, A., Weerkamp, F., Van Wamel, W.J., Heezius, E.C., Poppelier, M.J., Van Kessel, K.P., and van Strijp, J.A. (2004). Chemotaxis inhibitory protein of *Staphylococcus aureus*, a bacterial antiinflammatory agent. J Exp Med. *199*, 687–695.

Dempsey, R.M., Carroll, D., Kong, H., Higgins, L., Keane, C.T., and Coleman, D.C. (2005). Sau42I, a BcgI-like restriction-modification system encoded by the *Staphylococcus aureus* quadruple-converting phage Phi42. Microbiol. *151*, 1301–1311.

Deplano, A., Zekhnini, A., Allali, N., Couturier, M., and Struelens, M.J. (1997). Association of mutations in grlA and gyrA topoisomerase genes with resistance to ciprofloxacin in epidemic and sporadic isolates of methicillin-resistant *Staphylococcus aureus*. Antimicrob. Agents Chemother. *41*, 2023–2025.

Diep, B.A., Gill, S.R., Chang, R.F., Phan, T.H., Chen, J.H., Davidson, M.G., Lin, F., Lin, J., Carleton, H.A., Mongodin, E.F., Sensabaugh, G.F., and Perdreau-Remington, F. (2006). Complete genome sequence of USA300, an epidemic clone of community-acquired meticillin-resistant *Staphylococcus aureus*. Lancet *367*, 731–739.

Donovan, D.M., Dong, S., Garrett, W., Rousseau, G.M., Moineau, S., and Pritchard, D.G. (2006b). Peptidoglycan hydrolase fusions maintain their parental specificities. Appl. Environ. Microbiol. *72*, 2988–2996.

Donovan, D.M., Lardeo, M., and Foster-Frey, J. (2006a). Lysis of staphylococcal mastitis pathogens by bacteriophage phi11 endolysin. FEMS Microbiol. Lett. *265*, 133–139.

Dyer, D.W., Rock, M.I., Lee, C.Y., and Iandolo, J.J. (1985). Generation of transducing particles in *Staphylococcus aureus*. J. Bacteriol. *161*, 91–5.

Edgeworth, J.D., Yadegarfar, G., Pathak, S., Batra, R., Cockfield, J.D., Wyncoll, D., Beale, R., and Lindsay, J.A. (2007). An outbreak in an intensive care unit of a strain of methicillin-resistant *Staphylococcus aureus* sequence type 239 associated with an increased rate of vascular access device-related bacteremia. Clin. Infect. Dis. *44*, 493–501.

Elek, SD. (1959). Staphylococcus pyogenes. (E & S. Livingstone Ltd, Edinburgh and London).

Embleton, M.L., Nair, S.P., Heywood, W., Menon, D.C., Cookson, B.D., and Wilson, M. (2005). Development of a novel targeting system for lethal photosensitization of antibiotic-resistant strains of *Staphylococcus aureus*. Antimicrob. Agents Chemother. *49*, 3690–3696.

Ender, M., McCallum, N., Adhikari, R., and Berger-Bachi, B. (2004). Fitness cost of SCCmec and methicillin resistance levels in *Staphylococcus aureus*. Antimicrob. Agents Chemother. *48*, 2295–2297.

Enright, M.C., Day, N.P., Davies, C.E., Peacock, S.J., and Spratt, B.G. (2000). Multilocus sequence typing for characterization of methicillin-resistant and methicillin-susceptible clones of *Staphylococcus aureus*. J. Clin. Microbiol. *38*, 1008–1015.

Feil, E.J., Cooper, J.E., Grundmann, H., Robinson, D.A., Enright, M.C., Berendt, T., Peacock, S.J., Smith, J.M., Murphy, M., Spratt, B.G., Moore, C.E., and Day, N.P. (2003). How clonal is *Staphylococcus aureus?* J. Bacteriol. *185*, 3307–3316.

Fitzgerald, J.R., Monday, S.R., Foster, T.J., Bohach, G.A., Hartigan, P.J., Meaney, W.J., and Smyth, C.J. (2001). Characterization of a putative pathogenicity island from bovine *Staphylococcus aureus* encoding multiple superantigens. J. Bacteriol. *183*, 63–70.

Flannagan, S.E., Chow, J.W., Donabedian, S.M., Brown, W.J., Perri, M.B., Zervos, M.J., Ozawa, Y., and Clewell, D.B. (2003). Plasmid content of a vancomycin-resistant *Enterococcus faecalis* isolate from a patient also colonized by *Staphylococcus aureus* with a VanA phenotype. Antimicrob. Agents Chemother. *47*, 3954–3959.

Fujimoto-Nakamura, M., Ito, H., Oyamada, Y., Nishino, T., and Yamagishi, J. (2005). Accumulation of mutations in both gyrB and parE genes is associated with high-level resistance to novobiocin in *Staphylococcus aureus*. Antimicrob. Agents Chemother. *49*, 3810–3815.

Gill, S.R., Fouts, D.E., Archer, G.L., Mongodin, E.F., Deboy, R.T., Ravel, J., Paulsen, I.T., Kolonay, J.F., Brinkac, L., Beanan, M., Dodson, R.J., Daugherty, S.C., Madupu, R., Angiuoli, S.V., Durkin, A.S., Haft, D.H., Vamathevan, J., Khouri, H., Utterback, T., Lee, C., Dimitrov, G., Jiang, L., Qin, H., Weidman, J., Tran, K., Kang, K., Hance, I.R., Nelson, K.E., and Fraser, C.M. (2005). Insights on evolution of virulence and resistance from the complete genome analysis of an early methicillin-resistant *Staphylococcus aureus* strain and a biofilm-producing methicillin-resistant Staphylococcus epidermidis strain. J. Bacteriol. *187*, 2426–2438.

Goerke, C., Koller, J., and Wolz, C. (2006b). Ciprofloxacin and trimethoprim cause phage induction and virulence modulation in *Staphylococcus aureus*. Antimicrob. Agents Chemother. *50*, 171–177.

Goerke, C., Matias y Papenberg, S., Dasbach, S., Dietz, K., Ziebach, R., Kahl, B.C., and Wolz, C. (2004). Increased frequency of genomic alterations in *Staphylococcus aureus* during chronic infection is in part due to phage mobilization. J Infect Dis. *189*, 724–34.

Goerke, C., Wirtz, C., Fluckiger, U., and Wolz, C. (2006a). Extensive phage dynamics in *Staphylococcus aureus* contributes to adaptation to the human host during infection. Mol. Microbiol. *61*, 1673–1685.

Gomes, A.R., Westh, H., and de Lencastre, H. (2006). Origins and evolution of methicillin-resistant *Staphylococcus aureus* clonal lineages. Antimicrob. Agents Chemother. *50*, 3237–3244.

Gravet, A., Colin, D.A., Keller, D., Girardot, R., Monteil, H., and Prevost, G. (1998). Characterization of a novel structural member, LukE-LukD, of the bi-component staphylococcal leucotoxins family. FEBS Lett. *436*, 202–208.

Grundmeier, M., Hussain, M., Becker, P., Heilmann, C., Peters, G., and Sinha, B. (2004). Truncation of fibronectin-binding proteins in *Staphylococcus aureus* strain Newman leads to deficient adherence and host cell invasion due to loss of the cell wall anchor function. Infect. Immun. *72*, 7155–7163.

Hardy, K.J., Oppenheim, B.A., Gossain, S., Gao, F., and Hawkey, P.M. (2006). Use of variations in staphylococcal interspersed repeat units for molecular typing of methicillin-resistant *Staphylococcus aureus* strains. J. Clin. Microbiol. *44*, 271–273.

Harmsen, D., Claus, H., Witte, W., Rothganger, J., Claus, H., Turnwald, D., and Vogel, U. (2003). Typing of methicillin-resistant *Staphylococcus aureus* in a university hospital setting by using novel software for spa repeat determination and database management. J. Clin. Microbiol. *41*, 5442–5448.

Heusser, R., Ender, M., Berger-Bachi, B., and McCallum, N. (2007). Mosaic staphylococcal cassette chromosome mec containing two recombinase loci and a new mec complex, B2. Antimicrob. Agents Chemother. *51*, 390–393.

Holden, M.T., Feil, E.J., Lindsay, J.A., Peacock, S.J., Day, N.P., Enright, M.C., Foster, T.J., Moore, C.E., Hurst, L., Atkin, R., Barron, A., Bason, N., Bentley, S.D., Chillingworth, C., Chillingworth, T., Churcher, C., Clark, L., Corton, C., Cronin, A., Doggett, J., Dowd, L., Feltwell, T., Hance, Z., Harris, B., Hauser, H., Holroyd, S., Jagels, K., James, K.D., Lennard, N., Line, A., Mayes, R., Moule, S., Mungall, K., Ormond, D., Quail, M.A., Rabbinowitsch, E., Rutherford, K., Sanders, M., Sharp, S., Simmonds, M., Stevens, K., Whitehead, S., Barrell, B.G., Spratt, B.G., and Parkhill, J. (2004). Complete genomes of two clinical *Staphylococcus aureus* strains: evidence for the rapid evolution of virulence and drug resistance. Proc. Natl Acad. Sci. USA *101*, 9786–9791.

Iandolo, J.J., Worrell, V., Groicher, K.H., Qian, Y., Tian, R., Kenton, S., Dorman, A., Ji, H., Lin, S., Loh, P., Qi, S., Zhu, H., and Roe, B.A. (2002). Comparative analysis of the genomes of the temperate bacteriophages phi 11, phi 12 and phi 13 of *Staphylococcus aureus* 8325. Gene. *289*, 109–118.

Ince, D., and Hooper, D.C. (2003). Quinolone resistance due to reduced target enzyme expression. J. Bacteriol. *185*, 6883–6892.

Ito, T., Katayama, Y., and Hiramatsu, K. (1999), Cloning and nucleotide sequence determination of the entire mec DNA of pre-methicillin-resistant *Staphylococcus aureus* N315. Antimicrob. Agents Chemother. *43*, 1449–1458.

Ito, T., Katayama, Y., Asada, K., Mori, N., Tsutsumimoto, K., Tiensasitorn, C., and Hiramatsu, K. (2001). Structural comparison of three types of staphylococcal cassette chromosome mec integrated in the chromosome in methicillin-resistant *Staphylococcus aureus*. Antimicrob. Agents Chemother. *45*, 1323–1336.

Ito, T., Ma, X.X., Takeuchi, F., Okuma, K., Yuzawa, H., and Hiramatsu, K. (2004). Novel type V staphylococcal cassette chromosome mec driven by a novel cassette chromosome recombinase, ccrC. Antimicrob. Agents Chemother. *48*, 2637–2651.

Jansen, W.T., Beitsma, M.M., Koeman, C.J., van Wamel, W.J., Verhoef, J., and Fluit, A.C. (2006). Novel mobile variants of staphylococcal cassette chromosome mec in *Staphylococcus aureus*. Antimicrob. Agents Chemother. *50*, 2072–2078.

Jarraud, S., Peyrat, M.A., Lim, A., Tristan, A., Bes, M., Mougel, C., Etienne, J., Vandenesch, F., Bonneville, M., and Lina, G. (2001). egc, a highly prevalent operon of enterotoxin gene, forms a putative nursery of superantigens in *Staphylococcus aureus*. J. Immunol. *166*, 669–677.

Jin, T., Bokarewa, M., Foster, T., Mitchell, J., Higgins, J., and Tarkowski, A. (2004). *Staphylococcus aureus* resists human defensins by production of staphylokinase, a novel bacterial evasion mechanism. J. Immunol. *172*, 1169–1176.

Kahl, B.C., Duebbers, A., Lubritz, G., Haeberle, J., Koch, H.G., Ritzerfeld, B., Reilly, M., Harms, E., Proctor, R.A., Herrmann, M., and Peters, G. (2003). Population dynamics of persistent *Staphylococcus aureus* isolated from the airways of cystic fibrosis patients during a 6-year prospective study. J. Clin. Microbiol. *41*, 4424–4427.

Kaneko, J., Kimura, T., Narita, S., Tomita, T., and Kamio, Y. (1998). Complete nucleotide sequence and molecular characterization of the temperate staphylococcal bacteriophage phiPVL carrying Panton–Valentine leukocidin genes. Gene. *215*, 57–67.

Katayama, Y., Ito, T., and Hiramatsu, K. (2000). A new class of genetic element, staphylococcus cassette chromosome mec, encodes methicillin resistance in *Staphylococcus aureus*. Antimicrob. Agents Chemother. *44*, 1549–1555.

Katayama, Y., Takeuchi, F., Ito, T., Ma, X.X., Ui-Mizutani, Y., Kobayashi, I., and Hiramatsu, K. (2003). Identification in methicillin-susceptible Staphylococcus hominis of an active primordial mobile genetic element for the staphylococcal cassette chromosome mec of methicillin-resistant *Staphylococcus aureus*. J. Bacteriol. *185*, 2711–2722.

Khan, S.A. (2005). Plasmid rolling-circle replication: highlights of two decades of research. Plasmid. 53, 126–136.

Khan, S.A., and Novick, R.P. (1980). Terminal nucleotide sequences of Tn551, a transposon specifying erythromycin resistance in *Staphylococcus aureus*: homology with Tn3. Plasmid. *4*, 148–154.

Kluytmans-Vandenbergh, M.F., and Kluytmans, J.A. (2006). Community-acquired methicillin-resistant *Staphylococcus aureus*: current perspectives. Clin. Microbiol. Infect. *12 Suppl. 1*, 9–15.

Kuroda, M., Ohta, T., Uchiyama, I., Baba, T., Yuzawa, H., Kobayashi, I., Cui, L., Oguchi, A., Aoki, K., Nagai, Y., Lian, J., Ito, T., Kanamori, M., Matsumaru, H., Maruyama, A., Murakami, H., Hosoyama, A., Mizutani-Ui, Y., Takahashi, N.K., Sawano, T., Inoue, R., Kaito, C., Sekimizu, K., Hirakawa, H., Kuhara, S., Goto, S., Yabuzaki, J., Kanehisa, M., Yamashita, A., Oshima, K., Furuya, K., Yoshino, C., Shiba, T., Hattori, M., Ogasawara, N., Hayashi, H., Hiramatsu, K. (2001). Whole genome sequencing of meticillin-resistant *Staphylococcus aureus*. Lancet *357*, 1225–1240.

Kwan, T., Liu, J., DuBow, M., Gros, P., and Pelletier, J. (2005). The complete genomes and proteomes of 27 *Staphylococcus aureus* bacteriophages. Proc. Natl Acad. Sci. USA *102*, 5174–5179.

Kwong, S.M., Skurray, R.A., and Firth, N. (2004). *Staphylococcus aureus* multiresistance plasmid pSK41: analysis of the replication region, initiator protein binding and antisense RNA regulation. Mol. Microbiol. *51*, 497–509.

Kwong, S.M., Skurray, R.A., and Firth, N. (2006). Replication control of staphylococcal multiresistance plasmid pSK41: an antisense RNA mediates dual-level regulation of Rep expression. J. Bacteriol. *188*, 4404–4412.

Labandeira-Rey, M., Couzon, F., Boisset, S., Brown, E.L., Bes, M., Benito, Y., Barbu, E.M., Vazquez, V., Hook, M., Etienne, J., Vandenesch, F., and Bowden, M.G. (2007). *Staphylococcus aureus* Panton–Valentine leukocidin causes necrotizing pneumonia. Science. *315*, 1130–1133.

Langley, R., Wines, B., Willoughby, N., Basu, I., Proft, T., and Fraser, J.D. (2005). The staphylococcal superantigen-like protein 7 binds IgA and complement C5 and inhibits IgA-Fc alpha RI binding and serum killing of bacteria. J. Immunol. *174*, 2926–2933.

Lawrence, J.G., and Ochman, H. (1997). Amelioration of bacterial genomes: rates of change and exchange. J. Mol. Evol. *44*, 383–97.

Lee, C.Y., and Iandolo, J.J. (1986). Integration of staphylococcal phage L54a occurs by site-specific recombination: structural analysis of the attachment sites. Proc. Natl Acad. Sci. USA *83*, 5474–5478.

Lee, S.M., Ender, M., Adhikari, R., Smith, J.M.B., Berger-Bachi, B., and Cook, G.M. (2007). Fitness cost of staphylococcal cassette chromosome mec in methicillin-resistant *Staphylococcus aureus* by way of continuous culture. Antimicrob. Agents Chemother. *51*, 1497–1499.

Lewis, R.A., and Dyke, K.G. (2000). MecI represses synthesis from the beta-lactamase operon of *Staphylococcus aureus*. J. Antimicrob. Chemother. *45*, 139–144.

Lina, G., Durand, G., Berchich, C., Short, B., Meugnier, H., Vandenesch, F., Etienne, J., and Enright, M.C. (2006). Staphylococcal chromosome cassette evolution in *Staphylococcus aureus* inferred from ccr gene complex sequence typing analysis. Clin. Microbiol. Infect. *12,* 1175–1184.

Lina, G., Piemont, Y., Godail-Gamot, F., Bes, M., Peter, M.O., Gauduchon, V., Vandenesch, F., and Etienne, J. (1999). Involvement of Panton–Valentine leukocidin-producing *Staphylococcus aureus* in primary skin infections and pneumonia. Clin. Infect. Dis. *29,* 1128–1132.

Lindsay, J.A,, Moore, C.E., Day, N.P., Peacock, S.J., Witney, A.A., Stabler, R.A., Husain, S.E., Butcher, P.D., and Hinds, J. (2006). Microarrays reveal that each of the ten dominant lineages of *Staphylococcus aureus* has a unique combination of surface-associated and regulatory genes. J. Bacteriol. *188,* 669–676.

Lindsay, J.A., and Holden, M.T. (2004). *Staphylococcus aureus*: superbug, super genome? Trends Microbiol. *12,* 378–385.

Lindsay, J.A., and Holden, M.T.G. (2007) The staphylococci: a post-genomic view. In Bacterial Pathogenomics, Pallen MJ, Nelson KE, Preston GM, eds. (ASM Press, Washington, D.C.)

Lindsay, J.A., Ruzin, A., Ross, H., Kurepina, N., and Novick, R.P. (1998). The gene for toxic shock toxin is carried by a family of mobile pathogenicity islands in *Staphylococcus aureus*. Mol. Microbiol. *29,* 527–543.

Loessner, M.J., Gaeng, S., and Scherer, S. (1999). Evidence for a holin-like protein gene fully embedded out of frame in the endolysin gene of *Staphylococcus aureus* bacteriophage 187. J. Bacteriol. *181,* 4452–4460.

Lowbury, E.J., and Hood, A.M. (1953). The acquired resistance of *Staphylococcus aureus* to Bacteriophage. J. Gen. Microbiol. *9,* 524–535.

Luong, T.T., Ouyang, S., Bush, K., and Lee, C.Y. (2002). Type 1 capsule genes of *Staphylococcus aureus* are carried in a staphylococcal cassette chromosome genetic element. J. Bacteriol. *184,* 3623–3629.

Ma, X.X., Ito, T., Tiensasitorn, C., Jamklang, M., Chongtrakool, P., Boyle-Vavra, S., Daum, R.S., and Hiramatsu, K. (2002). Novel type of staphylococcal cassette chromosome mec identified in community-acquired methicillin-resistant *Staphylococcus aureus* strains. Antimicrob. Agents Chemother. *46,* 1147–1152.

Mahairas, G.G., Lyon, B.R., Skurray, R.A., and Pattee, P.A. (1989). Genetic analysis of *Staphylococcus aureus* with Tn4001. J. Bacteriol. *171,* 3968–3972.

Maiques, E., Ubeda, C., Campoy, S., Salvador, N., Lasa, I., Novick, R.P., Barbe, J., and Penades, J.R. beta-lactam antibiotics induce the SOS response and horizontal transfer of virulence factors in *Staphylococcus aureus*. J. Bacteriol. *188,* 2726–2729.

Maiques, E., Ubeda, C., Tormo, M.A., Ferrer, M.D., Lasa, I., Novick, R.P., and Penades, J.R. (2007). Role of staphylococcal phage and SaPI integrase in intra- and interspecies SaPI transfer. J. Bacteriol. *189,* 5608–5616.

Makris, G., Wright, J.D., Ingham, E., and Holland, K.T. (2004). The hyaluronate lyase of *Staphylococcus aureus* – a virulence factor? Microbiol. *150,* 2005–2013.

Matsuzaki, S., Rashel, M., Uchiyama, J., Sakurai, S., Ujihara, T., Kuroda, M., Ikeuchi, M., Tani, T., Fujieda, M., Wakiguchi, H., and Imai, S. (2005). Bacteriophage therapy: a revitalized therapy against bacterial infectious diseases. J. Infect. Chemother. 11, 211–219.

Matsuzaki, S., Yasuda, M., Nishikawa, H., Kuroda, M., Ujihara, T., Shuin, T., Shen, Y., Jin, Z., Fujimoto, S., Nasimuzzaman, M.D., Wakiguchi, H., Sugihara, S., Sugiura, T., Koda, S., Muraoka, A., and Imai, S. (2003). Experimental protection of mice against lethal *Staphylococcus aureus* infection by novel bacteriophage phi MR11. J. Infect. Dis. *187,* 613–24.

Merril, C.R., Biswas, B., Carlton, R., Jensen, N.C., Creed, G.J., Zullo, S., and Adhya, S. (1996). Long-circulating bacteriophage as antibacterial agents. Proc. Natl Acad. Sci. USA *93,* 3188–3192.

Milheirico, C., Oliveira, D.C., and de Lencastre, H. (2007). Multiplex PCR strategy for subtyping the staphylococcal cassette chromosome mec type IV in methicillin-resistant *Staphylococcus aureus*: 'SCCmec IV multiplex'. J. Antimicrob. Chemother. *60,* 42–48.

Moore, P.C., and Lindsay, J.A. (2001). Genetic variation among hospital isolates of methicillin-sensitive *Staphylococcus aureus*: evidence for horizontal transfer of virulence genes. J. Clin. Microbiol. 39, 2760–2767.

Moore, P.C., and Lindsay, J.A. (2002). Molecular characterisation of the dominant UK methicillin-resistant *Staphylococcus aureus* strains, EMRSA-15 and EMRSA-16. J. Med. Microbiol. *51,* 516–521.

Morfeldt, E., Taylor, D., von Gabain, A., and Arvidson, S. Activation of alpha-toxin translation in *Staphylococcus aureus* by the trans-encoded antisense RNA, RNAIII. EMBO J. 14, 4569–4577.

Murphy, E., Huwyler, L., and de Freire Bastos Mdo, C. (1985). Transposon Tn554: complete nucleotide sequence and isolation of transposition-defective and antibiotic-sensitive mutants. EMBO J. *4,* 3357–3365.

Murray, N.E. (2000). Type I restriction systems: sophisticated molecular machines (a legacy of Bertani and Weigle). Microbiol. Mol. Biol. Rev. *64,* 412–434.

Mwangi, M.M., Wu, S.W., Zhou, Y., Sieradzki, K., de Lencastre, H., Richardson, P., Bruce, D., Rubin, E., Myers, E., Siggia, E.D., and Tomasz, A. (2007). Tracking the *in vivo* evolution of multidrug resistance in *Staphylococcus aureus* by whole-genome sequencing. Proc. Natl Acad. Sci. USA *104,* 9451–9456.

Navaratna, M.A., Sahl, H.G., and Tagg, J.R. (1999). Identification of genes encoding two-component lantibiotic production in *Staphylococcus aureus* C55 and other phage group II *S. aureus* strains and demonstration of an association with the exfoliative toxin B gene. Infect. Immun. *67,* 4268–4271.

Navarre, W.W., Ton-That, H., Faull, K.F., and Schneewind, O. (1999). Multiple enzymatic activities of the murein hydrolase from staphylococcal phage phi11. Identification of a D-alanyl-glycine endopeptidase activity. J. Biol. Chem. *274,* 15847–15856.

Noble, W.C., Virani, Z., and Cree, R.G. (1992). Co-transfer of vancomycin and other resistance genes from *Enterococcus faecalis* NCTC 12201 to

Staphylococcus aureus. FEMS Microbiol. Lett. *72*, 195–198.

Nouwen, J., Boelens, H., van Belkum, A., and Verbrugh, H. (2004). Human factor in *Staphylococcus aureus* nasal carriage. Infect. Immun. *72*, 6685–6688.

Novick, RP. (1987). Plasmid incompatibility. Microbiol. Rev. *51*, 381–395

O'Neill, A.J., Larsen, A.R., Skov, R., Henriksen, A.S., and Chopra, I. (2007). Characterization of the epidemic European fusidic acid-resistant impetigo clone of *Staphylococcus aureus*. J. Clin. Microbiol. *45*, 1505–1510.

Patti, J.M., Allen, B.L., McGavin, M.J., and Hook, M. (1994). MSCRAMM-mediated adherence of microorganisms to host tissues. Annu. Rev. Microbiol. *48*, 585–617.

Pichon, C., and Felden, B. (2005). Small RNA genes expressed from *Staphylococcus aureus* genomic and pathogenicity islands with specific expression among pathogenic strains. Proc. Natl Acad. Sci. USA *102*, 14249–14254.

Pichon, C., and Felden, B. (2007). Proteins that interact with bacterial small RNA regulators. FEMS Microbiol. Rev. *31*, 614–625.

Popowicz, G.M., Dubin, G., Stec-Niemczyk, J., Czarny, A., Dubin, A., Potempa, J., and Holak, T.A. (2006). Functional and structural characterization of Spl proteases from *Staphylococcus aureus*. J. Mol. Biol. *358*, 270–279.

Projan, S.J. (2003). Why is big Pharma getting out of antibacterial drug discovery? Curr. Opin. Microbiol. *6*, 427–430.

Qi, W., Ender, M., O'Brien, F., Imhof, A., Ruef, C., McCallum, N., and Berger-Bachi, B. (2005). Molecular epidemiology of methicillin-resistant *Staphylococcus aureus* in Zurich, Switzerland (2003): prevalence of type IV SCCmec and a new SCCmec element associated with isolates from intravenous drug users. J. Clin. Microbiol. *43*, 5164–5170.

Resch, A., Fehrenbacher, B., Eisele, K., Schaller, M., and Gotz, F. (2005). Phage release from biofilm and planktonic *Staphylococcus aureus* cells. FEMS Microbiol. Lett. *252*, 89–96.

Robinson, D.A., and Enright, M.C. (2004). Evolution of *Staphylococcus aureus* by large chromosomal replacements. J. Bacteriol. 186, 1060–1064.

Roche, F.M., Meehan, M., and Foster, T.J. (2003). The *Staphylococcus aureus* surface protein SasG and its homologues promote bacterial adherence to human desquamated nasal epithelial cells. Microbiol. *149*, 2759–2767.

Rooijakkers, S.H., Ruyken, M., Roos, A., Daha, M.R., Presanis, J.S., Sim, R.B., van Wamel, W.J., van Kessel, K.P., and van Strijp, J.A. (2005a). Immune evasion by a staphylococcal complement inhibitor that acts on C3 convertases. Nat. Immunol. *6*, 920–927.

Rooijakkers, S.H., van Wamel, W.J., Ruyken, M., van Kessel, K.P. and van Strijp, J.A. (2005b). Antiopsonic properties of staphylokinase. Microbes Infect. *7*, 476–484.

Rountree, P.M. (1955). The role of divalent cations in the multiplication of staphylococcal bacteriophages. J. Gen. Microbiol *12*, 275–287.

Rowland, S.J., and Dyke, K.G. (1989). Characterization of the staphylococcal beta-lactamase transposon Tn552. EMBO J. *8*, 2761–2773.

Ruzin, A., Lindsay, J., and Novick, R.P. (2001). Molecular genetics of SaPI1 – a mobile pathogenicity island in *Staphylococcus aureus*. Mol. Microbiol. *41*, 365–377.

Sabat, A., Krzyszton-Russjan, J., Strzalka, W., Filipek, R., Kosowska, K., Hryniewicz, W., Travis, J., and Potempa, J. (2003). New method for typing *Staphylococcus aureus* strains: multiple-locus variable-number tandem repeat analysis of polymorphism and genetic relationships of clinical isolates. J Clin Microbiol. *41*, 1801–1804.

Sass, P., and Bierbaum, G. (2007). Lytic activity of recombinant bacteriophage phi11 and phi12 endolysins on whole cells and biofilms of *Staphylococcus aureus*. Appl. Environ. Microbiol. *73*, 347–352.

Shore, A., Rossney, A.S., Keane, C.T., Enright, M.C., and Coleman, D.C. (2005). Seven novel variants of the staphylococcal chromosomal cassette mec in methicillin-resistant *Staphylococcus aureus* isolates from Ireland. Antimicrob. Agents Chemother. *49*, 2070–2083.

Smith, E.M., Green, L.E., Medley, G.F., Bird, H.E., Fox, L.K., Schukken, Y.H., Kruze, J.V., Bradley, A.J., Zadoks, R.N., and Dowson, C.G. (2005). Multilocus sequence typing of intercontinental bovine *Staphylococcus aureus* isolates. J. Clin. Microbiol. *43*, 4737–4743.

Stahl, M.L., and Pattee, P.A. (1983). Confirmation of protoplast fusion-derived linkages in *Staphylococcus aureus* by transformation with protoplast DNA. J. Bacteriol. *154*, 406–412.

Sumby, P., and Waldor, M.K. (2003). Transcription of the toxin genes present within the Staphylococcal phage phiSa3ms is intimately linked with the phage's life cycle. J. Bacteriol. *185*, 6841–6851.

Sung, J.M., and Lindsay, J.A. (2007). *Staphylococcus aureus* strains that are hypersusceptible to resistance gene transfer from enterococci. Antimicrob. Agents Chemother. *51*, 2189–2191.

Takac, M., Witte, A., and Blasi, U. (2005). Functional analysis of the lysis genes of *Staphylococcus aureus* phage P68 in Escherichia coli. Microbiology. *151*, 2331–2342.

Tallent, S.M., Langston, T.B., Moran, R.G., and Christie, G.E. (2007). Transducing Particles of *Staphylococcus aureus* Pathogenicity Island SaPI1 are Comprised of Helper Phage-Encoded Proteins. J. Bacteriol. *189*, (e-version).

Ubeda, C., Barry, P., Penades, J.R., and Novick, R.P. (2007b) A pathogenicity island replicon in *Staphylococcus aureus* replicates as an unstable plasmid. Proc. Natl Acad. Sci. USA *104*, 14182–14188.

Ubeda, C., Maiques, E., Knecht, E., Lasa, I., Novick, R.P., and Penades, J.R. (2005). Antibiotic-induced SOS response promotes horizontal dissemination of pathogenicity island-encoded virulence factors in staphylococci. Mol. Microbiol. *56*, 836–844.

Ubeda, C., Maiques, E., Tormo, M.A., Campoy, S., Lasa, I., Barbe, J., Novick, R.P., and Penades, J.R. (2007a). SaPI operon I is required for SaPI packaging and is controlled by LexA. Mol. Microbiol. *65*, 41–50.

Ubeda, C., Tormo, M.A., Cucarella, C., Trotonda, P., Foster, T.J., Lasa, I., and Penades, J.R. (2003). Sip, an integrase protein with excision, circularization and integration activities, defines a new family of mobile *Staphylococcus aureus* pathogenicity islands. Mol. Microbiol. *49*, 193–210.

Valle, J., Vergara-Irigaray, M., Merino, N., Penades, J.R., and Lasa, I. (2007). sigmaB regulates IS256-mediated *Staphylococcus aureus* biofilm phenotypic variation. J. Bacteriol. *189*, 2886–2896.

van Wamel, W.J., Rooijakkers, S.H., Ruyken, M., van Kessel, K.P., and van Strijp, J.A. (2006). The innate immune modulators staphylococcal complement inhibitor and chemotaxis inhibitory protein of *Staphylococcus aureus* are located on beta-hemolysin-converting bacteriophages. J. Bacteriol. *188*, 1310–1315.

Vojtov, N., Ross, H.F., and Novick, R.P. (2002). Global repression of exotoxin synthesis by staphylococcal superantigens. Proc. Natl. Acad Sci. USA *99*, 10102–10107.

Voyich, J.M., Otto, M., Mathema, B., Braughton, K.R., Whitney, A.R., Welty, D., Long, R.D., Dorward, D.W., Gardner, D.J., Lina, G., Kreiswirth, B.N., and DeLeo, F.R. (2006). Is Panton–Valentine leukocidin the major virulence determinant in community-associated methicillin-resistant *Staphylococcus aureus* disease? J. Infect. Dis. *194*, 1761–1770.

Vybiral, D., Takac, M., Loessner, M., Witte, A., von Ahsen, U., and Blasi, U. (2003). Complete nucleotide sequence and molecular characterization of two lytic *Staphylococcus aureus* phages: 44AHJD and P68. FEMS Microbiol. Lett. *219*, 275–283.

Waldron, D.E., and Lindsay, J.A. (2006). Sau1: a novel lineage-specific type I restriction-modification system that blocks horizontal gene transfer into *Staphylococcus aureus* and between *S. aureus* isolates of different lineages. J. Bacteriol. *188*, 5578–5585.

Watanabe, S., Ito, T., Morimoto, Y., Takeuchi, F., and Hiramatsu, K. (2007). Precise excision and self-integration of a composite transposon as a model for spontaneous large-scale chromosome inversion/deletion of the Staphylococcus haemolyticus clinical strain JCSC1435. J. Bacteriol. *189*, 2921–2925.

Weigel, L.M., Clewell, D.B., Gill, S.R., Clark, N.C., McDougal, L.K., Flannagan, S.E., Kolonay, J.F., Shetty, J., Killgore, G.E., and Tenover, F.C. (2003). Genetic analysis of a high-level vancomycin-resistant isolate of *Staphylococcus aureus*. Science. 302, 1569–1571.

Werbick, C., Becker, K., Mellmann, A., Juuti, K.M., Eiff, C., Peters, G., Kuusela, P.I., Friedrich, A.W., and Sinha, B. (2007). Staphylococcal chromosomal cassette mec type I, spa type, and expression of Pls are determinants of reduced cellular invasiveness of methicillin-resistant *Staphylococcus aureus* isolates. J. Infect. Dis. *195*, 1678–1685.

Williams, R.E.O., and Rippon, J.E. (1952). Bacteriophage typing of *Staphylococcus aureus*. J. Hyg. (London) *50*, 320–353.

Wines, B.D., Willoughby, N., Fraser, J.D., and Hogarth, P.M. (2006). A competitive mechanism for staphylococcal toxin SSL7 inhibiting the leukocyte IgA receptor, Fc alphaRI, is revealed by SSL7 binding at the C alpha2/C alpha3 interface of IgA. J. Biol. Chem. *281*, 1389–1393.

Wu, S.W., de Lencastre, H., and Tomasz, A. (1999). The *Staphylococcus aureus* transposon Tn551: complete nucleotide sequence and transcriptional analysis of the expression of the erythromycin resistance gene. Microb. Drug Resist. 5, 1–7.

Wyman, L., Goering, R.V., and Novick, R.P. (1974). Genetic control of chromosomal and plasmid recombination in *Staphylococcus aureus*. Genetics. *76*, 681–702.

Yacoby, I., Bar, H., and Benhar, I. (2007). Targeted drug-carrying bacteriophages as antibacterial nanomedicines. Antimicrob. Agents Chemother. 51, 2156–2163.

Yamaguchi, T., Hayashi, T., Takami, H., Nakasone, K., Ohnishi, M., Nakayama, K., Yamada, S., Komatsuzawa, H., and Sugai, M. (2000). Phage conversion of exfoliative toxin A production in *Staphylococcus aureus*. Mol. Microbiol. *38*, 694–705.

Yamaguchi, T., Hayashi, T., Takami, H., Ohnishi, M., Murata, T., Nakayama, K., Asakawa, K., Ohara, M., Komatsuzawa, H., and Sugai, M. (2001). Complete nucleotide sequence of a *Staphylococcus aureus* exfoliative toxin B plasmid and identification of a novel ADP-ribosyltransferase, EDIN-C. Infect. Immun. *69*, 7760–7771.

Yarwood, J.M., McCormick, J.K., Paustian, M.L., Orwin, P.M., Kapur, V., and Schlievert, P.M. (2002). Characterization and expression analysis of *Staphylococcus aureus* pathogenicity island 3. Implications for the evolution of staphylococcal pathogenicity islands. J. Biol. Chem. *277*, 13138–13147.

Ye, Z.H., and Lee, C.Y. (1989). Nucleotide sequence and genetic characterization of staphylococcal bacteriophage L54a int and xis genes. J. Bacteriol. *171*, 4146–4153.

Ye, Z.H., and Lee, C.Y. (1993). Cloning, sequencing, and genetic characterization of regulatory genes, rinA and rinB, required for the activation of staphylococcal phage phi11 int expression. J. Bacteriol. *175*, 1095–1102.

Zhang, Y.Q., Ren, S.X., Li, H.L., Wang, Y.X., Fu, G., Yang, J., Qin, Z.Q., Miao, Y.G., Wang, W.Y., Chen, R.S., Shen, Y., Chen, Z., Yuan, Z.H., Zhao, G.P., Qu, D., Danchin, A., and Wen, Y.M. (2003). Genome-based analysis of virulence genes in a non-biofilm-forming Staphylococcus epidermidis strain (ATCC 12228). Mol. Microbiol. *49*, 1577–1593.

Ziebuhr, W., Krimmer, V., Rachid, S., Lossner, I., Gotz, F., and Hacker, J. (1999). A novel mechanism of phase variation of virulence in Staphylococcus epidermidis: evidence for control of the polysaccharide intercellular adhesin synthesis by alternating insertion and excision of the insertion sequence element IS256. Mol. Microbiol. *32*, 345–356.

Rapid Diagnosis and Typing of *Staphylococcus aureus*

4

Patrice Francois and Jacques Schrenzel

Abstract

Staphylococcus aureus is a major pathogen responsible for both nosocomial and community acquired infections. The severity of these infections varied from local benign wounds to severe systemic diseases. The situation is also complicated with emergence of bacterial resistance to common antibiotics, such as methicillin. Endemic strains of MRSA carrying multiple resistance determinants have become a worldwide nosocomial problem only in the early 1980s, carrying a threefold attributable cost and a threefold excess length of hospital stay when compared with methicillin-susceptible *S. aureus* bacteraemia. Recent genetic advances have enabled identification and characterization of clinical isolates in real-time. These tools support infection control strategies to limit bacterial spreading and ensure the appropriate use of diminishing antibiotics. They are also attractive for understanding the epidemiology of MRSA and the relationship between genome content and virulence.

Introduction

Staphylococcus aureus (*S. aureus*) is a frequent colonizer found in 20–30% of the general population without causing any clinical manifestation. Simultaneously, it represents a potential risk for colonized subjects and is the most common cause of nosocomial infections. Despite continuous medical progress, *S. aureus* remains a versatile pathogen (Sheagren, 1984a; Sheagren, 1984b) responsible for numerous diseases ranging from benign localized skin infections to severe systemic diseases such as endocarditis, septicaemia or osteomyelitis (Lowy, 1998; Waldvogel, 2000). *S. aureus* is also responsible for numerous chronic diseases such as osteomyelitis (Lew and Waldvogel, 2004), rhinosinusitis (Gittelman *et al.*, 1991), or otitis (Brook and Finegold, 1979). These chronic infections are difficult to eradicate and often relapse even after adapted and prolonged antibiotic therapy (Kauffman *et al.*, 1993; Powers *et al.*, 1990). This observation suggests that *S. aureus* is able to develop specific strategies to rapidly adapt to environmental conditions. Remarkably, such physiologic and genetic adaptations also rely on sequential acquisition of numerous antimicrobial resistance determinants, making *S. aureus* a formidable bacterial pathogen. The first *S. aureus* isolates displaying resistance to methicillin (MRSA) were reported in the early 1960s (Barber, 1961), only a few months after the introduction of the drug in human medicine. Endemic strains of MRSA carrying multiple resistance determinants have become a worldwide nosocomial problem only in the early 1980s (Hryniewicz, 1999). Healthcare centres represent a major concern as limiting MRSA spread appears particularly difficult to enforce, requiring application of elaborate and expensive infection control guidelines (Herwaldt, 1999; Pittet *et al.*, 1996; Pittet *et al.*, 2000). Difficulties in eradicating nosocomial MRSA infections may be explained by several factors including, the presence of an unknown hidden reservoir of MRSA carriers, the emergence of novel highly epidemic *S. aureus* clonotypes (such as EMRSA-15, EMRSA-16 (Aucken *et al.*, 2002; Cox *et al.*, 1995; Richardson and Reith, 1993) with unspecified selective advantages, and/or the failure to enforce infection control procedures. Some studies suggest the screening of high-risk patients for MRSA colonization is a cost-effective measure for limiting the spread of the organism in hospitals (Papia *et al.*, 1999). Thus, early and reliable detection of MRSA carriers appears crucial for infection control strategies. Furthermore, rapid diagnosis of MRSA infections ensures last-line antibiotics against MRSA (e.g. glycopeptides and oxazolidinones) are prescribed only when necessary (Sakoulas *et al.*, 2002).

MRSA has been considered as the prototype nosocomial pathogen for decades (Archer, 1998; Kreiswirth *et al.*, 1993), but several recent outbreaks in the community have been reported in young and healthy people (Campbell *et al.*, 2004; Jones *et al.*, 2002; Kazakova *et al.*, 2005) involving unusual MRSA strains harbouring toxins rarely present in strains responsible for nosocomial infections. These community-acquired strains (CA-MRSA) may lead to very severe infections or complications (Gillet *et al.*, 2002; Moran *et al.*, 2006) but remain generally susceptible to numerous non-beta-lactam antibiotics (Hiramatsu *et al.*, 2002). Taken together, these clinical observations suggest that the epidemiology of MRSA is rapidly changing (Carleton *et al.*, 2004; Chambers, 2001; Pan *et al.*, 2005) and that this evolution is not compatible with the time required to develop new antimicrobial drug classes. The bacterial genome and its plasticity support this evolution as it contains genes required to adapt to multiple environmental changes including the presence of antibiotics or disinfectants (Chapters 3 and 8). Exposed to such environmental modifications, adaptation occurs through genetic changes, within genes belonging to the core genome or acquired by an array of genetic events, such as horizontal transfer or various sequence modifications. Assessment of these genetic changes is now measurable using modern genotyping methods in the diagnostic and reference laboratory and provides important epidemiological information not only essential to improve our understanding of bacterial epidemiology and evolution but also to determine the relationship between genome content and virulence or pathogenicity of the bacterium.

MRSA: to screen or not to screen?

Extensive screening of MRSA carriers at hospital admission, despite its important cost, appears to have a major impact in reducing MRSA nosocomial infection rates, as recently shown by Wernitz and co-authors (Wernitz *et al.*, 2005). Indeed, MRSA carriage or colonization is a major risk factor for becoming infected. The preferred colonization sites are the nose, the throat, and the skin surface (Kluytmans *et al.*, 1997). The spread of MRSA occurs generally after contact with carriers (Grundmann *et al.*, 2005), or contact with 'MRSA reservoirs' (parts of which remain probably unknown). The spread of MRSA in health care centres is difficult to control and requires elaborate infection control guidelines (Chaix *et al.*, 1999; Cohen *et al.*, 1991; Coia *et al.*, 2007; Cosseron-Zerbib *et al.*, 1998; Harbarth *et al.*, 2000; Nettleman *et al.*, 1991; Papia *et al.*, 1999; Pittet *et al.*, 1996) including: (i) large-scale screening of suspected carriers (patients, staff, visitors), (ii) automated computerized alerts, (iii) specific recommendations for at-risk patients, such as decolonization and contact isolation (Harbarth and Pittet, 1998; Pittet *et al.*, 1996; Pittet and Waldvogel, 1997), and (iv) significant improvement of hand hygiene compliance (Pittet *et al.*, 2000). These measures, together with successful containment effort programs (Chaix *et al.*, 1999; Cohen *et al.*, 1991; Cosseron-Zerbib *et al.*, 1998; Harbarth *et al.*, 2000; Nettleman *et al.*, 1991; Papia *et al.*, 1999; Pittet *et al.*, 1996), prompt for screening high-risk patients even in a highly endemic setting (Rubinovitch and Pittet, 2001). Several international guidelines now recommend the screening of potential MRSA-positive patients at hospital admission (Anonymous, 1999; Ayliffe *et al.*, 1999; Muto *et al.*, 2003). However, despite intensive efforts in the application of such guidelines, MRSA spread remains difficult to control. In a concerted effort, major experts in the field edited guidelines under the auspices of the Society for Healthcare Epidemiology of America (SHEA) for preventing nosocomial transmission of resistant strains and clearly stated that 'active surveillance cultures are essential to identify the reservoir for spread of MRSA and make control possible using the CDC's long-recommended contact precautions'

(Muto *et al.*, 2003). Very recently, these decisions were reinforced by a joint position statement of the SHEA and the Association for Professionals in Infection Control and Epidemiology (APIC) who published a five points position consensus legislative mandate aiming at 'controlling antimicrobial-resistant pathogens through the use of active surveillance cultures to screen hospitalized patients' (Weber *et al.*, 2007). Despite this, there is still some debate about the cost effectiveness of screening and other guidelines are more circumspect (Coia *et al.*, 2006). In contrast, The Netherlands has a low endemic level of MRSA and implements a 'search and destroy' policy. All outside patients and staff are considered colonized until proven otherwise and subject to stringent and expensive infection control procedures. In this case, rapid MRSA screening is necessary to identify the non-carriers who can then have these procedures relaxed.

Culture-based screening methods

Mueller-Hinton broth supplemented by oxacillin (MHO) is still used to detect or confirm presence of MRSA from swabs sampled in surveillance programs. This medium appears in the guidelines for the prevention and control of antibiotic-resistant organisms from the CSLI recommendations (Clinical and Laboratory Standards Institute, 2005). Generally, after presumptive identification on MHO, presence of MRSA is confirmed using phenotypic assays such as coagulase test or Pastorex agglutination (agglutination test detecting the presence of the clumping factor and/or the protein A) and DNAse reaction on agar allowing the identification of *Staphylococcus aureus* specific factors (Compernolle *et al.*, 2007).

Solid media are now commonly used for MRSA isolation and identification, because they can be interpreted in approximately 24 hours. Agar plates provide numerous advantages, such as the possibility for microbiologists to detect the presence of relevant colony morphologies, isolate them by subplating, and assess their purity on isolation plates. Pure isolates are essential for further phenotypic testing, including speciation (when required), antimicrobial susceptibility testing, and genotyping. To date, numerous selective media containing β-lactam antibiotics and chromogenic substances are commercially available. The general principles are straightforward and consist in providing selective medium supplemented with: (i) Gram-negative growth inhibitor (required for samples containing mixed flora), (ii) antibiotic (allowing the selection of methicillin-resistant organisms) and (iii) chromogenic substrate for specific detection of growing *Staphylococcus aureus* colonies. ORSAB plates (Oxoid), a solid variant of the liquid mannitol-salt medium containing oxacillin and aniline blue, allows detection of mannitol-fermenting organisms as blue colonies, due to medium acidification. This medium appears adapted to high-risk populations (Simor *et al.*, 2001), but presents limitation for surveillance applications (Becker *et al.*, 2002) as some coagulase-negative staphylococci (mainly *S. haemolyticus*, a frequent skin colonizer) appear also blue (Becker *et al.*, 2002). Thus, the utilization of this plate requires additional tests for robust identification. In our institution, the implementation of a chromogenic-assay (ORSAB plates) together with different confirmatory assays (DNAse and Mueller-Hinton oxacillin broth in 2000, Vitek2 system in 2001, and optimization of the workflow in the laboratory in 2002) permitted shorter turn-around times for delivering results to clinicians. Yet, despite increasing workload and optimized procedures, average turn-around times decreased only from 96 to 72 h (Fig. 4.1). To date, ChromAgar/MRSA (ChromaAgar, Paris, France) stands among the most popular agar plates dedicated to the detection of MRSA. This medium where MRSA colonies appear mauve, while other bacteria display different colours, has been extensively tested and shows appreciable sensitivity and specificity (Diederen *et al.*, 2005; Kluytmans *et al.*, 2002). Among other chromogenic agar media showing interesting performance is MRSA ID (bioMérieux, France) which contains cefoxitin and a chromogenic substrate for detecting the α-glucosidase activity. Recent evaluation of this medium has shown improved performance when compared to ChromAgar/MRSA (Perry *et al.*, 2004). Comparable results were reported with MRSA Select (Bio-Rad) (www.rapidmicrobiology.com/PG/MRSA.php). For the direct identification of MRSA from blood

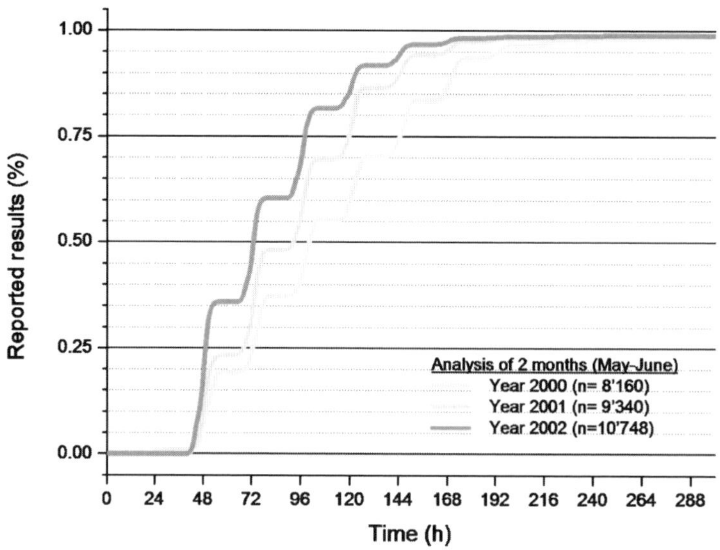

Figure 4.1 Identification of MRSA using culture-based methods. Evolution of time required for obtaining identification of MRSA from biological samples using chromogenic media. The first 24 hours corresponds to the growth of the organisms and cannot be reduced using selective media. Source: Central Laboratory of Bacteriology, Geneva University Hospitals.

cultures, this selective plate showed exquisite sensitivity and specificity (Pape *et al.*, 2006), but this type of sample is an exception since it typically contains a single bacterial species. Recently, our group has shown that reliable MRSA identification from complex flora-containing samples using chromogenic media is also affected by the choice of confirmatory tests, such as growth on Mueller Hinton oxacillin broth (6 mg/l) and a positive DNAse assay. Performances of selective media under real conditions of utilization are therefore heterogeneous, underlining the absence of a gold standard medium for MRSA screening (Cherkaoui *et al.*, 2007).

A rapid alternative to oxacillin susceptibility testing of a pure bacterial culture is direct or indirect particle agglutination assays using antibody-coated beads. For example, MRSA-Screen (Denka Seiken,Tokyo, Japan) provides the rapid, sensitive and specific immunodetection of pure MRSA culture by using anti-PBP2′ antibody-coated latex beads, and shows similar performance as standard oxacillin disc diffusion or oxacillin salt agar screening (Cavassini *et al.*, 1999; Hussain *et al.*, 2000). However, the immunodetection of MRSA based on PBP2′ expression fails to specifically identify MRSA in the presence of other methicillin-resistant staphylococcal species, organisms that are frequently recovered as commensals in mixed flora samples (Cavassini *et al.*, 1999). Indeed, the high level of homology of PBP2′ protein present in *S. aureus*, *S. epidermidis*, and potentially other coagulase-negative staphylococci (CNS) species (Ryffel *et al.*, 1990; Wielders *et al.*, 2001) precludes specific discrimination of methicillin-resistance between staphylococci. Furthermore, the reliability of latex agglutination based on detection of PBP2′ antigenic motif requires induction of its expression to reliably obtain detectable levels of the protein (Rohrer *et al.*, 2001).

All these media represent appreciable improvement of the current situation in the field of MRSA screening and identification, albeit these culture-based methods require at least 20–24 h to yield identification results. During this period of time, infection control measures cannot be optimally applied. Similarly, financial, logistical and psychological costs of cohorting patients before ruling out MRSA carriage are substantial and should not be overlooked. And in case of empirical treatment (treatment prior to diagnosis of the causative organism), options usually include glycopeptide prescription leading to important costs and suboptimal use of last barrier drugs.

Molecular methods for MRSA screening and identification

The *mecA* gene encoding for the low-affinity penicillin-binding protein PBP2′ is the genetic basis of methicillin-resistance in staphylococci. This gene is carried on a large mobile genetic element designated SCC*mec* (staphylococcal cassette chromosome *mec* (Katayama *et al.*, 2003)), which is invariably inserted into the *orfX* gene of the bacterial genome. Another constant element is the presence of site-specific cassette chromosome recombinase (*ccr*) responsible for the precise excision and integration of SCC*mec* within the bacterial chromosome (Katayama *et al.*, 2003). To date, six differently organized SCC*mec* elements have been characterized (Kondo *et al.*, 2007). Three types of SCC*mec* elements are typically found in HA-MRSA strains: (i) type I, a 34 kb element that was prevalent in MRSA isolates in the 1960s,(ii) type II, a 53 kb element that was identified in 1982 and is ubiquitous in Japan, Korea and the USA, and (iii) type III, the largest 67 kb element identified in 1985, currently prevalent in Germany, Austria, India, and other South Asian and Pacific areas (Hiramatsu *et al.*, 2001; Katayama *et al.*, 2003). In contrast to HA-MRSA, CA-MRSA isolates generally carry SCC*mec* type IV, V or VI elements, whose sizes are much smaller than those generally found in HA-MRSA (Ito *et al.*, 2003; Kondo *et al.*, 2007; Ma *et al.*, 2002). However, there are important exceptions, and some HA-MRSA carry SCC*mec* type IV (EMRSA-15), and CA-MRSA are becoming common in hospitals. The molecular structure of the recently described type V cassette (Ito *et al.*, 2004) encodes a new *ccrC* recombinase and does not contain any additional antibiotic resistance determinants; it is the smallest SCC*mec* cassette with a size <30 kb.

Since detection of the *mecA* gene in *S. aureus* represents the 'gold standard' for detecting methicillin resistance (Murakami *et al.*, 1991), several assays based on the direct detection of the *mecA* gene have been described. Some require the overnight culture and enrichment of clinical samples (148) making them less attractive as a rapid test. Others use chemiluminescent probes (Youmans *et al.*, 1993) or cycling probe technology on isolated strains (Hamels *et al.*, 2001). These techniques appear promising in terms of sensitivity but also of turn-around-time as they generally require only a few hours to provide results. However, any test that only detects *mecA* is not MRSA specific, as many CoNS also carry *mecA*.

Rohrer and colleagues reported the utilization of a duplex PCR recognizing the genes *mecA* and *femA* (a gene specific for and recovered in all *S. aureus* strains (Berger-Bachi *et al.*, 1989; Hurlimann-Dalel *et al.*, 1992) on a collection of isolated MRSA (Rohrer *et al.*, 2001). Other similar multiplexed PCR assays, using different targets (coagulase and *femA* genes) (Kearns *et al.*, 1999; Vannuffel *et al.*, 1998) or *S. aureus* toxins (Schmitz *et al.*, 1998), yielded promising results with good specificity. More recently, other promising results were obtained by using triplex PCR assays based on the detection of *S. aureus* rRNA, *mecA*, and *nuc* (Louie *et al.*, 2002;Maes *et al.*, 2002) or *clfA* genes (Mason *et al.*, 2001). However, all these method can produce false positives if the culture contains a mixture of methicillin-resistant CoNS and methicillin-sensitive *S. aureus*. Very recently, a smart approach to overcome this has been described that exploits the insertion of SCC*mec* into the specific *S. aureus* site *orfX*. Cuny and colleagues use a simplex PCR amplification using primers complementary to a conserved region of *orfX* (*S. aureus* specific) and a combination of SCC*mec* specific primers representing each of the major types, thus amplifying a region spanning the *S. aureus* SCC*mec* insertion site (Cuny and Witte, 2005). Overall, PCR-based methods perform extremely well when applied to isolated strains or samples containing organisms in pure cultures; however this is not rapid enough for direct MRSA screening.

Multiplexed and real-time qPCR techniques

Fluorescently labelled detection of PCR products (Lee *et al.*, 1999) combines accurate identification with very low limits of detection, close to a single gene copy per sample. These techniques provided a significant technological advantage for the rapid and large-scale identification of various microorganisms (Brancart *et al.*, 2005; Corless *et al.*, 2001; Gieseler *et al.*, 2005; He and Jiang, 2005; Hsu *et al.*, 2005; Lee *et al.*, 2005; Mackay *et al.*, 2002; Svenstrup *et al.*, 2005). Molecular methods outperform conventional detection strategies by providing rapid and sensitive detection, and avoiding the use of toxic acrylamide gels. Using the *nuc* gene as target, Fang and Hedin reported a fast screening and identification assay applicable to isolated bacteria (Fang and Hedin, 2003). In this field, our group reported the use of a novel immuno-qPCR procedure allowing rapid detection of MRSA from mixed flora samples (Francois *et al.*, 2003). The procedure consists in a direct one-step enrichment of *S. aureus* present in either nasal or inguinal swabs, requiring 45 min, followed by DNA extraction of immunocaptured bacteria and their identification by a triplex qPCR. The specificity of MRSA identification is based on the quantitative correlation of the *mecA* gene and that of the *S. aureus* -specific *femA* signal, a probe that does not cross-react with other bacterial species, including *S. epidermidis*. This assay allows detection and identification of MRSA in less than 6 h after sample collection, thus allowing same day identification. We have successfully applied this method to the rapid screening of patients admitted to the ICU and shown that infection control measures could be implemented earlier than with optimized culture-based methods (22.2 versus 93.1 hours, respectively). This molecular screening strategy resulted in a 70% decrease in MRSA infections in the MICU, after coupling assay implementation to strict infection control measures (Harbarth *et al.*, 2006; Masuet-Aumatell *et al.*, 2004). Results obtained in this study showed that a drastic reduction of infections was obtained without change in colonization pressure, thus indicating that rapid screening and isolation measures can decrease the transmission rate.

The other major development was the introduction of methods designed to detect MRSA directly from patient swabs without prior cultivation. Using the smart method of specific primer pairs in the SCC*mec* and *orfX* regions generating products that span the insertion site, Huletsky and colleagues reported a very fast assay allowing direct identification of MRSA in 2 h, even in the presence of other staphylococcal species (Huletsky *et al.*, 2004). This method is now commercially available from BD GeneOhm and its performance has been evaluated by several groups. The test appears robust for the direct identification of MRSA in samples originating from nose or groins swabs, showing sensitivities of 89%, 82% and 87% for the nose, groin, or combined nose-groin swabs, respectively (Bishop *et al.*, 2006), using enrichment cultures (i.e. with backup broth) as gold standard. This observation was further supported by de San and colleagues, who compared the BD-Geneohm MRSA assay to enrichment culture (de San *et al.*, 2007). The sensitivity and specificity of the PCR assay were 81% and 97%, respectively, and its positive and negative predictive values were 75% and 98%, respectively. The BD-Geneohm assay was more sensitive on swabs from nares (90.6%) than from other body sites (76.5%, $P < 0.01$) (de San *et al.*, 2007). However, another recent study reports a moderate sensitivity of 76% in nares or rectum swabs and only 56% in other sites including open wounds and line exit sites (Zhang *et al.*, 2007), even using charcoal-based swabs. Furthermore, the emergence of new SCC*mec* elements warrants further modifications of the design and revalidation of the whole assay as new molecular variants continuously emerge (Berglund *et al.*, 2005). Donnio and colleagues nicely described that the partial excision of the SCC*mec* cassette occurs not infrequently in MSSA strains, and might compromise the efficiency of this strategy (Donnio *et al.*, 2005). There are also SCC elements described that do not contain the *mecA* gene, and that can cross-react with the assay (Holden *et al.*, PNAS, 2004). From a practical standpoint, the rapid delivery of the swabs to the laboratory remains a challenging task. In our institution, times from screening to arrival of the sample in the laboratory range from 1.1 to 12.1 h (Harbarth *et al.*, 2006; Masuet-Aumatell *et al.*, 2004). A careful implementation of the method is also warranted to balance the convenience of a batch procedure with acceptable delays for performing the assay. This can be achieved by defining PCR run times, implementing software for master mix preparation coupled to sample tracking and transmission of the results to clinicians. Overall, a short training period is warranted for implementing this procedure within routine laboratories. The cost of the reagents ranges from 10 to 40 US$ and they can prove cost-effective, at least in defined clinical settings (Harbarth *et al.*, 2006; Masuet-Aumatell *et al.*, 2004).

Molecular epidemiology of MRSA

Molecular genotyping assays permit the detection and delineation of outbreaks. The rapid implementation of such assays focuses the efforts of the infection control team on more targeted interventions. The population structure of *S. aureus* is now known to be clonal with dominant lineages that have evolved independently and mobile genetic elements encoding virulence and resistance genes that move into and out of strains (Chapter 3). Genotyping methods in a diagnostic setting need to be discriminatory enough to make clinical judgments, but also cheap, rapid, easy to interpret and simple to perform. Genotyping methods can either rely on non-cognate approaches that generate patterns (e.g. PFGE-Pulsed-Field Gel Electrophoresis or RAPD-Random Amplification of Polymorphic DNA) or focus on genetic variations in well-defined molecular targets (e.g. MLST-Multi-Locus Sequence Typing or MLVA-Multi-Locus Variation Analysis). In addition, other approaches can specifically focus on a given gene, for providing a less discriminative, but more rapid, epidemiologic information. Among these targets, the analysis of toxin(s) or exotoxin(s) content has been proposed as a typing tool (Aguiar-Alves *et al.*, 2006; Nishi *et al.*, 2002) and showed appreciable power of discrimination in selected populations. The method chosen should be dependent on the question asked, and the typical local *S. aureus* population.

Recent efforts in the field of high-throughput sequencing yielded to the release of numerous bacterial genome sequences. To date, nine fully sequenced strains of *Staphylococcus aureus* are publicly available (Baba *et al.*, 2002; Gill *et al.*, 2005; Holden *et al.*, 2004; Kuroda *et al.*, 2001) providing opportunities for searching conserved or variable targets within the genome of MRSA, thus allowing improved molecular identification and characterization (Charbonnier *et al.*, 2005; Couzinet *et al.*, 2005). For example, the markers described above (e.g. *mecA* and SCC*mec* organization) are important epidemiological indicators of strain origin (Dufour *et al.*, 2002; Naimi *et al.*, 2003). Benefiting from this knowledge, our group recently developed a series of molecular assays allowing us to decipher SCC*mec* element types (Francois *et al.*, 2004), global regulator *agr* type (Francois *et al.*, 2006), toxin content, and genotyping by MLVA (Francois *et al.*, 2005; Francois *et al.*, 2008). This panel of tests was used to evaluate the trends in our local epidemiology over a 13-year period. To evaluate population dynamics of methicillin-resistant *Staphylococcus aureus* (MRSA) strains differing from the multiresistant endemic clone responsible for the majority of hospital-acquired infections, we identified a collection of 151 strains showing numerous antibiotic susceptibilities. Surprisingly, our collection contained an extraordinary diversity of strains in terms of genome content, including commonly isolated European isolates. Fifty-eight (58%) of strains contained one toxin (either, one exfoliatin, the Panton–Valentine leukocidin (PVL) or the staphylococcal toxic shock-syndrome toxin) while 42% of strains were devoid of any toxin. The large majority (82%) harboured a SCC*mec* IV or V cassette whereas 65% harboured at least one important toxin. The most frequently isolated strains were ST80 and ST5 containing PVL and ST88 harbouring the exfoliatin A toxin. Remarkably, a very homogeneous cluster of strains harbouring the PVL was found specifically in patients originating from Kosovo (ST152). Other strains showed new profiles with previously uncharacterized genome contents, such as PVL-negative MLST type 1 with a SCC*mec* V cassette, or one isolate harbouring two important toxins. Finally, the oldest strain isolated from our institution showing common features to community-onset MRSA was a PVL-positive ST80 strain with a SCC*mec* IV cassette; it was identified as early as in 1994 from a wound of patient originating from Libya. Our observations suggest that the epidemiology of MRSA is rapidly evolving and that our area contains a wide variety of genetic backgrounds.

Thus, our situation is clearly different from that observed in other countries such as USA, and is characterized by numerous events of importation (Fig. 4.2). In addition none of our CA-MRSA was responsible for necrotizing pneumonia and they were rarely responsible for severe infections (Francois *et al.*, 2005; Francois, *et al.*, 2008).

The capacity to continuously monitor molecular changes over time is invaluable for:

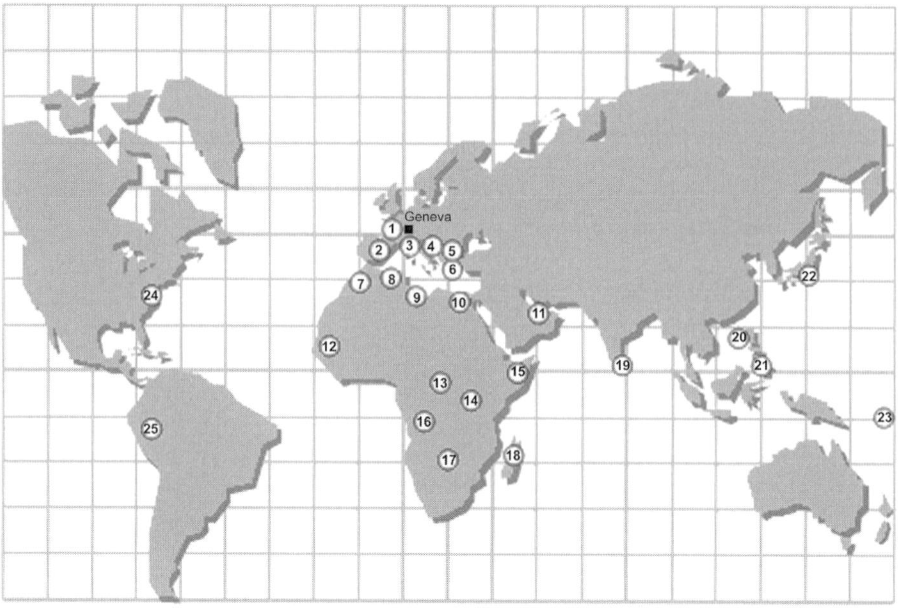

Figure 4.2 Geographical origin of CA-MRSA. Direct origin or recent travel destinations of patients detected with CA-MRSA since 2002. Circles show the location of 1: France, 2: Spain, 3: Italia, 4: Kosovo, 5: Bulgaria, 6: Greece, 7: Morocco, 8: Algeria, 9: Libya, 10: Egypt, 11: Qatar, 12: Senegal, 13: Central African Republic, 14: Uganda, 15: Somalia, 16: Angola, 17: Botswana, 18: Madagascar, 19: Sri Lanka, 20: Taiwan, 21: Philippines, 22: Japan, 23: French Polynesia, 24: New York city, 25: Bolivia. Geneva appears as a black square (courtesy of Dr S. Harbarth).

i) focusing infection control interventions on specific subgroups of patients that would remain undetected otherwise and might thus spread novel MRSA genotypes into hospitals (e.g. SCC*mec* strain variants that are undetected by current commercial assays) and ii) correlating clinical presentations with new genetic backgrounds to uncover significant epidemiological changes, as those witnessed during the emergence of CA-MRSA.

Pulsed-field gel electrophoresis (PFGE) is still the most widely used method for micro-epidemiology (local outbreak) studies. The method relies on the direct digestion of genomic DNA by a restriction enzyme (typically *Sma*I), followed by separation of fragments by gel electrophoresis, and has necessitated protocol standardization (Murchan *et al.*, 2003; Van Belkum *et al.*, 1998). Although tedious, costly (Van Belkum *et al.*, 1998) and technically challenging (Murchan *et al.*, 2003), this method can detect some MGE movements. However, PFGE is not appropriate for lineage identification. Multilocus sequence typing (MLST) is an alternative typing method more popular in epidemiological typing (Enright *et al.*, 2000). MLST basically consists in amplification and sequencing of amplicons (usually 450–500 bp in length) of a series of genes (7 genes for *Staphylococcus aureus*) encoding for central metabolic compounds that evolve slowly, and comparing to an international database (www.mlst.net). The method is expensive and provides genetic information only in highly conserved genes but not MGE, therefore limiting its discriminatory power between related strains. The method is more suited to the investigation of bacterial phylogeny and evolution of population lineages than to the typing of strains in hospital outbreaks.

Recently, different low-cost and reduced turn-around time strategies were reported to study strain relatedness or outbreaks of strain collections, such as Random Amplification of Polymorphic DNA (RAPD) (Luijendijk *et al.*, 1996), Amplified Fragment Length Polymorphism (AFLP) (Van Leeuwen *et al.*, 1996), multi-locus variation analysis (MLVA) (Sabat *et al.*, 2003) or *spa* typing (Koreen *et al.*, 2004). AFLP is a whole genome typing method that scans for polymorphism in restriction sites and nucleotides bordering these sites. It documents nucleotide sequence variation, insertions and deletions across genomes. Recently, MLVA has been introduced as a typing method for a large number of bacterial pathogens (Schouls *et al.*, 2006; Tenover *et al.*, 2007; van Belkum, 2007). In MLVA, the variability in the numbers of short tandem repeated within a limited number of genes is utilized to create DNA fingerprints for epidemiological studies (Tenover *et al.*, 2007) (Fig. 4.3). *Spa* typing is technically easier to perform, although it does not provide similar levels of discrimination than AFLP or MLVA. Assays based on the amplification of a limited number of genomic sequences are all potentially amenable to high-throughput for providing real-time epidemiology, as recently shown by Melles and co-workers (Melles *et al.*, 2004). This study relying on >800 clinical isolates or carriage strains allowed the authors to finely dissect the bacterial population composition using previously well-characterized epidemiological situations.

Another important contribution was brought by Kuhn *et al* who developed an original sequence-based analysis of genes belonging either to the core adhesin or accessory gene categories (Kuhn *et al.*, 2006). This method, potentially suitable for a diagnostic setting, exploits the fact that each lineage is highly independent. Independence is due to the role of a lineage specific restriction-modification system (Waldron and Lindsay, 2006). This study clearly demonstrated that Sau1, a restriction-modification system was able to block acquisition of DNA not only from exogenous origin but also from other *S. aureus* belonging to other lineages. This system probably slows down the evolution of new strains or lineages (Feil *et al.*, 2003;Feil and Enright, 2004). Conveniently, the Sau1 system is a useful marker of independent lineages and a simple PCR test based on Sau1 variation can be used to identify *S. aureus* lineages (Cockfield *et al.*, 2007). This method is simple, fast and inexpensive for typing.

Many of the methods described only identify lineage but not the various MGE which often encode important resistance and virulence genes. The only methods available for assessing them comprehensively are sequencing or multi-strain microarray but neither are currently suitable for the diagnostic setting. Antibiotic resistance patterns can be useful in some situations when the lineages are known. Some toxins are also useful markers and can be assessed using PCR.

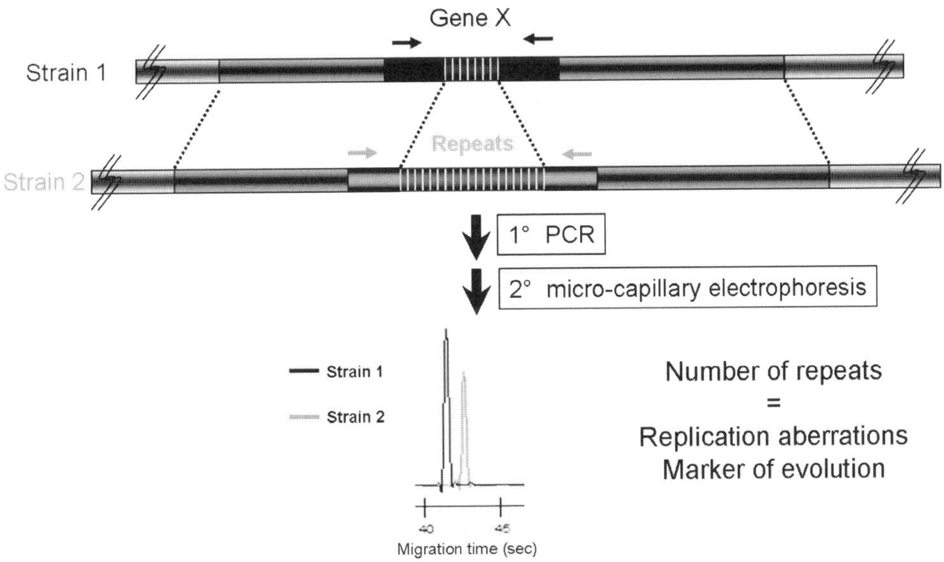

Figure 4.3 Microcapillary electrophoresis system. MLVA and other techniques relying on DNA fragments size separation are amenable to high-throughput format using capillary electrophoretic system, avoiding utilization of agarose gels and intercalating agents. In this example, PCR amplicons (1) generated using primers designed in conserved regions flanking repeated motives are separated using capillary electrophoresis system (2). Gene X containing a higher number of repeats in strains 2 (grey peak) displays a migration time higher than that obtained for strain 1 (black peak).

Markers of unique or virulent strains

Some strains are known to be particularly virulent, and it can be useful to identify a rapid test to identify them. Panton–Valentine leukocidin (PVL) is a toxin carried by many types of MRSA, and is associated with recurrent skin and soft tissue infections or lethal necrotizing pneumonia (Gillet *et al.*, 2002). Its role as a marker of CA-MRSA is increasingly debated (Karden-Lilja *et al.*, 2007). Staphylococcal superantigens (SAg) are secreted toxins that induce a strong activation of large T-cell subpopulations which can result in toxic shock (Holtfreter *et al.*, 2007). Certain SAg genes are associated with particular clonal lineages (Booth *et al.*, 2001; Jarraud *et al.*, 2002; Moore *et al.*, 2001; Peacock *et al.*, 2002; Van Belkum *et al.*, 2006). Indeed, most of the 19 described *S. aureus* superantigens, the enterotoxins SEA to SEE, SEG to SER, SEU, and the toxic shock syndrome toxin 1 (TSST-1), are encoded on phages and pathogenicity islands (Lindsay and Holden, 2006).

Tenover and colleagues recently showed that USA300, a strain responsible for the vast majority of community-acquired MRSA infections in the USA (e.g. 97% in skin and soft tissue infections (Moran *et al.*, 2006) contains an Arginine Catabolic Mobile Element (ACME) adjacent to SCC*mec* at *orfX* (Goering *et al.*, 2007). This location appears specifically associated with strains harbouring a SCC*mec* type IVa element and was never found in MSSA strains. It thus appears as potentially useful for typing purpose. From a biological standpoint, this observation suggests a role of SCC*mec* recombinase genes in ACME integration.

In 2003, Day and colleagues identified a link between specific MLST profiles and strain virulence (Day *et al.*, 2001) using a large collection of *S. aureus* recovered from asymptomatic nasal carriers and from episodes of invasive infections in the community. However, results of this seminal study were retracted after careful inspection of sequences by the authors (Day *et al.*, 2002). Further investigation of these strains using a seven-strain *S. aureus* microarray (including most known toxins), confirmed a lot of variation between isolates but no difference between the carriage and invasive strains (Lindsay *et al.*, 2006). However, hospitals may be different. Specific clones of MRSA in the hospital setting have been shown to cause different clinical disease, including some associated with vascular access devices (Edgeworth *et al.*, 2007). Thus, the quest for linking specific genes or genotypic markers to a defined clinical outcome remains an open and very desirable goal. In the future, this specific topic will benefit from the considerable amount of information resulting from the utilization of new molecular techniques and especially related to high-throughput sequencing and arrays.

Conclusions

Infections due to MRSA are frequent and represent an economical burden, requiring utilization of last barrier drugs. Thus, rapid detection and identification of MRSA is an absolute prerequisite to adopt prompt isolation measures. Until recently, microbiological methods dedicated to MRSA identification were based on the utilization of selective growth media, which are time-consuming and preclude same-day diagnosis. For more than one decade, nucleic acid-based identification assays have demonstrated their usefulness and robustness for the detection of hardly cultivable, non-cultivable and even killed microorganisms, as well as for the identification of specific pathogens against the background of a complex microflora. The current view is still that molecular methods are used to supplement, but not to replace cultures. MRSA molecular detection nicely illustrates this paradigm: it provides early warning but cultures are still required for further antimicrobial susceptibility testing or epidemiological typing. Molecular assays based on target nucleic acid amplification, and especially real-time PCR, have proven rapid, affordable and successful in terms of sensitivity and specificity. Current challenges for MRSA screening are centred on the selection of the most appropriate assay, both in terms of feasibility (costs, technical expertise) and assay performance. One has to be especially careful when embarking on detection strategies that are based on mobile and highly variable genetic regions, such as the SCC*mec* insertion site. Indeed, iterative changes in the detection protocol to adapt for emerging variants might not only affect the performance of the assay but also open unpredictable and systematic breaches in the infection control programme. To date, with several decades of hindsight, we can firmly state that molecular variations will continuously emerge, but it is currently impossible to predict where these changes will emerge. As the molecular epidemiology may substantially vary from country to country, and possibly also between different regions, this underlines the importance of having an epidemiologic molecular surveillance, ideally as close as possible to our own lab practice.

Future

Detection of amplification reactions is still improving and future developments coupled to the parallelism of hybridization techniques might provide more broadly usable tools. Such strategies might be adapted to identify bacteria as well as to provide their genetically encoded antibiotic resistance and virulence determinants.

Determination of strain relatedness using genotyping methods allows documenting the spread of strains and potential outbreaks, helping to adapt infection control measures. Discrimination power of molecular methods is dependent on the number and quality of tested parameters. Currently, most genotyping methods rely on a limited number of targets (e.g. PFGE or MLVA). Microarrays covering whole genomes provide extensive – but still partial – catalogue of genome content during a single experiment. This limits the potential of the technique to discover rare or recently acquired clusters of genes still not characterized. High-throughput sequencing represents the most advanced approach, but is still costly and not yet adapted to a routine utilization. However, the added value of such tool will prove considerable in infectious diseases diagnostic. Unravelling bacterial pathogenic potential as well as all genetically encoded resistance determinants will impact drug prescription and patient monitoring. Additionally, assessing epidemiology in real-time will likely contribute to improve and standardize infection control strategies.

Acknowledgements

This work was supported by grants from the Swiss National Science Foundation nos. 3100A0–112370/1, COST C05.0103 (JS), 3100A0-116075 (PF) and from the University of Geneva Hospitals Quality Improvement Research Program.

References

Aguiar-Alves, F., Medeiros, F., Fernandes, O., Gudziki Pereira, R.M., Perdreau-Remington, F., and Riley, L.W. (2006). New *Staphylococcus aureus* genotyping method based on exotoxin (set) genes. J. Clin. Microbiol. *44*, 2728–2732.

Anonymous (1999). Komimission für Krankenhaushygiene und Infektionspravention. Empfehlung zur Prävention und Kontrolle von Methicillin-resistenten *Staphylococcus aureus*-Stämmen (MRSA) in Krankenhäusern und anderen medizinischen Einrichtungen. Bundesgesundheitsbl Gesundheitsforsch Gesundheitsschutz *42*, 954–958.

Archer, G.L. (1998). *Staphylococcus aureus*: a well-armed pathogen. Clin. Infect. Dis. *26*, 1179–1181.

Aucken, H.M., Ganner, M., Murchan, S., Cookson, B.D., and Johnson, A.P. (2002). A new UK strain of epidemic methicillin-resistant *Staphylococcus aureus* (EMRSA-17) resistant to multiple antibiotics. J. Antimicrob. Chemother. *50*, 171–175.

Ayliffe, G.A., Buckles, A., Casewell, M.W., Cookson, B.D., Cox, R.A., Duckworth, G.J., French, G.L., Griffiths-Jones, A., Heathcock, R., Humphreys, H., Keane, C.T., Marples, R.R., Shanson, D.C., Slack, R., and Tebbs, E. (1999). Revised guidelines for control of MRSA: applying appropriately-based recommendations. J. Hosp. Infect. *43*, 315–316.

Baba, T., Takeuchi, F., Kuroda, M., Yuzawa, H., Aoki, K., Oguchi, A., Nagai, Y., Iwama, N., Asano, K., Naimi, T., Kuroda, H., Cui, L., Yamamoto, K., and Hiramatsu, K. (2002). Genome and virulence determinants of high virulence community-acquired MRSA. Lancet 359, 1819–1827.

Barber, M. (1961). Methicillin-resistant staphylococci. J. Clin. Pathol. *14*, 385–393.

Becker, A., Forster, D.H., and Kniehl, E. (2002). Oxacillin resistance screening agar base for detection of methicillin-resistant *Staphylococcus aureus*. J. Clin. Microbiol. *40*, 4400–4401.

Berger-Bachi, B., Barberis-Maino, L., Strassle, A., and Kayser, F.H. (1989). FemA, a host-mediated factor essential for methicillin resistance in *Staphylococcus aureus*: molecular cloning and characterization. Mol. Gen. Genet. *219*, 263–269.

Berglund, C., Molling, P., Sjoberg, L., and Soderquist, B. (2005). Predominance of staphylococcal cassette chromosome mec (SCCmec) type IV among methicillin-resistant *Staphylococcus aureus* (MRSA) in a Swedish county and presence of unknown SCC*mec* types with Panton–Valentine leukocidin genes. Clin. Microbiol. Infect. *11*, 447–456.

Bishop, E.J., Grabsch, E.A., Ballard, S.A., Mayall, B., Xie, S., Martin, R., and Grayson, M.L. (2006). Concurrent analysis of nose and groin swab specimens by the IDI-MRSA PCR assay is comparable to analysis by individual-specimen PCR and routine culture assays for detection of colonization by methicillin-resistant *Staphylococcus aureus*. J. Clin. Microbiol. *44*, 2904–2908.

Booth, M.C., Pence, L.M., Mahasreshti, P., Callegan, M.C., and Gilmore, M.S. (2001). Clonal associations among *Staphylococcus aureus* isolates from various sites of infection. Infect. Immun. *69*, 345–352.

Brancart, F., Rodriguez-Villalobos, H., Fonteyne, P.A., Peres-Bota, D., and Liesnard, C. (2005). Quantitative TaqMan PCR for detection of *Pneumocystis jiroveci*. J. Microbiol. Methods *61*, 381–387.

Brook, I., and Finegold, S.M. (1979). Bacteriology of chronic otitis media. JAMA *241*, 487–488.

Campbell, K.M., Vaughn, A.F., Russell, K.L., Smith, B., Jimenez, D.L., Barrozo, C.P., Minarcik, J.R., Crum, N.F., and Ryan, M.A. (2004). Risk factors for community-associated methicillin-resistant *Staphylococcus aureus* infections in an outbreak of disease among military trainees in San Diego, California, in 2002. J. Clin. Microbiol. *42*, 4050–4053.

Carleton, H.A., Diep, B.A., Charlebois, E.D., Sensabaugh, G.F., and Perdreau-Remington, F. (2004). Community-adapted methicillin-resistant *Staphylococcus aureus* (MRSA): population dynamics of an expanding community reservoir of MRSA. J. Infect. Dis. *190*, 1730–1738.

Cavassini, M., Wenger, A., Jaton, K., Blanc, D.S., and Bille, J. (1999). Evaluation of MRSA-Screen, a simple anti-PBP 2a slide latex agglutination kit, for rapid detection of methicillin resistance in *Staphylococcus aureus*. J. Clin. Microbiol. *37*, 1591–1594.

Chaix, C., Durand-Zaleski, I., Alberti, C., and Brun-Buisson, C. (1999). Control of endemic methicillin-resistant *Staphylococcus aureus*: a cost-benefit analysis in an intensive care unit. JAMA *282*, 1745–1751.

Chambers, H.F. (2001). The Changing Epidemiology of *Staphylococcus aureus* ? Emerg. Infect. Dis. *7*, 178–182.

Charbonnier, Y., Gettler, B.M., Francois, P., Bento, M., Renzoni, A., Vaudaux, P., Schlegel, W., and Schrenzel, J. (2005). A generic approach for the design of whole-genome oligoarrays, validated for genomotyping, deletion mapping and gene expression analysis on *Staphylococcus aureus*. BMC Genomics 6, 95.

Cherkaoui, A., Renzi, G., Francois, P., and Schrenzel, J. (2007). Comparison of four chromogenic media for culture-based screening of meticillin-resistant *Staphylococcus aureus*. J. Med. Microbiol. 56, 500–503.

Clinical and Laboratory Standards Institute (2005). Performance standards for antimicrobial susceptibility testing: 15th informational supplement M100-S15. Clinical and Laboratory Standards Institute *Wayne. PA*.

Cohen, S.H., Morita, M.M., and Bradford, M. (1991). A seven-year experience with methicillin-resistant *Staphylococcus aureus*. Am. J. Med. *91*, 233S-237S.

Coia, J.E., Duckworth, G.J., Edwards, D.I., Farrington, M., Fry, C., Humphreys, H., Mallaghan, C., and Tucker, D.R. (2007). Guidelines for the control and prevention of meticillin-resistant *Staphylococcus aureus* (MRSA) in healthcare facilities. J. Hosp. Infect. *63 Suppl 1*, S1–44.

Compernolle, V., Verschraegen, G., and Claeys, G. (2007). Combined use of Pastorex Staph-Plus and either of two new chromogenic agars, MRSA ID and CHROMagar MRSA, for detection of methicillin-resistant *Staphylococcus aureus*. J. Clin. Microbiol. *45*, 154–158.

Corless,C.E., Guiver,M., Borrow,R., Edwards-Jones,V., Fox,A.J., and Kaczmarski,E.B. (2001). Simultaneous detection of *Neisseria meningitidis*, *Haemophilus influenzae*, and *Streptococcus pneumoniae* in suspected cases of meningitis and septicemia using real-time PCR. J. Clin. Microbiol. *39*, 1553–1558.

Cosseron-Zerbib, M., Roque Afonso, A.M., Naas, T., Durand, P., Meyer, L., Costa, Y., el Helali, N., Huault, G., and Nordmann, P. (1998). A control programme for MRSA (methicillin-resistant *Staphylococcus aureus*) containment in a paediatric intensive care unit: evaluation and impact on infections caused by other micro-organisms. J. Hosp. Infect. *40*, 225–235.

Couzinet, S., Yugueros, J., Barras, C., Visomblin, N., Francois, P., Lacroix, B., Vernet, G., Lew, D., Troesch, A., Schrenzel, J., and Jay, C. (2005). Evaluation of a high-density oligonucleotide array for characterization of *grlA*, *grlB*, *gyrA* and *gyrB* mutations in fluoroquinolone resistant *Staphylococcus aureus* isolates. J. Microbiol. Methods *60*, 275–279.

Cox, R.A., Conquest, C., Mallaghan, C., and Marples, R.R. (1995). A major outbreak of methicillin-resistant *Staphylococcus aureus* caused by a new phage-type (EMRSA-16). J. Hosp. Infect. *29*, 87–106.

Cuny, C., and Witte, W. (2005). PCR for the identification of methicillin-resistant *Staphylococcus aureus* (MRSA) strains using a single primer pair specific for SCC*mec* elements and the neighbouring chromosome-borne *orfX*. Clin. Microbiol. Infect. *11*, 834–837.

Day, N.P., Moore, C.E., Enright, M.C., Berendt, A.P., Smith, J.M., Murphy, M.F., Peacock, S.J., Spratt, B.G., and Feil, E.J. (2002). Retraction. Science *295*, 971.

Day, N.P., Moore, C.E., Enright, M.C., Berendt, A.R., Smith, J.M., Murphy, M.F., Peacock, S.J., Spratt, B.G., and Feil, E.J. (2001). A link between virulence and ecological abundance in natural populations of *Staphylococcus aureus*. Science *292*, 114–116.

de San, N., Denis, O., Gasasira, M.F., De Mendonca, R., Nonhoff, C., and Struelens, M.J. (2007). Controlled evaluation of the IDI-MRSA assay for detection of colonization by methicillin-resistant *Staphylococcus aureus* in diverse mucocutaneous specimens. J. Clin. Microbiol. *45*, 1098–1101.

Diederen, B., van Duijn, I., Van Belkum, A., Willemse, P., van Keulen, P., and Kluytmans, J. (2005). Performance of CHROMagar MRSA medium for detection of methicillin-resistant *Staphylococcus aureus*. J. Clin. Microbiol. *43*, 1925–1927.

Donnio, P.Y., Oliveira, D.C., Faria, N.A., Wilhelm, N., Le Coustumier, A., and de Lencastre, H. (2005). Partial Excision of the Chromosomal Cassette Containing the Methicillin Resistance Determinant Results in Methicillin-Susceptible *Staphylococcus aureus*. J. Clin. Microbiol. *43*, 4191–4193.

Dufour, P., Gillet, Y., Bes, M., Lina, G., Vandenesch, F., Floret, D., Etienne, J., and Richet, H. (2002). Community-acquired methicillin-resistant *Staphylococcus aureus* infections in France: emergence of a single clone that produces Panton–Valentine leukocidin. Clin. Infect. Dis. 35, 819–824.

Enright, M.C., Day, N.P., Davies, C.E., Peacock, S.J., and Spratt, B.G. (2000). Multilocus sequence typing for characterization of methicillin- resistant and methicillin-susceptible clones of *Staphylococcus aureus*. J. Clin. Microbiol. 38, 1008–1015.

Fang, H., and Hedin, G. (2003). Rapid screening and identification of methicillin-resistant *Staphylococcus aureus* from clinical samples by selective-broth and real-time PCR assay. J. Clin. Microbiol. *41*, 2894–2899.

Feil, E.J., Cooper, J.E., Grundmann, H., Robinson, D.A., Enright, M.C., Berendt, T., Peacock, S.J., Smith, J.M., Murphy, M., Spratt, B.G., Moore, C.E., and Day, N.P. (2003). How clonal is *Staphylococcus aureus?* J. Bacteriol. *185*, 3307–3316.

Feil, E.J., and Enright, M.C. (2004). Analyses of clonality and the evolution of bacterial pathogens. Curr. Opin. Microbiol. *7*, 308–313.

Francois, P., Harbarth, S., Huyghe, A., Renzi, G., Bento, M., Gervaix, A., Pittet, D., and Schrenzel, J. (2008). Community-associated MRSA, Geneva, Switzerland, 1993–2005. Emerg. Infecto. Dis. In press.

Francois, P., Huyghe, A., Charbonnier, Y., Bento, M., Herzig, S., Topolski, I., Fleury, B., Lew, D., Vaudaux, P., Harbarth, S., Van Leeuwen, W., Van Belkum, A., Blanc, D.S., Pittet, D., and Schrenzel, J. (2005). Use of an automated multiple-locus, variable-number tandem repeat-based method for rapid and high-throughput genotyping of *Staphylococcus aureus* isolates. J. Clin. Microbiol. *43*, 3346–3355.

Francois, P., Koessler, T., Huyghe, A., Harbarth, S., Bento, M., Lew, D.P., Etienne, J., Pittet, D., and Schrenzel, J. (2006). Rapid *Staphylococcus aureus agr* Type Determination by A Novel Multiplex Real-Time qPCR Assay. J. Clin. Microbiol. *44*, 1892–1895.

Francois, P., Pittet, D., Bento, M., Pepey, B., Vaudaux, P., Lew, D., and Schrenzel, J. (2003). Rapid detection of methicillin-resistant *Staphylococcus aureus* directly from sterile or nonsterile clinical samples by a new molecular assay. J. Clin. Microbiol. *41*, 254–260.

Francois, P., Renzi, G., Pittet, D., Bento, M., Lew, D., Harbarth, S., Vaudaux, P., and Schrenzel, J. (2004). A novel multiplex real-time PCR assay for rapid typing of major staphylococcal cassette chromosome *mec* elements. J. Clin. Microbiol. *42*, 3309–3312.

Gieseler, S., Konig, B., Konig, W., and Backert, S. (2005). Strain-specific expression profiles of virulence genes in *Helicobacter pylori* during infection of gastric epithelial cells and granulocytes. Microbes. Infect. *7*, 437–447.

Gill, S.R., Fouts, D.E., Archer, G.L., Mongodin, E.F., Deboy, R.T., Ravel, J., Paulsen, I.T., Kolonay, J.F., Brinkac, L., Beanan, M., Dodson, R.J., Daugherty, S.C., Madupu, R., Angiuoli, S.V., Durkin, A.S., Haft, D.H., Vamathevan, J., Khouri, H., Utterback, T., Lee, C., Dimitrov, G., Jiang, L., Qin, H., Weidman, J., Tran, K., Kang, K., Hance, I.R., Nelson, K.E., and Fraser, C.M. (2005). Insights on evolution of virulence and resistance from the complete genome analysis of an early methicillin-resistant *Staphylococcus aureus* strain and a biofilm-producing methicillin-resistant *Staphylococcus epidermidis* strain. J. Bacteriol. *187*, 2426–2438.

Gillet, Y., Issartel, B., Vanhems, P., Fournet, J.C., Lina, G., Bes, M., Vandenesch, F., Piemont, Y., Brousse, N., Floret, D., and Etienne, J. (2002). Association between *Staphylococcus aureus* strains carrying gene for Panton–Valentine leukocidin and highly lethal necrotising pneumonia in young immunocompetent patients. Lancet *359*, 753–759.

Gittelman, P.D., Jacobs, J.B., Lebowitz, A.S., and Tierno, P.M., Jr. (1991). *Staphylococcus aureus* nasal carriage in patients with rhinosinusitis. Laryngoscope *101*, 733–737.

Goering, R.V., McDougal, L.K., Fosheim, G.E., Bonnstetter, K.K., Wolter, D.J., and Tenover, F.C. (2007). Epidemiologic distribution of the arginine catabolic mobile element (ACME) Among selected methicillin-resistant and methicillin-susceptible *Staphylococcus aureus* isolates. J. Clin. Microbiol. 45, 1981–1984.

Grundmann, H., Barwolff, S., Tami, A., Behnke, M., Schwab, F., Geffers, C., Halle, E., Gobel, U.B., Schiller, R., Jonas, D., Klare, I., Weist, K., Witte, W., Beck-Beilecke, K., Schumacher, M., Ruden, H., and Gastmeier, P. (2005). How many infections are caused by patient-to-patient transmission in intensive care units? Crit. Care Med. 33, 946–951.

Hamels, S., Gala, J.L., Dufour, S., Vannuffel, P., Zammatteo, N., and Remacle, J. (2001). Consensus PCR and microarray for diagnosis of the genus *Staphylococcus*, species, and methicillin resistance. Biotechniques *31*, 1364–2.

Harbarth, S., Martin, Y., Rohner, P., Henry, N., Auckenthaler, R., and Pittet, D. (2000). Effect of delayed infection control measures on a hospital outbreak of methicillin-resistant *Staphylococcus aureus*. J. Hosp. Infect. *46*, 43–49.

Harbarth, S., Masuet-Aumatell, C., Schrenzel, J., Francois, P., Akakpo, C., Renzi, G., Pugin, J., Ricou, B., and Pittet, D. (2006). Evaluation of rapid screening and pre-emptive contact isolation for detecting and controlling methicillin-resistant *Staphylococcus aureus* in critical care: an interventional cohort study. Crit. Care *10*, -R25.

Harbarth, S., and Pittet, D. (1998). MRSA-a European currency of Infection control. Quarterly Journal of Medicine *91*, 519–521.

He, J.W., and Jiang, S. (2005). Quantification of enterococci and human adenoviruses in environmental samples by real-time PCR. Appl. Environ. Microbiol. *71*, 2250–2255.

Herwaldt, L.A. (1999). Control of methicillin-resistant *Staphylococcus aureus* in the hospital setting. Am. J. Med. *106*, 11S–18S.

Hiramatsu, K., Cui, L., Kuroda, M., and Ito, T. (2001). The emergence and evolution of methicillin-resistant *Staphylococcus aureus*. Trends Microbiol. 9, 486–493.

Hiramatsu, K., Okuma, K., Ma, X.X., Yamamoto, M., Hori, S., and Kapi, M. (2002). New trends in *Staphylococcus aureus* infections: glycopeptide resistance in hospital and methicillin resistance in the community. Curr. Opin. Infect. Dis. *15*, 407–413.

Holden, M.T., Feil, E.J., Lindsay, J.A., Peacock, S.J., Day, N.P., Enright, M.C., Foster, T.J., Moore, C.E., Hurst, L., Atkin, R., Barron, A., Bason, N., Bentley, S.D., Chillingworth, C., Chillingworth, T., Churcher, C., Clark, L., Corton, C., Cronin, A., Doggett, J., Dowd, L., Feltwell, T., Hance, Z., Harris, B., Hauser, H., Holroyd, S., Jagels, K., James, K.D., Lennard, N., Line, A., Mayes, R., Moule, S., Mungall, K., Ormond, D., Quail, M.A., Rabbinowitsch, E., Rutherford, K., Sanders, M., Sharp, S., Simmonds, M., Stevens, K., Whitehead, S., Barrell, B.G., Spratt, B.G., and Parkhill, J. (2004). Complete genomes of two clinical *Staphylococcus aureus* strains: Evidence for the rapid evolution of virulence and drug resistance. Proc. Natl. Acad. Sci. U. S. A *101*, 9786–9791.

Holtfreter, S., Grumann, D., Schmudde, M., Nguyen, H.T., Eichler, P., Strommenger, B., Kopron, K., Kolata, J., Giedrys-Kalemba, S., Steinmetz, I., Witte, W., and Broker, B.M. (2007). Clonal Distribution of Superantigen Genes in Clinical *Staphylococcus aureus* Isolates. J. Clin. Microbiol. *45*, 2669–2680.

Hryniewicz, W. (1999). Epidemiology of MRSA. Infection *27 Suppl 2*, S13–S16.

Hsu, C.F., Tsai, T.Y., and Pan, T.M. (2005). Use of the duplex TaqMan PCR system for detection of Shiga-like toxin-producing *Escherichia coli* O157. J. Clin. Microbiol. *43*, 2668–2673.

Huletsky, A., Giroux, R., Rossbach, V., Gagnon, M., Vaillancourt, M., Bernier, M., Gagnon, F., Truchon, K., Bastien, M., Picard, F.J., Van Belkum, A., Ouellette, M., Roy, P.H., and Bergeron, M.G. (2004). New real-time PCR assay for rapid detection of methicillin-resistant *Staphylococcus aureus* directly from specimens containing a mixture of staphylococci. J. Clin. Microbiol. *42*, 1875–1884.

Hurlimann-Dalel, R.L., Ryffel, C., Kayser, F.H., and Berger-Bachi, B. (1992). Survey of the methicillin resistance-associated genes *mecA*, *mecR1-mecI*, and *femA-femB* in clinical isolates of methicillin-resistant *Staphylococcus aureus*. Antimicrob. Agents Chemother. *36*, 2617–2621.

Hussain, Z., Stoakes, L., Garrow, S., Longo, S., Fitzgerald, V., and Lannigan, R. (2000). Rapid detection of mecA-positive and mecA-negative coagulase-negative staphylococci by an anti-penicillin binding protein 2a slide latex agglutination test. J. Clin. Microbiol. *38*, 2051–2054.

Ito, T., Ma, X.X., Takeuchi, F., Okuma, K., Yuzawa, H., and Hiramatsu, K. (2004). Novel type V staphylococcal cassette chromosome mec driven by a novel cassette chromosome recombinase, *ccrC*. Antimicrob. Agents Chemother. *48*, 2637–2651.

Ito, T., Okuma, K., Ma, X.X., Yuzawa, H., and Hiramatsu, K. (2003). Insights on antibiotic resistance of *Staphylococcus aureus* from its whole genome: genomic island SCC. Drug Resist. Updat. *6*, 41–52.

Jarraud, S., Mougel, C., Thioulouse, J., Lina, G., Meugnier, H., Forey, F., Nesme, X., Etienne, J., and Vandenesch, F. (2002). Relationships between *Staphylococcus aureus* genetic background, virulence factors, *agr* groups (alleles), and human disease. Infect. Immun. *70*, 631–641.

Jones, M.E., Mayfield, D.C., Thornsberry, C., Karlowsky, J.A., Sahm, D.F., and Peterson, D. (2002). Prevalence of oxacillin resistance in *Staphylococcus aureus* among inpatients and outpatients in the United States during 2000. Antimicrob. Agents Chemother. *46*, 3104–3105.

Karden-Lilja, M., Ibrahem, S., Vuopio-Varkila, J., Salmenlinna, S., Lyytikainen, O., Siira, L., and Virolainen, A. (2007). Panton–Valentine leukocidin genes and staphylococcal chromosomal cassette *mec* types amongst Finnish community-acquired methicillin-resistant *Staphylococcus aureus* strains, 1997–1999. Eur. J. Clin. Microbiol. Infect. Dis.

Katayama, Y., Takeuchi, F., Ito, T., Ma, X.X., Ui-Mizutani, Y., Kobayashi, I., and Hiramatsu, K. (2003). Identification in methicillin-susceptible *Staphylococcus hominis* of an active primordial mobile genetic element for the staphylococcal cassette chromosome mec of methicillin-resistant *Staphylococcus aureus*. J. Bacteriol. *185*, 2711–2722.

Kauffman, C.A., Terpenning, M.S., He, X., Zarins, L.T., Ramsey, M.A., Jorgensen, K.A., Sottile, W.S., and Bradley, S.F. (1993). Attempts to eradicate methicillin-resistant *Staphylococcus aureus* from a long-term-care facility with the use of mupirocin ointment. Am. J. Med. *94*, 371–378.

Kazakova, S.V., Hageman, J.C., Matava, M., Srinivasan, A., Phelan, L., Garfinkel, B., Boo, T., McAllister, S., Anderson, J., Jensen, B., Dodson, D., Lonsway, D., McDougal, L.K., Arduino, M., Fraser, V.J., Killgore, G., Tenover, F.C., Cody, S., and Jernigan, D.B. (2005). A clone of methicillin-resistant *Staphylococcus aureus* among professional football players. N. Engl. J. Med. *352*, 468–475.

Kearns, A.M., Seiders, P.R., Wheeler, J., Freeman, R., and Steward, M. (1999). Rapid detection of methicillin-resistant staphylococci by multiplex PCR. J. Hosp. Infect. *43*, 33–37.

Kluytmans, J., Van Belkum, A., and Verbrugh, H. (1997). Nasal carriage of *Staphylococcus aureus*: epidemiology, underlying mechanisms, and associated risks. Clin. Microbiol. Rev. *10*, 505–520.

Kluytmans, J., van Griethuysen, A., Willemse, P., and van Keulen, P. (2002). Performance of CHROMagar selective medium and oxacillin resistance screening agar base for identifying *Staphylococcus aureus* and detecting methicillin resistance. J. Clin. Microbiol. *40*, 2480–2482.

Kondo, Y., Ito, T., Ma, X.X., Watanabe, S., Kreiswirth, B.N., Etienne, J., and Hiramatsu, K. (2007). Combination of multiplex PCRs for SCC*mec* type assignment: Rapid identification system for *mec*, *ccr*, and major differences in junkyard regions. Antimicrob. Agents Chemother. *51*, 264–274.

Koreen, L., Ramaswamy, S.V., Graviss, E.A., Naidich, S., Musser, J.M., and Kreiswirth, B.N. (2004). spa typing method for discriminating among *Staphylococcus aureus* isolates: implications for use of a single marker to detect genetic micro- and macrovariation. J. Clin. Microbiol. *42*, 792–799.

Kreiswirth, B., Kornblum, J., Arbeit, R.D., Eisner, W., Maslow, J.N., McGeer, A., Low, D.E., and Novick, R.P. (1993). Evidence for a clonal origin of methicillin resistance in *Staphylococcus aureus*. Science *259*, 227–230.

Kuhn, G., Francioli, P., and Blanc, D.S. (2006). Evidence for clonal evolution among highly polymorphic genes in methicillin-resistant *Staphylococcus aureus*. J. Bacteriol. *188*, 169–178.

Kuroda, M., Ohta, T., Uchiyama, I., Baba, T., Yuzawa, H., Kobayashi, I., Cui, L., Oguchi, A., Aoki, K., Nagai, Y., Lian, J., Ito, T., Kanamori, M., Matsumaru, H., Maruyama, A., Murakami, H., Hosoyama, A., Mizutani-Ui, Y., Takahashi, N.K., Sawano, T., Inoue, R., Kaito, C., Sekimizu, K., Hirakawa, H., Kuhara, S., Goto, S., Yabuzaki, J., Kanehisa, M., Yamashita, A., Oshima, K., Furuya, K., Yoshino, C., Shiba, T., Hattori, M., Ogasawara, N., Hayashi, H., and Hiramatsu, K. (2001). Whole genome sequencing of meticillin-resistant *Staphylococcus aureus*. Lancet 357, 1225–1240.

Lee, L.G., Livak, K.J., Mullah, B., Graham, R.J., Vinayak, R.S., and Woudenberg, T.M. (1999). Seven-color, homogeneous detection of six PCR products. Biotechniques *27*, 342–349.

Lee, S.O., Na, D.K., Kim, C.M., Li, Y.H., Cho, Y.H., Park, J.H., Lee, J.H., Eo, S.K., Klein, T.A., and Chae, J.S. (2005). Identification and prevalence of *Ehrlichia chaffeensis* infection in *Haemaphysalis longicornis* ticks from Korea by PCR, sequencing and phylogenetic analysis based on 16S rRNA gene. J. Vet. Sci. 6, 151–155.

Lew, D.P., and Waldvogel, F.A. (2004). Osteomyelitis. Lancet *364*, 369–379.

Lindsay, J.A., and Holden, M.T. (2006). Understanding the rise of the superbug: investigation of the evolution and genomic variation of *Staphylococcus aureus*. Funct. Integr. Genomics 6, 186–201.

Louie, L., Goodfellow, J., Mathieu, P., Glatt, A., Louie, M., and Simor, A.E. (2002). Rapid detection of methicillin-resistant staphylococci from blood culture bottles by using a multiplex PCR assay. J. Clin. Microbiol. *40*, 2786–2790.

Lowy, F.D. (1998). *Staphylococcus aureus* infections. N. Engl. J. Med. 339, 520–532.

Luijendijk, A., Van Belkum, A., Verbrugh, H., and Kluytmans, J. (1996). Comparison of five tests for identification of *Staphylococcus aureus* from clinical samples. J. Clin. Microbiol. 34, 2267–2269.

Ma, X.X., Ito, T., Tiensasitorn, C., Jamklang, M., Chongtrakool, P., Boyle-Vavra, S., Daum, R.S., and Hiramatsu, K. (2002). Novel type of staphylococcal cassette chromosome mec identified in community-acquired methicillin-resistant *Staphylococcus aureus* strains. Antimicrob. Agents Chemother. *46*, 1147–1152.

Mackay, I.M., Arden, K.E., and Nitsche, A. (2002). Real-time PCR in virology. Nucleic Acids Res. *30*, 1292–1305.

Maes, N., Magdalena, J., Rottiers, S., De Gheldre, Y., and Struelens, M.J. (2002). Evaluation of a triplex PCR assay to discriminate *Staphylococcus aureus* from coagulase-negative staphylococci and determine methicillin resistance from blood cultures. J. Clin. Microbiol. *40*, 1514–1517.

Mason, W.J., Blevins, J.S., Beenken, K., Wibowo, N., Ojha, N., and Smeltzer, M.S. (2001). Multiplex PCR protocol for the diagnosis of staphylococcal infection. J. Clin. Microbiol. *39*, 3332–3338.

Masuet-Aumatell, C., Pittet, D., Schrenzel, J., Akapko, C., Renzi, G., Ricou, B., Pugin, J., and Harbarth, S. Evaluation of a rapid screening strategy for detecting methicillin-resistant *Staphylococcus aureus* (MRSA) in critical care. 44th Interscience Conference on Antimicrobial Agents and Chemotherapy Poster D57, Washington DC, USA 2004.

Melles, D.C., Gorkink, R.F., Boelens, H.A., Snijders, S.V., Peeters, J.K., Moorhouse, M.J., van der Spek, P.J., van Leeuwen, W.B., Simons, G., Verbrugh, H.A., and Van Belkum, A. (2004). Natural population dynamics and expansion of pathogenic clones of *Staphylococcus aureus*. J. Clin. Invest *114*, 1732–1740.

Moore, J.E., Millar, B.C., Yongmin, X., Woodford, N., Vincent, S., Goldsmith, C.E., McClurg, R.B., Crowe, M., Hone, R., and Murphy, P.G. (2001). A rapid molecular assay for the detection of antibiotic resistance determinants in causal agents of infective endocarditis. J. Appl. Microbiol. *90*, 719–726.

Moran, G.J., Krishnadasan, A., Gorwitz, R.J., Fosheim, G.E., McDougal, L.K., Carey, R.B., and Talan, D.A. (2006). Methicillin-resistant *S. aureus* infections among patients in the emergency department. N. Engl. J. Med. 355, 666–674.

Murakami, K., Minamide, W., Wada, K., Nakamura, E., Teraoka, H., and Watanabe, S. (1991). Identification of methicillin-resistant strains of staphylococci by polymerase chain reaction. J. Clin. Microbiol. *29*, 2240–2244.

Murchan, S., Kaufmann, M.E., Deplano, A., de Ryck, R., Struelens, M., Zinn, C.E., Fussing, V., Salmenlinna, S., Vuopio-Varkila, J., El Solh, N., Cuny, C., Witte, W., Tassios, P.T., Legakis, N., Van Leeuwen, W., Van Belkum, A., Vindel, A., Laconcha, I., Garaizar, J., Haeggman, S., Olsson-Liljequist, B., Ransjo, U., Coombes, G., and Cookson, B. (2003). Harmonization of pulsed-field gel electrophoresis protocols for epidemiological typing of strains of methicillin-resistant *Staphylococcus aureus*: a single approach developed by consensus in 10 European laboratories and its application for tracing the spread of related strains. J. Clin. Microbiol. *41*, 1574–1585.

Muto, C.A., Jernigan, J.A., Ostrowsky, B.E., Richet, H.M., Jarvis, W.R., Boyce, J.M., and Farr, B.M. (2003). SHEA guideline for preventing nosocomial transmission of multidrug-resistant strains of *Staphylococcus aureus* and *enterococcus*. Infect. Control Hosp. Epidemiol. *24*, 362–386.

Naimi, T.S., LeDell, K.H., Como-Sabetti, K., Borchardt, S.M., Boxrud, D.J., Etienne, J., Johnson, S.K., Vandenesch, F., Fridkin, S., O'Boyle, C., Danila, R.N., and Lynfield, R. (2003). Comparison of community- and health care-associated methicillin-resistant *Staphylococcus aureus* infection. JAMA *290*, 2976–2984.

Nettleman, M.D., Trilla, A., Fredrickson, M., and Pfaller, M. (1991). Assigning responsibility: using feedback to achieve sustained control of methicillin-resistant *Staphylococcus aureus*. Am. J. Med. *91*, 228S-232S.

Nishi, J., Yoshinaga, M., Miyanohara, H., Kawahara, M., Kawabata, M., Motoya, T., Owaki, T., Oiso, S., Kawakami, M., Kamewari, S., Koyama, Y., Wakimoto, N., Tokuda, K., Manago, K., and Maruyama, I. (2002). An epidemiologic survey of methicillin-resistant *Staphylococcus aureus* by combined use of mec-HVR genotyping and toxin genotyping in a university hospital in Japan. Infect. Control Hosp. Epidemiol. *23*, 506–510.

Pan, E.S., Diep, B.A., Charlebois, E.D., Auerswald, C., Carleton, H.A., Sensabaugh, G.F., and Perdreau-Remington, F. (2005). Population dynamics of nasal strains of methicillin-resistant *Staphylococcus aureus* – and their relation to community-associated disease activity. J. Infect. Dis. *192*, 811–818.

Pape, J., Wadlin, J., and Nachamkin, I. (2006). Use of BBL CHROMagar MRSA medium for identification of methicillin-resistant *Staphylococcus aureus* directly from blood cultures. J. Clin. Microbiol. *44*, 2575–2576.

Papia, G., Louie, M., Tralla, A., Johnson, C., Collins, V., and Simor, A.E. (1999). Screening high-risk patients for methicillin-resistant *Staphylococcus aureus* on admission to the hospital: is it cost effective? Infect. Control Hosp. Epidemiol. *20*, 473–477.

Peacock, S.J., Moore, C.E., Justice, A., Kantzanou, M., Story, L., Mackie, K., O'Neill, G., and Day, N.P. (2002). Virulent combinations of adhesin and toxin genes in natural populations of *Staphylococcus aureus*. Infect. Immun. *70*, 4987–4996.

Perry, J.D., Davies, A., Butterworth, L.A., Hopley, A.L., Nicholson, A., and Gould, F.K. (2004). Development and evaluation of a chromogenic agar medium for methicillin-resistant *Staphylococcus aureus*. J. Clin. Microbiol. *42*, 4519–4523.

Pittet, D., Hugonnet, S., Harbarth, S., Mourouga, P., Sauvan, V., Touveneau, S., and Perneger, T.V. (2000). Effectiveness of a hospital-wide programme to improve compliance with hand hygiene. Infection Control Programme. Lancet *356*, 1307–1312.

Pittet, D., Safran, E., Harbarth, S., Borst, F., Copin, P., Rohner, P., Scherrer, J.R., and Auckenthaler, R. (1996). Automatic alerts for methicillin-resistant *Staphylococcus aureus* surveillance and control: role of a hospital information system. Infect. Control Hosp. Epidemiol. *17*, 496–502.

Pittet, D., and Waldvogel, F.A. (1997). To control or not control colonization with MRSA... that's the question! Quarterly Journal of Medicine *90*, 239–241.

Powers, K.A., Terpenning, M.S., Voice, R.A., and Kauffman, C.A. (1990). Prosthetic joint infections in the elderly. Am. J. Med. *88*, 9N–13N.

Richardson, J.F., and Reith, S. (1993). Characterization of a strain of methicillin-resistant *Staphylococcus aureus* (EMRSA-15) by conventional and molecular methods. J. Hosp. Infect. *25*, 45–52.

Rohrer, S., Tschierske, M., Zbinden, R., and Berger-Bachi, B. (2001). Improved methods for detection of methicillin-resistant *Staphylococcus aureus*. Eur. J. Clin. Microbiol. Infect. Dis. *20*, 267–270.

Rubinovitch, B., and Pittet, D. (2001). Screening for methicillin-resistant *Staphylococcus aureus* in the endemic hospital: what have we learned? J. Hosp. Infect. *47*, 9–18.

Ryffel, C., Tesch, W., Birch-Machin, I., Reynolds, P.E., Barberis-Maino, L., Kayser, F.H., and Berger-Bachi, B. (1990). Sequence comparison of *mecA* genes isolated from methicillin-resistant *Staphylococcus aureus* and *Staphylococcus epidermidis*. Gene *94*, 137–138.

Sabat, A., Krzyszton-Russjan, J., Strzalka, W., Filipek, R., Kosowska, K., Hryniewicz, W., Travis, J., and Potempa, J. (2003). New method for typing *Staphylococcus aureus* strains: multiple-locus variable-number tandem repeat analysis of polymorphism and genetic relationships of clinical isolates. J. Clin. Microbiol. *41*, 1801–1804.

Sakoulas, G., Eliopoulos, G.M., Moellering, R.C., Jr., Wennersten, C., Venkataraman, L., Novick, R.P., and Gold, H.S. (2002). Accessory gene regulator (agr) locus in geographically diverse *Staphylococcus aureus* isolates with reduced susceptibility to vancomycin. Antimicrob. Agents Chemother. *46*, 1492–1502.

Schmitz, F.J., Steiert, M., Hofmann, B., Verhoef, J., Hadding, U., Heinz, H.P., and Kohrer, K. (1998). Development of a multiplex-PCR for direct detection of the genes for enterotoxin B and C, and toxic shock syndrome toxin- in *Staphylococcus aureus* isolates. J. Med. Microbiol. *47*, 335–340.

Sheagren, J.N. (1984a). *Staphylococcus aureus*. The persistent pathogen (first of two parts). N. Engl. J. Med. *310*, 1368–1373.

Sheagren, J.N. (1984b). *Staphylococcus aureus*. The persistent pathogen (second of two parts). N. Engl. J. Med. *310*, 1437–1442.

Simor, A.E., Goodfellow, J., Louie, L., and Louie, M. (2001). Evaluation of new medium, oxacillin resistance screening agar base, for the detection of methicillin-resistant *Staphylococcus aureus* from clinical specimens. J. Clin. Microbiol. *39*, 3422.

Svenstrup, H.F., Jensen, J.S., Bjornelius, E., Lidbrink, P., Birkelund, S., and Christiansen, G. (2005). Development of a quantitative real-time PCR assay for detection of *Mycoplasma genitalium*. J. Clin. Microbiol. *43*, 3121–3128.

Van Belkum, A., Melles, D.C., Snijders, S.V., van Leeuwen, W.B., Wertheim, H.F., Nouwen, J.L., Verbrugh, H.A., and Etienne, J. (2006). Clonal distribution and differential occurrence of the enterotoxin gene cluster, egc, in carriage- versus bacteremia-associated isolates of *Staphylococcus aureus*. J. Clin. Microbiol. *44*, 1555–1557.

Van Belkum, A., Van Leeuwen, W., Kaufmann, M.E., Cookson, B., Forey, F., Etienne, J., Goering, R., Tenover, F., Steward, C., O'Brien, F., Grubb, W., Tassios, P., Legakis, N., Morvan, A., El Solh, N., de Ryck, R., Struelens, M., Salmenlinna, S., Vuopio-Varkila, J., Kooistra, M., Talens, A., Witte, W., and Verbrugh, H. (1998). Assessment of resolution and intercenter reproducibility of results of genotyping *Staphylococcus aureus* by pulsed-field gel electrophoresis of SmaI macrorestriction fragments: a multicenter study. J. Clin. Microbiol. *36*, 1653–1659.

Van Leeuwen, W., Sijmons, M., Sluijs, J., Verbrugh, H., and Van Belkum, A. (1996). On the nature and use of randomly amplified DNA from *Staphylococcus aureus*. J. Clin. Microbiol. *34*, 2770–2777.

Vannuffel, P., Laterre, P.F., Bouyer, M., Gigi, J., Vandercam, B., Reynaert, M., and Gala, J.L. (1998). Rapid and specific molecular identification of methicillin-resistant *Staphylococcus aureus* in endotracheal aspirates from mechanically ventilated patients. J. Clin. Microbiol. *36*, 2366–2368.

Waldron, D.E., and Lindsay, J.A. (2006). Sau1: a novel lineage-specific type I restriction-modification system that blocks horizontal gene transfer into *Staphylococcus aureus* and between *S. aureus* isolates of different lineages. J. Bacteriol. *188*, 5578–5585.

Waldvogel, F.A. (2000). *Staphylococcus aureus* (Including Staphylococcal Toxic Shock). In Principles and Practice of Infectious Diseases, G.L. Mandell, J.E. Bennet, and R. Dolin, eds. (New-York: Churchill Livingstone), pp. 2069–2092.

Weber, S.G., Huang, S.S., Oriola, S., Huskins, W.C., Noskin, G.A., Harriman, K., Olmsted, R.N., Bonten, M., Lundstrom, T., Climo, M.W., Roghmann, M.C., Murphy, C.L., and Karchmer, T.B. (2007). Legislative mandates for use of active surveillance cultures to screen for methicillin-resistant *Staphylococcus aureus* and vancomycin-resistant enterococci: Position statement from the Joint SHEA and APIC Task Force. Am. J. Infect. Control 35, 73–85.

Wernitz, M.H., Swidsinski, S., Weist, K., Sohr, D., Witte, W., Franke, K.P., Roloff, D., Ruden, H., and Veit, S.K. (2005). Effectiveness of a hospital-wide selective screening programme for methicillin-resistant *Staphylococcus aureus* (MRSA) carriers at hospital admission to prevent hospital-acquired MRSA infections. Clin. Microbiol. Infect. *11*, 457–465.

Wielders, C.L., Vriens, M.R., Brisse, S., Graaf-Miltenburg, L.A., Troelstra, A., Fleer, A., Schmitz, F.J., Verhoef, J., and Fluit, A.C. (2001). Evidence for in-vivo transfer of *mecA* DNA between strains of *Staphylococcus aureus*. Lancet 357, 1674–1675.

Youmans, G.R., Davis, T.E., and Fuller, D.D. (1993). Use of chemiluminescent DNA probes in the rapid detection of oxacillin resistance in clinically isolated strains of *Staphylococcus aureus*. Diagn. Microbiol. Infect. Dis. *16*, 99–104.

Zhang, S.X., Drews, S.J., Tomassi, J., and Katz, K.C. (2007). Comparison of two versions of the IDI-MRSA assay using charcoal swabs for prospective nasal and nonnasal surveillance Samples. J. Clin. Microbiol. *45*, 2278–2280.

Genetic Manipulation of *Staphylococcus aureus*

5

Peter J. McNamara

Abstract

Genetic manipulation of *S. aureus* has provided an avenue for the discovery of new approaches to treat *S. aureus* infections, therapies that are based on a molecular understanding of virulence and pathogenesis. This chapter provides basic information on strains of *S. aureus* and the tools and techniques that have been successfully employed to engineer their genome. The topics that are covered include: choosing a strain for study; the cultivation and husbandry of *S. aureus*; vectors and their uses; methods to introduce DNA into *S. aureus*, and methods for the mutagenesis of *S. aureus*. With the information in this chapter, it is possible to design experiments to examine a broad range of biological phenomena that will contribute to the understanding of the biology of *S. aureus*, and the factors that make these bacteria such formidable pathogens.

Introduction

Studies of the molecular genetics of *Staphylococcus aureus* have centred on genes whose products confer virulence. As a consequence of *S. aureus* virulence to human health, from Sir Alexander Ogston's isolation of the organism from a surgical abscess in 1880 to the current spread of community-acquired multiple drug-resistant strains, a natural rationalization has been available for the study of this organism. While the effect of a particular gene and its products on virulence might seem limiting for understanding the basic biology of an organism, many of the cellular processes of *S. aureus* have been addressed. In addition to work on the specific adhesons, toxins, and enzymes that mediate interactions between *S. aureus* and a host, nucleic acid metabolism, the genetic organization of the genome, the nature of plasmids, phage, and transposable elements, the physiology of metabolic processes, the detection and response to stimuli, and the regulation of genes and their products have been explored. In examining staphylococcal virulence, much has been learned about various strains of *S. aureus* and many techniques have been developed for their genetic manipulation. Compared to laboratory strains of *Escherichia coli*, and many other pathogens, *S. aureus* can be difficult to genetically manipulate because of its intrinsic restriction modification systems, low transduction frequency and low efficiency of recombination. Methodologies have been developed to circumvent the problems encountered in working with *S. aureus*. This chapter discusses the choice of a strain for study, and current techniques for their cultivation and husbandry. This chapter also provides useful methods for engineering of the *S. aureus* genome.

Most of the methodologies developed for study of the structure and function of genes at a molecular level have been adapted for use in *S. aureus*. Forward genetics, an experimental methodology that begins with creating mutants and screening for those with a desired phenotype, which is then followed by identification of the mutated genes and determining their function, is a mainstay of *S. aureus* research. Tools are available to create *S. aureus* harbouring a transposon library; a large pool of bacteria, each with a unique chromosomal mutation that can be screened or selected for isolates that have lost (or gained) a desired phenotype. In addition, some transposons have been modified to carry reporter genes that can be used to identify genes that are expressed under specific environmental or genetic conditions, broadening the scope of biological phenomenon that can be examined. Genes identified by transposon mutagenesis can be cloned into a variety of plasmids designed for complementation or other studies of gene function. Reverse genetics, a methodology where the function of a gene is determined by creating a mutant bacteria, has been employed with *S. aureus*. Sophisticated tools for allelic exchange, a technique that can be used to make specific genetic mutations, have been developed and are now routinely used by many laboratories. Tools for forward and reverse genetic studies of essential genes, DNA that encodes proteins that are required for viable bacteria, have been developed for *S. aureus*. These methodologies are based on targetrons, iRNA, or the ectopic expression of genes in a corresponding null mutant. Using all the tools available to manipulate the genome, the function of an increasing number of staphylococcal genes is being experimentally established. These studies are providing a picture of the genetic underpinnings and their relationships for all the biological processes associated with *S. aureus*.

Strains of *S. aureus*: history, physiology and genomic content

The choice an appropriate strain of *S. aureus* for a particular line of research can be problematic. With the exception of lines of inquiry into specific clinical isolates (e.g. the community acquired *S. aureus* strain, USA300), investigators must be aware that genetic and physiological differences among strains can influence their experimental results and the generality of their conclusions. Furthermore, practical considerations such as phage or plasmid content, antibiotic resistance profiles, or the availability of specific genome sequences may limit the methodologies that can be used for the molecular genetics of a particular strain of *S. aureus* (i.e. it can be difficult to genetically manipulate many clinical isolates).

Historically, the justification of *S. aureus* research dictated that strains chosen for study displayed a pathogenic potential. This potential was typically manifest as an ability to effectively elaborate a specific virulence factor or to homogeneously expresses resistance to an antibiotic. However, some strains were chosen for further study solely because they were available. These strains often provided the reservoir for bacteriophages that were used in an early typing scheme developed for epidemiological monitoring (Table 5.1). Many of these prototype strains are still studied today and are available from the American Type Culture Collection (http://www.atcc.org) or the Network on Antimicrobial Resistance in *Staphylococcus aureus* (http://www.narsa.net).

To date derivatives of strain NCTC 8325 (a.k.a. 8325) have been the subject of the most detailed genetic studies of *S. aureus*. The most useful isolate, 8325–4 (a.k.a. RN450, RN6390, and RN6390B), was created by curing three prophages (Φ11, Φ12, and Φ13) from a spontaneous streptomycin-resistant mutant of 8325 (Novick, 1967). *S. aureus* 8325 and its derivatives gained favour as the host strains for molecular genetic experiments due to the relative ease of their genetic manipulation and the phenotypic consistency among their progeny. One strain in particular, RN4220 a chemical mutant of 8325-4, can accept plasmid DNA isolated from *E. coli* an important intermediate step in establishing sophisticated constructs in *S. aureus* (Kreiswirth 1983; Waldron & Lindsay 2006).

The relevance of studies using 8325 derivatives to decipher the molecular biology of pathogenesis is complicated by recent finding showing that these strains are genetically and physiologically different than various clinical isolates. In the eyes of many investigators, a specific mutation in 8325 that negatively affects the activity of the alternative *S. aureus* sigma factor (σ^B) has rendered this strain and its derivatives less desirable for study. Compared to clinical isolates, the mutation affecting σ^B in 8325 reduces its level of stationary phase growth (Shaw *et al.*, 2006), decreases its pigmentation (Horsburgh *et al.*, 2002), alters its stress responses (Chan *et al.*, 1998; Gertz *et al.*, 2000; Kullik *et al.*, 1997; Kupferwasser *et al*, 2003) and reduces its pathogenic potential (Bischoff *et al*, 2001; Bischoff *et al.*, 2004; Cheung *et al.*, 1999; Deora *et al.*, 1997;

Table 5.1 Prototypical strains of *S. aureus*

Strain	Phenotype
8325–4	Derivative of NCTC8325 devoid of bacteriophage and plasmids
Becker	Capsule producer
COL	Methicillin resistance
Cowan	Protein A producer
Foggi	Nuclease producer
FRI1187	TSST-1 producer
MN8	TSST-1 producer
NCTC 8325	Propagating strain for typing phage 47, staphylokinase producer
Newman	Fibrinogen-binding protein clumping factor producer
Reynolds	Type 5 capsular polysaccharide producer
S6	Enterotoxin B hyper-producer
TC82	β-Toxin producer
TC128	δ-Toxin producer
V8	Serine protease producer

Gertz *et al.*, 2000; Horsburgh *et al.*, 2002; Kullik *et al.*, 1998; Nair *et al.*, 2003; Renzoni *et al.*, 2004; Shaw *et al.*, 2004; Ziebandt *et al.*, 2001). As one might expect, these traits are common to 8325-derived strains. For one isolate of 8325–4, the mutation affecting σ^B activity has been seamlessly repaired (Horsburgh *et al.*, 2002). This strain, *S. aureus* SH1000, has been used by investigators to extend ongoing lines of research, although there is not universal agreement of the relevance of SH1000 to the study of *S. aureus* as a pathogen.

To foster the use of clinical isolates in *S. aureus* research, the genetic content of 8325–4 was compared to a recent clinical isolate, *S. aureus* strain UAMS-1 (Cassat *et al.*, 2005). In this comparison, genetic differences were observed. It was found that the clinical isolate encodes *cna* (the gene encoding an adhesin), but lacks *isaB* (a gene encoding an immuno-dominant antigen), *sarT* and *sarU* (two genes encoding MarR-family transcriptional regulators), and *sasG* (a surface adhesin). A different genetic content for virulence factors and their regulators among strains is to be expected (Iandolo, 1989), and these differences appear to be significant in terms of pathogenesis. Furthermore, mutations in regulatory genes created in 8325-4, other prototypical isolates, and newly isolated clinical strains can result in dissimilar phenotypes (Cassat *et al.*, 2005). Genomic sequence comparisons of *S. aureus* have shown that chromosomal variations play an important role in the evolution of this pathogen (Lindsay and Holden, 2006), a finding that validates studying different strains.

Physiologically, 8325 and its derivatives may display activities deviating from those seen in recently isolated clinical strains. For example, the 8325–4 isolate RN6390 has reduced aconitase activity compared to clinical isolates of *S. aureus* (Somerville *et al.*, 2003). From the limited data obtained using *S. aureus* and by analogy to other organisms, the observed reduction in aconitase activity may affect pathogenesis by known links between the tricarboxylic acid cycle, virulence factor production, and bacterial survival (Dassy *et al.*, 1996; Gerard *et al.*, 2002; Lorenz *et al.*, 2001; McKinney *et al.*, 2000; Somerville *et al.*, 1999; Somerville *et al.*, 2002; Sonenshein *et al.*, 2002).

Given the differences among strains and the varied importance of these differences to a specific line of inquiry, practical considerations often take precedence over other criteria when selecting a strain of *S. aureus* for molecular genetic experiments. The ability to meaningfully genetically manipulate bacteria presumes that the target strain encodes the desired molecules, manifests the desired biological activity or trait, and is amenable to the techniques needed for genetic manipulation. The latter point is considered below.

When a suitable strain is found, one must consider its independently replicating and chromosomally integrated plasmid content, as well as their complement of lysogenic bacteriophage. In addition to complicating the results from such basic methodologies as chromosomal DNA isolation, restriction mapping, or *trans*-complementation of a mutation, plasmids frequently confer antibiotic resistance to potential host strains. This intrinsic resistance can severely limit the scope of genetic manipulation of *S. aureus*, an organism with very few useful selectable genetic markers (Table 5.2). Bacteriophage content can be a further consideration for the choice of a strain. Most strains of *S. aureus* harbour multiple bacteriophage, typically they are integrated into specific sites within the genome (Lee *et al.*, 1988). Experiments requiring bacteriophage mediated transduction of phenotypic traits (an intermediate step in many methodologies for genome manipulation) can be influenced by the bacteriophage content of the target cell. While a number of similar generalized transducing bacteriophage are available for mobilization of staphylococcal markers, susceptibility of recipient strains to commonly used transducing bacteriophages is not universal (Novick, 1967). At least in part, the mechanism of bacteriophage immunity involves the nature of intrinsic host prophage. Thus, strains lacking bacteriophage are more likely to undergo transduction with commonly available bacteriophage than strains with intrinsic bacteriophage.

Another consideration of bacteriophage content of strains is lysogenic conversion, the gain and/or loss of specific phenotypic traits that is mediated by the chromosomal integration of a bacteriophage that may carrying an accessory gene or genes. While lysogenic conversion can occur silently (Jan *et al.*, 1972; Lee *et al.*, 1985; Pattee, 1990), this process can disrupt a normally expressed chromosomal gene. Multiple conversions due to site specific integration of bacteriophage within the structural gene for β-toxin with the concomitant introduction of genes encoding a variety of virulence factors has been well documented (Betley *et al.*, 1885; Coleman *et al.*, 1989; De Haas *et al.*, 2004; Iandolo *et al.*, 2002; Rooijakkers *et al.*, 2005; Winkler *et al.*, 1965). Awareness of the capacity for lysogenic conversion may help explain aberrant or unexpected results obtained after transduction or transformation of certain strains of *S. aureus*.

A further consideration of the bacteriophage content of *S. aureus* is that prophage have been directly shown to influence biological processes, including pathogenesis. In an example that is poignant for staphylococcal research, the pathogenic potential of *S. aureus* can be dependent on its bacteriophage content. *S. aureus* strain Newman is normally lysogenized by four bacteriophage: ΦNM1, ΦNM2, ΦNM3, and ΦNM4. Strains of Newman lacking either ΦNM1 or ΦNM3, or cured of ΦNM2 and ΦNM4, all displayed virulence defects in a murine model of infection. In this same animal model, Newman that lacks all four bacteriophage was shown to be avirulent (Bae *et al.*, 2006).

Until recently, no strain of *S. aureus* was completely characterized with regard to its genetic content. As late as 1990, only about a hundred phenotypic markers, many of which were silent transposon insertions, had been mapped by classical genetics means and pulse-field gel electrophoresis analysis (Pattee, 1990). Since the beginning of this decade, the genomic DNA sequences (chromosomal and plasmid) of a number of strains have been completely elucidated, annotated, and placed in the pubic domain (Chapter 1). Notably, the genomic sequence of two prototypical strains, 8325 and COL, are available. These strains, or derivatives of these strains, are often preferred by researchers for experiments that require extensive manipulation of the staphylococcal genome.

S. aureus: growth media and husbandry

S. aureus can be grown using a variety of media at temperatures up to 45–50°C, with 35–37°C being optimal. Various media for the selection, routine cultivation, and nutrient-defined growth of *S. aureus* are available. Staphylococci can be enriched from biological samples containing a mixed population of bacteria on selective/differential medium including mannitol salt agar or Baird-Parker Agar (Baird-Parker, 1962). Newer formulations of selective/differential media, such as DOTEMSA (Zadik *et al.*, 2001) and CHROMagar Staph aureus (Samra *et al.*, 2004)

Table 5.2 Commonly used antibiotic resistance markers for *S. aureus*

Antibiotic markers							
Gene	Resistance[2]	Activity	Resistance Level (μg/ml)	Selective Cconcentration (μg/ml)	Inducible expression	Lag time (h)	GenBank accession no.
aacA-aphD	Gm^r, Km^r, Tm^r	Phospho- and acetyltransfrease	10	6	–	0	AY541446
BlaIR1Z	Pc^r	β-Lactmase/inactivation	0.1	0.1	+	0	AB245469
cat	Cm^r	Acetylase	10	5	+	0	NC_002013
ermB	Em^r	Ribosome methylase	20	5	–	0	Y13600
ermC	Em^r	Ribosome methylase	>1000	5	+ and –	0	NC_005908 Y15273 Y15274 Y17294 Y18018
linA	Lm^r	Adenylylation	64	5	–	2	AY541446
tetK	Tc^r	Efflux pump	25	3–5	+	1	S67449
tetM	Tc^r	Altered ribosomes	25–65	5	–	0	M21136

Modified from Novick, 1991.

[2]Gm^r, gentamycin resistance; Km^r, kanamycin resistance; Tm^r, tobramycin resistance, Pc^r, penicillin resistance; Cm^r, chloramphenicol resistance, Em^r, erythromycin resistance; Lm^r, lincomycin resistance; Tc^r, tetracycline resistance.

are also commonly used. Commercially available (e.g. Difco Laboratories, Oxoid, BBL) blends of Brain Heart Infusion, Tryptic Soy, Columbia, Mueller–Hinton, Penassay (Antibiotic Medium #3) and nutrient broths or semi-solid media can be used in the routine cultivation of *S. aureus*. CY medium [10 g/l casamino acids, 10 g/l yeast extract, 5 g/l glucose, 5.9 g/l NaCl with 40 ml/l 1.5 M β-glycerophosphate] has also been used for the routine cultivation. Several synthetic growth media have been developed for *S. aureus* (Charles *et al.*, 1976; Laue *et al.*, 1968; Patte *et al.*, 1975; Stringfellow *et al.*, 1991; Wu *et al.*, 1971). Each formulation of chemically defined medium has specific intrinsic characteristics including composition, ultraviolet absorptivity, and a variable ability to promote growth or support production of a desired molecule. For a particular line of inquiry, either rational or empirical choices may be required for the selection of an appropriate synthetic medium.

As with other genetic stock organisms, strains of *S. aureus* can be maintained for prolonged periods of time as viable bacteria frozen in media containing 30% vol./vol. glycerol at −80°C. With both clinical and laboratory isolates, care should be taken to select low-passage organisms to prevent the loss of desired traits (Björklind *et al.*, 1980; Somerville *et al.*, 2002). Strains should be checked for desired traits after freezing because there have been reports of genotypic and phenotypic changes displayed by frozen stocks upon cultivation (Novick *et al.*, 1993).

Nucleic acid isolation and the curing of mobile genetic elements

In addition to the chromosome, the genomic content of *S. aureus* can include plasmids, bacteriophage, transposons, insertion sequences, and other mobile and non-mobile elements. The *S. aureus* genome contains on the order of 2.7–2.9 mega-base pairs of chromosomal DNA with an approximate G/C content of 32.8%. The emerging picture from the publicly available chromosomal genomic sequences is that of a large highly conserved core set of genes interspersed with strain-specific segments of genetic diversity (Baba *et al.*, 2002; Holden *et al.*, 2004; for additional information see Chapters 1 and 3). The content of integrated plasmids and resident bacteriophage, and to a lesser degree other mobile elements, have been characterized in several of the sequenced strains. Many on-line tools are available for analysis of whole genome sequences. These tools can be found at The Institute for Genomic Research (TIGR) Comprehensive Microbial Resource homepage (http://cmr.tigr.org) and at the National Center for Biological Information (NCBI) Entrez Genomes homepage (http://www.ncbi.nlm.nih.gov).

The isolation of whole-cell DNA

Genetic manipulation of an organism often begins with the isolation of high-quality whole-cell DNA. Whole-cell DNA can be isolated from *S. aureus* using a number of procedures; here, three procedures are described. The first procedure is for the small-scale isolation of DNA and can be used to isolate nucleic acids from large numbers of samples. The second and third procedures are for the large-scale isolation of DNA from a limited number of samples. The large scale procedures differ in that one procedure uses phenol chloroform purification and the other uses a cesium chloride step gradient to purify the DNA.

Until recently, it has been difficult to reliably isolate small quantities of pure whole-cell DNA from *S. aureus*. Newer DNA purification matrices, available in kit form, have made the small-scale purification of chromosomal DNA from *S. aureus* routine. While several commercially available kits for the isolation of whole-cell DNA can be used, the QIAamp DNA Blood Mini or DNeasy Kits, (Qiagen, GmbH, Germany) are recommended. In using either kit, follow the protocol for the isolation of DNA from Gram-positive bacteria. To facilitate lysis of the bacteria, 5×10^9 *S. aureus* are suspended in 200 ml of 20 mM Tris-HCl (pH 8.0), 2 mM EDTA, 1.2% Triton, 200 μg/ml lysostaphin for 30 min at 37°C. Lysostaphin is a glycylglycine endopeptidase which specifically cleaves the pentaglycine cross-bridges found in the staphylococcal peptidoglycan and is commercially available from Sigma Chemicals (St. Louis, MO, USA) or AMBI Products (Lawrence, NY, USA). The expected yield from this procedure is approximately 10–20 μg DNA. However, to obtain pure DNA, DNase-free RNase treatment of samples is necessary because the Qiagen kits purify total nucleic acids. At any rate, the resulting DNA is sufficiently pure for cloning, transformation PCR applications, and hybridization methodologies.

A large-scale whole-cell DNA isolation procedure has been described by Novick (1991). This procedure lyses *S. aureus* with lysostaphin and in a classical manner, purifies whole-cell DNA by phenol extraction and ethanol precipitation. The main advantages of this procedure are the relatively short period of time it takes to complete, and that no ultracentrifugation is involved. A 20 ml culture of *S. aureus* in CY medium (see *S. aureus*: Growth Media and Husbandry) is grown to a Klett value of 250 (optical density at 540 nm of 0.5). The cells are harvested by centrifugation and the pellet is washed in TES Buffer [20 mM Tris-HCl (pH 7.5), 50 mM NaCl, 10 mM EDTA]. The pellet is then suspended in 5 ml of TES buffer containing 25 μg/ml lysostaphin and incubated at 37°C for 1 h. After, 500 μg/ml pronase is added and the cells are further incubated for an additional 1 h at 37°C. The DNA is extracted with equal volumes of buffer saturated phenol, then with 25:1 (vol./vol.) chloroform-isoamyl alcohol (Sambrook *et al.*, 1998). The DNA is then precipitated by the addition of one-tenth volume of 3 M sodium acetate (pH 5.4) and two volumes of ice-cold ethanol. Contaminating salts are then removed using a 70% ethanol wash. The DNA is pelleted, briefly dried, suspended in 1 ml of TE buffer [10 mM Tris-HCl (pH 7.5), 1 mM EDTA] containing 0.2 units RNase, and incubated for 1 h. at 37°C. The DNA is again ethanol precipitated, and then it is suspended in 0.5 ml of TE. This procedure yields milligram amounts of pure DNA.

Perhaps the best procedure for the isolation of a large amount of very pure whole-cell DNA from *S. aureus* was described by Dyer and Iandolo (Dyer *et al.*, 1993). In this procedure (Table 5.3), the *S. aureus* cell wall and membrane is disrupted using lysostaphin and sarkosyl. Then a chaotropic agent is added to facilitate the solubilization of contaminating proteins, and the DNA is purified on a caesium chloride step gradient. A step gradient can be readily made by using a pipette to add the 5.7 M CsCl solution directly to the bottom of a centrifuge tube containing the 2.85 M CsCl solution. As the 5.7 M CsCl is added, the lighter 2.85 M CsCl solution will float on top of the heavier liquid, creating a discrete interface between the two concentrations of salts. The DNA is precipitated using ice-cold ethanol. As the solution warms, the DNA traps bubbles and floats to the top of the container. Occasionally, the precipitated DNA does not float to the top of the alcohol solution. If the DNA solution does not float, remove the DNA from the ethanol, dry briefly, and suspend in 5 ml of TE [10 mM Tris-HCl (pH 7.5), 1 mM EDTA) containing 10 μg/ml Proteinase K. The DNA solution is incubated at 37°C for 2 h overnight and then ethanol is precipitated. Up to 2 mg of DNA that is sufficiently pure for any subsequent manipulation can be obtained using this procedure.

Plasmids in *S. aureus*

Most strains of *S. aureus* contain native plasmids. These plasmids have been catalogued into four general classes labelled I through IV (Novick *et al.*, 1989; Firth *et al.*, 2000). Class I plasmids are small (< 5 kb), present in multiple (15 to 60) copies per cell, and generally encode resistance to a single antibiotic or heavy metal. Class II plasmids are 10 kb to 30 kb in size, have a low copy number (4 to 6 per cell), and can encode inducible resistance to β-lactam and macrolide antibiotics, as well as assorted heavy metals. Class III plasmids are large (30 to 60 kb), low copy (4 to 6 per cell), encode multiple resistance markers, and in general are capable of conjugative transfer among bacteria. Class II and III plasmids often harbour composite transposons, and many of their resistance markers can mobilize from the plasmid backbone to other replicating deoxyribonucleic acids. Finally, plasmids whose biology remains to be elucidated have been placed into class IV.

The biochemistry of staphylococcal plasmids has been extensively studied and excellent reviews have been published (Firth *et al.*, 2000; Lyon *et al.*, 1987; Novick, 1987; Novick, 1989; Russell, 1997; Thomas, 1988). The reader is referred to these articles for information on the genetic content, structural relationships, and mechanisms of replication for specific plasmids. In addition, the genetics of antibiotic and heavy metal resistance markers has been reviewed (Novick *et al.*, 1968;

Table 5.3 The Isolation of chromosomal DNA from *S. aureus*

1	Pellet cells from a 20 ml overnight culture of *S. aureus* for 15 min at 4°C at 5000 × *g*.
2	Wash the pellet in 30 ml TES Buffer [20 mM Tris-HCl (pH 7.5), 50 mM NaCl, 10 mM EDTA]. Centrifuge as in step 1.
3	Suspend the pellet in 5 ml High Salt TES Buffer [20 mM Tris-HCl (pH 7.5), 2.5 M NaCl, 10 mM EDTA].
4	Transfer the culture to a 50 ml flask and add lysostaphin to final concentration of 100 µg/ml. Initially incubate the flask with gentle rocking at 37°C for 30 min. and then transfer the solution to 55°C for an additional 15 min.
5	Add 0.5 ml of 20% sarcosyl and gently agitate the solution to facilitate lysis.
6	Add 3.69 g guanidine HCl (a final concentration of 7 M) and incubate the flasks at 60°C for 1 h or until the solution has cleared.
7	Set up a CsCl step gradient in tubes for the Beckman SW-41 rotor (or an equivalent swinging bucket rotor) using 2 ml of 2.85 M CsCl layered on top of 2 ml of 5.7 M CsCl.
8	Add 1.66 ml 18 MΩ water to the DNA-containing solution and overlay CsCl gradient with 9 ml of the solution.
9	Centrifuge at 30,000 rpm at 20°C for 24 h.
10	Clamp the centrifuge tube in a ring stand. Remove about 5 ml of solution from the top of the tube (do not disturb the CsCl layers). Poke a hole in bottom of tube with 18 gauge needle and collect the viscous fraction (1–2 ml).
11	Increase the volume of the viscous fraction to 5 ml with 18 MΩ and pour the DNA solution into a 25 ml graduated cylinder containing 20 ml of ice-cold 95% Ethanol. As solution warms, bubbles collect in the DNA allowing it to float to the top.
12	Retrieve DNA using pasteur pipette with hook on end, dip in 70% ethanol, dry briefly, and resuspend in 0.2–0.5 ml TE [10 mM Tris-HCl (pH 7.5); 1 mM EDTA].

Lyon *et al.*, 1987). Several class I plasmids and elements from class II plasmids have been modified for the genetic manipulation of *S. aureus* (These plasmids are discussed under Engineered cloning and shuttle plasmids).

The isolation of plasmid DNA

Plasmids can be isolated from *S. aureus* by alkaline lysis using procedures that have been modified to allow for disruption of the staphylococcal cell wall. Most plasmids can be isolated in closed covalently closed circularized (CCC) form; however, other forms are often isolated in plasmid-dependent manner in varying amounts.

Commercially available plasmid DNA mini-, midi-, and maxi-preparation kits can be used to isolate plasmid DNA if *S. aureus* is incubated with lysostaphin prior to lysis of the bacteria with detergent. Many investigators routinely use Qiagen Plasmid Kits (Qiagen, GmbH, Germany). In using these kits, the bacteria are harvested by centrifugation and suspended in Qiagen P1 Buffer. To the cell solution 100 µg/ml lysostaphin is added, and the cell wall is digested during a 30 min incubation at 37°C. After the lysostaphin step, the cells are lysed under alkaline conditions, the solution is neutralized, and the DNA is purified as recommended by the manufacturer. Small-scale plasmid preparations are suitable for PCR and restriction endonuclease analysis, subcloning and transformation, as well as other DNA modifying procedures. Midi- and maxi-preparations can be used to increase the amount of isolated plasmid.

Plasmids can be readily isolated using a small-scale alkaline lysis procedure (Table 5.4). In this procedure, the bacteria are treated with lysostaphin in the presence of sucrose. Sucrose is used to osmotically stabilize the bacteria, preventing premature lysis and the release of intracellular proteases that can degrade lysostaphin. Most alkaline lysis mini-preparations include a phenol purification step. Although sometimes unnecessary, this step can be quickly accomplished and increases the likelihood that the purified DNA will be amenable to further manipulation.

Occasionally, it is necessary to obtain relatively large amounts of CCC plasmid DNA

Table 5.4. *S. aureus* alkaline lysis plasmid preparation

1	Pellet cells from 2–10 ml overnight culture of *S. aureus* for 2 min in a microfuge or for larger volumes 15 min. at 4°C at 5,000 × *g* in a clinical centrifuge.
2	Remove all media and suspend the pellet in 1 ml 3% NaCl. If necessary transfer to an Eppendorf tube.
3	Microfuge 2 min., discard the NaCl solution, and suspend the pellet in 150 µl SET [20 mM Tris-HCl (pH 7.0), 50 mM EDTA, 0.58 M sucrose]. Add 15 µl of 10 µg/ml lysostaphin and incubate for 30 min at 37°C.
4	Add 300 µl of freshly made Lysis Solution [0.1 N NaOH, 1% sodium dodecyl sulfate]. Mix and let stand on ice 5 min.
5	Add 225 µl 1.5 M potassium acetate (pH 4.8). Mix and place on ice for 5 min.
6	Microfuge 10 min in a cold microfuge. Transfer the supernatant to a new tube and extract with 500 µl of phenol–chloroform–isoamyl alcohol (25:24:1 vol:vol:vol). Vortex to make an emulsion.
7	Microfuge 1 min in a room temperature microfuge and transfer the supernatant to a new tube.
8	Add 2 volumes of 100% ethanol. Mix well and let stand at room temperature 10 min.
9	Microfuge 10 min in a room temperature microfuge and remove the supernatant.
10	Add 1 ml 70% ethanol. Vortex briefly and repeat step 9.
11	Remove the supernatant and dry the pellet. Suspend the pellet in 25 µl TE [10 mM Tris-HCl (pH 7.5), 1 mM EDTA].

(e.g. preparation of plasmid stocks) or to obtain a modest amount of a low copy number plasmid. In our laboratory, we use a classical large-scale alkaline lysis plasmid preparation where the DNA is purified by sedimentation on a cesium gradient. While time consuming and involving ultracentrifugation, our procedure routinely yields between 100 µg to 1 mg of pure CCC DNA. For *S. aureus*, one litre of bacteria containing the plasmid of interest is divided between four 250 ml bottles and harvested by centrifuge at 5,000 × *g* for 15 min. at 4°C. The supernatant fluids are discarded and the bacterial pellets washed in a total of 400 ml of 3% NaCl. The 3% NaCl solution is discarded, and the bacterial pellets are suspended to a final volume of 36 ml in SET [20 mM Tris-HCl (pH 7.5), 2.5 M NaCl, 10 mM EDTA]. The bacterial suspension is then evenly divided among four 50 ml Erylenmeyer flasks and 90 µl of a 10 mg/ml solution of lysostaphin is added to each flask. The flasks are then incubated with gentle stirring for 30 min. at 37°C. After the incubation period, 18 ml of freshly made lysis solution (0.1 N NaOH, 1% sodium dodecyl sulfate) is slowly mixed into each flask. The flasks are incubated on ice for 10 min, then 13.5 ml of ice-cold 1.5 M potassium acetate (pH 4.8) is added to each flask, the mixture is shaken, and allowed to stand on ice for ≥10 min. The content of each flask is transferred to a SW-28 centrifuge tube (Beckmann Coulter, Palo Alto, CA) and centrifuged at 141,000 × *g* for ≥1 h at 4°C. The plasmid DNA in the supernatant fluids is recovered into four centrifuge tubes and precipitated for 30 min. at room temperature using 0.6 volumes isopropanol. The DNA is recovered by centrifugation at 12,000 × *g* for 30 min at 26°C. The pellets are suspended in 70% ethanol and collected by centrifugation, again at 12,000 × *g* for 30 min at 26°C. Each pellet is suspended in 5 ml of TE [10 mM Tris-HCl (pH 7.5), 1 mM EDTA], and 2.5 ml of 7.5 M ammonium acetate is added to each tube. The solution is centrifuged at room temperature for 10 min at 10,000 × *g* and supernatant is collected into a clean tube and precipitated by the adding of 2 volumes of ethanol and placing the tubes at −20°C for ≥ 30 min. The precipitate is collected by centrifugation, briefly dried, and suspended in a total of 4 ml TE. A total of 8.8 g of CsCl is dissolved in the solution and 800 µl of a 10 mg/ml solution of ethidium bromide is added. The solution is placed in 0.5" × 2' ultracentrifuge tubes. The tubes are heat sealed and then centrifuged at the maximum speed for 4.5 h at 20°C using a VTi65 rotor (Beckman Coulter, Fullerton, CA, USA). The tubes are vented at the top, and the lower CCC plasmid band, visualized by irradiation with UV light, is removed from the gradient using a 16-Ga needle. The ethidium bromide is removed from the solution by repeated extraction with 1-butanol saturated with HPLC-grade water (18 MΩ). Finally, the plasmid solution is dialysed overnight against 2 litres of TE (pH 8.0).

The elimination of plasmids from *S. aureus*

The genetic manipulation of bacteria is greatly facilitated by using strains that are initially plasmid-free and sensitive to antibiotics or other selective agents. As previously noted, most clinical isolates of *S. aureus* harbour plasmids, and therefore, the first step in genetic engineering of clinical isolates is often to cure these strains of their resident plasmids. Plasmids have been eliminated from *S. aureus* by growth of the bacteria at an elevated temperature (i.e. 42–43°C) (Udo *et al.*, 2001), or by growth in the presence of a range of concentrations of ethidium bromide (Dajani *et al.*, 1974; Silhavy *et al.*, 1984), or by using acridine orange (Dajani *et al.*, 1974; Silhavy *et al.*, 1984), L-ascorbic acid (Amabile-Cuevas, 1988; Amabile-Cuevas *et al.*, 1991), or trace levels of sodium dodecyl sulfate (Sonatein *et al.*, 1972). The regeneration of protoplasts (see Protoplast transformation of *S. aureus*) has also been shown to be useful in eliminating class I and class II plasmids from *S. aureus* (Abdell-Galal *et al.*, 1988; Gruss *et al.*, 1986). Finally, plasmids can be reliably eliminated from *S. aureus* using a second plasmid that carries a unique selectable marker and an incompatible thermo-sensitive replicon. The resident plasmid is eliminated by displacement, selecting for a newly introduced marker. The introduced plasmid is then eliminated by curing at the non-permissive replication temperature (Novick, 1967; Novick, 1987). Several of the referenced methods involve the use of known mutagens or are generally perceived to be mutagenic. As a result, plasmid-cured strains should be examined, at least as much as possible, for concomitant mutations. In addition, staphylococcal plasmids can encode resistance to ethidium bromide (Heir *et al.*, 1999), lysostaphin (Grundling *et al.*, 2006), as well as other antibiotics that may be required for a given plasmid-curing methodology. Therefore, for specific strains the most expedient plasmid-curing technique may need to be empirically determined.

The isolation of total RNA from *S. aureus*

RNA is commonly isolated for the analysis of strains of *S. aureus*. Procedures for the chemical extraction [e.g. hot phenol, RNA Bee (IsoTex Diagnostics, Inc. Friendswood, TX, USA), or Trizol (Invitrogen, Madison, WI, USA)] of RNA have been adapted for use with *S. aureus*. However, most investigators use commercially available RNA isolation kits. The RNAeasy Kit (Qiagen, GmbH, Germany) can be used according to the manufacturer's instructions, although these instructions must be modified so that the RNA is collected from at least 20 Units of cells as measured at an optical density 600 nM using a path length of 1 cm. The bacteria are incubated in 100 μl of TE (pH 7.5) containing 100 μg/ml lysostaphin at 37°C for 30 min. to facilitate the initial lysis step. The fastest, most reliable, and easiest procedure to isolate RNA uses the FastRNA Blue Kit (QIOgene, Irvine, CA, USA). The only disadvantage of this technique is that it requires a mechanical cell disruptor (Thermo Electron Corporation, Waltham, MA). Regardless of the RNA isolation procedure that is used, RNA from multiple samples of bacteria isolated over time can be processed together by stabilizing the RNA using RNAprotect (Qiagen) or the phenol/guanidine isothiocyanate solutions (RNA BEE or Trizol). These solutions eliminate the need for freezing cell pellets in liquid nitrogen or a dry ice/acetone bath. Stabilization of RNA in samples is important because the half-life of many bacterial transcripts is very short, sometimes less than one minute.

Bacteriophage in *S. aureus*

Temperate phages of the family Siphoviridae play an important role in the biology of *S. aureus*, and most wild-type strains of *S. aureus* are lysogenized with bacteriophage. *S. aureus* bacteriophage nomenclature can be confusing. The bacteriophage have been grouped by serotype (A, B, and F), by *Hin*dIII-restriction endonuclease digest analysis of genomic DNA of lysogenic bacteria of the International Typing Set (A, Ba, Bb, Bc, F), and by host restriction-modification-dependent lytic groups (I, II, III, and miscellaneous {IV}). Over thirty complete genomic sequences of *S. aureus* bacteriophages have been determined, with the work of Kwan *et al.* accounting for of 27 sequences (Kwan *et al.*, 2005). This information has been used to confirm the three group classification system that was deduced by classical methods (Kwan *et al.*, 2005). Furthermore these data showed that more than half of the annotated bacteriophage genes encode proteins of unknown function. Despite this fact, the Kwan study, as well as data from other sequenced *S. aureus* bacteriophage genomes, gives a molecular description of the architecture of *S. aureus* bacteriophages beyond their being circularly permuted, terminally redundant, and flush ended. *S. aureus* bacteriophage have not yet been modified for use as cloning vectors, nor has *in vitro* packaging of DNA using *S. aureus* bacteriophage proteins been developed.

Within the *S. aureus* genome, prophages are often resident at specific chromosomal sites and they can be responsible for lysogenic conversion – the bacteria either gaining or losing a specific phenotype or multiple phenotypes. For example, it has long been known that a staphylococcal phage disrupts the structural gene encoding glycerol ester hydrolase (*geh*), causing loss of lipase activity (Rosendal *et al.*, 1964). In another example, exfoliative A and Panton–Valentine leucocidin activity is conferred to *S. aureus* by bacteriophages that carry either *eta* or *lukS/lukF-PV* (Yamaguchi *et al.*, 2000; Vijver *et al.*, 1972). More complex patterns of lysogenic conversion have also been documented. For instance, a single type of bacteriophage is responsible for the introduction of *entA* (*sea*) and *sak* and the gain of enterotoxin A and staphylokinase activities. The integration site for this same bacteriophage is within *hlb*, and lysogenization confers the loss of β-toxin activity (Coleman *et al.*, 1989). In addition to virulence determinants, *S. aureus* bacteriophage can mobilize class I plasmids and transposable elements by incorporating them within their genome (for more information see Chapter 4).

The isolation of bacteriophage DNA

The isolation of *S. aureus* bacteriophage DNA is a two-step process. In the first step, virus particles are purified from a liquid culture of bacteria lysogenized with the bacteriophage of interest. In the second step, DNA is purified from the isolated virus particles. To isolate virus particles, an overnight culture of bacteria is used for a 1:100 inoculation of 500 ml of Tryptic Soy broth supplemented with $CaCl_2$ (5 mM final concentration). The bacteria are grown at 37°C with rotary aeration to an optical density at 600 nm (1 cm path length) of 1.5. When the bacteria have reached the required density, 1 μg mitomycin C per ml of culture is added, and bacterial cultivation is continued for an additional 5 h. Bacteriophage-containing supernatant fluids are recovered by centrifugation at 15,000 × *g* for 20 min at 4°C. The cleared lysate is then treated with 10 μg/ml each of DNase I and RNase A for 1 h at room temperature. After, the lysate is passed through a 0.22 μm cellulose acetate filter, and polyethylene glycol 6000 and NaCl are added to final concentrations of 10% and 0.7 M, respectively. Viral particles are allowed to precipitate for 12 h at 4°C after which they are collected by centrifugation at 3,000 × *g* for 20 min. The bacteriophage are suspended in SM Phage buffer without gelatin [50 mM Tris-Cl (pH 7.0), 0.1 M NaCl, 8 mM $MgSO_4$, 5 mM $CaCl_2$]. The bacteriophage solution is extracted with an equal volume of chloroform, and the bacteriophage are harvested by ultracentrifugation at 35,000 × g for 2 h and the pellet is suspended in TE (pH 7.5) with 50 mg/ml of Proteinase K. Next, sodium dodecyl sulfate is added to final concentration of 0.5% and the suspension is incubated for 1 h at 65°C to disrupt the viral particles. After, bacteriophage DNA can be purified by phenol extraction and ethanol precipitation using standard methods (Ausubel *et al.*, 2007).

Eliminating bacteriophage from *S. aureus* using ultraviolet light

Curing *S. aureus* of chromosomally resident bacteriophage has been used in experiments to determine the phenotypic effect of prophages, as well as for providing strains that are amenable to generalized transduction. To cure *S. aureus*, bacteriophage-containing bacteria are grown from a 1:100 dilution of an overnight culture to an optical density at 660 nm (1 cm path length) of 0.4–0.5 in 25 ml aliquots of Tryptic

Soy broth. One 25 ml sample of bacteria will be required for each dosage of ultraviolet light. Typically aliquots of bacteria receive one of ten different light dosages, a number that is generally sufficient to obtain bacteriophage-free bacteria. The cells are harvested by centrifugation at 5,000 × *g* for 15 min at 4°C using an SS34 rotor (Thermo Electron Corporation, Waltham, MA, USA). The cell pellet is suspended in 12.5 ml of 0.15 M NaCl, and the solution is placed inside a covered sterile glass Petri dish. With the cover removed from the Petri dish, the cell solution is irradiated for different amounts of time using a 15 W germicidal lamp placed 50 cm from the sample. The delivered dosage is approximately 1,200 erg/mm^2/min. Serial dilutions of each sample are plated on Tryptic Soy agar and grown overnight. *S. aureus* from plates with a 1–10% survival rate after irradiation are patched onto fresh Tryptic Soy agar plates supplemented with $CaCl_2$ (5 mM final concentration). These bacteria are overlaid with 4 ml of Tryptic Soy soft agar (3 g/L Bacto-agar with $CaCl_2$) containing 200 µl of a 1:3 dilution of an overnight culture of a bacteriophage-susceptible strain of *S. aureus*. Typically, *S. aureus* strains RN4220 (Kreiswirth 1983; Waldron and Lindsay 2006) or 80CR3 (Stobberingh and Winkler, 1977) are used because they are restriction and modification deficient and are sensitive to a wide range bacteriophage types. After overnight growth at 37°C, plates are scored for those colonies that do not have a clear zone above them. These colonies are composed of bacteria that are potentially cured of bacteriophage.

Introduction of plasmid DNA into *S. aureus*

DNA can be introduced into *S. aureus* by electroporation, protoplast transformation, bacteriophage-mediated transduction, conjugation, or by natural competence. As with other microorganisms, host restriction barriers play a role in acceptance of foreign DNA by *S. aureus*. Elucidation of the sequence of the *S. aureus* genome has allowed for the identification of several restriction and/or modification enzymes (see REBASE (http://rebase.neb.com/rebase/rebase.html) for a current listing of restriction endonucleases and methylases (Roberts *et al.*, 2007)). In *S. aureus*, few of these enzymes have been studied in detail; however, it is known that the genome encodes both type I and II restriction-modification (R-M) systems (Godany *et al.*, 2004; Sjöström *et al.*, 1978; Waldron *et al.*, 2006). To circumvent the restriction systems, strains of *S. aureus* (*e.g.* RN4220 or 80CR3) have been developed that accept foreign DNA, including shuttle vectors propagated in *E. coli* (Chapter 3). Typically, plasmid purified DNA is introduced into *S. aureus* by electroporation, although protoplast transformation remains a viable technique. Class III plasmids (plasmids encoding *tra* functions) can be mobilized into *S. aureus* by conjugation. In *S. aureus* conjugation occurs at low frequency and only with a restricted set of vectors. This technique has largely been supplanted by the use of generalized vectors, electroporation, and transduction. Therefore, conjugation will not be covered in this chapter, but information on conjugation in *S. aureus* can be obtained in a review by Novick (1991). Furthermore, unlike the case of *Bacillus subtilis*, natural competence *S. aureus* has not been developed as a viable technique for the introduction of DNA. Therefore, competence is not covered in this chapter. Finally, plasmids are routinely transferred at high-frequency among strains by *S. aureus* by transduction. This important technique is covered under its own heading (see Bacteriophage-mediated transduction of *S. aureus*, below).

Electroporation of *S. aureus*

Electroporation uses a large electric pulse to disturb the cell membrane, allowing for the passage of polar molecules such as DNA. The ease, rapidity, and efficiency of electroporation have made it the preferred method for the introduction of plasmids into *S. aureus*. Several procedures for the electroporation of *S. aureus* have been published (Augustin *et al.*, 1990; Kraemer *et al.*, 1990; Lee, 1995; Schenk *et al.*, 1992). In each case, the basic protocol involves washing mid-exponential phase *S. aureus* of salts using a sucrose or glycerol solution and concentrating the bacteria to 1–3 × 10^{10} cfu/ml. The bacteria can either be used immediately or stored frozen at –80°C for later use. Electroporators and electroporation cuvettes are commercially available from several vendors. The settings for electric pulse necessary to transform *S. aureus* are machine (e.g. fixed versus adjustable resistance) and cuvette specific (i.e. 0.1 versus 0.2 cm gap), but generally fall into the range of 100 ohm resistance, 25 μF capacitance, and 2–2.5 kV charging capacity. After a voltage has been passed through *S. aureus*, the bacteria are placed in an isotonic medium and allowed to recover. Detailed instructions for electroporation of *S. aureus* based on the report of Krammer and Iandolo (Krammer *et al.*, 1990) is presented in Table 5.5. For small high-copy number CCC vectors, this procedure routinely results in 10^5–10^6 transformants per μg DNA. The use of a different growth medium, without a post-electroporation recovery media, has been reported to produce transformation efficiencies of up to 4×10^8 transformants per μg DNA (Schenk *et al.*, 1995). While the outlined procedure for electro-transformation of *S. aureus* is sufficient for use with most staphylococcal class I plasmids and plasmid constructs, the modifications of Schrenk *et al.* (1995) may be useful in the transformation of class III plasmids, or other large low-copy number plasmids. There is considerable speculation as to the factors that affect the electro-transformation efficiency of *S. aureus*. The bacterial growth medium and the size, copy number, and resistance marker of the transforming plasmid all appear to play a role.

Protoplast transformation of *S. aureus*

Prior to the routine use of electroporation, plasmid DNA was introduced into *S. aureus* using a modified version of the protoplast transformation procedure originally described by Chang and Cohen for use with *Bacillus subtilis* (Chang *et al.*, 1979). For *S. aureus*, protoplasts are prepared

Table 5.5. Preparation of electro-competent *S. aureus* and electroporation of plasmid DNA

Preparation of cells

1 Inoculate 10 ml of Tryptic Soy broth with a single colony of *S. aureus.* Grow overnight at 37°C with shaking (250 rpm).

2 Inoculate 200 ml of Tryptic Soy broth with 4 ml of the overnight culture. Grow to an OD_{660} of 0.4 at 37°C with shaking (250 rpm).

3 Harvest the cell by centrifugation at 5,000 × *g* for 20 min at 4°C.

4 Suspend the pellet in 10 ml ice-cold 0.5 M Sucrose using a sterile pipette. After the cells are resuspended, add 190 ml sterile ice-cold 0.5 M sucrose. Centrifuge as in step 3.

5 Suspend the pellet in 10 ml ice-cold 0.5 M sucrose using a sterile pipette. After the cells are suspended, add 90 ml sterile ice-cold 0.5 M sucrose and let stand on ice 30 min. Centrifuge as in step 3.

6 Suspend the pellet in 10 ml ice-cold 0.5 M sucrose using a sterile pipette. After the cells are suspended, transfer to a 25 ml Corex tube. Wash the bottle with an additional 10 ml sterile ice-cold 0.5 M Sucrose, and add the cells to the Corex tube. Centrifuge as in step 3.

7 Suspend the pellet in 300 μl of ice-cold 0.5 M sucrose. Aliquot 80 μl of cells into microfuge tubes and either use immediately or store frozen at –80°C.

Electroporation of plasmid DNA

1 Mix 80 μl of the cells with 0.1–1 μg of plasmid DNA.

2 Set the Electoporator to the following settings: 100 ohms resistance, 25 mF capacitance, and a charging voltage of 2.5 kV.

3 Place 80 μl of the cell/DNA mixture into a disposable cuvette (2.0 mm gap). Place the cuvette into the reaction chamber/electrode and pulse the cuvette.

4 Immediately add 920 μl of SMMP {5.5 parts SMM Buffer [1 M sucrose, 0.04 M maleic acid, 0.04 M $MgCl_2$, (pH 6.5)], 4 parts 7% Panassy Broth, 0.5 parts 10% BSA} to the cuvette and transfer the broth to a polypropylene tube.

5 Plate directly on selective media or after incubation at 37°C with shaking to allow for marker expression.

with lysostaphin in a stabilizing medium, and transformation is accomplished by mixing the bacteria with a solution of polyethylene glycol and DNA. The transformed bacteria are then plated on an isotonic media to allow for regeneration of the cell wall and plasma membrane.

To prepare protoplasts of *S. aureus*, a 10 ml starter culture of the DNA-recipient strain is grown overnight in 2× Penassay Broth (Antibiotic Medium Number 3; Difco, Detroit, MI). [Note for this procedure all glassware must be detergent-free.] The overnight culture is then used to inoculate 40 ml of 2× Penassay Broth to an OD_{540} (1 cm path length) of 0.05–0.07 units. The bacteria are grown with aeration until the culture reaches an OD_{540} of 0.3–0.4 units. The cells are divided into 6 ml aliquot, and placed in a plastic Falcon tube (#2027). The cells are harvested by centrifugation at 1.25 × *g* for 5 min. at room temperature. The supernatant is then discarded, and the cell pellets are washed in 5 ml SMMP [5.5 parts SMM Buffer {1 M sucrose, 0.04 M maleic acid, 0.04 M $MgCl_2$, (pH 6.5)}, 4 parts 7% Penassy Broth, 0.5 parts 10% BSA] and collected by centrifugation as above. After the wash step, the cell pellets are again suspended in 5 ml SMMP and lysostaphin is added to a final concentration of 30 μg/ml. To monitor the formation of protoplasts, at five minute intervals, harvest a 100 μl aliquot of cells from one tube. To the harvested cells, add 50 μl of HPLC-grade water and then 50 μl of a 2% solution of sodium dodecyl sulfate. If the cell pellet disappears in the presence of detergent and the solution becomes viscous, then the cells are protoplasts. Remove the lysostaphin in the cell solutions by harvesting the cells in an SS34 rotor as before, and wash the pellets in 5 ml SMMP. After centrifugation to collect the cell, the supernatant is decanted and each pellet is suspended in 0.5 SMMP. The protoplasts can be used immediately or stored overnight at room temperature.

To transform *S. aureus* protoplasts, an equal volume of 2 × SMM [1 M sucrose, 0.04 M maleic acid, 0.04 M $MgCl_2$, (pH 6.5)] is added to 0.1–1 μg plasmid DNA. Up to 100 μl of the DNA-SMM solution can be added to each tube of protoplasts. Next, 1.5 ml of PEG/SMM (0.125 mM Polyethylene glycol 8000 in 1 × SMM, heat to dissolve, sterilize by autoclaving for 8 min) is added to each tube, and the cell-containing solution is mixed by gentle inversion for 2 min. After, 5 ml of SMMP is added to each tube, which are then mixed by inversion. Cells are collected by room temperature centrifugation at 1.25 × *g*. The liquid contents of the tubes are carefully decanted. The tubes are drained by inverting them over a Kimwipe, and as much excess fluid as possible is removed, being careful not to disturb the pellet. Suspend the cells in 1 ml of SMMP and incubate the solution at 32 °C for 2 h. If necessary, add an inducing concentration of antibiotic, and grow at 32°C for an additional hour. Plate 100 μl aliquots of the cells onto DM3 regeneration plates and incubate at 32 °C for 3–5 days. To make DM3 plates, dissolve 67.54 g disodium succinic acid hexahydrate in 300 ml HPLC-grade water, adjust the pH to 7.3 with HCl, bring the solution volume to 350 ml, add 4.25 g Bacto-agar, and sterilize. After, equilibrate medium's components to 65 °C add: 50 ml of 5% casamino acids, 12.5 ml 20% dextrose, 25 ml 10% yeast extract, 50 ml 5% phosphate salts (20 mM K_2HPO_4, 10 mM KH_2PO_4), and 5 ml 1 M $MgCl_2$. Slowly swirl the mixture to combine the ingredients. Add 5 ml of room temperature 5% bovine serum albumin and any required antibiotics. Again, gently swirl to mix, and place the medium at 65°C until no surface bubbles are visible. Carefully swirl the flask to mix its contents and pipette 20–25 ml of the medium per standard-sized plate.

Plasmid vectors for use in *S. aureus*

A wide variety of naturally occurring and engineered plasmids are available for manipulation of the *S. aureus* genome. There are general cloning vectors *E. coli*/*S. aureus* shuttle vectors, expression vectors, and reporter systems. (Transposons and their delivery systems will be covered in the section on Transposon mutagenesis.) Recently, many of the *S. aureus* vectors that have been used for the last 20 or 30 years have been re-engineered, creating modern families of well-designed and easy-to-use plasmids. These plasmids often encode for replication in both *E. coli* and *S. aureus*. The shuttle plasmids allow construction of recombinant plasmids in *E. coli*, an organism that can be easily and efficiently manipulated, rather than *S. aureus*, an organism that is more difficult to manipulate. For selection within the host strain, these plasmids carry genes conferring resistance to various antibiotics and/or heavy metals. Often two antibiotic markers are carried on shuttle vectors, one for *E. coli* and another for *S. aureus*. Many of the engineered vectors contain a multiple cloning site to facilitate insertion of DNA sequences. Still other plasmids harbour elements to perform specific tasks: transposons for mutagenesis, reporters to assay promoter activity, or inducible promoters to control gene expression. Commonly used plasmid are summarized in Table 5.6.

General cloning and shuttle vectors

Four *S. aureus* class I plasmids, pC194 (GenBank accession NC_002013), pE194 (V01278), and pT181 (NC_001393), pUB110 (X03885), or derivatives of these plasmids, have been popular *S. aureus* vectors. These plasmids were isolated over 30 years ago, and through the years have been extensively studied (Chopra *et al.*, 1973; Horinouchi, *et al.* ,1972; Iordanescu, 1975, Iordanescu, 1976, Iordanescu *et al.*, 1978; reviewed in Novick, 1989). While pC194 and pE194 have been shown to be naturally occurring shuttles, they both replicate poorly in *E. coli* and lack convenient restriction endonuclease sites for cloning. Most investigators prefer to use engineered *E.*

Table 5.6. *S. aureus* shuttle plasmids

Type	Examples	Description	Uses
Native	pC194 pE194	Unmodified naturally occuring plasmids with a limited number of cloning sites. They replicate poorly in *E. coli*	Cloning vectors that can be used for complementation
Modified	pSK265 pCU1	Naturally occurring plasmids. Modified to contain a multiple cloning site	Generalized cloning vectors that can be used for complementation
Low copy number	pSK5630	Plasmids maintained in *S. aureus* at 10 copies per cell	To express a known gene in *S. aureus* and identify its phenotype. Expression of toxic products
Integration	pCL55 pCL84	Plasmids that integrate at specific *S. aureus* bacteriophage attachment sites, incapable of cytoplasmic replication in *S. aureus*	Use for single-copy gene expression
Temperature sensitive	pSPT181 pBT1	Plasmids that replicate in *S. aureus* at 32°C, but not 42°C	Used in allelic exchange and for plasmid curing
Inducible promoter	pCX15 pSA1 pP*spac* pSK236	Plasmids that carry a promoter that can be activated or repressed by the addition of specific compounds to the growth medium	To artificially regulate expression of a known gene in *S. aureus.* These plasmids are useful for identifying genes that are essential for *S. aureus* growth and expressing genes that encode toxic products
Reporter	pSK5805 pSB197 pALC1484	Plasmids that carry a reporter gene or genes downstream of a multiple cloning site	Detects the activity of a cloned promoter. Can be used under various growth conditions to assess promoter activation or represssion

Shuttle plasmids are frequently used to manipulate *S. aureus*. While these vectors have been designed to meet a variety of experimental needs, they all share the ability to replicate in *E. coli* and *S. aureus* and carry resistance marker or markers for the selection in both species. Shuttle vectors allow for creating constructs for *S. aureus* in *E. coli.* Ligation reactions are typically transformed first into *E. coli*. This step takes advantage of the high transformation/electroporation frequency associated with *E. coli*, but not *S. aureus*. The plasmid is then grown in *E. coli* under selection, purified, and used to electroporate into *S. aureus*.

coli/S. aureus shuttle plasmids with multiple cloning sites and that allow for the efficient propagation of the plasmid in both organisms. From a practical point of view, genetic constructs intended for use in *S. aureus* can be created in *E. coli*, a strategy that overcomes the problem of the poor transformation efficiency of *S. aureus* with low concentrations of *in vitro* ligated DNA. Shuttle plasmids can be quickly and efficiently isolated, modified, and transformed into *E. coli*. The modified plasmid is then isolated from *E. coli*, and CCC DNA is used to transform a restriction-minus strain of *S. aureus*. The plasmid can be isolated from the specialized strains of *S. aureus* and used to transform the strain of interest. While a cumbersome process, this strategy is often the quickest route to cloning genes into *S. aureus*.

The 2.9-kb plasmid pC194 can be propagated in *E. coli* and *S. aureus*; however, the plasmid is poorly replicated by *E. coli*. This plasmid confers chloramphenicol-induced resistance to chloramphenicol to both organisms (Goze *et al.*, 1980; Horinouchi *et al.*, 1982). In *S. aureus*, a 30 min. induction and selection of chloramphenicol resistance requires a medium containing 5 μg/ml and 25 μg/ml chloramphenicol, respectively. For *E. coli*, the 30 min. induction and selection use 0.5 μg/ml and 20 μg/ml chloramphenicol, respectively. Plasmid pC194 copy number and replication temperature sensitivity mutants have been isolated (Gruss *et al.*, 1987; Iordanescu, 1983). Copy number mutants are useful for either overexpressing genes or cloning genes that express a toxic product. Temperature-sensitive mutants replicate at low temperature, but cannot replicate at high temperatures (see Temperature-sensitive vectors, below).

Plasmid pE194, another small *S. aureus* R-plasmid, encodes erythromycin-inducible resistance to macrolide, lincosamide, and streptogramin type B antibiotics (Horinouchi *et al.*, 1972). Like pC194, pE194 replicates in both *E. coli* and *S. aureus*, although pE194 is inefficiently replicated in *E. coli* (Hardy *et al.*, 1982; Sozhamannan *et al.*, 1990). Plasmid pE194 is naturally temperature-sensitive for replication, and this feature has been used by investigators in the study of *S. aureus* (Scheer-Abramowitz *et al.*, 1981). Further modifications to pE194 include mutations that increase the temperature sensitivity (super-thermosensitive mutants) of the plasmid replicon, and mutations that alter the copy number of the plasmid (Gruss *et al.*, 1987; Villafane *et al.*, 1987). These derivative plasmids have been exploited for use in *S. aureus* and other Gram-positive organisms.

The *S. aureus* plasmid pT181 encodes inducible resistance to tetracycline. This plasmid contains four open reading frames (Kahn *et al.*, 1983). One of the open reading frames is required for plasmid replication (*repC*), two are required for expression of tetracycline resistance (*tetA* and *tetB*), and the fourth (*pre*) encodes a *recA*-independent recombinase. Interestingly, pT181 resides in the *S. aureus* chromosome in strains with type III staphylococcal cassette chromosome (SCC*mec*) (Ito *et al.*, 2001). Although the mechanism of insertion is unknown, it appears likely that insertion sequences rather than plasmid encoded elements are involved because in these strains, pT181 is flanked by the insertion element IS*431* (Ito *et al.*, 2001). In any event, the *pre* open reading frame is considered dispensable in that plasmid replication and selection are unaffected by its deletion. Making use of this observation, investigators have used the unique *Hin*dIII sites within *pre* as cloning sites. Mutants of pT181 with a copy number of 800 to 1,000 per cell (rather than the normal 20 to 25) have been isolated, as well as mutations that confer temperature-sensitive replication to pT181 (Iordanescu, 1983; Novick *et al.*, 1982; Wang *et al.*, 1991).

In addition to pC194, pE194, and pT181, plasmid pUB110 has been used for cloning in *S. aureus*. The nucleotide sequence of pUB110 is known and has been shown to contain, in addition to a gene that is required for resistance to kanamycin or neomycin (*neo*), four open reading frames that encode proteins of more than 80 amino acids. These open reading frames are labelled α, β, δ and γ (McKenzie *et al.*, 1986). The α open reading frame encodes an essential replication protein, while β, δ and γ are considered non-essential and they can be disrupted by cloned elements without an obvious effect on plasmid replication or selection. Plasmid pUB110 can be used in other Gram-positive organisms, and many pUB110-derivative plasmids that are useful for cloning in *S. aureus* have been constructed for the genetic manipulation of *B. subtilis*.

Engineered cloning and shuttle plasmids

In recent years, investigators have created a variety of shuttle vectors based on the replicons and resistance determinants from staphylococcal plasmids. In comparison to the earlier modifications to pC194, pE194, pT181, and pUB110, the most recently created shuttle vectors contain a greater number of unique cloning sites to facilitate the cloning of insert DNA, are more compact, are more structurally and segregationally stable, and are more versatile.

Many iterations of pC194 are available as cloning vectors. Jones and Khan added unique cloning sites by inserting the commercially available cloning vector pUC18 (GenBank accession number L09136) polylinker at the *Hin*dIII site within pC194, creating plasmid pSK265 (Jones *et al.*, 1986). With this plasmid, the copy number increased from the 40 to 90 copies per cell, due to an uncharacterized mutation (Novick, 1991).

While pC194 naturally replicates and confers chloramphenicol resistance to *E. coli* and *S. aureus*, it is inefficiently replicated in *E. coli*, forcing most investigators to transform pC194-derivative plasmids directly into *S. aureus*. The transformation of ligation products into *S. aureus* produces few recombinant bacteria. Furthermore, analysis of the transformed plasmids is usually with DNA purified from *S. aureus*. Again, this process is more cumbersome than procedures to isolate plasmids from *E. coli*. To increase the ease of use of pC194, fusion *E. coli*/*S. aureus* shuttle vectors have been constructed. One example, pCU1, has been shown to be stably replicated in both hosts while carrying DNA inserts of less than 8 kb (Augustin *et al.*, 1992). Another popular pC194-based shuttle is plasmid pSK236, which has plasmid pUC19 cloned into the *Hin*dIII site of pC194 (Gaskill *et al.*, 1988). Yet another readily available pC194-fusion plasmid is pMK4 (Sullivan, 1984). Similar plasmids have been constructed using pE194 and various *E. coli* vectors (Murray *et al.*, 1996). Despite the introduction of more efficient vectors, pC194 and pE194-derivatives remain popular among *S. aureus* researchers.

A comprehensive set of shuttle vectors has been described by Charpentier *et al.* (Charpentier *et al.*, 2004). This family of vectors (the pCN-family and plasmids pRN145, pRN146, and pRN8298) are composed of PCR-generated modular elements that confer a variety of attributes. One can select among replicons with very-low (five per cell), low (20–25 per cell), or high (300–400 per cell) copy numbers. In addition, there is a cassette that confers temperature-sensitive replication. Other modules confer resistance in *S. aureus* to erythromycin, tetracycline, chloramphenicol, kanamycin, or spectinomycin. These plasmids also include a module containing the ColE1 replicon and *blaZ* for efficient replication and selection and *E. coli*. Within this group of plasmids, other modules allow for high-frequency transduction, site-specific integration into the *S. aureus* chromosome, constitutive or inducible ectopic expression of cloned products, and for the use of the *Vibrio fisheri luxAB* or *Aequorea victoria* green fluorescent protein as reporters. In all, 28 plasmids, each consisting of a different combination of cassettes that confer different functions, are available.

In a less comprehensive manner, other investigators have created very useful *E. coli*/*S. aureus* shuttle vectors that fulfill specific needs (e.g. low copy number vectors, vectors that can be used for gene expression, vectors that contain reporter constructs, or chromosomal integration vectors). Some of these plasmids have been widely distributed, and their stability and functionality has been demonstrated by a number of laboratories.

Low copy number vectors

Grkovic *et al.* reported on the construction of a series of low copy number *E. coli*/*S. aureus* shuttle vectors (Grkovic *et al.*, 2003). In part, these vectors are based on pSK1, a low copy number *S. aureus* vector that undergoes the theta mode rather than rolling circle mode replication (simultaneous replications of both strands of DNA versus leading and lagging strand DNA synthesis). Most *S. aureus* vectors are based on plasmids that utilize rolling circle replication, the primary type of replication seen with staphylococcal plasmids. Recombinant plasmids that utilize the theta mode of replication are more structurally stabile than plasmids that replicate using a rolling circle mechanism. In addition, plasmids can harbour genes that ensure the efficient partitioning of plasmids into each daughter cell that is formed during division thus increasing their stability. The *E. coli/S. aureus* vector, pSK5630, was created to combine these desirable attributes. It contains the *rep* and *par* loci, which encode the replication and partitioning functions of pSK1. Plasmid pSK5630 also contains a ColE1 origin of replication and *blaZ*, for efficient replication and selection in *E. coli*. In addition, this plasmid has a multiple cloning site within *lacZ*α to facilitate the insertion and the identification of recombinant plasmids in *lacZ*ΔM15 *E. coli* host strains. Plasmid pSK5630 was used as the backbone in the construction of transcriptional and translational promoter fusion vectors that use β-lactamase as a reporter.

Integration vectors

Site-specific single-copy *E. coli/S. aureus* integration/shuttle vectors are often used to express genes that confer a dosage-sensitive phenotype. While low copy number shuttle plasmids can be used, plasmids have been engineered to integrate as a single-copy into specific sites within the *S. aureus* chromosome. One of these plasmids is pCL55 (Lee *et al.*, 1991). In *E. coli*, this plasmid confers resistance to ampicillin and contains the replicative functions from plasmid pBR322. Unique *Eco*RI, *Sal*I, *Kpn*I, *Sma*I, and *Bam*HI sites are available to accept cloned DNA. In *S. aureus*, pCL55 is incapable of autonomous cytoplasmic replication; instead this plasmid carries the gene encoding the bacteriophage L54a integrase (*int*) and attachment site (*attP*). Expression of integrase allows for the site-specific integration of the vector into the bacterial bacteriophage attachment site within the chromosomally encoded gene for glycerol ester hydrolase (*geh*). Thus, the vector is replicated as part of the chromosome when *S. aureus* divides. For selection in *S. aureus*, pCL55 carries *cat* from pC194, which in single copy on the chromosome confers resistance to 5 μg/ml chloramphenicol. Due to the low expression of *cat*, pCL55 cannot be used in all strains of *S. aureus*. The authors also report on the construction of plasmid pCL84. This vector confers tetracycline resistance to *S. aureus* and contains *attP*, but lacks a bacteriophage integrase gene. To establish this plasmid in *S. aureus*, it can be introduced in strain CYL316, which carries multiple copies of a constitutively expressed allele of *int*. While the authors report an increase in the transformation efficiency of *S. aureus* with pCL84 versus pCL55, the difference may be attributed to the efficiency of the selectable marker on the plasmid. Other site-specific single-copy *E. coli/S. aureus* integration/shuttle vectors that use the bacteriophage ϕ11 *int* and *attP* have also been developed (Dr Chia Lee, personal communication). Plasmid pCL74 contains unique restriction sites for cloning, propagates in *E. coli* via a pSC101 origin of replication and confers resistance to 25–50 μg/ml spectinomycin in *E. coli* (Note: some common *E. coli* host strains are resistant to spectinomycin, while others, like strain LE392, are spectinomycin-sensitive.). Single-copy integrates can readily be selected in *S. aureus* using tetracycline.

Temperature-sensitive vectors

A number of *E. coli/S. aureus* shuttle vectors that are temperature-sensitive for replication in *S. aureus* have been reported (Janzon *et al.*, 1990). The vectors are primarily used for curing native plasmids, the delivery of transposons, and allelic exchange, a technique for site-specific mutagenesis of chromosomal genes (see The elimination of plasmids from *S. aureus* and Allelic replacement). One vector, pSPT181, has been successfully used in a number of laboratories. Plasmid pSPT181 was constructed by fusing the *E. coli* vector pSP64 with the staphylococcal vector pRN8103 (a pT181-derivative) at their *Hin*dIII sites. This cloning resulted in loss of a *Hin*dIII fragment within *pre* on pRN8103. The resulting plasmid can be selected on media containing ampicillin for *E. coli* or on a medium supplemented with 10 μg/ml tetracycline for *S. aureus*. In *S. aureus*, the plasmid is capable of autonomous replication in bacteria grown at 32°C, but not at 42°C. The plasmid contains unique *Pst*I, *Bam*HI, *Sma*I, and *Xma*I cloning sites. Brückner reported on two additional *E. coli/S. aureus* vectors that are temperature-sensitive for replication in *S. aureus*, pBT1 and pBT2 (Brückner, 1997). These vectors are fusions of the region of pBR322 that encodes for replication and ampicillin selection in *E. coli* and pTV1(ts) which provides for chloramphenicol selection and temperature-sensitive replication in *S. aureus*. The two vectors differ in that pBT2 contains a multiple cloning site allowing for the use of *Eco*R1, *Sst*I, *Kpn*I, *Sma*I, *Bam*HI, *Xba*I, *Sal*I, *Pst*I, and *Hin*dIII, while pBT1 contains a limited number of useful cloning sites.

Inducible expression vectors and their promoters

Several inducible-expression plasmids are available for use in *S. aureus*. The earliest example of controlled expression of *S. aureus* genes was based on the β-lactamase promoter, its *trans*-encoded regulator, and the ampicillin analogue inducer, carboxyphenlybenzyl aminopenicillic acid (Vandenessch *et al.*, 1991). This early system lacks generality in that two plasmids are required: one large plasmid that encodes several resistance determinants and their regulators, and a second plasmid that contains a promoter fragment with two-thirds of the structural gene for β-lactamase and a single restriction site for creating promoter fusions. In the last 15 years, more convenient and universally applicable vectors have been adapted for use in *S. aureus* or constructed *de novo*.

Two vectors, pCX15 and pCX26 use the xylose-inducible, glucose-repressible, xylose operon promoter from *S. xylosus* (Wieland, 1995). A promoterless gene can be inserted into this vector in a manner that confers carbohydrate-controlled expression. Plasmid pCX15 is used for genes with a ribosomal binding site, and pCX26 for genes without a ribosomal binding site. While the effect of glucose was not quantified, one may be able to expect a 40-fold increase in gene expression upon xylose-induction. Other plasmids that are designed to place genes under xylose control are available. An *E. coli/S. aureus* shuttle/controlled gene expression vector containing *xylR* and the *xylA* promoter, pSA1, was created by Tegmark *et al.* (2000). Another plasmid, pEPSA5, is a well-designed *E. coli/S. aureus* shuttle vector that carries an optimized bacteriophage T5 PN25 promoter under the control of *xylR* (Forsyth *et al.*, 2002).

Workers in *B. subtilis* created a inducible promoter, P*spac*, that has been used control gene expression in *S. aureus*. P*spac* is a hybrid promoter consisting of *E. coli lac* operator sequences fused to the −35 and −10 regions of the *B. subtilis* bacteriophage SPO-1 promoter (Yansura *et al.*, 1984). Transcription from P*spac* is repressed in alleles expressing *lacI*. Induction of P*spac* occurs when isopropyl-β-D-thiogalactopyranoside (IPTG) is added to the growth medium. In the case of *S. aureus*, *lacI* is expressed from a constitutively expressed penicillinase operon promoter from *Bacillus licheniformis* (P*penP*). Genes under P*spac* control have been cloned into either shuttle or integration plasmids. In *S. aureus* these constructs have successfully been used to study essential genes (Luong *et al.*, 2006; Pinho *et al.*, 2003; Sun *et al.*, 2005; Zheng *et al.*, 2005; Zheng *et al.*, 2007) and in experiments that require the controlled over-expression of genes (Puyang *et al.*, 2003). Unfortunately, several investigators have observed the regulation of P*spac* is not absolute in *S. aureus*. A high level of basal promoter activity limits the number applications for which this system is appropriate (Bateman, *et al.*, 2001; Charpentier *et al.*, 2007; Jana *et al.*, 2000).

An inducible promoter was created from the *xyl* promoter and sequences from *E. coli* transposon *Tn*10-encoded *tetRA* (Geissendorfer *et al.*, 1990). This composite promoter, residing on plasmid pWH353, has *tetR* operator sequences that are normally associated with *tetA* placed in between the −35 and −10 regions of *B. subtilis xyl* promoter. In the presence of the *tetR* product, a repressor that recognizes the operator sequence, transcription of genes under the control of hybrid promoter are not expressed. When tetracycline or anhydrotetracycline (analogue of tetracycline, anhydrotetracycline, which is a poor antibiotic, but a strong inducer of the *tet* operon) is added to cells expressing *tetR*, the hybrid promoter is active due to the antibiotic-induced allosteric inactivation of the repressor and dissociation of the repressor-operator DNA complex. Under induced conditions, genes under the control of the *xylE-tetR* operator are induced 100-fold. Several investigators have used this inducible promoter in *S. aureus*, including in experiments where a regulated gene was tested in an animal model of staphylococcal disease (Bankosky *et* *al.*, 2000; Ji *et al.*, 1999, Bateman *et al.*, 2001). In each case, the investigators could demonstrate very low basal levels of expression of the regulated genes in the absence of tetracycline, and a dose-dependent response for induction of the regulated gene. Bateman *et al.* created plasmid pSK236 that contains *tetR* and the *xyl/tetA* operator promoter from pWH353 cloned into the *Pst*I and *Sma*I sites of pSK236 (see Engineered cloning and shuttle plasmids).

A third inducible expression system exploited is P*cad-cadC* from the *S. aureus* plasmid pI258. The *cad* operon of pI258 confers inducible resistance to heavy metals (Cd^{2+} > Bi^{3+} and Pb^{2+} >> Co^{2+} and Zn^{2+}) (Yoon *et al.*, 1991). Resistance is mediated by a P-type ATPase pump encoded by *CadA*. The promoter is induced by dissociation of a *cadC*-encoded repressor from *cad* operator sequences binding the appropriate heavy metal (Endo *et al.*, 1995; Ye *et al.*, 2005; Yoon *et al.*, 1991). Several constructs of the modular system of vectors described by Charpentier *et al.* carry a P*cad-cadC*. Using luciferase or β-lactamase reporters, low levels of basal promoter activity were observed. These same constructs also produced a dose-dependent expression of the reporter gene with 1 to 15 μM $CdCl_2$, concentrations that do not affect bacterial growth (Charpentier *et al.*, 2004).

Reporters and their activity assays

Reporter genes, genes whose products are easily identified and quantified, have been adapted for use in *S. aureus*. The reader is referred to review articles for the uses and limitation of each type of reporter gene (Schenborn *et al.*, 1999; Wood, 1995). In *S. aureus*, reporter genes have been used in the study of promoters (Clarke *et al.*, 2001; Tseng *et al.*, 2005), in the analysis of gene expression (Bargonetti *et al.*, 1993; Cheung *et al.*, 1998; Luong *et al.*, 2002; Luong *et al.*, 2006), and in specialized methodologies such as the yeast hybrid system (Rohrer, 2003; Yan *et al.*, 2001). (The referenced articles provide examples of the use of reporters in *S. aureus*, and do not constitute a comprehensive list of publications.) The most commonly used reporters in *S. aureus* are promoterless alleles of β-lactamase (*blaZ*) from plasmid pI258, chloramphenicol acetyltransferase (*cat*) from pC194, lucerificase from bacterial or insect sources, or a derivative of green fluorescent protein. The choice of a reporter is dictated by the nature of its intended use. *BlaZ* is a secreted product and is not considered to be appropriate for protein or bacterial localization studies. Instead, it has found a role in experiments aimed at understanding gene expression and regulation as a transcriptional or translational fusion with a gene of interest. Similarly, *cat* is used to study gene expression and regulation; however, the *cat* gene product is not secreted and assays for *cat* activity require lysis of the bacteria. Luciferase is a very sensitive reporter because of the lack of background luminescence in *S. aureus*. In addition, luciferase has a short half-life, and it can be used to examine the kinetics of gene expression. However, luciferase has an energy requirement, limiting its use in microscopic localization studies of fixed samples (Hill, 1994). In *S. aureus*, the green fluorescent protein reporter is less sensitive than a luciferase, and due to the long half-life of the protein, it cannot be used to monitor promoter activity in real-time. However, the green fluorescent protein can be used in retrospective studies of promoter activity and for photonic imaging of fixed samples.

β-lactamase reporters and assays

After the initial use of reporters in *E. coli*, investigators of Gram-positive organisms began to construct and use such systems. β-lactamase, encoded by *blaZ*, was initially chosen as a reporter. Several *blaZ* reporter systems are now available, from early multi-copy plasmid constructs to several modern low and moderate copy shuttle vectors. The first *S. aureus* reporter plasmids, pWN1818 and pWN1819, were based on a pUC18/pC194 *E. coli/S. aureus* shuttle vector. These vectors carry an allele of *blaZ* that allow for either a translational or transcriptional fusion between a cloned gene and a promoterless allele of *blaZ* (Wang *et al.*, 1987). These *blaZ* alleles have been used in a number of other shuttle vectors including: the *S. aureus* low-copy number plasmids pSK5645, pSK5805, and pCN41 (Grkovic *et al.*, 2003; Charpentier *et al.*, 2004). It is preferable to use reporters on low copy number vectors with genes that are normally present on the chromosome in a single copy.

Plate assays have been developed for detecting β-lactamase activity. These assays have been described in detail (Novick, 1991), but for the reader's convenience are reiterated here. One assay detects β-lactamase using starch–iodine indicator plates. This assay is based on the hydrolyses of the β-lactam ring in penicillin, which results in the production of penicilloic acid. This acid hydrolyses starch, a reaction made visible by staining the plates with iodine. In this assay, β-lactamase producing colonies are surrounded by a clear zone. Plates are prepared using the desired media supplemented with 0.2% potato starch. After growth of the colonies, the plates are flooded with a freshly made 1% (wt/vol.) penicillin G and incubated for 15 min at 37°C. The residual penicillin is decanted, and a solution of 80 mM I_2 and 3.2 M KI is poured onto the plate. The plates can be read instantaneously. The concentrations of starch and iodine can be altered to adjust the sensitivity of the assay. As an alternative, acidification of the medium surrounding β-lactamase producing colonies can also be detected using the indicator N-phenyl-1-naphthylamine 4-azo-*o*-carbobenzene (PNCB). A 0.25% solution of PNCB in N,N-dimethylformamide and 5.8 mM NaOH is used to cover the bacterial colonies on a plate. The plate is incubated for 30 min at room temperature, at which time the excess dye is decanted. The plate is then flooded with a solution of freshly made 1% (wt/vol.) penicillin G. β-Lactamase producing colonies develop a dark red colour in seconds to several minutes depending on the concentration of the enzyme.

Measurement of β-lactamase activity associated with *S. aureus* grown in broth can be measured using nitrocefin, a chromogenic substrate of β-lactamase (Novick, 1991). In wells of a microtitre plate, 50 µl of 50 mM potassium phosphate buffer (pH 5.9) are added. These wells are used to make serial 1:2 or 1:3 dilutions of test samples, starting with 25 µl of sample in the first well. If live bacteria are tested, a growth inhibitor (i.e. 0.1 M NaN_3) must be added to the wells. To the diluted samples, 50 ml of a 0.1 mM solution of nitrocefin that consists of 5.16 mg nitrocefin in 100 ml of 50 mM potassium phosphate buffer (pH 7.0) is added. The plates are read immediately at 486 mm and at 10 min intervals for 1 h. For each dilution series, data points with an optical density below 0.2 are plotted and the slope of the curve is determined for the linear portion of the curve. Activities are reported as units/ml = [(slope) (V_d)/[$(\mathcal{E}_m)$ (l) (s)], where the slope is in absorbance units/h, V is the volume of the reaction (0.1 ml), d is the dilution factor, $\mathcal{E}_m$ is the millimolar extinction coefficient for nitrocefin (20,500 M^{-1} cm^{-1} at 486 nm), l is the path length in cm (0.3 for round bottom microtitre plates containing 0.1 ml of sample), and s the amount of sample in the first well. For simplicity, kits for the measurement of β-lactamase activity are commercially available from Fluka (Test Kit 40561, Sigma-Aldrich, St. Louis, MO, USA). In addition, larger scale test tube β-lactamase assays have been reported (O'Callaghan *et al.*, 1972).

Chloramphenicol acetyltransferase reporters and assays

Plasmids pBR322 and pUB110 based *E. coli/S. aureus* shuttle vectors that contain a promoterless *cat*, ribosomal binding site, and a multiple cloning site have been described (Band *et al.*, 1983; Brückner, 1992; Hudson *et al.*, 1986).

Chloramphenicol acetyltransferase (CAT) activity can be spectrophotometrically detected by measuring the chloramphenicol-dependent appearance of coenzyme A using acetyl coenzyme A as substrate. A detectable change occurs by reaction of 5,5′ dithiobis-2-nitrobenzoic acid (DTNB, Ellman's reagent) with the free sufhydryl group of coenzyme A, which releases thiolate ions that produce a yellow colour (Shaw, 1976). A microtitre plate assay for measuring CAT activity in *S. aureus* has been reported by Tseung *et al.*, and is presented with slight modifications (Tseung *et al.*, 2005). Briefly, an overnight culture of bacteria is used to inoculate 10 ml of a pre-warmed medium, to an optical density at 540 nm (OD_{540}) of 0.2. The cultures are cultivated at 37°C with rotary aeration until an OD_{540} of 2.5. A 5 ml aliquot of culture harvested by centrifugation at 5,000 × *g* for 10 min, and the supernatant is decanted. The cell pellet is washed in 5 ml TE buffer (pH 8.0), harvested by centrifugation, and suspended in 1 ml TE buffer. Cells are transferred to a Fast-Prep Blue tube (QBIOgene) that contains 0.1 mm silica beads, and disrupted at 6.0 m/s for 40 s using a

Fast-Prep FP cell homogenizer (Thermo Savant, Carlsbad, CA, USA). The lysates are cleared by centrifugation at 2,000 × *g* for 10 min at 4 °C, the supernatant is recovered, and may be used immediately or stored at –70 °C. Samples are analysed in duplicate; therefore for each assay, two 25 μl aliquots of cell lysates, serial dilutions of cell lysates, or water (as a control) are added to 37.5 μl of 0.4% DTNB in 100 mM Tris-HCl (pH 8.0). Then, 7.5 μl of 5 mM acetyl coenzyme A and 180 μl of distilled water are added to each well. The mixtures are then incubated at 37°C for 10 min. Then 5 pmol of chloramphenicol (in 10 μl of 70% ethanol) is added to each well. Measurements are made at an optical density at 412 nm. The results are normalized to the total protein concentration of the cell extracts or dry cell weight of the cells (Tseng *et al.*, 2005). CAT activity is calculated as the change in absorbance per minute, divided by 13.6 (the molar extinction value for DTNB) and the amount of protein added or dry weight of the bacteria. These values are expressed as nM chloramphenicol acetylated per minute per mg protein.

Luciferase reporters and their applications

Bioluminescence assays that measure luciferase activity have been used to measure gene expression in *S. aureus* (Corbisier *et al.*, 1993; Giachino *et al.*, 2001; Sheehan *et al.*, 1992). In addition, luciferase activity has been used to estimate the number and cellular state of bacteria exposed to antibiotics (McWalter, 1984; Tenhami *et al.*, 2001) and for specialized transcription/translation assays (Murray *et al.*, 2001). Alleles of *luxAB* from *Vibrio harveyi* (GenBank accession number E15888) or *Vibrio fischeri* (E16961), and firefly *luc* (U47123) have been expressed in *S. aureus* (Corbisier *et al.*, 1993; Giachino *et al.*, 2001; Charpenter *et al.*, 2004). At least two *E. coli/S. aureus* shuttle vectors that carry promoterless alleles of the genes encoding luciferase activity have been constructed: plasmid pC101 (Corbisier *et al.*, 1993), and pSB197 (Sheehan *et al.*, 1992).

The use of *luxAB* requires exogenous long-chain aldehyde for bioluminescence, while *luxABCDE* confers bioluminescence without the addition of long-chain aldehydes. The metabolism for long-chain aldehydes provides the energy for bioluminescence. *luxCDE* encodes a transferase, synthetase, and reductase that synthesize and recycle long-chain aldehydes, eliminating the need for an exogenous energy supply. Luciferase activity is measured with a luminometer and the assays for use with *S. aureus* have been reported (Corbisier *et al.*, 1993; Giachino *et al.*, 2001; Timmins *et al.*, 1996). The advantages of this assay are that it is extremely quick, simple and sensitive.

Green fluorescent protein reporters and their applications

The detection of the chemiluminescent green fluorescent protein has been used by a number of investigators to study gene activation and bacterial localization in models of *S. aureus* infections or within mammalian cells using time-lapse confocal scanning laser microscopy and/or flow cytotometry (Leid *et al.*, 2002; van Wamel *et al.*, 2002; Xiong *et al.*, 2002, Xiong *et al.*, 2004, Xiong *et al.*, 2006; Yarwood Xiong *et al.*, 2004). These studies used one of two variants of the gene encoding the green fluorescent protein (*gfp*), the FACS-optimized variant, *gfpmut2* (Cormack *et al.*, 1996; Xiong *et al.*, 1996), or the red-shifted bacteria-optimized variant, *gfp*uvr (Clontech, Palo Alto, CA). A pSK236-based plasmid, pALC1484, has been created to generate transcriptional fusions between a promoter of interest and *gfpuvr* (Wolz *et al.*, 2000). The activity of the green fluorescent protein can be measured by application-specific fluoremetry using microplate fluorimeters, single tube fluorometers, or systems that can detect fluorescence through tissues of live animals. In addition, fluorescence activated cell sorters can be used to isolate bacteria expressing a reporter from a mixed population of cells. In addition fluorescent microscopes can detect individual cells expressing the reporter.

Duel reporters and their applications

A luciferase/green fluorescent protein tandem reporter under the control of a growth-dependent promoter has been used to monitor bacterial internalization, replication, and gene expression within mammalian cells (Qazi *et al.*, 2001; Sunderland *et al.*, 1995). Furthermore, a duel reporter system with *gfp-lux* has been used to study the internalization of *S. aureus* and the response of these bacteria to antibiotics, with bacterial localization and enumeration determined using the green fluorescent protein and the metabolic state of the bacteria monitored using luciferase (Qazi *et al.*, 2004). Dual reporter systems have the advantage in that the green fluorescent protein can be used to examine chemically fixed specimens, while living specimens can be examined using the *lux* reporter, which is more sensitive than the green fluorescent protein. In these dual reporter systems, *gfp* and either *luxAB* or the entire *lux* operon (*luxABCDE*) is used.

Bacteriophage-mediated transduction of *S. aureus*

Bacteriophage-mediated transduction is commonly used to manipulate the *S. aureus* genome. Viral transducing particles can be used to quickly and easily transfer plasmid- or chromosomally encoded markers between strains of *S. aureus*. One limitation of transduction in *S. aureus*, particularly of chromosomally encoded markers, is that *S. aureus* bacteriophage package and transfer large segments of DNA (i.e. 43–46 kb). An example of this limitation is seen when transducing a specific mutation which may be physically linked to undesired secondary mutations.

Nearly all the International Typing Set group B bacteriophage can be used for transduction, but investigators have preferred to use Φ11 (Iandolo *et al.*, 2002), 80α – a hybrid of Φ11 and 80 (Novick, 1963) – and L54a (Lee *et al.*, 1985) in their experiments. One advantage of the use of bacteriophage L54a is that unintentionally formed lysogenic strains can be identified by screening bacteria for the loss of lipase activity, which can be visualized as clear zones surrounding colonies growing on egg yolk agar.

Transduction can involve several steps: the isolation of bacteriophage, the making of a high-titre lysate, determinating the plaque forming units within a lysate, the making of a transducing lysate, and finally, the transfer of a plasmid- or chromosome-encoded marker into a susceptible strain. These experiments can take from overnight to several days depending on the availability of high-titre bacteriophage. Plastic tubes seem to lower viral titres, and many investigators use only glass tubes in their work with bacteriophage.

Induction of prophage from a single colony of a *S. aureus*

As with most temperate bacteriophage, prophage induction – the beginning of a cycle that multiplies the bacteriophage genome and releases viral particles – can be initiated by exposure of lysogenic bacteria to a range of mitomycin C or ultraviolet light (Novick, 1963; Sjöström *et al.*, 1974; Wilkinson *et al.*, 1987). In one commonly used method for isolating bacteriophage from lysogenic *S. aureus*, the bacteria are grown with rotary aeration in 110 ml Tryptic Soy broth containing $CaCl_2$ at 37°C to an optical density of 1.5 at 600 nm (1 cm path length). Ten different concentrations of mitomycin C ranging from 0.1 to 1 μg/ml are added to 10 ml aliquots of the bacterial culture with one concentration of mitomycin C used per tube. One tube is left as an untreated control. The cultures are incubated as before for 6 h. After, the lysate is cleared by centrifugation and sterilized by passage through a 0.22 μm cellulose acetate filter. The bacteriophage content of the Mitomycin C-induced lysates are then compared by spotting 10 μl of serial dilutions [made in SM buffer (50 mM Tris-Cl buffer {pH 7.0}, 0.1 M NaCl, 8 mM $MgSO_4$, 5 mM $CaCl_2$, 0.01% gelatin)] of each lysate grown on a lawn of susceptible bacteria. After overnight growth at 37°C, the lysate that clears of the test bacteria at the greatest dilution can be used to obtain a high-titre lysate by the plate amplification method.

Preparation of high-titre bacteriophage lysates and transduction

In general, the titre of *S. aureus* bacteriophage grown in liquid media is lower than can be achieved on plates. Therefore, high-titre lysates are created using a plate amplification method. Initially bacteriophage-susceptible cells are streaked across the entire surface of a plate containing a solid medium and grown overnight. The bacteria are harvested in 5 ml of Tryptic Soy broth with 5 mM $CaCl_2$ using a sterile policeman or inoculating loop. The harvested bacteria are diluted 1:3 in Tryptic Soy broth with $CaCl_2$, and 100 μl aliquots are added to a series of tubes. While working with bacteriophage is aided by an empirical knowledge of viral and bacterial titres, generally, six tubes are sufficient to obtain a multiplicity of infection that amplifies the bacteriophage. One hundred microlitres of undiluted or 1:10 serially diluted bacteriophage (the bacteriophage can be diluted in Tryptic Soy broth with $CaCl_2$) are added to bacteria-containing tubes. Then 4 ml of cooled, but not solidified, Tryptic Soy soft agar (Tryptic Soy broth with $CaCl_2$ and 0.5% Bacto Agar) is added to each tube and each mixture is poured over a fresh Tryptic Soy agar plate supplemented with $CaCl_2$. The soft agar is allowed to gel and the plates are incubated top-side-up overnight at the appropriate temperature for growth of the recipient strain. Typically, a range of lysis of the susceptible bacteria are seen on the plates, from complete lysis to a limited number of distinct plaques. If no plate has completely cleared the bacteria, plates can be incubated at 4°C or 48°C until clearing occurs (Lundholm *et al.*, 2005). The lowest dilution of bacteriophage that completely clears the plate is harvested in 5 ml of Tryptic Soy broth with $CaCl_2$ into a glass centrifuge tube. The Tryptic Soy broth and soft agar mixture is placed on a rocker at 4°C for 3–5 h. After, the debris is removed by centrifugation at 4°C at 8,500 × *g* for 15 min. The supernatant is sterilized by three passages through 0.22 mm cellulose acetate syringe filters and stored refrigerated at 4°C.

Transduction

A procedure for bacteriophage-mediated transduction of chromosomal- or plasmid-encoded markers between strains of *S. aureus* is presented in Table 5.7. This procedure works best with high-titre bacteriophage (10^{10}–10^{12} plaque-forming units/ml), especially when mobilizing chromosomally encoded markers. To prevent transformed bacteria from becoming lysogens, a low multiplicity-of-infection is desirable. This requires enumeration of both the bacteriophage in lysates and the bacteria to be transformed. As with other bacteriophage, the number of virions can be determined from plates of susceptible bacteria infected with serially diluted bacteriophage. Essentially, the procedure is the same as described above; however, for high-titre lysates, it is often necessary to plate 10–12 1:10 serial dilutions of a stock lysate. The number of bacterial cells can be estimated by comparison of the optical density of the recipient cells to a graph of optical density versus viable counts.

Mutagenesis of *S. aureus*

As with many organisms, the study of the molecular genetics of *S. aureus* has relied on mutagenesis. There are a few current reports of the chemical mutagenesis of *S. aureus*, with this technique having reached its zenith in *S. aureus* almost a half a century ago. Chemical mutagenesis has been largely supplanted by transposon mutagenesis. While transposon mutagenesis remains a useful technique, especially for generating libraries containing thousands of mutants, directed mutagenesis by allelic replacement is now the most commonly used technique to alter the *S. aureus* genome.

Table 5.7. Transduction of *S. aureus*

1	Inoculate the entire surface of a Tryptic Soy agar plate, supplemented with antibiotics as necessary, with the recipient strain. The plate is incubated overnight under appropriate conditions.
2	Harvest the recipient strain in 5 ml Tryptic Soy broth supplemented with $CaCl_2$ (5 mM final concentration).
3	Label screw-capped 15 ml Cortex centrifuge tubes 'control' and 'transduction'. Add 0.5 ml of the cells to each tube.
4	To the tube labelled 'transduction' add up to 0.5 ml of bacteriophage to achieve a multiplicity of infection between 0.1 to 1.0. Add Tryptic Soy broth with $CaCl_2$ to a final volume of 2.0 ml.
5	To the tube labelled 'control' add 1.5 ml Tryptic Soy broth with $CaCl_2$.
6	Incubate both tubes at 37°C with shaking for 20 min. Induce antibiotic resistance if necessary.
7	Add 1.0 ml of ice-cold 0.02 M sodium citrate to each tube, and pellet the cells by centrifugation at 5,000 × *g* at 4°C for 20 min.
8	Resuspend the pellets in 1 ml of 0.02 M sodium citrate and plate 100 µl aliquots on Tryptic Soy broth with $CaCl_2$ and any required antibiotic(s). Grow the bacteria under appropriate conditions for 12–48 h.

Chemical mutagenesis

Currently, the whole-cell chemical mutagenesis of *S. aureus* is almost exclusively used to examine the emergence of resistance to antibiotics (Gustafson *et al.*, 1992; Markham *et al.*, 1999; Silverman *et al.*, 2001). The limited use of this technique is due to the difficulty of identifying the genetic lesion associated with a given phenotype. While chemically induced lesions are routinely defined in other organisms including *B. subtilis*, in *S. aureus* very few have been genetically or physically mapped or identified by complementation and DNA sequencing. This is not surprising since the original genetic linkage map of *S. aureus* was constructed after the isolation of most of the chemical mutants and because historically, complementation screens of genomic DNA libraries were often fruitless due to the low transformation frequency of *S. aureus* (Pattee, 1990). Furthermore, while chemical mutagenesis has been successfully used to study certain phenotypes (Barnes *et al.*, 1969; Good *et al.*, 1972), there are examples of chemical mutants with a specific phenotype that were incorrectly ascribed to a structural gene (Phonimdaeng *et al.*, 1990). This was due, at least for most of the mutations, to an inability to transfer the lesion to a clean genetic background because the mutations were not tagged with a selectable marker.

Miller has published a comprehensive guide to the chemical mutagenesis of *E. coli* along with a discussion of the advantages and disadvantages of many chemical mutagens (Miller, 1972). These classical procedures can be used with *S. aureus* by modifying the medium for bacterial growth.

Transposon mutagenesis

The use of transposons for the disruption of genes has grown in sophistication as investigators have responded to their experimental needs. Composite transposons including Tn*551*, Tn*917* and Tn*917*-derivatives, Tn*916*, Tn*918*, and Tn*4001* have all been used to identify genes that confer specific phenotypes. These class II transposable elements reside on an assortment of delivery vectors and encode a variety of antibiotic resistance markers. Two advantages of the use of transposons are that mutations are genetically tagged, making the target gene easy to identify, and that the mutation is readily verified by transduction element-encoded antibiotic resistance. This latter advantage is a necessary step in transposon mutagenesis to demonstrate that a transposon insertion is genetically linked to an observed phenotype (see McNamara *et al.*, 1998).

Transposon Tn*551* (GenBank accession number Y13600) is a 5.2-kb element from the *S. aureus* plasmid pI258. This transposon carries a constitutively expressed determinant that confers resistance to macrolide-lincosamide-streptogramin antibiotics. A temperature-sensitive version of pI258, pRN3208, has been used to deliver Tn*551* to the chromosome of *S. aureus* and a detailed procedure has been described (Kornblum *et al.*, 1986). Using plasmid pRN3208, transposition frequencies of between 1 in 6 × 10^3 to 1 in 10^5 mutants to total bacteria can be expected. Preferential regions for Tn*551* insertions in the chromosome have been noted (Pattee, 1981).

Transposons Tn*917* (GenBank accession number M11180) and Tn*551* are approximately 99% identical in their native form. Functionally, Tn*917* differs from Tn*551* in that macrolide–lincosamide–streptogramin resistance is inducible (Horinouchi *et al.*, 1983) and that Tn*917* preferentially inserts into different preferential regions within the chromosome than Tn*551*. Philip Youngman and co-workers have created a series of Tn*917*-derivatives and a pC194-based delivery system (Perkins *et al.*, 1986; Youngman *et al.*, 1984a, Youngman *et al.*, 1984b). While developed for use in *B. subtilis*, several of these plasmids have been used in *S. aureus* to create genetically tagged mutations. These plasmids include the original Tn*917* vector, pTV1, pTV20, a Tn917/pBR322 hybrid used to ease the cloning of the transposon and flanking DNA for later identification of insertion sites, and pTV32 which carries Tn917*lac*, a reporter that can be used to detect expressed promoters. A review that includes a theoretical examination of transposon mutagenesis, plasmid maps of Tn*917* and its derivatives, a discussion of the use of pTV series plasmids, and detailed procedures has been published (Youngman *et al.*, 1987).

Other transposon have been used in the genetic analysis of *S. aureus*. The conjugative transposon Tn*918* confers resistance to tetracycline and is transferred to *S. aureus* by membrane-filter mating, typically using a strain of *Enterococcus* (*Streptococcus*) *faecalis* carrying a plasmid with a hitchhiking copy of Tn*918*. The plasmid is readily transferred from *E. faecalis* to *S. aureus* where it cannot replicate. Transposition occurs when Tn*918* excises from the plasmid and inserts into the *S. aureus* chromosome. A variety of mutants have been made using Tn*918* (Albus *et al.*, 1991; Greene *et al.*, 1996; Nakao *et al.*, 2000). Two other transposons, Tn*916* (GenBAnk accession number AF343837) and Tn*4001* were important to creating the original genetic map of *S. aureus* (Pattee, 1990). Tn*916* is a conjugal tetracycline-resistance-encoding transposon originally isolated from *E. faecalis* (Clewell *et al.*, 1986). Tn*4001* is a gentamicin, tobramycin and kanamycin resistance-encoding transposon from *S. aureus* (Lyon *et al.*, 1984). Until the use of Tn*916* and Tn*4001*, information contained on the *S. aureus* genetic linkage map consisted of auxotrophic markers, genes that conferred a phenotype and silent insertions of Tn*551*. The addition of Tn*916* and Tn*4001* provided selectable markers other than Tn*551*-encoded erythromycin resistance that eased multifactor transformation analysis. Tn*916* residing on the conjugative plasmids pAD1::Tn*916* and pAD5::Tn*916* can be introduced to *S. aureus* by intergeneric protoplasting or membrane-filter mating; However, the transformation efficiency is low, and using the published experimental procedures, many investigators have concluded that Tn*916* is a marginally helpful genetic tool in *S. aureus* (Jones *et al.*, 1987). For Tn*4001* delivery of the transposon to the chromosome uses thermosensitive replication mutant derivatives of plasmids pI258 or pII147 (Mahairas *et al.*, 1989). While Tn*4001* remains a useful tool, reversion frequencies of Tn*4001* mutants were reported as being between 1 in 10^6 or 10^7, higher than those observed with Tn*551* and Tn*917*.

Finally, a transposon named *bursa aurealis* has been engineered for the random mutagenesis of the *S. aureus* genome. This transposon is based on a synthetic transposon, himar 1. Himar 1 is composed of *mariner* elements from the horn fly and the lacewing (Lampe *et al.*, 1996; Robertson *et al.*, 1995). This element requires only its self-encoded transposase and a dinucleotide 'AT' sequence in the recipient cell's DNA for transposition. These two features make himar 1 ideal as a tool of the mutagenesis of low 'GC'-content organisms like *S. aureus*.

To create *Bursa aurealis* Bae *et al.* further modified himar1 and created a delivery system for use in *S. aureus* creating *bursa aurealis* (Bae *et al.*, 2004). They constructed plasmid pBursa, which includes a temperature-sensitive replicon for *S. aureus*, a chloramphenicol resistance marker, and a modified version of himar 1. Between the terminal inverted repeats of the element are: a promoterless allele of the gene encoding the green fluorescent protein, the *E. coli* R6K origin of replication, and a *S. aureus* erythromycin resistant determinant. Plasmid pBursa can be used to identify promoters that are expressed under selected growth conditions. Bae *et al.* also constructed a second plasmid, pFA545. This plasmid contains the mariner transposase expressed from the xylose-inducible promoter. This system is like the minitransposons of *E. coi* where the transposon integrates without its transposase preventing further transposition events. *S. aureus* chromosomal inserts of *bursa aurealis* can be selected for by erythromycin resistance at elevated growth temperatures. The green fluorescent protein is expressed upon transposition of the element downstream of a nearby promoter, and theoretically, the element-encoded R6K origin of replication can be used to rescue transposon-flanking chromosomal DNA into *E. coi* λ lysogens, although in the original report the flanking DNA was identified by sequencing inverse-PCR products. *Bursa aurealis* has now been used for the mutagenesis of *S. aureus* in a number of recent studies (Burtz *et al.*, 2005; Gründling *et al.*, 2006; Wardenburg *et al.*, 2007; Bae *et al.*, 2006).

Signature-tagged mutagenesis and *in vivo* expression technology

Signature-tagged mutagenesis (STM) is an approach for the identification of mutants in mixed populations of bacteria (For a complete review of STM see Mazurkiewicz *et al.*, 2006). The original application of this technique was to identify bacterial virulence factors by analysis of 'dropout' mutants (i.e. mutants that fail to survive) that are selected against during infection in an animal model (Hensel, 1995). Importantly, STM uses a small number of animals. The key feature of this technique is the methodology used for tracking mutaginized genes. Each library mutation is tagged with a unique index sequence, a molecular bar code. Bacteria with a transposon harbouring a specific bar code can be identified by DNA hybridizations. STM uses pooled mutants grown within a host. After growth *in vivo*, viable bacteria are collected and their index regions are amplified and compared to those within the pool of mutants. This process allows the identification of mutants that dropped out during the selection. Transposon-interrupted genes can then be identified by comparing the sequence of DNA flanking the transposon to the sequence of genomic DNA. Using modified versions of Tn*917*, STM has been used to identify genes that are essential for *S. aureus* pathogenesis by selection in a murine model of bacteraemia (Mei *et al.*, 1997) and a lethality model for *Caenorhabditis elegans* (Begun *et al.*, 2005).

In a related technique, *in vivo* expression technology (IVET), genes that play a role in pathogenesis can be identified by looking for promoters that are inactive during *in vitro* growth, but induced within the host. This technique uses promoter traps, which are promoterless genes that express an easily identifiable phenotype when downstream of a heterologous promoter. IVET technology has been adapted for use in *S. aureus* by Lowe *et al.* (1998). This procedure uses a strain of *S. aureus* with a chromosomally integrated copy of a kanamycin cassette. This cassette is flanked by direct repeat sequences that are recognized by the site-specific resolvase encoded by *tnpR* from the transposon γδ. The investigators created plasmid pESAL2, which can accept staphylococcal DNA fragments upstream of a promoterless allele of *tnpR*. Promoter fragments that allow for expression of *tnpR* induce resolvase-mediated excision of the kanamycin cassette and the loss of kanamycin resistance, providing a negative marker of gene expression. By screening *S. aureus* harbouring a library of fragments in pESAL2 for strong expression in a murine model of abscessation, but minimal expression *in vitro*, virulence-associated genes were identified. For further information on IVET, the reader is referred to a concise review that outlines variations of technique and their advantages and limitations by Angelichio and Camilli (2002).

Allelic replacement

Among the first reports of targeted gene replacement in prokaryotes was that of Ruvkun *et al.*, who, working with *Escherichia coli*, transferred a *Tn*5 knock out of a cloned gene into the chromosomally encoded wild-type copy of the gene (Ruvkun *et al.*, 1981). Allelic replacement been consistently improved and its use has been expanded to a wide variety of organisms. For *S. aureus*, O'Reilly *et al.* were the first to use allelic replacement to create a mutation in a chromosomally encoded copy of a structural gene (O'Reilly *et al.*, 1986). Since that initial report, many *S. aureus* genes have been insertionally inactivated using a similar methodology. Allelic replacement does have limitations. First, as with other gene inactivation methodologies, there is a risk of introducing unintended polar effects; typically effects on genes downstream of the point of the mutation. These effects can be mitigated by carefully creating mutant alleles of the target genes. For example, it may be necessary to design a 'hit and run' knockout allele of a target gene where a promoter is placed at the end of the transposon to express downstream genes (Thourp *et al.*, 2001; Xia *et al.*, 1999). Second, at least in *E. coli*, there is a risk of creating unwanted secondary mutations at locations that are physically distinct from the targeted gene (Johnson *et al.*, 2003).

In *S. aureus* allelic replacement (Fig. 5.1) mutants are usually created by cloning the specific gene of interest and then inserting a selectable marker into the gene. From a practical perspective, approximately 1-kb of *S. aureus* DNA flanking gene-disrupting marker is required, although longer stretches of DNA appear to increase the efficiency of allelic replacement. Rarely, non-cassette-tagged mutations (*e.g.* small deletions) have been created by screening for a visible phenotype (McNamara *et al.*, 2000). There are new modifications to the basic technique that allow for the isolation of non-cassette-tagged mutations that do not confer a visible phenotype (discussed below). Regardless of the specific mutation, the altered gene is transferred into an allele replacement vector. These vectors are

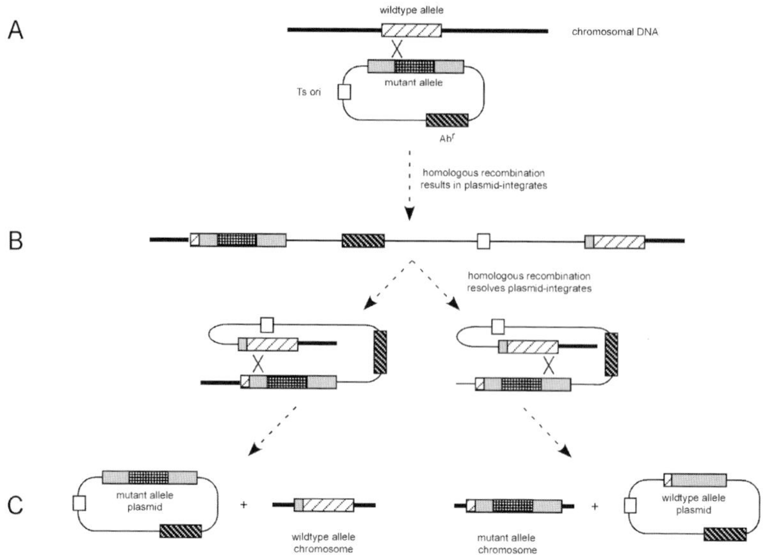

Figure 5.1 Allelic exchange mutagenesis. Chromosomal DNA is depicted as thick lines. Plasmid DNA is depicted as thin lines. Genetic features are depicted as rectangles or boxes that are differentiated by shading or a pattern. Dashed arrows point to the results of recombination events. (A) A chromosomally encoded wild-type copy of a target gene (white rectangle with stripes) contains DNA that is homologous to a cloned mutant allele (grey rectangles) that is disrupted using a unique selectable marker (dotted rectangle). The mutant allele of the target gene is carried on a plasmid with a temperature-sensitive origin of replication (white box) and a second selectable marker (black rectangle with thin white stripes). (B) Bacteria mediated homologous recombination, the location denoted with a large cross (X), allows for the chromosomal integration of the plasmid containing the mutant allele. (C) A second recombination event excises a plasmid. The resulting products are either the original DNA elements or a strain which now has the mutant allele of the target gene within the chromosome the wild-type allele on the plasmid. The products are determined by the location of the recombination events. If the second crossover occurs on the opposite side of the target gene-disrupting marker, the chromosome of the resulting bacteria will contain the mutant allele of the target gene.

typically *E. coli*/*S. aureus* shuttle plasmids that carry a unique marker or markers for selection in *E. coli* and/or *S. aureus*, and a temperature-sensitive replicon for *S. aureus* (see section on temperature-sensitive vectors). The vector with the mutated gene is introduced into *S. aureus*, usually by electroporation, and the bacteria are propagated at the permissive temperature for vector replication selecting for the phenotype conferred by the marker that resides on the backbone of the plasmid. After introduction of the vector, the cultivation temperature is raised to the non-permissive temperature and autonomous plasmid replication is blocked. Plasmids that have integrated into the chromosome are selected using the marker on the gene-interrupting cassette. Integration of the allele replacement vector relies on homologous recombination between the cloned gene and its counterpart in the *S. aureus* genome. Plasmid integration results in a bacterial chromosome harbouring both mutant and wild-type alleles of the cloned gene separated by vector sequences. Plasmid sequences and one copy of the target gene are then resolved from the chromosome by serial passage of the bacteria at the non-permissive replication temperature in media without antibiotic supplementation. Growth of the bacteria under these conditions allows for a second recombinational event. The resolution of the plasmid-integrates involves recombination and relies on the homologous DNA surrounding the wild-type and mutant allele of the target gene. The final outcome of the second recombinational event are bacteria with a chromosomal copy of either a mutant or wild-type allele of the target gene. Mutant bacteria are cultivated on a solid medium at the non-permissive temperature for plasmid replication, selecting for the phenotype conferred by the gene-disrupting marker. The resulting colonies are then replica plated to solid media containing the antibiotic that corresponds to the vector-encoded marker.

Bacteria that grow on media that is selective for the target gene-associated marker, but not on media that selects for the vector-encoded marker, are potential mutants. Screening for a phenotype other than that of the gene-disrupting cassette is very helpful in the isolation of the desired mutant. Potential mutant strains are verified by analysis of the chromosome (e.g. Southern or PCR analysis). Transduction of the marker within the disrupted gene into a strain with a clean genetic background is recommended prior to extensive analysis of the phenotypic effect of the mutation.

The efficiency of identifying engineered strains of *S. aureus* created by allelic replacement is greatly enhanced by employing either a screen to identify plasmid-free bacteria or by counter selection techniques designed to eliminate non-altered bacteria. Arnaud *et al.* created the allelic replacement vector pMAD that carries a 'universal' system for screening for the loss of vector sequences from Gram-positive bacteria. Plasmid pMAD encodes a constitutively expressed thermostable β-galactosidase that can be used in combination with the chromogenic substrate X-gal (5-bromo-4-chloro-3-indoyl-β-D-galactopyranoside) to identify plasmid-free bacteria. With this plasmid, plasmid-free bacterial colonies appear white, while colonies of bacteria which retain plasmid sequences appear blue. While the use of pMAD does not select for allelic replacement mutants, the ability to screen for bacteria that have lost plasmid sequence increases the efficiency of mutant isolation when compared to the classical method of allelic replacement.

In a second system (Bae *et al.*, 2006), a PCR amplified mutated gene is inserted into an *E. coli*/*S. aureus* shuttle vector that carries a temperature-sensitive *S. aureus* replicon, pKOR1. Plasmid pKOR1 allows for cloning by *in vitro* site-specific recombination using components from bacteriophage λ. In addition, successful cloning of the mutated gene disrupts a pKOR1-encoded copy of *ccdB*, a gene encoding a cytotoxin for *E. coli*, providing positive selection for the desired construct. The chromosomal integration of the allele replacement plasmid and resolution of plasmid-integrates are bacterial cultivation temperature-dependent events. Counter selection uses pKOR1-encoded anhydrotetracycline-induced antisense *secY* RNA to inhibit the growth of bacteria with cytoplasmic or chromosomally integrated copies of the allelic replacement vector, thus eliminating background colonies. Allele replacement can be confirmed by Southern analysis or by sequencing mutation-encoding PCR amplified DNA as appropriate.

The third system, based on the work of Pósfai *et al.*, uses restriction endonuclease counter-selection of the allelic exchange vector and chromosomal plasmid-integrates (McNamara, unpublished data; Pósfai *et al.*; 1999). The restriction enzyme used by this system is the *Saccharomyces cerevisiae* mitochondrial DNA-encoded meganuclease, I-*Sce*-I. I-*Sce*-I recognizes a specific 18-bp target sequence not normally found in the genome of *E. coli* or *S. aureus*. Cleavage of DNA by I-*Sce*-I generates double-stranded breaks that abolish cytoplasmic plasmid replication and, in the case of chromosomal cleavage, are bactericidal. The system uses two *E. coli*/*S. aureus* shuttle vectors that encode temperature-sensitive replicons for *S. aureus*. The first plasmid, pJM930, is pBT2 that contains an I-*Sce*-I recognition sequence cloned into the *Kpn*I site (although other vectors can be used because an I-Sce-I site can easily be added to the end of any *S. aureus* gene by PCR). The second plasmid, pJM918, is pSPT181(Ts) that constitutively expresses I-*sce*I. Plasmid pJM930 with the mutant allele of the target gene is introduced into *S. aureus*. Plasmid-integrates are selected by cultivation of the bacteria at 42°C in the presence of 10 μg/ml tetracycline. Resolution of the plasmid-integrates is accomplished by overnight growth of bacteria at 42°C in medium that is free of antibiotics. A lysate *S. aureus* strain PM928 (*S. aureus* RN4220 harbouring pJM918) is used to transduce chloramphenicol resistance and I-*Sce*-I activity to a pool of the resolved bacteria. After growth at 32°C, the bacteria are harvested and pJM928 is eliminated by growth bacteria at 42°C on solid media lacking antibiotics. Again, allele replacement can be verified by Southern analysis or by sequencing mutation-encoding PCR amplified DNA as required.

Regardless of the technique, the counter selection techniques are sufficiently robust to routinely isolate mutants without the use of a gene-disrupting marker-encoding cassette (non-tagged mutations). While a direct comparison is not possible because different target genes encoded on variously sized fragments of DNA were used, it appears that both antisense *secY* RNA selection and I-*Sce*-I counter selection are equally efficient (Bae *et al.*, 2006; McNamara, unpublished data). Interestingly, I-*Sce*-I counter selection was efficient enough to aid in the isolation of a point mutation (i.e. a single base-pair substitution) in the chromosomal copy of the gene encoding the *S. aureus* homologues of histone-like protein HU encoded by *hsa* (McNamara, unpublished data). After I-*Sce*-I counter selection, mutations in *hsa* were isolated at frequencies of 70%. The isolation of non-tagged mutations is desirable for the generation of non-polar chromosomal mutants, conditionally silent mutations, and for conserving the limited number of antibiotic markers for use in further manipulation of the mutant strains. However, there may be a limitation in that it is difficult to mobilize these mutations to a clean genetic background unless there is a nearby marker.

Essential genes and conditional mutants

Genes can encode for proteins that are absolutely required for the replication and growth of bacteria or that are indispensable to metabolic processes under certain conditions. Bacteria with null mutations in essential genes are non-viable. For *S. aureus*, large-scale searches have been used to identify essential genes (Forsyth *et al.*, 2002). An online database of such genes is maintained by the National Microbial Pathogen Data Resource (www.nmpdr.org) (McNeil *et al.*, 2007) and DEG (http://tubic.tju.edu.cn/deg/). Essential genes are potential targets for novel antimicrobial therapies. Several strategies have been employed for the identification and/or genetic analysis of essential genes. One strategy involves the creation of conditional mutants; a second strategy involves the regulated *trans*-expression of the essential gene in a corresponding mutant; a third strategy involves the use of targetrons; a fourth strategy uses the controlled expression of antisense RNA that is complementary to target gene mRNA sequences. Each of these methodologies has advantages and disadvantages. For example, chemically induced conditional mutants can be quickly isolated, but their the genetic lesions are difficult to identify. If a gene is known to be essential, the time consuming clonings that are required for the controlled expression of a wild-type allele of the gene in a corresponding mutant background my be deemed worthwhile. Targetrons may be used to conditionally mutagenize any gene; however, the mutant phenotype is only expressed at elevated temperatures. Antisense RNA has been shown to only repress, not eliminate gene function, but this technique is suitable for high-throughput studies. Thus, chosen methodology for the examination of essential genes is likely to be determined by the nature of the encoded protein and the goals of the experimenter.

Early studies of essential genes relied on the isolation of chemically induced conditional mutants. For example, in one study temperature-sensitive lethal mutants were generated and phenotypically characterized (Good *et al.*, 1970). At the non-permissive temperature, death of these bacteria could be prevented by NaCl and the bacteria were shown to accumulate cell wall precursors. This study and similar studies (Tomasz *et al.*, 1970), were the introduction to the genetics of the synthesis and degradation of the *S. aureus* cell wall. With better and more direct approaches now available, the techniques used in these studies and the mutants that were created, are mostly of historical interest.

For *S. aureus*, Luong *et al.* described a methodology for the identification and analysis of essential genes that involves the regulated expression of a wild-type allele of a gene in a cognate mutant (Luong *et al.*, 2000). In this case the target gene was *murE*, which encodes protein involved in cell wall biosynthesis. Their methodology was to place *murE* under the control of P*spac* promoter at its chromosomal locus. The temperature-sensitive vector pCL52.2 carrying a P*spac-murE* construct was integrated into the *S. aureus* chromosome creating a strain of bacteria with *MUTE* under the control of LacI. Constitutively expressed copies of the *lacI* repressor were introduced on the multicopy plasmid, pMJ8426. By demonstrating IPTG-dependent growth, *murE* was shown to be an essential gene.

Since this original report, several examples of this methodology have been published. In a recent example, a single copy of *murF*, another gene involved in cell wall biosynthesis and cell division, was placed under the control of P*spac* promoter (Sobral *et al.*, 2006). By demonstrating IPTG-dependent growth, *murF* was shown to be an essential gene. Furthermore, at suboptimal concentrations the investigators could examine the affect of *murF* on growth, cell wall synthesis, and the antibiotic susceptibility of *S. aureus*. Two plasmids were constructed for controlled expression of *MUTE*. The first was plasmid pMGPI, an *S. aureus* integration vector with an ampicillin-resistance determinant and origin of replication from *E. coli*, the P*spac* promoter, and a constitutively expressed allele of *lacI*. However, as seen by other investigators, a single copy of *lacI* did not lead to total repression of P*spac* (see section on Inducible expression vectors and their promoters). Therefore, a second plasmid (pMGPII), an *E. coli*/*S. aureus* shuttle vector with the *lacI* gene, was used to provide multiple copies *lacI* and tighter regulation of P*spac*. Other genes involved in cell wall metabolism have been examined using vectors and methodologies that are similar to those that have been reported (Zheng *et al.*, 2005).

Yao *et al.* developed a methodology for the inactivation of genes using a modified version of the *Lactococcus lactis* L1.LtrB intron (Yao *et al.*, 2006). L1.LtrB is a class II intron or targetron (reviewed in Lambowitz *et al.*, 2004, and Lambowitz *et al.*, 2005), which is a naturally mobile, temperature-sensitive, catalytic RNA. Targetrons are RNAs that catalyse a self-splicing reaction that precisely excises themselves from a precursor messenger RNA. Furthermore, these RNAs can ligate the ends of their flanking exons reconstituting the original precursor message. Targetrons have an intron-encoded protein (IEP) that, in addition to aiding in splicing, confers reverse transcriptase and endonuclease activities. Using these later two functions, targetrons can mobilize themselves as part of a IEP-intron ribonucleoparticle. As mobile elements, targetrons insert into DNA based upon their ability to recognize and pair with 13- to 16-nt target DNA sequence. By altering their target-recognition sequence it has been shown to be possible to 'retarget' a targetron to insert at a desired sequence (Lambowitz, 2004, 2005). For essential genes, in order for the targetron to excise and restore a functional target gene mRNA, the targetron must integrate into the DNA in the same orientation as the target gene; otherwise, targetron IEP is not expressed.

As with other bacteria, targetrons are delivered to *S. aureus* on plasmids. These vectors carry a modified version of the L1.LtrB intron, L1.LtrB-ΔORF, in which the IEP has been deleted from the element and moved downstream of the element's 3′-exon. Furthermore, to retarget the element to a chromosomal site specified by three small regions of DNA named IBS1, IBS2, and δ′, corresponding changes are made in the targetron sequence at EBS1, EBS2, and δ. A computer algorithm finds 'best-match' for IEP binding to the target sequence and then designs primers to modify the targetron so that there are optimal target and intron IBS/EBS and δ′/δ base-parings. The required modifications to the targetron are generated by PCR and the modified L1.LtrB-ΔORF is exchanged for the unmodified element in the delivery vector. When delivered to *S. aureus*, the targetron specifically inserts within the gene of interest. Fortunately, the splicing functions encoded by the targetron are temperature-sensitive; therefore, to express a mutant phenotype, the bacteria encoding targeted essential genes can be grown at 43°C. This technique was used by Yao *et al.* to create a conditional mutation in *hsa*.

Essential genes are also studied by using antisense RNA methodologies. Messenger RNA (single-stranded protein-coding sense RNA) can duplex with RNA having a complementary sequence, an antisense strand. Duplex formation between sense and antisense RNA physically hinders ribosomal binding and translation of the messenger RNA. For *S. aureus*, the first demonstration of the use of antisense technology was for decreasing expression of α-toxin (Kernodle *et al.*, 1997). In this report, a 600-bp fragment encoding the gene for α-toxin was inverted and cloned behind its native promoter. When introduced into *S. aureus* on an *E. coli*/*S. aureus* shuttle vector, a 16-fold decrease in α-toxin activity and reduce a virulence in an animal model was seen with transformed bacteria as compared to their controls. This, while *hla* is a non-essential gene *in vitro*, *hla* activity appears to aid the bacteria in an infection. Ji *et al.* advanced the technology by demonstration that *hla* expression could be controlled by induction of antisense *hla* RNA (Ji *et al.*, 1999). In these experiments, an inverted DNA fragment from *hla* was cloned onto a pUC19/pE194 *E. coli*/*S. aureus* shuttle plasmid behind the inducible *tet* promoter (*Pxyl/tetO*). When the construct was introduced into *S. aureus* and the antisense *hla* RNA was expressed, the investigators saw a 14-fold decrease in *in vitro* *hla* expression and attenuation of the bacteria in an murine model of infection. Antisense technology has been used in *S. aureus* for genome-wide searches of essential genes.

In addition to the study of the activity of specific genes, antisense RNA technology can be used to catalogue genes associated with lethal or growth-deficient phenotypes. A library of bacteria that each contain a conditionally expressed, random, small, antisense-oriented DNA fragment can be used to screen for growth under induced or non-induced conditions. If a colony of bacteria is viable in the absence of the inducer and non-viable under induced conditions, its antisense-orientated fragment is inferred to down-regulate a specific essential gene. Sequencing of the insert and searches of the genomic sequence database can then be used to identify the gene. For *S. aureus*, in the first of two independent studies, a random library of 200- to 800-bp chromosomal DNA fragments were cloned into a vector under the control of the *tet* promoter (Ji *et al.*, 2001). The plasmids were transformed in *S, aureus*, and the resulting bacteria replica platted on media with or without the induction. The twenty-thousand bacteria were screened for growth, with 3% displaying a growth-defective or lethal phenotype on plates supplemented with anhydrotetracyclin. The DNA sequence of single-insert antisense-oriented clones were identified (the correct orientation of the insert was present in one-third of the bacteria), and a list of 150 genes that confer a lethal or growth impaired phenotype was compiled. In a second similar report, a corresponding list of essential genes was compiled from analysis of a small DNA fragment shotgun library whose expression was controlled by the *xyl* promoter (Forsyth, 2002).

Future directions

The development of new vectors and methodologies for examining *S. aureus* has and will continue to be indespensible to our understanding of this organism and its relationship with its hosts. While native plasmids have been greatly improved for use in *S. aureus* in terms of ease of use and versatility, *S. aureus* research could benefit from engineered bacteriophage vectors. One possible improvement would be construction of a bacteriophage that mobilizes one to two kilobase pairs of DNA. This vector that would make transduction of chromosomal markers more meaningful. Furthermore, the development of *in vitro* packaging of DNA could provide an avenue of very high-frequency transformation of *S. aureus*. The ultimate expression of engineering of the *S. aureus* genome would be the direct transformation and chromosomal integration of PCR products. It is possible that in the future this technique could be developed.

Perhaps the biggest challenge for *S. aureus* researchers lies in the distribution of plasmid constructs and strains. This challenge will be magnified by the retirement of the generation of scientists that originally created many of the *S. aureus* constructs. While other organisms have specific repositories for such materials, workers in *S. aureus* generally rely on obtaining publicly documented materials directly from their original source. The American Type Culture Collection and the Network on Antimicrobial Resistance in *Staphylococcus aureus* (NARSA) have some of the prototypical strains and current clinical isolates of *S. aureus*. These collections include many strains with valuable plasmids and plasmid constructs. In addition, the *Bacillus subtilis* Genetic Stock Center (http://www.bgsc.org) has a few of the original *S. aureus* plasmids. However, some collections of historical and practical importance (i.e. strains from Dr Merlin Bergdoll of The Food Research Institute, Madison, WI, USA, or Dr Peter Pattee of Iowa State University) have been lost or are only preserved in the private collections of individuals.

Conclusions

Many tools and techniques are available for manipulating the *S. aureus* genome. These methodologies represent the application of over half a century of accumulated knowledge of bacterial genetics, the collective efforts of workers in the field, and employees of corporations that have advanced the reagents, genetic systems, and instrumentation used by biologists. Most experimental strategies and techniques that have been employed with other bacteria have been adapted for use in *S. aureus*. As our knowledge of the basic biology of *S. aureus* increases, primarily through our understanding of the pathogenesis of this organism, there will undoubtedly be advances that facilitate the engineering of the *S. aureus* genome.

Web resources

American Type Culture Collection
http://www.atcc.org
Source of general strains

Network on Antimicrobial Resistance in *Staphylococcus aureus*
http://www.narsa.net
Source of *S. aureus* strains

Bacillus subtilis Genetic Stock Center
http://www.bgsc.org
Source of *B. subtilis* strains

The Institute for Genome Research
http://cmr.tigr.org
Genome and proteome data for *S. aureus* and other bacteria

GenomeNet
http://www.genome.jp
S. aureus genome, proteome and metabolic data

The National Center for Biological Information (NCBI)
http://www.ncbi.nlm.nih.gov
S. aureus genome, proteome and metabolic data

National Microbial Pathogen Data Resource
www.nmpdr.org
Genome and proteome data for *S. aureus* and other bacteria. Lists of essential genes

REBASE
http://rebase.neb.com/rebase/rebase.html
Information on restriction enzymes and modification systems, including those encoded by *S. aureus* (Roberts, 2007)

Database of Essential Genes (DEG)
http://tubic.tju.edu.cn/deg/

References

Abdalla-Galal, S., Ramuz, M., and Schmitt-Slomska, J. (1988). Plasmid curing in *Staphylococcus aureus* by antibiotics affecting the bacterial cell wall. FEMS Microbiol. Lett. *49*, 500–504.

Albus, A., Arbeit, R.D., and Lee, J.C. (1991). Virulence of *Staphylococcus aureus* mutants altered in type 5 capsule production. Infect. Immun. *59*, 1008–1014.

Amabile-Cuevas, C.F. (1988). Loss of penicillinase plasmids of *Staphylococcus aureus* after treatment with L-ascorbic acid. Mutat. Res. *207*, 107–109.

Amabile-Cuevas, C.F., Pina-Zentella, R., and Wah-Laborde, M.E. (1991). Decreased resistance to antibiotics and plasmid loss in plasmid-carrying strains of *Staphylococcus aureus* treated with ascorbic acid. Mutat. Res. *264*, 119–125.

Angelichio, M.J., and Camilli, A. (2002). *In vivo* expression technology. Infect. Immun. *70*, 6518–6523.

Arnaud, M., Chastanet, A., and Débarbouillé, M. (2004). New vector for efficient allelic replacement in naturally nontransformable, low-GC-content, gram-positive bacteria. Appl. Environ. Microbiol. *70*, 6887–6891.

Augustun, J., Rosenstein, R., Wieland, B., Schneider, U., Schell, N., Engelke, G., Entian, K.D., and Götzm, F. (1992). Genetic analysis of epidermin biosynthetic genes and epidermin-negative mutants of *Staphylococcus epidermidis*. Eur. J. Biochem. *204*, 1149–1154.

Ausubel, F.A., Brent, R., Kingston, R.E., Moore, D.D., Steidman, J.G., Smith, J.A., and Struhl, K. (2007). Current Protocols in Molecular Biology. (New York: Greene Publishing and Wiley Interscience).

Augustin, J., and Götz, F. (1990). Transformation of *Staphylococcus epidermidis* and other staphylococcal species with plasmid DNA by electroporation. FEMS Microbiol. Lett. *66*, 203–208.

Baba, T., Takeuchi, F., Kuroda, M., Yuzawa, H., Aoki, K., Oguchi, A., Nagai, Y., Iwama, N., Asano, K., Naimi, T., Kuroda, H., Cui, L., Yamamoto, K., and Hiramatsu, K. (2002). Genome and virulence determinants of high virulence community-acquired MRSA. Lancet 359, 1819–1827.

Bae, T., Baba, T., Hiramatsu, K., and Schneewind, O. (2006). Prophages of *Staphylococcus aureus* Newman and their contribution to virulence. Molec. Microbiol. *62*, 1035–1047.

Bae, T., Banger, A.K., Wallace, A., Glass, E.M., Åslund, F., and Schneewind, O., and Missiakas, D.M. (2004). *Staphylococcus aureus* virulence genes identified by *bursa aurealis* mutagenesis and nematode killing. Proc. Natl. Acad. Sci. USA *101*, 12312–12317.

Bae, T., and Schneewind, O. (2006). Allelic replacement in *Staphylococcus aureus* with inducible counter-selection. Plasmid 55, 58–63.

Baird-Parker, A.C. (1962). An improved diagnostic and selective medium for isolating coagulase positive staphylococci. J. Appl. Bact. *25*, 12–19.

Band, L., Yansura, D.G., and Henner, D.J. (1983). Construction of a vector for cloning promoters in *Bacillus subtilis*. Gene 26, 313–315.

Bargonetti, J., Wang, P.Z., and Novick, R.P. (1993). Measurement of gene expression by translational coupling: effect of copy mutations on pT181 initiator synthesis. EMBO J. *12*, 3659–67.

Barnes, I.J., Bondi, A., and Moat A.G. (1969). Biochemical characterization of lysine auxotrophs of *Staphylococcus aureus*. J. Bacteriol. *99*, 169–174.

Bateman, B.T., Donegan, N.P.. Jarry, T.M., Palma, M., and Cheung, A.L. (2001). Evaluation of a tetracycline-inducible promoter in *Staphylococcus aureus in vitro* and *in vivo* and its application in demonstrating the role of *sigB* in microcolony formation. Infect. Immun. *69*, 7851–7857.

Begun, J., Sifri, C.D., Goldman, S., Calderwood, S.B., and Ausubel, F.M. (2005). *Staphylococcus aureus* virulence factors identified by using a high-throughput *Caenorhabditis elegans*-killing model. Infect Immun. *73*, 872–877.

Betley, M.J., and Mekalanos, J.J. 1985. Staphylococcal enterotoxin A is encoded by phage. Science *229*, 185–187.

Bischoff, M., Entenza, J.M., and Giachino, P. (2001). Influence of a functional *sigB* operon on the global regulators *sar* and *agr* in *Staphylococcus aureus*. J. Bacteriol. *183*, 5171–5179.

Bischoff, M., Dunmanm P., Kormanecm, J., Macapagal, D., Murphy, E., Mounts, W., Berger-Bächi, B., and Projan, S. (2004). Microarray-based analysis of the *Staphylococcus aureus* σ^B regulon. J. Bacteriol. *186*, 4085–4099.

Björklind, A., and Arvidson S. (1980). Mutants of *Staphylococcus aureus* affected in the regulation of exoprotein synthesis. FEMS Microbiol. Lett. *7*, 203–206.

Brückner, R. (1992). A series of shuttle vectors for *Bacillus subtilis* and *Escherichia coli*. Gene. *122*, 187–192.

Brückner, R. (1997). Gene replacement in *Staphylococcus carnosis* and *Staphylococcus xylosus*. FEMS Microbiol. Lett. *151*, 1–8.

Burts, M.L., Williams, W.A., DeBordm K., and Missiakasm D.M. (2005). EsxA and EsxB are secreted by an ESAT-6-like system that is required for the pathogenesis of *Staphylococcus aureus* infections. Proc. Natl. Acad. Sci. USA *102*, 1169–1174.

Cassat, J.E., Dunman, P.M., McAleese, F., Murphy, E., Projan, S.J., and Smeltzer, M.S. (2005). Comparative genomics of *Staphylococcus aureus* musculoskeletal isolates. J. Bacteriol. *187*, 576–592.

Cassat, J., Dunman, P.M., Murphym E., Projanm S.J., Beenken, K.E., Palm, K.J., Yang, S.J., Rice, K.C., Bayles, K.W., and Smeltzer, M.S. (2006). Transcriptional profiling of a *Staphylococcus aureus* clinical isolate and its isogenic *agr* and *sarA* mutants reveals global differences in comparison to the laboratory strain RN6390. Microbiology *152*, 3075–90.

Chan, P.F., Foster, S.J., Ingham, E. and Clements, M.O. (1998). The *Staphylococcus aureus* alternative sigma factor σ^B controls the environmental stress response but not starvation survival or pathogenicity in a mouse abscess model. J. Bacteriol. *180*, 6082–6089.

Chang, S., and Cohen, S.N. (1979). High frequency transformation of *Bacillus subtilis* protoplasts by plasmid DNA. Mol. Gen. Genet. *168*, 111–115.

Charles, B.G., and Rawal, B.D. (1976). Cultivation of *Staphylococcus aureus* in a synthetic medium of low ultraviolet absorptivity. Antimicrob. Agents Chemother. *10*, 915–917.

Charpentier, E., Anton, A.I., Barry, P., Alfonso, B., Fang. Y., and Novick, R.P. (2004). Novel cassette-based shuttle vector system for gram-positive bacteria. Appl. Environ. Microbiol. *70*, 6076–6085.

Cheung, A.L., Chien, Y.T., and Bayer, A.S. (1999). Hyperproduction of alpha-hemolysin in a *sigB* mutant is associated with elevated SarA expression in *Staphylococcus aureus*. Infect. Immun. 67, 1331– 1337.

Cheung, A.L., Nast, C.C., and Bayer, A.S. (1998). Selective activation of *sar* promoters with the use of green fluorescent protein transcriptional fusions as the detection system in the rabbit endocarditis model. Infect. Immun. *66*, 5988–5993.

Chopra, I., Bennett, P.M, and Lacey, R.W. (1973). A variety of staphylococcal plasmids present as multiple copies. J. Gen. Microbiol. *79*, 343–345.

Clarke, S.R., and Dyke, K.G. (2001). The signal transducer (BlaRI) and the repressor (BlaI) of the *Staphylococcus aureus* beta-lactamase operon are inducible. Microbiology 147, 803–810.

Clewell, D.B., and Gawron-Burke, C. (1986). Conjugative transposons and dissemination of antibiotic resistance in streptococci. Annu. Rev. Microbiol. *40*, 635–659.

Coleman, D.C., Sullivan, D.J., Russell, R.J., Arbuthnott, J.P., Carey, B.F., and Pomeroy, H.M. (1989). *Staphylococcus aureus* bacteriophages mediating the simultaneous lysogenic conversion of beta-lysin, staphylokinase and enterotoxin A: molecular mechanism of triple conversion. J. Gen. Microbiol. *135*, 1679–1697.

Corbisier, P., Ji, G., Nuyts, G., Mergeay, M., and Silver, S. (1993). *luxAB* gene fusions with the arsenic and cadmium resistance operons of *Staphylococcus aureus* plasmid pI258. FEMS Microbiol. Lett. *110*, 231–238.

Cormack, B.P., Valdivia, R.H., and Falkow, S. (1996). FACS-optimized mutants of the green fluorescent protein (GFP). Gene *173*, 33–38.

Dajani, A.S., and Taube, Z. (1974). Plasmid-mediated production of staphylococcin in bacteriophage type 71 *Staphylococcus aureus*. Antimicrob. Agents Chemother. 5, 594–598.

Dassy, B., and Fourniern J.M. (1996). Respiratory activity is essential for post-exponential-phase production of type 5 capsular polysaccharide by *Staphylococcus aureus*. Infect. Immun. *64*, 2408–2414.

Deora, R., Tseng, T., and Misra, T.K. (1997). Alternative transcription factor sigmaSB of *Staphylococcus aureus*: characterization and role in transcription of the global regulatory locus *sar*. J. Bacteriol. *179*, 6355–6359.

Diep, B.A., Gill, S.R., Chang, R.F., Phan, T.H., Chen, J.H., Davidson, M.G., Lin, F., Lin, J., Carleton, H.A., Mongodin, E.F., Sensabaugh, G.F., and Perdreau-Remington, F. (2006). Complete genome sequence of USA300, an epidemic clone of community-acquired meticillin-resistant *Staphylococcus aureus*. Lancet *367*, 731–739.

Dyer, D.W., and Iandolo, J.J. (1983). Rapid isolation of DNA from *Staphylococcus aureus*. Appl. Environ. Microbiol. *46*, 283–285.

Endo, G., and Silver, S. (1995). CadC, the transcriptional regulatory protein of the cadmium resistance system of *Staphylococcus aureus* plasmid pI258. J. Bacteriol. *177*, 4437–4441.

Firth, N., Apisiridej, S., Berg, T., O'Rourke, B.A., Curnock, S., Dyke, K.G, and Skurray, R.A. (2000). Replication of staphylococcal multiresistance plasmids. J. Bacteriol. 182, 2170–2178.

Forsyth, R.A., Haselbeck, R.J., Ohlsen, K.L., Yamamoto, R.T., Xu, H., Trawick, J.D., Wall, D., Wang, L., Brown-Driver, V., Froelich, J.M., C, K.G., King, P., McCarthy, M., Malone, C., Misiner, B., Robbins, D., Tan, Z., Zhu, Z.Y., Carr, G., Mosca, D.A., Zamudio, C., Foulkes, J.G., and Zyskind, J.W. (2002). A genome-wide strategy for the identification of essential genes in *Staphylococcus aureus*. Mol. Microbiol. *43*, 1387–1400.

Gaskill, M.E., and Khan, S.A. (1988). Regulation of the enterotoxin B gene in *Staphylococcus aureus*. J. Biol. Chem. *263*, 6276–6280.

Geissendorfer, M., and Hillen, W. (1990). Regulated expression of heterologous genes in *Bacillus subtilis* using the *Tn*10 encoded *tet* regulatory elements. Appl. Microbiol. Biotechnol. 33, 657–663.

Gerard, H.C., Freise, J., Wang, Z., Roberts, G., Rudy, D., Krauss-Opatz, B., Kohler, L., Zeidler, H., Schumacher, H.R., Whittum-Hudson, J.A., and Hudson, A.P. (2002). *Chlamydia trachomatis* genes whose products are related to energy metabolism are expressed differentially in active vs. persistent infection. Microbes Infect. *4*, 13–22.

Gertz, S., Engelmann, S., Schmid, R., Ziebandt, A-K., Tischerm K., Scharf, C., Hacker, J., and Hecker, M. (2000). Characterization of the σ^B regulon in *Staphylococcus aureus*. J. Bacteriol. *182*, 6983–6991.

Giachino, P., Engelmann, S., and Bischoff, M. (2001). σ^B activity depends on RsbU in *Staphylococcus aureus*. J. Bacteriol. *183*, 1843–1852.

Godány, A., Bukovská, G., Farkasovská, J., Brnáková, Z., Dmitriev, A., Tkacikova, E., Ayele, T., and Mikula, I. (2004). Characterization of a complex restriction-modification system detected in *Staphylococcus aureus* and *Streptococcus agalactiae* strains isolated from infections of domestic animals. Folia. Microbiol. (Praha). *49*, 307–314.

Good, C.M., and Tipper, D.J. (1972). Conditional Mutants of *Staphylococcus aureus* defective in cell wall precursor synthesis. J. Bacteriol. *111*, 231–241.

Goze, A., and Ehrlich, S.D. (1980). Replication of plasmid from *Staphylococcus aureus* in *Escherichia coli*. Proc. Natl. Acad. Sci. USA *77*, 7333–7337.

Greene, C., Vaudaux, P.E., Francois, P., Proctor, R.A., McDevitt, D., and Foster, T.J. (1996). A low-fibronectin-binding mutant of *Staphylococcus aureus* 879R4S has Tn*918* inserted into its single *fnb* gene. Microbiology *142*, 2153–2160.

Grkovic, S., Brown, M.H., Hardie, K.M., Firth, N., and Skurray, R.A. (2003). Stable low-copy-number *Staphylococcus aureus* shuttle vectors. Microbiology *149*, 785–94.

Grundling, A., Missiakas, D.M., and Schneewind, O. (2006). *Staphylococcus aureus* mutants with increased lysostaphin resistance. J. Bacteriol. *188*, 6286–6297.

Gruss, A., and Novick, R. (1986). Plasmid instability in regenerating protoplasts of *Staphylococcus aureus* is caused by aberrant cell division. J. Bacteriol. *165*, 878–883.

Gruss, A.D., Ross, H.F., and Novick, R.P. (1987). Functional analysis of a palindromic sequence required for normal replication of several staphylococcal plasmids. Proc. Natl. Acad. Sci. USA *84*, 2165–2169.

Gustafson, J.E., Berger-Bachi, B., Strassle, A., and Wilkinson, B.J. (1992). Autolysis of methicillin-resistant and -susceptible *Staphylococcus aureus*. Antimicrob. Agents Chemother. 36, 566–572.

Hardy, K., and Haefeli, C. (1982). Expression in *Escherichia coli* of a staphylococcal gene for resistance to macrolide, lincosamide, and streptogramin type B antibiotics. J. Bacteriol. 152, 524–526.

Heir, E., Sundheim, G., and Holck, A.L. (1999). The *qacG* gene on plasmid pST94 confers resistance to quaternary ammonium compounds in staphylococci isolated from the food industry. J. Appl. Microbiol. *86*, 378–388.

Hensel, M., Shea, J.E., Gleeson, C., Jones, M.D., Dalton, E., and Holden, D.W. (1995) Simultaneous identification of bacterial virulence genes by negative selection. Science *269*, 400–403.

Hill, P.J., Eberl, L., Molin, S., and Stewart G.S.A.B. (1994). Imaging bioluminescence in single bacteria, In Bioluminescence and chemiluminescence: fundamental and applied aspects., A.K., Campbell, L.J. Kricka, and P.E. Stanley, ed. (John Wiley & Sons, Chichester, UK) pp. 629–632.

Holden, M.T., Feil, E.J., Lindsay, J.A., Peacock, S.J., Day, N.P., Enright, M.C., Foster, T.J., Moore, C.E., Hurst, L., Atkin, R., Barron, A., Bason, N., Bentley, S.D., Chillingworth, C., Chillingworth, T., Churcher, C., Clark, L., Corton, C., Cronin, A., Doggett, J., Dowd, L., Feltwell, T., Hance, Z., Harris, B., Hauser, H., Holroyd, S., Jagels, K., James, K.D., Lennard, N., Line, A., Mayes, R., Moule, S., Mungall, K., Ormond, D., Quail, M.A., Rabbinowitsch, E., Rutherford, K., Sanders, M., Sharp, S., Simmonds, M., Stevens, K., Whitehead, S., Barrell, B.G., Spratt, B.G., and

Parkhill, J. (2004). Complete genomes of two clinical strains: evidence for the rapid evolution of virulence and drug resistance. Proc. Natl. Acad. Sci. USA *101*, 9786–9791.

Horinouchi, S., Byeon, W-H., and Weisblum, B. (1983). A complex attenuator regulates inducible resistance to macrolides, lincosamides, and streptogramin type B antibiotics in *Streptococcus sanguis*. J. Bacteriol. *154*, 1252–1262.

Horinouchi, S., and Weisblum, B. (1982). Nucleotide sequence and functional map of pC194, a plasmid that specifies inducible chloramphenicol resistance. J. Bacteriol. *150*, 815–825.

Horsburgh, M.J., Aish, J.L., White, I.J., Shaw, L., Lithgow, J.K., and Foster, S.J. (2002). sigmaB modulates virulence determinant expression and stress resistance: characterization of a functional *rsbU* strain derived from *Staphylococcus aureus* 8325-4. J. Bacteriol. *184*, 5457–5467.

Hudson, M.C., and Stewart, G.C.. (1986). Differential utilization of *Staphylococcus aureus* promoter sequences by *Escherichia coli* and *Bacillus subtilis*. Gene. 48, 93–100.

Iandolo, J.J. (1989). Genetic analysis of extracellular toxins of *Staphylococcus aureus*. Annu. Rev. Microbiol. *43*, 375–402.

Iandolo, J.J., Worrell, V., Groicher, K.H., Qian, Y., Tian, R.Y., Kenton, S., Dorman, A., Jia, H-G., Lin, S., Loh, P., Qi, S., Zhu, H., and Roe, B.A. (2002). Comparative analysis of the genomes of the temperate bacteriophages phi 11, phi 12 and phi 13 of *Staphylococcus aureus* 8325. Gene *289*, 109–118.

Iordanescu, S. (1975). Recombinant plasmid obtained from two different, compatible staphylococcal plasmids. J. Bacteriol. *124*, 597–601.

Iordanescu, S. (1976). Three distinct plasmids originating in the same *Staphylococcus aureus* strain. Arch. Roum. Pathol. Exp. Microbiol. 35, 111–118.

Iordanescu, S. (1983). *Staphylococcus aureus* chromosomal mutation specifically affecting the copy number of Inc3 plasmids. Plasmid. *10*, 130–137.

Iordanescu, S., and Surdeanu, M. (1983). Isolation and complementation of temperature-sensitive replication mutants of *Staphylococcus*. Mol. Gen. Genet. *191*, 201–206.

Iordanescu, S., Surdeanu, M., Della Latta, P., and Novick, R.P. (1978). Incompatibility and molecular relationships between small staphylococcal plasmids carrying the same resistance marker. Plasmid *1*, 468–479.

Ito, T., Katayama, Y., Asada, K., Mori, N., Tsutsumimoto, K., Tiensasitorn, C., and Hiramatsu, K. (2001). Structural comparison of three types of staphylococcal cassette chromosome *mec* integrated in the chromosome in methicillin-resistant *Staphylococcus aureus*. Antimicrob. Agents Chemother. *45*, 1323–1336.

Jana, M., Luong, T.T., Komatsuzawa, H., Shigeta, M., and Lee C.Y. (2000). A method for demonstrating gene essentiality in *Staphylococcus aureus*. Plasmid *44*, 100–104.

Janzon, L., and Arvidson, S. (1990). The role of the delta-lysin gene (*hld*) in the regulation of virulence genes by the accessory gene regulator (*agr*) in *Staphylococcus aureus*. EMBO J. *9*, 1391–1399.

Ji, Y., Marra, A., Rosenberg, M., and Woodnut, G. (1999). Regulated antisense RNA eliminates alpha-toxin virulence in *Staphylococcus aureus* infection. J. Bacteriol. *181*, 6585–6590.

Ji, Y., Zhang, B., Van Horn, S.F., Warren, P., Woodnutt, G., Burnham M.K.R., and Rosenberg, M. (2001). Identification of critical Staphylococcal genes using conditional phenotypes generated by antisense RNA. Science *293*, 2266–2269.

Johnson, J.R., Lockman, H.A., Owens, K., Jelacic, S., and Tarr, P.I. (2003). High-frequency secondary mutations after suicide-driven allelic exchange mutagenesis in extraintestinal pathogenic *Escherichia coli*. J. Bacteriol. *185*, 5301–5305.

Jones, C.L., and Khan, S. A. (1986). Nucleotide sequence of the enterotoxin B gene from *Staphylococcus aureus*. J. Bacteriol. *166*, 29–33.

Jones, J.M., Yost, S.C., and Pattee, P.A. (1987). Transfer of the conjugal tetracycline resistance transposon Tn*916* from *Streptococcus faecalis* to *Staphylococcus aureus* and identification of some insertion sites in the staphylococcal chromosome. J. Bacteriol. *169*, 2121–2131.

Kernodle, D.S., Voladri, R.K., Menzies, B.E., Hager, C.C., and Edwards K.M. (1997). Expression of an antisense *hla* fragment in *Staphylococcus aureus* reduces alpha-toxin production *in vitro* and attenuates lethal activity in a murine model. Infect. Immun. *65*, 179–184.

Khan, S., and Novick, R. (1983). Complete nucleotide sequence of pT181, a tetracycline resistance plasmid from *Staphylococcus aureus*. Plasmid *10*, 251–259.

Kornblum, J., Hartman, B.J., Novick, R.P., and Tomasz, A. (1986). Conversion of a homogeneously methicillin-resistant strain of *Staphylococcus aureus* to heterogeneous resistance by Tn*551*-mediated insertional inactivation. Eur. J. Clin. Microbiol. *5*, 714–718.

Kraemer, G.R., and Iandolo, J.J. (1990). High-frequency transformation of *Staphylococcus aureus* by electroporation. Curr. Microbiol. *21*, 373–376.

Kullik, I., and Giachino, P. (1997). The alternative sigma factor σ^B in *Staphylococcus aureus*: regulation of the *sigB* operon in response to growth phase and heat shock. Arch. Microbiol. *167*, 151–159.

Kullik, I., Giachino, P., and Fuchs, T. (1998). Deletion of the alternative sigma factor σ^B in *Staphylococcus aureus* reveals its function as a global regulator of virulence genes. J. Bacteriol. *180*, 4814–4820.

Kupferwasser, L.I, Yeaman, M.R., Nast, C.C., Kupferwasser, D., Xiong, Y.Q., Palma, M., Cheung, A.L., and Bayer, A.S. (2003). Salicylic acid attenuates virulence in endovascular infections by targeting global regulatory pathways in *Staphylococcus aureus*. J. Clin. Invest. *112*, 222–233.

Kwan, T., Liu, J., Dubow, M., Gros, P., and Pelletier, J. (2005). The complete genomes and proteomes of 27 *Staphylococcus aureus* bacteriophages. Proc. Natl. Acad. Sci. USA *102*, 5174–5179.

Lambowitz, A.M., Mohr, G., and Zimmerly, S. (2005). Group II intron homing endonucleases: Ribonucleoprotein complexes with programmable target specificity. In Homing endonucleases and inteins, M. Belfort, B.L. Stoddard, D.W. Wood, and V.

Derbyshire, ed. Nucleic acids and molecular biology. (Springer-Verlag, Heidelberg, Germany), vol. 16, pp. 121–145.

Lambowitz, A.M., and Zimmerly, S. (2004). Mobile Group II introns. Ann. Rev. Genet. *38*, 1–35.

Lampe, D.J., Churchill, M.E., Robertson, H.M. (1996). A purified mariner transposase is sufficient to mediate transposition *in vitro*. EMBO J. *15*, 5470–5479.

Laue, P., and MacDonald, R.E. (1968). Growth stimulation of *Staphylococcus aureus* by L-cystine or L-djenkolic acid. J. Bacteriol. *95*, 2420–2421.

Lee, C.Y., Buranenm S.L., and Ye, Z.H. (1991). Construction of single-copy integration vectors for *Staphylococcus aureus*. Gene *103*, 101–105.

Lee, C.Y., and Iandolo, J.J. (1985). Mechanism of bacteriophage conversion of lipase activity in *Staphylococcus aureus*. J. Bacteriol. *164*, 288–293.

Lee, J.C. (1995). Electrotransformation of staphylococci. Methods Mol. Biol. *47*, 209–216.

Leid, J.G., Shirtliff, M.E., Costerton, J.W., and Stoodley, P. (2002). Human leukocytes adhere to, penetrate, and respond to *Staphylococcus aureus* biofilms. Infect. Immun. 70, 6339–6345.

Lindsay, J.A., and Holden, M.T. (2006). Understanding the rise of the superbug: investigation of the evolution and genomic variation of *Staphylococcus aureus*. Funct. Integr. Genomics. 6,186–201.

Lorenz, M.C., and Fink, G.R. (2001). The glyoxylate cycle is required for fungal virulence. Nature *412*, 83–86.

Lowe A.M., Beattie, D.T., and Deresiewicz R.L. (1998). Identification of novel staphylococcal virulence genes by *in vivo* expression technology. Mol. Microbiol. *27*, 967–976.

Luong, T., and Lee, C.Y. (2006). The *arl* locus positively regulates *Staphylococcus aureus* type 5 capsule via an *mgrA*-dependent pathway. Microbiology *152*, 3123–3131.

Luong, T., Sau, I.S., Gomez, M., Lee, J.C., and Lee, C.Y. (2002). Regulation of *Staphylococcus aureus* capsular polysaccharide expression by *agr* and *sarA*. Infect. Immun. *70*, 444–450.

Lundholm, M., and Bergendahl, B. (2005). Heat treatment to increase phage typability of *staphylococcus aureus*. Eur. J. Clin. Microbiol. Infect. Dis. *7*, 300–302

Lyon, B.R., May, J.W., and Skurray, R.A. (1984). Tn*4001*: a gentamicin and kanamycin resistance transposon in *Staphylococcus aureus*. Mol. Gen. Genet. *193*, 554–555.

Lyon, B.R., and Skurray, R. (1987). Antimicrobial resistance of *Staphylococcus aureus*: genetic basis. Microbiol. Rev. *51*, 88–134.

Mahairas, G.G., Lyon, R.B., and Skurray, R.A., and Pattee, P.A. (1989). Genetic analysis of *Staphylococcus aureus* with Tn*4001*. J. Bacteriol. *171*, 3968–3972.

Markham, P.N., Westhaus, E., Klyachko, K., Johnson, M.E., and Neyfakh, A.A. (1999). Multiple novel inhibitors of the NorA multidrug transporter of *Staphylococcus aureus*. Antimicrob. Agents Chemother. *43*, 2404–2408.

Mazurkiewicz, P., Tang, C.M., Boone, C., and Holden, D.W. Signature-tagged mutagenesis: barcoding mutants for genome-wide screens. Nature Rev. Genet. *7*, 929–939.

McKenzie, T., Hoshino, T., Tanaka, T., and Sueoka, N. (1986). The nucleotide sequence of pUB110: some salient features in relatyion to replication and its regulation. Plasmid *15*, 93–103.

McKinney, J.D., Honer zu Bentrup, K., Munoz-Elias, E.J., Miczak, A., Chen, B., Chan, W.T., Swenson, D., Sacchettini, J.C., Jacobs, R.W., and Russell, D.G. (2000). Persistence of *Mycobacterium tuberculosis* in macrophages and mice requires the glyoxylate shunt enzyme isocitrate lyase. Nature *406*, 735–738.

McNamara, P.J., and Iandolo, J.J. (1998). Genetic instability of the global regulator *agr* explains the phenotype of the *xpr* mutant *Staphylococcus aureus* KSI9051. J. Bacteriol. *180*, 2609–2615.

McNamara, P.J., Milligan-Monroe, K.C., Khalili, S., and Proctor, R.A. (2000). Identification, cloning, and initial characterization of *rot*, a locus encoding a regulator of virulence factor expression in *Staphylococcus aureus*. J. Bacteriol. *182*, 3197–203.

McNeil, L.K., Reich, C., Aziz, R.K., Bartels, D., Cohoon, M., Disz, T., Edwards, R.A., Gerdes, S.Y., Hwang, K., Kubal, M., Margaryan, G.R., Meyer, F., Mihalo, W., Olsen, G.J., Olson, R., Osterman, A.L., Paarmann, D., Paczian, T., Parrello, B., Pusch, G.D., Rodionov, D.A., Shi, X., Vassieva, O., Vonstein, V., Zagnitko, O.P., Xia, F., Zinner, J., Overbeek, R., and Stevens, R. (2007). The National Microbial Pathogen Database Resource (NMPDR): A genomics platform based on subsystem annotation. Nucleic Acids Res. 35, D347–D353.

McWalter, P.W. (1984). Determination of susceptibility of *Staphylococcus aureus* to methicillin by luciferin-luciferase assay of bacterial adenosine triphosphate. J. Appl. Bacteriol. 56, 145–50.

Mei, J. M., Nourbakhsh, F., Ford, C. W., and Holden, D. W. (1997). Identification of *Staphylococcus aureus* virulence genes in a murine model of bacteraemia using signature-tagged mutagenesis. Mol. Microbiol. *26*, 399–407.

Miller, J.H. (1972). Experiments in molecular genetics (New York: Cold Spring Harbor Laboratory Press).

Murray, D.L., Earhart, C.A., Mitchell, D.T., Ohlendorf, D.H., Novick, R.P., and Schlievert, P.M. (1996). Localization of biologically important regions on toxic shock syndrome toxin 1. Infect. Immun. *64*, 371–374.

Murray, R.W., Melchior, E.P., Hagadorn, J.C., and Marotti, K.R. (2001). *Staphylococcus aureus* cell extract transcription-translation assay: firefly luciferase reporter system for evaluating protein translation inhibitors. Antimicrob. Agents Chemother. *45*, 1900–1904.

Nair, S.P., Bischoff, M., Senn, M.M., Berger-Bächi, B. (2003). The σ^B regulon influences internalization of *Staphylococcus aureus* by osteoblasts. Infect. Immun. *71*, 4167–4170.

Nakao, A., Imai, S., and Takano, T. (2000). Transposon-mediated insertional mutagenesis of the D-alanyl-lipoteichoic acid *(dlt)* operon raises methicillin resistance in *Staphylococcus aureus*. Res. Microbiol. *151*, 823–829.

Novick, R.P. (1963). Analysis by transduction of muataions affecting penicillinase formation in *Staphylococcus aureus*. J. Gen. Microbiol. 33, 121–136.

Novick, R. (1967). Properties of a cryptic high-frequency transducing phage in *Staphylococcus aureus*. Virology 33, 155–166.

Novick, R.P. (1987). Plasmid Incompatibility. Microbiol. Rev. *51*, 381–395.

Novick, R.P. (1989). Staphylococcal plasmids and their replication. Annu. Rev. Microbiol. *43*, 537–560.

Novick, R.P. (1991). Genetic systems in staphylococci. Methods Enzymol. *204*, 587–633.

Novick, R.P, Adler, G.K., Majumder, S., Khan, S.A., Carleton, S., Rosenblum, W.D., and Iordanescu, S. (1982). Coding sequence for the pT181 *repC* product: a plasmid-coded protein uniquely required for replication. Proc. Natl. Acad. Sci. USA *79*, 4108–4112.

Novick, R.P., Ross, H.F., Projan, S.J., Kornblum, J., Kreiswirth, G., and Moghazeh, S. (1993). Synthesis of staphylococcal virulence factors is controlled by a regulatory molecule. EMBO J. *12*, 3967–397.

Novick, R.P., and Roth, C. (1968). Plasmid-linked resistance to inorganic salts in *Staphylococcus aureus*. J. Bacteriol. 95, 1335–1342.

Novick, R.P., and Schwesinger, M. (1976). Independence of plasmid incompatibility and replication control functions in *Staphylococcus aureus*. Nature *262*, 623–626.

O'Callaghan, C.H., Morris, A., Kirby, S.M., Shingler, A.H. (1972). Novel method for detection of beta-lactamases by using a chromogenic cephalosporin substrate. Antimicrob. Agents Chemother. *1*, 283–288.

O'Reilly, M., de Azavedo, J.C.S., Kennedy, S., and Foster, T.J. (1986). Inactivation of the alpha-haemolysin gene of *Staphylococcus aureus* by site-directed mutagenesis and studies on the expression of its haemolysins. Microb. Pathogenen. *1*, 125–138.

Pattee, P.A. (1981). Distribution of Tn*551* insertion sites responsible for auxotrophy on the *Staphylococcus aureus* chromosome. J. Bacteriol. *145*, 479–488.

Pattee, P.A. (1990). *Staphylococcus aureus*. In Genetic maps of complex genomes, 5th edition, S.J. O'Brien, ed. (New York: Cold Spring Harbor Laboratory Press), pp. 2.22–2.27.

Pattee, P.A., Lee, H.C., and Bannantine, J.P. (1990). Genetic and physical mapping of the chromosome of *Staphylococcus aureus*. In Molecular biology of the staphylococci. R.P. Novick, ed. (New York: VCH Publishers), pp. 41–58.

Perkins, J.B., and Youngman, P.J. (1986). Construction and properties of *Tn917-lac*, a transposon derivative that mediates transcriptional gene fusions in *Bacillus subtilis*. Proc. Natl. Acad. Sci. USA *83*, 140–144.

Phonimdaeng, P., O'Reilly, M., Nowlan, P., Bramley, A.J., and Foster, T.J. (1990). The coagulase of *Staphylococcus aureus* 8325–4. Sequence analysis and virulence of site-specific coagulase-deficient mutants. Molec. Microbiol. *4*, 393–404.

Pinho, M.G., Filipe, S.R., de Lencastre, H., and Tomasz, A. (2001). Complementation of the essential peptidoglycan transpeptidase function of penicillin-binding protein 2 (PBP2) by the drug resistance protein PBP2A in *Staphylococcus aureus*. J. Bacteriol. *183*, 6525–6531.

Pinho, M.G., Errington, J. (2003). Dispersed mode of *Staphylococcus aureus* cell wall synthesis in the absence of the division machinery. Mol. Microbiol. *50*, 871–881.

Pósfai, G., Kolisnychenko, V., Bereczki, Z., and Blattner, F.R. (1999). Markerless gene replacement in *Escherichia coli* stimulated by a double-strand break in the chromosome. Nucleic Acids Res. *27*, 4409–4415.

Puyang, X., Fan, J., Schumacher, T., Borsari, B., Davey, M., and Ling, L.L. (2003). Over-expression of *mur*G in *S. aureus* does not alter the Ramoplanin MIC. 43rd Annual Interscience Conference on Antimicrobial Agents. Chicago, IL, Sept. 14–17.

Qazi, S.N., Counil, E., Morrissey, J., Rees, C.E., Cockayne, A., Winzer, K., Chan, W.C., Williams, P., and Hill, P.J. (2001). *agr* expression precedes escape of internalized *Staphylococcus aureus* from the host endosome. Infect. Immun. *69*, 7074–7082.

Qazi, S.N.A., Harrison, S.E., Self, T., Williams, P., and Hill, P.J. (2004). Real-time monitoring of intracellular *Staphylococcus aureus* replication. J. Bacteriol. *186*, 1065–1077.

Renzoni, A., Francois, P., Li, D., Kelley, W.L., Lew, D.P., Vaudaux, P., Schrenzel J. (2004). Modulation of fibronectin adhesins and other virulence factors in a teicoplanin-resistant derivative of methicillin-resistant *Staphylococcus aureus*. Antimicrob. Agents Chemother. *48*, 2958–2965.

Roberts, R.J., Vincze, T., Posfai, J., and Macelis, D. (2007). REBASE: restriction enzymes and modifiction. Nucleic Acids Res. 35, D265–D270.

Robertson, H.M., and Lampe, D.J. (1995). Recent horizontal transfer of a mariner transposable element among and between diptera and neuroptera. Mol. Biol. Evol. *12*, 850–862.

Rohrer, S., and Berger-Bächi, B. (2003). Application of a bacterial two-hybrid system for the analysis of protein–protein interactions between FemABX family proteins. Microbiol. *149*, 2733–2738.

Rosendal, K., Bulow, P., and Jessen, O. (1964). Lysogenic conversion in *Staphylococcus aureus*, to a change in the production of extracellular 'tween'-splitting enzyme. Nature *204*, 1222–1223.

Russell, A.D. (1997). Plasmids and bacterial resistance to biocides. J. Appl. Microbiol. 83, 155–165.

Ruvkun, G.B., and Ausubel, F.M. (1981). A general method for site-directed mutagenesis in prokaryotes. Nature *289*, 85–88.

Sambrook, J., Fritsch, E.F., and Maniatis, T. (1989) Molecular Cloning: a Laboratory Manual (New York: Cold Spring Harbor Laboratory Press).

Samra, Z., Ofir, O,. and Bahar, J. (2004). Optimal detection of *Staphylococcus aureus* from clinical specimens using a new chromogenic medium. Diag. Microbiol. Infect. Dis. *49*, 243–247.

Scheer-Abramowitz, J., Gryczan, T., and Dubnau, D. (1981). Origin and mode of replication of plasmids pE194 and pUB110. Plasmid 6, 67–77.

Schenborn, E., and Groskreutz, D. (1999). Reporter gene vectors and assays. Mol. Biotechnol. *13*, 29–44.

Schenk, S., and Laddaga, R.A. (1992). Improved method for electroporation of *Staphylococcus aureus*. FEMS Microbiol. Lett. *73*, 133–138.

Shaw, W.V. (1976). Chloramphenicol acetyltransferase from chloramphenicol-resistant bacteria. Meth. Enzymol. *43*, 737–756.

Shaw, L., Golonka, E., Potempa, J., and Foster, S.J. (2004). The role and regulation of the extracellular proteases of *Staphylococcus aureus*. Microbiology *150*, 217–28.

Shaw, L.N., Aish, J., Davenport, J.E., Brown, M.C., Lithgow, J.K., Simmonite, K., Crossley, H., Travis, J., Potempa, J., and Foster S.J. (2006). Investigations into sigmaB-modulated regulatory pathways governing extracellular virulence determinant production in *Staphylococcus aureus*. J. Bacteriol. *88*, 6070–6080.

Sheehan, B.J., Foster, T.J., Dorman, C.J., Park, S., and Stewart, G.S. (1992). Osmotic and growth-phase dependent regulation of the *eta* gene of *Staphylococcus aureus*: a role for DNA supercoiling. Mol. Gen. Genet. *232*, 49–57.

Silhavy, T.J., Berman, M.L., and Enquist, L.W. (1984). Experiments with Gene Fusions (New York: Cold Spring Harbor Laboratory Press).

Silverman, J.A., Oliver, N., Andrew, T., and Li, T. (2001). Resistance Studies with Daptomycin. Antimicrob. Agents Chemother. *45*, 1799–1802.

Sjöström, J.E., Löfdahl, S., and Philipson, L. (1978). Biological characteristics of a type I restriction-modification system in *Staphylococcus aureus*. J. Bacteriol. *133*, 1144–1149.

Sobral, R.G., Ludovice, A.M., de Lencastre, H., and Tomasz, A. (2006). Role of *murF* in cell wall biosynthesis: isolation and characterization of a *murF* conditional mutant of *Staphylococcus aureus*. J. Bacteriol. *188*, 2543–2553.

Somerville, G.A., Beres, S.B., Fitzgerald, J.R., DeLeo, F.R., Cole, R.L., Hoff, J.S., and Musser, J.M. (2002). *In vitro* serial passage of *Staphylococcus aureus*: changes in physiology, virulence factor production, and *agr* nucleotide sequence. J. Bacteriol. *184*, 1430–1437.

Somerville, G., Mikoryak, C.A., and Reitzer, L. (1999). Physiological characterization of *Pseudomonas aeruginosa* during exotoxin A synthesis: glutamate, iron limitation, and aconitase activity. J. Bacteriol. *181*, 1072–1078.

Somerville, G.A., Said-Salim, B., Wickman, J.M., Raffel, S.J., Kreiswirth, B.N., and Musser, J.M. (2003). Correlation of acetate catabolism and growth yield in *Staphylococcus aureus*: implications for host–pathogen interactions. Infect. Immun. *71*, 4724–4732.

Sonenshein, A.L. (2002). The Krebs citric acid cycle. In *Bacillus subtilis* and its Closest Relatives: from Genes to Cells. A.L. Sonenshein, J.A. Hoch, and R. Losick ed. (Washington, DC: ASM Press), pp. 151–162.

Sonstein, S.A., and Baldwin, J.N. (1972). Loss of the penicillinase plasmid after treatment of *Staphylococcus aureus* with sodium dodecyl sulfate. J. Bacteriol. *109*, 262–265.

Sozhamannan, S., Dabert, P., Moretto, V., Ehrlich, S.D., and Gruss, A. (1990). Plus-origin mapping of single-stranded DNA plasmid pE194 and nick site homologies with other plasmids. J. Bacteriol. *172*, 4543–4548.

Stringfellow, W.T., Dassy, B., Lie, M., and Fournier, J.M. (1991). *Staphylococcus aureus* growth and type 5 capsular polysaccharide production in synthetic media. Appl. Environ. Microbiol. *57*, 618–621.

Stobberingh, E. E., and K. C. Winkler. 1977. Restriction-deficient mutants of *Staphylococcus aureus*. J. Gen. Microbiol. *99*, 359–367.

Sullivan, M.A., Yasbin, R.E., and Young, F.E. (1984). New shuttle vectors for *Bacillus subtilis* and *Escherichia coli* which allow rapid detection of inserted fragments. Gene *29*, 21–26.

Sunderland, J., Brenwald, N., Smith, E., Andrews, J.M., Wise, R., and Honeybourne, D. (1995). Enumeration of phagocytosed *Staphylococcus aureus* inside cells of the mouse macrophage cell line J774 by bioluminescence, fluorescence and viable counting techniques. J. Biolumin. Chemilumin. *10*, 291–299.

Sun, J., Zheng, L., Landwehr, C., Yang, J., and Ji, Y. (2005). Identification of a novel essential two-component signal transduction system, YhcSR, in *Staphylococcus aureus*. J. Bacteriol. *187*, 7876–7880.

Tegmark, K., Karlsson, A., and Arvidson, S. (2000). Identification and characterization of SarH1, a new global regulator of virulence gene expression in *Staphylococcus aureus*. Mol. Microbiol. *37*, 398–409.

Tenhami, M., Hakkila, K., and Karp, M. (2001). Measurement of effects of antibiotics in bioluminescent *Staphylococcus aureus* RN4220. Antimicrob. Agents Chemother. *45*, 3456–3461.

Thomas, C.M. (1988). Recent studies on the control of plasmid replication. Biochim. Biophys. Acta. *949*, 253–263.

Throup, J.P., Zappacosta, F., Lunsford, R.D., Annan, R.S., Carr, S.A., Lonsdale, J.T., Bryant, A.P., McDevitt, D., Rosenberg, M., and Burnham, M.K.R. (2001). The *srhSR* gene pair from *Staphylococcus aureus*: genomic and proteomic approaches to the identification and characterization of gene function. Biochemistry *40*, 10392–10401.

Tomasz, A., Albinom A., and Zanatim E. (1970). Multiple antibiotic resistance in a bacterium with suppressed autolytic system. Nature *227*, 138–140.

Timmins, B.S., Farrell, A.M., Stewart, G.S.A.B., and Holland, K.T. (1996). Construction and use of a luciferase based reporter system for detection of toxic shock syndrome Toxin-1 production by *Staphylococcus aureus*. Zbl. Bakt. Suppl. 28, 67–68.

Tseng, C.W., and Stewart, G.C. (2005). Rot repression of enterotoxin B expression in *Staphylococcus aureus*. J. Bacteriol. *187*, 5301–5309.

Tseng, C.W., Zhang, S., and Stewart, G.C. (2004). Accessory gene regulator control of staphyloccoccal enterotoxin D gene expression. J. Bacteriol. *186*, 1793–1801.

Udo, E.E., Jacob, L.E., and Mathew, B. (2001). Genetic analysis of methicillin-resistant *Staphylococcus aureus* expressing high- and low-level mupirocin resistance. J. Med. Microbiol. *50*, 909–915.

Vijver, J.C.M., Es-Boon, M., and Michel, M.F. (1972). Lysogenic Conversion in *Staphylococcus aureus* to leucocidin production. J. Virol. *10*, 318–319.

Vandenesch, F., Kornblum, J., and Novick, R.P. (1991). A temporal signal, independent of *agr*, is required for *hla* but not *spa* transcription in *Staphylococcus aureus*. J. Bacteriol. *173*, 6313–6320.

van Wamel, W., Xiong, Y-Q., Bayer, A.S., Yeaman, M.R., Nast, C.C., and Cheung, A.L. (2002). Regulation of *Staphylococcus aureus* type 5 capsular polysaccharides by *agr* and *sarA in vitro* and in an experimental endocarditis model. Microb. Pathogen. *33*, 73–79.

Villafane, R., Bechhofer, D., Naraymn, C., and Dubnau, D. (1987). Replication control genes of plasmid pE194. J. Bacteriol. *169*, 4822–4829.

Waldron, D.E., and Lindsay, J.A. (2006). Sau1: a novel lineage-specific type I restriction-modification system that blocks horizontal gene transfer into *Staphylococcus aureus* and between *S. aureus* isolates of different lineages. J. Bacteriol. *188*, 5578–5585.

Wang, P.Z., Henriquez, V.B., Projan, S.J., Iordanescu, S., and Novick, R.P. (1991). The effect of plasmid copy number mutations on pT181 replication initiator protein expression. Plasmid 25, 198–207.

Wang, P.Z., Projan, S.J., Leason, K.R., and Novick, R.P. (1897). Translational fusion with a secretory enzyme as an indicator. J. Bacteriol. *169*, 3082–3087.

Wardenburg, J.B., Patel, R.J., and Schneewind, O. (2007). Surface proteins and exotoxins are required for the pathogenesis of *Staphylococcus aureus* pneumonia. Infect. Immun. *75*, 1040–1044.

Wieland, K-P., Wieland, B., and Götz, F. (1995). A promoter-screening plasmid and xylose-inducible, glucose-repressible expression vectors for *Staphylococcus carnosus*. Gene *158*, 91–96.

Winkler, K.C., de Waart, J., and Grootsen, C. (1965). Lysogenic conversion of staphylococci to loss of beta-toxin. J. Gen. Microbiol. *39*, 321–333.

Wolz, C., Pöhlmann-Dietze, P., Steinhuber, A., Chien, Y-T., Manna, A., van Wamel, W., and Cheung, A. (2000). Agr-independent regulation of fibronectin-binding protein(s) by the regulatory locus *sar in Staphylococcus aureus*. Molec. Microbiol. 36, 230–243.

Wood, K.V. (1995). Marker proteins for gene expression. Curr. Opin. Biotechnol. *6*, 50–58.

Wu, C-H., and Bergdoll, M.S. (1971). Stimulation of enterotoxin B production. II. Synthetic medium for staphylococcal growth and enterotoxin B production. Infect. Immun. 3, 784–792.

Xia, M., Lunsford, R.D., McDevitt, D., and Iordanescu, S. (1999). Rapid method for the identification of essential genes in *Staphylococcus aureus*. Plasmid *42*, 144–149.

Xiong, Y-Q., Bayer, A.B., Yeaman, M.R., van Wamel, W., Manna, A.C., and Cheung, A.L. (2004). Impacts of *sarA* and *agr* in *Staphylococcus aureus* strain Newman on fibronectin-binding Protein A gene expression and fibronectin adherence capacity *in vitro* and in experimental infective endocarditis Infect. Immun. *72*, 1832–1836.

Xiong, Y-Q., van Wamel, W., Nast, C.C., Yeaman, M.R., Cheung, A.L., and Bayer, A.S. (2002). Activation and transcriptional interaction between *agr* RNAII and RNAIII in *Staphylococcus aureus in vitro* and in an experimental endocarditis model. J. Infect. Dis. *186*, 668–677.

Xiong, Y-Q., Willard, J., Yeaman, M.R., Cheung, A.L., and Bayer, A.S. (2006). Regulation of *Staphylococcus aureus* alpha-toxin gene (*hla*) expression by *agr*, *sarA*, and *sae in vitro* and in experimental infective endocarditis. J. Infect. Dis. *194*, 1267–1275.

Yamaguchi, T., Hayashi, T., Takami, H., Nakasone, K., Ohnishi, M., Nakayama, K., Yamada, S., Komatsuzawa, H., and Sugai, M. (2000). Phage conversion of exfoliative toxin A production in *Staphylococcus aureus* Mol. Microbiol. *38*, 694–705.

Yan, K., Sossong, T.M., and Payne, D.J. (2001). Regions of FtsZ important for self-interaction in *Staphylococcus aureus*. Biochem. Biophys. Res. Commun. *4*, 515–518.

Yarwood, J.M., Bartels, D.J., Volper, E.M., Greenberg, E.P. (2004). Quorum Sensing in *Staphylococcus aureus* Biofilms. J. Bacteriol. *186*, 1838–1850.

Yansura, D.G., and Henner, D.J. (1984). Use of the *Escherichia coli lac* repressor and operator to control gene expression in *Bacillus subtilis*. Proc. Natl. Acad. Sci. USA *81*, 439–443.

Yao, J., Zhong, J., Fang, Y., Geisinger, E., Novick, R.P., and Lambowitz, A.M. (2006). Use of targetrons to disrupt essential and nonessential genes in *Staphylococcus aureus* reveals temperature sensitivity of Ll.LtrB group II intron splicing. RNA *12*, 1271–1281.

Ye, J., Kandegedara, A., Martin, P., and Rosen, B.P. (2005). Crystal structure of the *Staphylococcus aureus* pI258 CadC Cd(II)/Pb(II)/Zn(II)-responsive repressor. J. Bacteriol. *187*, 4214–4221.

Yoon, K.P., Misra, T.K., and Silver, S. (1991). Regulation of the *cadA* cadmium resistance determinant of *Staphylococcus aureus* plasmid pI258. J. Bacteriol. *173*, 7643–7649.

Yoon, K.P., and Silver, S. (1991). A second gene in the *Staphylococcus aureus cadA* cadmium resistance determinant of plasmid pI258. J. Bacteriol. *173*, 7636–7642.

Youngman, P., Perkins, J.B., and Losick, R. (1984). A novel method for the rapid cloning in *Escherichia coli* of *Bacillus subtilis* chromosomal DNA adjacent to Tn*917* insertions. Mol. Gen. Genet. 195, 424–433.

Youngman, P.J., Perkins, J.B., and Sandman, K. (1984b). New genetic methods, molecular cloning strategies, and gene fusion techniques for *Bacillus subtilis* which take advantage of Tn*917* insertional mutagenesis. In Genetics and biotechnology of bacilli, J.A. Hoch, and A.T. Ganesan ed. (New York: Academic Press) pp. 103–111.

Zadik, P.M., Davies, S., Whittaker, S., and Mason, C. (2001). Evaluation of a new selective medium for methicillin-resistant *Staphyloccocus aureus*. J. Med. Microbiol. *50*, 476–479.

Zheng, L., Yang, J., Landwehr, C., Fan, F., and Ji, Y. (2005). Identification of an essential glycoprotease in *Staphylococcus aureus*. FEMS Microbiol. Lett. *245*, 279- 285.

Zheng, L., Yu, C., Bayles, K., Lasa, I., and Ji, Y. (2007). Conditional mutation of an essential putative glycoprotease eliminates autolysis in *Staphylococcus aureus*. J. Bacteriol. *189*, 2734–2742.

Zhang, L., Fan, F., Palmer, L.M., Lonetto, M.A., Petit, C., Voelker, L.L., St. John, A., Bankosky, B., Rosenberg, M., and McDevitt, D. (2000). Regulated gene expression in *Staphylococcus aureus* for identifying conditional lethal phenotypes and antibiotic mode of action. Gene. *255*, 297–300.

Ziebandt, A.K., Weber, H., Rudolph, J., Schmid, R., Höper, D., Engelmann, S., and Hecker, M. (2001). Extracelellular proteins of *Staphylococcus aureus* and the role of SarA and sigma B. Proteomics *1*, 480–493.

Global Regulators of *Staphylococcus aureus* Virulence Genes

6

Bénédicte Fournier

Abstract

Staphylococci cause a great diversity of infections and have thus developed sophisticated mechanisms for eliciting infection in different environments. Staphylococcal infection requires the production of various virulence factors. The expression of these virulence factor genes is co-ordinated by global regulators. These regulators help bacteria to adapt to a hostile environment by producing factors enabling them to survive and subsequently to cause infection at the appropriate time. Several of these global virulence regulators, such as the Agr system, Sar and Sae, have been well characterized. Others, such as the Arl system, Sar homologues (Rot, MgrA, SarS, SarR, SarT, SarU, SarV, SarX, SarZ and TcaR), the Srr system and TRAP, require further study to determine their exact role in the virulence regulon. Several proteins and regulators with primary functions other than the regulation of virulence, such as Clp proteins, HtrA, MsrR, aconitase and CcpA, are also involved in regulating virulence, often through interactions with major virulence regulators. Other proteins, such as SvrA, Msa, CfvA and CfvB, regulate virulence, but their main function remains unknown.

Introduction

Staphylococcus aureus is a major human and animal pathogen responsible for a wide diversity of infections. It remains an important cause of hospital- and community-acquired infections. *S. aureus* survives in the nares – its main reservoir – but may also be found in axillae, perineum, vagina and pharynx. About 30% of healthy adults carry *S. aureus* in the nares (Lowy, 1998; Tenover and Gaynes, 2000). The implantation of foreign objects, such as catheters, increases the risk of developing staphylococcal infection (Lowy, 1998). Staphylococcal infections can usually be subdivided into three main groups: superficial lesions (e.g. skin and soft tissue infections such as abscess or furuncles), invasive infections (e.g. osteomyelitis, endocarditis, pneumonia or septicaemia), and toxinoses (e.g. food poisoning, scalded-skin syndrome, toxic shock syndrome) (Novick, 2000; Tenover and Gaynes, 2000).

Invasive infections are usually a complication of bacteraemia in patients presenting multiple risk factors for infection. Bacteraemia may lead to endocarditis, metastatic infection (e.g. osteomyelitis, arthritis and sepsis) (Lowy, 1998). Endothelial cells are crucial in the infection process, because staphylococci bind to these cells, which subsequently take them up by phagocytosis. *S. aureus* is usually extracellular, but it may survive in cells, including endothelial and epithelial cells, osteoblasts and neutrophils (Buggy *et al.*, 1984; Hamill *et al.*, 1986; Vann and Proctor, 1987; Hudson *et al.*, 1995; Almeida *et al.*, 1996; Beekhuizen *et al.*, 1997; Bayles *et al.*, 1998; Menzies and Kourteva, 1998; Gresham *et al.*, 2000; Kahl *et al.*, 2000; Lowy, 2000a; Jett and Gilmore, 2002; Hess *et al.*, 2003). Phagocytes take the bacteria up into a plasma membrane-derived vacuole called the phagosome. Phagosomes then undergo maturation into endosomes and lysosomes with degradative properties (Vieira *et al.*, 2002). Staphylococci may escape from the endosomes, replicating in the cytoplasm of the cells (Almeida *et al.*, 1996; Bayles *et al.*, 1998; Kahl *et al.*, 2000). The survival and replication of staphylococci in cells may protect these bacteria against host responses and antibiotic effects (Craven and Anderson, 1979; Bayles *et al.*, 1998; Lowy, 2000a). Indeed, relapse after antibiotic treatment frequently occurs in staphylococcal infections (Lowy, 2000a). The ability of staphylococci to survive in cells accounts for several pathogenetic features, including the ability to cause endocarditis or metastatic infections in particular (Lowy, 2000b). Furthermore, the intracellular environment may also be responsible for the selection of small-colony variants. These bacteria are auxotrophic for menadione, thymidine or haemin, lack α-toxin and persist for longer in cells (Vesga *et al.*, 1996; Lowy, 2000a; Kahl *et al.*, 2005). The internalization of *S. aureus* in endothelial cells triggers a series of cellular changes, including an increase in cytokine production (Lowy, 2000b).

Toxinoses are different from classical staphylococcal infections in that they are caused by pyrogenic-toxin superantigens, usually encoded by genes present in mobile genetic elements. Superantigens are extracellular molecules with a high potency for stimulating immune cells. Indeed, some pathogens avoid destruction by the immune system by secreting superantigens that bind at high frequency to lymphocyte antigen receptors (Goodyear and Silverman, 2004). These toxins associate with major histocompatibility complex class II molecules present on antigen-presenting cells (outside the classical antigen-binding region). The binding of this complex to T-cell receptors results in the non-specific activation of T lymphocytes and cytokine release (Alouf and Muller-Alouf, 2003). The massive release of inflammatory mediators is responsible for the toxic shock syndrome induced by toxic shock syndrome toxin-1 (see section on toxin superantigens) (Lowy, 1998).

S. aureus virulence factors

The induction of infection requires the production of virulence factors. However, only certain toxinoses caused by specific toxins can induce disease alone. Other virulence factors present in almost all strains seem to play a role in one or more of the steps leading to infection, but none of these factors has been found to be essential for infection. Host invasion by *S. aureus* and the subsequent establishment of infection involve several essential steps: binding to the cell surface or other host components, escape from the host immune system, dissemination and tissue invasion (Projan and Novick, 1997). Virulence factors have therefore been classified according to their role during the infection process. Some are bound to the cell wall of *S. aureus*, whereas others are secreted into the extracellular medium. Large number of virulence factors have been described and considerable progress towards understanding the role of staphylococcal virulence factors in infection has been made in recent years, particularly as concerns the mechanism used by *S. aureus* to circumvent host defences (Foster, 2005; Rooijakkers *et al.*, 2005b).

Staphylococci cause a wide diversity of infections and have thus developed sophisticated mechanisms for eliciting infection in different environments. The expression of the virulence factor genes involved in these mechanisms is coordinated by global regulators. These regulators help bacteria to adapt to a hostile environment by producing factors enabling the bacteria to survive and to cause subsequent infection.

MSCRAMMS (microbial surface components recognizing adhesive matrix molecules)

MSCRAMMS are proteins anchored to the peptidoglycan by a covalent bond. They are responsible for attaching the bacterium to various extracellular matrices, such as collagen, fibronectin or vitronectin. These cell wall-associated proteins are mostly produced during the exponential growth phase when bacteria are grown *in vitro*. Many of these proteins have been shown to bind to host components (Foster and McDevitt, 1994; Foster and Hook, 1998). Fibronectin-binding proteins FnbpA and FnbpB, encoded by *fnbA* and *fnbB*, respectively, bind fibronectin (Flock *et al.*, 1987; Signas *et al.*, 1989; Jonsson *et al.*, 1991; Schwarz-Linek *et al.*, 2003). FnbPs have recently been shown to be involved in cell invasion (Dziewanowska *et al.*, 1999; Lammers *et al.*, 1999), and this factor is necessary and sufficient for invasion in some cell lines (Sinha *et al.*,

2000; Ahmed *et al.*, 2001; Massey *et al.*, 2001). Indeed, once FnbPs have bound fibronectin, they bind integrin α5β1 – the cell fibronectin receptor – facilitating uptake into the cell (Sinha *et al.*, 1999; Dziewanowska *et al.*, 2000; Massey *et al.*, 2001). FnBPs also mediate adhesion to platelets via fibronectin and fibrinogen (Heilmann *et al.*, 2004). They have also been shown to bind elastin (Roche *et al.*, 2004). ClfA and ClfB (Clf for 'clumping factor') bind fibrinogen and type I cytokeratin 10, leading to platelet aggregation (McDevitt *et al.*, 1994; Bayer *et al.*, 1995; Ni Eidhin *et al.*, 1998; O'Brien *et al.*, 2002a; O'Brien *et al.*, 2002b; Walsh *et al.*, 2004). ClfA protects *S. aureus* against phagocytosis by macrophages (Palmqvist *et al.*, 2004). Cna (collagen adhesin) binds collagen (Patti *et al.*, 1992; Gillaspy *et al.*, 1997; Gillaspy *et al.*, 1998).

Protein A (Spa)

Protein A is also a cell wall-associated protein that interacts with several host factors (immunoglobulins G, A and E, platelets) (Lofdahl *et al.*, 1983). Protein A binds to the Fc of IgG. Immunoglobulin G is an opsonin, similar to C3 product (see section on the capsule). Opsonins are molecules that associate with bacteria, favouring the recognition of bacteria by neutrophils, resulting in phagocytosis. Protein A binding to IgG attenuates the phagocytosis mediated by Fc (Peterson *et al.*, 1977). Indeed, protein A has an antiphagocytic effect (Gemmell *et al.*, 1991). It also modulates platelet aggregation (O'Brien *et al.*, 2002a). Protein A binding to von Willebrand factor may be involved in the induction of endovascular disease by *S. aureus* (Hartleib *et al.*, 2000). Furthermore, protein A has been shown to sensitize B cells, thereby increasing their recognition of TLR2 ligands (Bekeredjian-Ding *et al.*, 2007). Finally, protein A binds TNFR-1 (tumour necrosis factor receptor 1), and is thus a crucial virulence factor for staphylococcal pneumonia (Gomez *et al.*, 2004).

S. aureus promotes the apoptosis of mammalian cells. Apoptosis is a well-characterized, highly controlled mechanism involving the activation of a cascade of intracellular events, leading to the death of unwanted cells. It is often observed in phagocytic cells, such as monocytes and macrophages (Menzies and Kourteva, 1998). Caspases are intracellular cysteine proteases responsible for causing apoptotic cell death. They are activated by the intrinsic mitochondrial pathway or by an extrinsic death receptor pathway (Haslinger *et al.*, 2003). *S. aureus* seems to trigger apoptosis by several different mechanisms (see the section on cytotoxins, and α-toxin in particular) (Baran *et al.*, 1996; Bayles *et al.*, 1998; Bantel *et al.*, 2001; Essmann *et al.*, 2003; Haslinger *et al.*, 2003). Several *S. aureus* virulence factors are involved in apoptosis. Protein A binds to IgM associated with B cells, inducing proliferation by acting like a superantigen. It subsequently induces apoptosis and the depletion of cells crucial for host defence (Goodyear and Silverman, 2004). Protein A therefore has several, highly diverse functions and contributes to several steps in the infection process.

Capsule

Most *S. aureus* bacteria produce a capsule (Lee *et al.*, 1994; Lin *et al.*, 1994; Sau and Lee, 1996; Ouyang and Lee, 1997; O'Riordan and Lee, 2004). In total, 11 serotypes have been identified. Serotypes 5 and 8 account for 75–85% of all staphylococcal strains involved in human infections. Their microcapsules are produced in small amounts and are not visible on negative staining, yet still play a role in protection from the innate defences including complement activation. Three different pathways (classical, lectin and alternative), depending on the mode of recognition, can be triggered and lead to the activation of C3 convertases, which cleave C3 into C3a and C3b (Fujita, 2002). C3b covalently binds to the bacterial pathogen by opsonization. Bacteria are recognized by receptors of phagocytic cells specific for complement that subsequently induce opsonophagocytosis (Rooijakkers *et al.*, 2005a). Staphylococcal capsules increase resistance to opsonophagocytosis and prevent staphylococci from being killed by neutrophils or macrophages (Karakawa *et al.*, 1988; Nilsson *et al.*, 1997; Thakker *et al.*, 1998; Cunnion *et al.*, 2001; Kampen *et al.*, 2005; Watts *et al.*, 2005). The expression of capsule synthesis genes is enhanced by the internalization of staphylococci in neutrophils (Voyich *et al.*, 2005). Finally, the staphylococcal capsule is a potent activator of human CD4+ T cells (Tzianabos *et al.*, 2001).

Extracellular adherence protein (Eap)

This protein, also known as Map (major histocompatibility complex class II analogous protein) is secreted (McGavin *et al.*, 1993; Palma *et al.*, 1999). It binds to several plasma proteins (fibrinogen, fibronectin, prothrombin, vitronectin), fibroblasts and epithelial cells and to itself, promoting aggregation (Palma *et al.*, 1999; Chavakis *et al.*, 2002; Hussain *et al.*, 2002). Eap is also involved in inhibiting leukocyte migration. Mac-1 and LFA-1, receptors present on the surface of leukocytes, interact with the ICAM-1 present on endothelial cells. Mac-1 can also associate with fibrinogen. During pathogen infection, leukocytes migrate from blood to the infection site. This migration is due to the association of leukocyte receptors, such as Mac-1 or LFA-1, with the ICAM-1 of endothelial cells. Eap interacts with the ICAM-1 present on endothelial cells and with fibrinogen. The presence of Eap therefore decreases neutrophil migration and inhibits the binding of staphylococci to endothelial cells via ICAM-1, subsequently preventing leukocyte extravasation (Chavakis *et al.*, 2002; Harraghy *et al.*, 2003; Haggar *et al.*, 2004). Furthermore, Eap binds to T cells, impairing their function and proliferation. It is involved in the internalization of staphylococci in fibroblast and epithelial cells (Lee *et al.*, 2002).

Extracellular fibrinogen-binding protein (Efb)

Efb is a secreted protein that binds fibrinogen. It prevents platelet aggregation and may be involved in the pathogenesis of wound infections due to *S. aureus* (Boden and Flock, 1994; Palma *et al.*, 1996; Palma *et al.*, 2001). Efb has been shown to bind C3 and abolishes the classical and alternative pathways leading to complement activation. It also abolishes complement-mediated opsonophagocytosis (Lee *et al.*, 2004a; Lee *et al.*, 2004b). Its structure has recently been studied. Efb binds to the native form of C3, triggering conformational changes in C3 that abolish its activation by enzymes (Hammel *et al.*, 2007).

Staphylokinase (Sak)

Staphylokinase is a secreted protein encoded by prophages (Sako and Tsuchida, 1983; van Wamel *et al.*, 2006). It converts the human plasminogen bound to the bacterial surface to plasmin. Sak has been shown to have anti-opsonic properties. Sak-activated plasmin is a potent serine protease that cleaves and inactivates both immunoglobulin G and human C3b, thereby inhibiting phagocytosis (Rooijakkers *et al.*, 2005c). Plasmin may also cleave the fibrin, collagen and elastin present around the site of infection, facilitating the entry of *S. aureus* into host tissues (Bokarewa *et al.*, 2006). Several antimicrobial peptides are detected in the respiratory tract. One such peptide, cathelicidin, is present before the release of others, such as α-defensins, by polymorphonuclear cells (PMNs). Staphylokinase binds directly to cathelicidin and the resulting complex has enhanced plasminogen activator activity (Braff *et al.*, 2007). Sak then triggers the release of defensins from neutrophils, binds to these defensins and represses their bactericidal effects (Jin *et al.*, 2004; Bokarewa *et al.*, 2006).

CHIPS (chemotaxis inhibitory protein of *S. aureus*)

CHIPS is a small secreted protein, 14.1 kDa in size, encoded by the *chp* gene. This gene has been detected in an immune evasion cluster (IEC) present in a β-haemolysin-converting bacteriophage (van Wamel *et al.*, 2006). CHIPS represses the neutrophil and monocyte chemotaxis induced by C5a (activated complement) and N-formyl peptide, but not the chemotaxis elicited by IL-8, a neutrophil chemoattractant (Veldkamp *et al.*, 2000; de Haas *et al.*, 2004). Indeed, CHIPS specifically binds to human neutrophils and monocytes on C5a receptors (C5aR and C5L2) and the formylated peptide receptor (FPR), thereby inhibiting the induction of these receptors by their respective ligands (de Haas *et al.*, 2004; Haas *et al.*, 2005; Wright *et al.*, 2007). Furthermore, the intraperitoneal injection of CHIPS in mice attenuates the neutrophil recruitment triggered by C5a (de Haas *et al.*, 2004), suggesting that this protein plays an important role in host immune defences.

SCIN (staphylococcal complement inhibitor)

SCIN is a small secreted protein, 9.8 kDa in size, encoded by *scn* and forming part of the same IEC as CHIPS. SCIN inhibits the three complement pathways. C3 convertases involved in C3 activation (see section on capsule) consist of a complex of two molecules, C4b2a and C3bBb. SCIN interacts with these complexes, abolishing their function. SCIN thus inhibits C3b binding to target bacteria and the subsequent phagocytosis of these bacteria (Rooijakkers *et al.*, 2005a). Thus, Eap, Efb, Sak, CHIPS and SCIN are interesting factors helping staphylococci to circumvent host defences.

Proteases

Several proteases are secreted by *Staphylococcus*. SspA is a serine protease first isolated from the V8 strain. The *sspA* gene is part of the *sspABC* operon (Rice *et al.*, 2001). SspB or staphopain B is a cysteine protease and SspC is a cysteine proteinase inhibitor also known as staphostatin B (Massimi *et al.*, 2002; Rzychon *et al.*, 2003). Staphopain A (ScpA) is a cysteine protease encoded by a gene of the *scpAB* operon. ScpB or staphostatin A is a cysteine protease inhibitor (Rzychon *et al.*, 2003). Aur (for aureolysin) is a metalloprotease (Arvidson, 2000; Shaw *et al.*, 2004). SplA-BCDE and F are also serine proteases, encoded by the same operon (Arvidson, 2000; Reed *et al.*, 2001; Shaw *et al.*, 2004). These proteins have been implicated principally in the degradation or activation of external staphylococcal proteins (Drapeau, 1978; McAleese *et al.*, 2001; Massimi *et al.*, 2002; Rzychon *et al.*, 2003). However they are also involved in the infection process. Indeed, some of these proteases have been implicated in the degradation of host components, such as immunoglobulins, antimicrobial peptides (e.g. peptide LL37 synthesized by the host) or plasma proteins (e.g. α-1-anti-chymotrypsin) (Potempa *et al.*, 1986; Potempa *et al.*, 1991; Goguen *et al.*, 1995; Projan and Novick, 1997; Ulvatne *et al.*, 2002; Sieprawska-Lupa *et al.*, 2004). They may also be involved in tissue degradation (Lowy, 2000a). Indeed, cysteine proteases have been implicated in elastin degradation (Potempa *et al.*, 1988). However, these proteases may also induce the activation of host components, as in the case of SspB, which cleaves chemerin. Chemerin is a low-biological activity chemoattractant found in blood. Cleavage of the C-terminal region of this protein results in an active chemoattractant that triggers the migration of plasmacytoid dendritic cells and macrophages to the site of infection. Thus, SspB is probably involved in the inflammatory response observed during staphylococcal infection (Kulig *et al.*, 2007). The role of proteases in staphylococcal virulence is complex and involves several of their functions.

Toxin superantigen

The staphylococcal superantigens are enterotoxins, toxic shock syndrome toxin-1 and exfoliative toxins (Kreiswirth *et al.*, 1983; Alouf and Muller-Alouf, 2003; Baker and Acharya, 2004). The genes encoding these virulence factors are carried by accessory genetic elements, such as phages, plasmids or staphylococcal pathogenicity islands (SaPI) (Betley and Mekalanos, 1985; Lindsay *et al.*, 1998; Yamaguchi *et al.*, 2001). Enterotoxins cause food poisoning. A large number of enterotoxins (A to E and G to R) have been characterized based on serological or amino-acid sequence differences (Betley and Mekalanos, 1985; Jones and Khan, 1986; Betley and Mekalanos, 1988; Couch *et al.*, 1988; Bayles and Iandolo, 1989; Hovde *et al.*, 1990; Su and Wong, 1995; Zhang *et al.*, 1998). Toxic shock syndrome toxin-1 (Tst) is responsible for toxic shock syndrome associated with menstruation in 90% of cases (Kreiswirth, 1989; Lowy, 2000a). It has also been implicated in *S. aureus*-induced apoptosis (Hofer *et al.*, 1996). Exfoliative toxin A and exfoliative toxin B, encoded by *eta* and *etb*, respectively, may cause a spectrum of damage, from localized lesions to extensive exfoliation, as observed in staphylococcal scalded-skin syndrome (Lee *et al.*, 1987; Bohach and Foster, 2000; Ladhani, 2003). These toxins display some similarity to serine proteases, but have no protease activity. It has therefore been suggested that, after recognizing and binding their target (desmoglein-1, a glycoprotein involved in cell-cell adhesion in the epidermis) these proteins undergo a change in conformation, inducing protease activity and the degradation of desmoglein and exfoliation (Ladhani, 2003).

Cytotoxins

These cytolytic toxins are secreted by *S. aureus* and induce the lysis of polymorphonuclear leukocytes, monocytes and/or erythrocytes (Song *et al.*, 1996). The α-haemolysin or α-toxin (Hla), β-haemolysin (Hlb), δ-haemolysin (Hld) and leukocidins of this group of toxins are involved in the tissue invasion phase (Kielian *et al.*, 2001).

α-haemolysin, which targets red blood cells, mononuclear immune cells, platelets, endothelial and epithelial cells, has been well studied (Gray and Kehoe, 1984; Patel *et al.*, 1987; Bramley *et al.*, 1989; Bhakdi and Tranum-Jensen, 1991; Callegan *et al.*, 1994; Bayer *et al.*, 1997; Bohach and Foster, 2000). α-toxin is present in the culture supernatant as a monomer. On interacting with the target cytoplasmic membrane, this toxin associates into hexamers or heptamers to form a pore in the membrane. Membrane damage then leads to cell lysis (Bhakdi and Tranum-Jensen, 1991; Bohach and Foster, 2000). α-toxin also induces the release of various proinflammatory cytokines, such as IL-1β, IL-6, IL-8, without cell lysis (Bhakdi *et al.*, 1989; Dragneva *et al.*, 2001; Kielian *et al.*, 2001). α-toxin is necessary for the process of apoptosis induced by *S. aureus* (Vann and Proctor, 1988; Jonas *et al.*, 1994; Menzies and Kourteva, 1998; Bantel *et al.*, 2001; Haslinger *et al.*, 2003; Haslinger-Loffler *et al.*, 2005). It induces caspase 8 and 9 via the mitochondrial cytochrome c pathway or the death receptor pathway, depending on the type of cell (Bantel *et al.*, 2001; Haslinger *et al.*, 2003). α-Toxin-induced cell death also seems to be triggered by a caspase-independent pathway resulting in necrosis (Essmann *et al.*, 2003). α-Toxin may be involved in the internalization of *S. aureus* by host cells. Indeed, the attenuation of α-toxin production increases staphylococcal adhesion to cells and enhances internalization by cells (Liang and Ji, 2006). As α-toxin has been shown to interact with β1 integrin, a cell receptor for fibronectin (Isberg and Leong, 1990), competition between FnbPA and α-toxin for β1 integrin may modulate the internalization of staphylococci (Liang and Ji, 2006). Thus, α-toxin is a major virulence factor of *S. aureus*.

β-Haemolysin differs from α-toxin in that it causes the hot-cold lysis of erythrocytes, for which a period in the cold is required (Projan *et al.*, 1989; Bohach and Foster, 2000). It is a sphingomyelase that degrades the sphingomyelin present in the cytoplasmic membrane (Bohach and Foster, 2000). Mammalian cell surfaces are processed by ectodomain shedding, a proteolytic event leading to the release of proteins bound to the cell surface into the extracellular medium. This mechanism is also tightly modulated by various extracellular molecules and several intracellular signalling pathways. Ectodomain shedding is a regulatory process that allows the cell surface to adapt rapidly to stimuli. It is also involved in staphylococcal pathogenesis (Park *et al.*, 2004). A substantial number of proteins, including cytokines, are shed by metalloproteases known as sheddases or secretases. Syndecan-1, a major heparan sulphate proteoglycan, binds and regulates various components. It can be shed and its soluble form abolishes certain interactions at the cell surface. Indeed, shed syndecan-1 usually inhibits the excess of proinflammatory mediators, which may be deleterious to the host. Syndecan-4 is also a heparan sulphate proteoglycan (Bernfield *et al.*, 1999). HB-EGF (heparin-binding epidermal growth factor) belongs to the epidermal growth factor family and is involved in wound repair and staphylococcal infection (Lemjabbar and Basbaum, 2002). *S. aureus* has been shown to stimulate the shedding of FasL, syndecans-1 and -4 and HB-EGF (Baran *et al.*, 2001; Park *et al.*, 2004). α- and β-haemolysins stimulate syndecan-1 shedding. β-haemolysins induce syndecan-4 and HB-EGF shedding, whereas α-toxin does not (Park *et al.*, 2004).

γ-haemolysin and leukocidins are two-component compounds, consisting of one S-class and one F-class component. They bind sequentially to cells, resulting in synergistic effects (Kamio *et al.*, 1993; Choorit *et al.*, 1995; Bohach and Foster, 2000). They also form a pore resulting in leukocyte lysis (Jayasinghe and Bayley, 2005; Joubert *et al.*, 2006). Panton–Valentine leukocidin (PVL) lyses polymorphonuclear cells and monocytes/macrophages (Finck-Barbancon *et al.*, 1993) and causes necrotizing pneumonia (Labandeira-Rey *et al.*, 2007). Until recently, PVL was rare and associated with about 2% of clinical strains (Prevost *et al.*, 1995a). However, an increase in the frequency of highly lethal necrotizing pneumonia due to PVL was observed in community-acquired infections (Gillet *et al.*, 2002). The two proteins constituting Panton–Valentine leukocidin are encoded by the *lukS-PV* (or *lukM-PV*) and *lukF-PV* genes, which are usually organized into an operon (Rahman *et al.*, 1992; Prevost *et al.*, 1995b). γ-haemolysins also consist of two components encoded by the *hlgACB* operon: HlgB (class F) and HlgC (class S) or HlgA (class S) (Cooney *et al.*, 1993; Kamio *et al.*, 1993; Supersac *et al.*, 1993). Leucotoxins are more leucotoxic and dermonecrotic than γ-haemolysins. In contrast, γ-haemolysins lyse rabbit erythrocytes (Prevost *et al.*, 1995b).

Lipases

Several lipases are detected in the supernatant of *S. aureus* cultures. Glycerolester hydrolase, encoded by the *geh* gene, hydrolyses long-chain and water-soluble triacylglycerols (Lee and Iandolo, 1986; Arvidson, 2000). An esterase encoded by the *lip* gene cleaves short-chain triacylglycerols. The role of this esterase in virulence remains unclear. Lipase has been shown to be strongly chemotactic to granulocytes and to decrease the phagocytic killing of these cells (Rollof *et al.*, 1988). It has been suggested that the fatty acids produced by triacylglycerol hydrolysis may modulate the immune system (Arvidson, 2000). Two phospholipase C proteins have been described: the sphingomyelinase β-haemolysin (see section on cytotoxins) and the phosphatidylinositol-specific phospholipase C (PI-PLC) encoded by *plc*. PI-PLC cleaves the glycan-phosphatidylinositol moiety of proteins anchored in the membrane of host cells. This enzyme has been identified as a potential virulence factor, but its role in virulence remains unknown (Marques *et al.*, 1989). Fatty acid modifying enzyme (FAME) catalyses the esterification of long-chain free fatty acids to generate cholesterol. These fatty acids, found in abscesses, are known to be toxic to *S. aureus*. FAME thus abolishes the bactericidal effect of these lipids, enabling the bacterium to invade abscesses (Chamberlain and Imanoel, 1996).

Coagulase

Coagulase is not associated with the cell wall, but it is bound to the surface of *S. aureus* (Kaida *et al.*, 1989). It associates with prothrombin and forms the staphylothrombin responsible for the conversion of fibrinogen to fibrin, leading to the coagulation of serum (Lowy, 2000a). Its role in staphylococcal infections remains unclear. Coagulase seems to play a crucial role in experimental blood-borne pneumonia (Sawai *et al.*, 1997). In contrast, it seems to have no effect in experimental endocarditis, whereas it contributes to acute bacterial endocarditis (Baddour *et al.*, 1994; Moreillon *et al.*, 1995; Panizzi *et al.*, 2004). The deletion of *coa* has no effect in subcutaneous and intramammary mouse models (Phonimdaeng *et al.*, 1990).

The Agr system

The Agr system (accessory gene regulator) has been studied in considerable detail since its discovery over a decade ago. A mutant carrying a Tn*551* insertion in this locus was first recovered in 1982, based on its *hla*-negative phenotype (Mallonee *et al.*, 1982). In 1986, Recsei *et al.* showed that this locus probably encodes a regulator active in *trans* and required for the transcription of virulence factor genes, such as *tst*, *spa*, *hla*, *hlb* and *hld* (Recsei *et al.*, 1986). Indeed, the Agr system regulates the expression of numerous virulence genes. Its mechanism has been well characterized. An RNA molecule, which is produced in response to quorum sensing, regulates gene translation. This RNA has a striking feature: it may act as a repressor but also as an activator (see section on RNAIII). Interestingly, four component systems with structures similar to that of Agr are present in other species such as *Listeria monocytogenes* or *Enterococcus faecalis* (Autret *et al.*, 2003; Nakayama *et al.*, 2006), suggesting that this system is crucial for bacteria.

Structure of the *agr* locus

The *agr* locus was cloned and sequenced (Peng *et al.*, 1988) and found to encode a two-component system. However, the regulation of this system is much more complicated than that of a classical two-component system. Indeed, the *agr* locus is partly constituted by an operon of four genes, *agrBDCA*. The transcription of this operon produces an RNA molecule, RNAII. Another RNA molecule, RNAIII, is produced by transcription in the opposite direction, overlapping the *hld* gene encoding the δ-toxin (Novick *et al.*, 1993; Morfeldt *et al.*, 1995). Deletion of this region results in an *agr*$^-$ phenotype, demonstrating the importance of this region for Agr activity (Janzon and Arvidson, 1990). Thus, two different and divergent transcripts, RNAII and RNAIII, play a crucial role in the regulation of *agr*. These transcripts are initiated from two promoters,

P2 and P3, respectively (Fig. 6.1A) (Novick *et al.*, 1995). The Agr system has been shown to respond to quorum sensing (Ji *et al.*, 1995).

RNAIII

RNAIII is a long (0.5 kb) RNA molecule, the transcription of which is positively regulated by the proteins encoded by the *agrBDCA* operon. RNAIII is not produced in an *agrA* mutant (Janzon *et al.*, 1989). Its transcription increases during growth, reaching a maximum in the post-exponential phase (Vandenesch *et al.*, 1991). It has a long half-life, dependent on growth phase – 20 minutes in the exponential growth phase and 11 minutes in the post-exponential phase (Janzon *et al.*, 1989). *hld* is a short open reading frame encoding a 26 aa haemolysin secreted without a signal peptide. The first question posed concerned whether Hld was responsible for the regulatory function of this DNA region. Deletion or mutation of the *hld* gene prevented the production of Hld and its release into the culture supernatant, but did not causes an agr^- phenotype (Janzon and Arvidson, 1990; Novick *et al.*, 1993), suggesting that *hld* is not responsible for Agr activity. The transcription initiation site of *hld* is 188 bp downstream from the initiation site of RNAIII, and this RNA may be responsible for Agr activity (Janzon and Arvidson, 1990). Indeed, the introduction of a plasmid containing the RNAIII sequence under the control of an inducible promoter in an agr^- strain is sufficient to induce an agr^+ phenotype, whereas the introduction of a similar plasmid carrying RNAII does not complement an agr^- phenotype (Novick *et al.*, 1993). This strongly suggests that RNAIII itself is the regulatory effector of the Agr system. The secondary structure of RNAIII is well known and consists of 14 hairpin structures with three long-range interactions bringing the 5′ end near the 3′ end (Fig. 6.1B). Hairpins 1, 7, 13 and 14 are highly conserved among RNAIIIs from various staphylococcal species, suggesting that several hairpins may have a common, major function in RNAIII (Benito *et al.*, 2000). It has been suggested that hairpin 1, located at the 5′ end of RNAIII, may stabilize RNAIII (Benito *et al.*, 2000). In contrast, hairpins 6, 8, 9, 11 and 12 diverge considerably between species (Vandenesch *et al.*, 1993; Tegmark *et al.*, 1998; Benito *et al.*, 2000). RNAIIIs from various coagulase-negative species (*Staphylococcus epidermidis*, *Staphylococcus warneri*, *Staphylococcus simulans* or *Staphylococcus lugdunensis*) complement an agr^- phenotype in *S. aureus*. Indeed, RNAIII from *S. epidermidis*, *S. warneri* and *S. simulans* completely restores the repression of *spa* transcription and the activation of *hla* transcription. In contrast, the RNAIII of *S. lugdunensis* increases *hla* expression but does not suppress *spa* expression (Benito *et al.*, 1998; Tegmark *et al.*, 1998). This suggests that the function of RNAIII is common to different species but that some domains of RNAIII have specific, independent functions. The translation of δ-haemolysin *in vivo* does not occur until 1 h after the synthesis of RNAIII, and deletion of the 3′ half of RNAIII restores prompt translation after RNAIII synthesis. It has therefore been suggested that RNAIII may present different conformational forms, one of which may inhibit Hld translation. Benito *et al.* showed that ribosomes bind to the ribosome site of *hld in vitro*, suggesting that the *in vitro* form of RNAIII is responsible for Hld translation (Benito *et al.*, 2000). Comparison of the overall pattern of nucleotide reactivity to chemical probes shows that *in vivo* and *in vitro* forms are very similar, with the exception of the region containing the ribosome-binding site, which is not cleaved *in vivo*. The ribosome-binding site of *hld* is therefore probably protected *in vivo* and an unknown factor is probably required to unfold RNAIII and subsequently elicit Hld translation (Balaban and Novick, 1995b; Benito *et al.*, 2000). RNAIII acts at both the transcriptional and translational levels (Novick *et al.*, 1993). The regulation of RNAIII at the post-transcriptional level has been extensively studied. However, the mechanism by which RNAIII modulates transcription remains vague.

The regulation of *hla* by *agr* is complex. In an RNAIII mutant, *hla* transcript levels are one fifth to one tenth those in the wild-type, whereas protein levels are one seventieth those in the wild-type, suggesting that *hla* is regulated mostly at the post-transcriptional level (Novick *et al.*, 1993). Indeed, the *hla* mRNA presents an unusually long 5′ untranslated region (UTR) of 330 nt. This 5′UTR forms a hairpin containing the Shine–Dalgarno sequence, which blocks

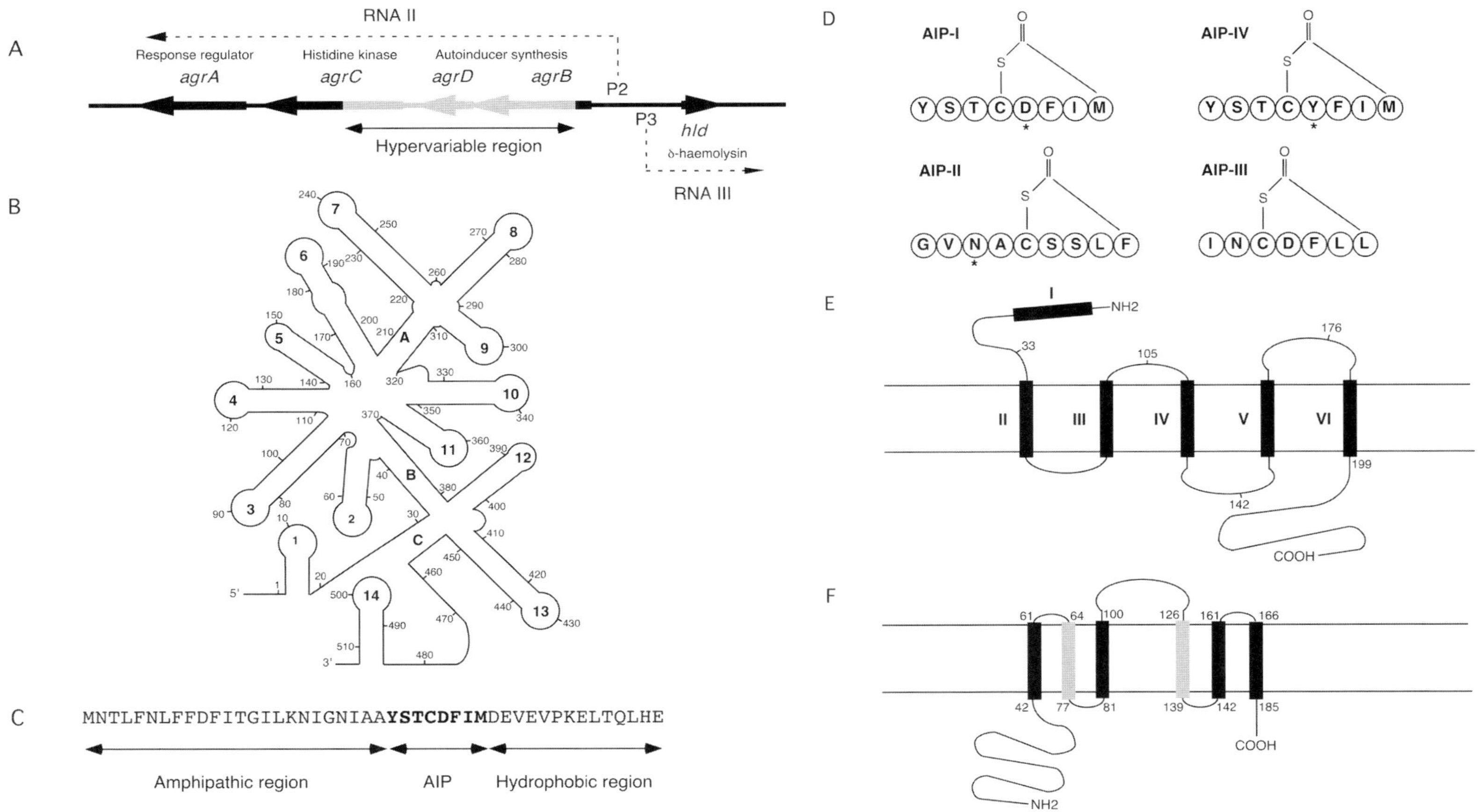

Figure 6.1 The *agr* locus. (A) The *agr* locus, with RNAII and RNAIII synthesis from promoters P2 and P3, as described by Jarraud *et al.* (2000). Dotted lines represent the transcripts. (B) The secondary structure of RNAIII, composed of 14 hairpins (the numbers of the different hairpins are indicated) with three long-range interactions indicated as A, B and C, as described by Benito *et al*. (2000). (C) The amino acid sequence of group I AgrD, as described by Zhang *et al*. (2004). The AIP sequence is indicated in bold typeface. (D) The structure of the different AIP groups, as described by Lyon *et al*. (2002b). Amino acids are circled. The thioester bond between the carboxy group of the peptide and the sulfhydryl group of the cysteine is also indicated. Asterisks indicate crucial amino acids. (E) The structure of AgrC. The amino acid residues are numbered as described by Lina *et al*. (1998). F. The structure of group I AgrB. The amino acid residues are numbered as described by Zhang *et al.* (2002). Hydrophobic transmembrane domains are indicated in black, whereas hydrophilic domains are indicated in grey.

translation. Deletion of the first 95 nt of RNAIII prevents Hla production, suggesting that the 5′ terminus is responsible for the activation of Hla production (Novick *et al.*, 1993). Indeed, RNAIII (residues 15 to 87) binds to the 5′UTR of *hla* mRNA (from 117 to 200 nt). When RNAIII is associated with *hla* mRNA, the G residue of the Shine–Dalgarno sequence is cleaved by RNase T1, which cleaves single-strand RNA on the 3′ side of a G nucleotide. In contrast, RNaseT1 cannot cleave this DNA region in the absence of RNAIII. The binding of *hla* mRNA to RNAIII frees the ribosome-binding site for translation. Thus, base-pairing between the two RNAs inhibits the formation of a secondary structure containing the ribosome-binding site sequence and repressing Hla translation (Morfeldt *et al.*, 1995).

The mechanism by which RNAIII regulates *spa* expression has also been elucidated. Several studies have shown that the 3′ end of RNAIII is necessary and sufficient to repress *spa* mRNA production (Novick *et al.*, 1993; Benito *et al.*, 1998; Benito *et al.*, 2000). Indeed, deletion of the 3′ sequence corresponding to the last five amino acids is sufficient to abolish the effects of RNAIII on *spa* transcription (Novick *et al.*, 1993). A chimera composed of the 5′ half of RNAIII from *S. lugdunensis* and the 3′ half of RNAIII from *S. aureus* represses *spa* transcription. In contrast, RNAIII from *S. lugdunensis* does not repress *spa* mRNA, suggesting that the 3′ end of the RNAIII of *S. aureus* is essential for the repression of *spa* transcription (Benito *et al.*, 1998). The subcloning of various regions of RNAIII showed that residues between 391 and 514, corresponding to part of hairpin 12, hairpin 13, hairpin 14 and residues 403 to 455, corresponding to hairpin 13 repress *spa* mRNA. In contrast, residues between 484 and 514, corresponding to hairpin 14, only slightly decreased *spa* mRNA levels (Fig. 6.1B). These findings suggest that hairpin 13 is essential for the repression of *spa* mRNA (Benito *et al.*, 2000). RNAIII residues 385 to 445 are complementary to *spa* mRNA nucleotides 2 to 58 (Huntzinger *et al.*, 2005). Thus, RNAIII binds to *spa* mRNA in this region. Indeed, a mutated RNAIII lacking hairpin 13 or residues between 430 to 437 cannot bind *spa* mRNA (Huntzinger *et al.*, 2005). The *spa* mRNA contains two hairpins, the first of which includes the ribosome-binding site. When RNAIII is associated with hairpin 1 of the *spa* mRNA, the translational initiation complex cannot form. In contrast, in the presence of a mutant RNAIII lacking the complementary region, ribosomes can bind to *spa* mRNA. The binding of RNAIII to *spa* mRNA is rapid, as required for antisense regulation. The rate of binding is within the range of various complementary antisense-target RNAs. Thus, the primary effect of the binding of RNAIII to *spa* mRNA is the inhibition of ribosome binding (Huntzinger *et al.*, 2005). A mutant strain lacking the RNase III gene (*rnc*) has been shown to display *spa* expression similar to that of an *agr*$^-$ strain, suggesting that RNase III may be involved in Agr regulation. The half-life of *spa* mRNA depends on genetic background. In an *agr*$^+$ strain, the half-life is so short that the molecule is not detected. In *agr*$^-$ or *rnc*$^-$ strains, the half-life is much longer, at 15 to 32 minutes. Thus, the rapid degradation of *spa* mRNA depends on its association with RNAIII. Indeed, hairpin 2 of *spa* mRNA is the only target of RNase III in the absence of RNAIII. In contrast, positions 5, 31 and 39, present in the complementary region of the *spa* mRNA, are cleaved in the presence of RNA III. Thus, RNAIII has the secondary effect of increasing *spa* mRNA degradation by RNase III, probably by inhibiting translation (Huntzinger *et al.*, 2005). RNAIII also binds to the 5′UTR region of the gene of the fibrinogen-binding protein, SA1000, repressing its translation. *SA1000* mRNA has a structure similar to that of *spa* (Boisset *et al.*, 2007). RNAIII is thus a fascinating regulator of virulence that operates mainly through the post-transcriptional modulation of gene expression based on the secondary structures of RNA molecules.

AgrD and the agr pheromone

Regulation of the expression of the *agr* locus depends on production of the four proteins encoded by RNAII and, subsequently, on P2 promoter activity. Indeed, the deletion of each of these protein genes abolishes P2 and P3 activity (Novick *et al.*, 1995). This suggests that the *agr* locus displays autocatalytic feedback. AgrA and AgrC are similar to the response regulator and histidine kinase components, respectively, of a two-component system (Novick *et al.*, 1995; Lina *et al.*, 1998). In the presence of an external signal, histidine kinase undergoes autophosphorylation at a crucial histidine residue. The phosphoryl group is subsequently transferred to the response regulator. This phosphorylation modifies the conformation of the response regulator and, thus, the binding of the regulator to the promoter region of its target genes. Indeed, phosphorylation of the response regulator increases or represses target gene expression (West and Stock, 2001).

AgrD is a 46- to 47-amino acid peptide (Fig. 6.1C). During the exponential growth phase, this propeptide is cleaved by two proteolytic digestions to produce a peptide of seven to nine residues (Ji *et al.*, 1997). This peptide undergoes post-translational modifications to generate the autoinducing peptide (AIP), which is then secreted into the extracellular medium. The modification and secretion of AgrD is promoted by AgrB (see section on AgrB below). When AIP reaches a threshold concentration directly correlated to the density of bacteria, AgrC is activated (see section on AgrC below), triggering the autoinduction of *agrBDCA* and the synthesis of RNAIII (Ji *et al.*, 1995). Thus, RNAII and RNAIII are mostly expressed during the post-exponential growth phase. AgrD is an integral membrane protein that is not extracted in NaCl or $NaCO_3$ (Zhang *et al.*, 2004). It consists of an N-terminal amphipathic domain with an α-helix, AIP and a hydrophobic C-terminal region (Fig. 6.1C). The N-terminal region is vital for the anchoring of AgrD in the membrane, whereas the C-terminal region is detected in the cytoplasm. The association of AgrD with the membrane is a major step in processing by AgrB. Indeed, AgrD not bound to the membrane is not modified by AgrB. However, the N-terminal region does not promote binding to AgrB (Zhang *et al.*, 2004). Interestingly, no transmembrane helix domain that could account for the interaction of AgrD with the membrane is present in the N-terminal region, but the α-helical amphipathic motif is clearly involved in membrane association. Indeed, the replacement of this N-terminal region by an artificial amphipathic sequence of 11 amino acids results in properties similar to those of the native AgrD. However, the replacement of the N-terminal region by the α-helix of the transmembrane domains of the *Escherichia coli* signal peptidase resulted in the anchorage of AgrD to the membrane but not its processing by AgrB (Zhang *et al.*, 2004). This suggests that AgrD is probably located in the inner leaflet of the membrane. Thus, the amphipathic N-terminal region plays a crucial role in AgrD processing as the membrane-anchoring agent required for AgrB activity. The mechanism by which AgrB binds to AgrD remains unknown.

In 1997, Ji *et al.* showed that the *agr* locus of laboratory strain RN6390B was inhibited by the culture supernatants of other strains, such as RN6596, whereas it was activated by its own supernatant and by the culture supernatants of several different strains, such as SA502A (Ji *et al.*, 1997). They defined three groups of staphylococci, with each group showing mutual activation of the *agr* locus by the AIPs of its members but inhibition by the AIPs of the other two groups. A fourth group was subsequently described (Jarraud *et al.*, 2000; McDowell *et al.*, 2001; Otto *et al.*, 2001). Sequencing of the *agr* region in two or three members of each group revealed the sequence of the *agr* locus to be highly conserved, except for the region corresponding to amino acid 34 of AgrB to amino acid 205 of AgrC (Fig. 6.1A) (Ji *et al.*, 1997). This region is conserved within individual groups but differs considerably between groups. AIP sequences were determined by sequencing AIPs purified from culture supernatants or by assessing the efficiency of a synthetic peptide. A new method based on targeted multistage mass spectrometry can be used for the direct sequencing of peptides in the culture supernatant (Kalkum *et al.*, 2003). AIPs from groups I, II and III have highly divergent sequences, whereas group IV AIPs differ from group I AIPs by only one amino acid (Fig. 6.1D). AgrC from group IV is activated by the AIP of its own group, weakly activated by AIP I and inhibited by AIP II and AIP III. AIP IV strongly activates AgrC IV, but inhibits AgrC II and III (Jarraud *et al.*, 2000). Its effects on AgrC I remain unclear: two groups have reported that AIP IV inhibits AgrC I (McDowell *et al.*, 2001; Goerke *et al.*, 2003), whereas another group found that AIP IV activated AgrC I (Jarraud *et al.*, 2000; Lyon *et al.*, 2002b). Interestingly, several recombinants of *agr* have been observed: *agr I/IV* is composed of a characteristic sequence from *agr I* running from the 5′ end of *agrB* to *agrD*, a unique sequence including part of *agrB* and the 5′ end of *agrC* and an *agr IV*-specific sequence from the 3′ end of *agrC* to *agrA* (Robinson *et al.*, 2005). Another *agr* locus, *agrIc*, is specific to group I in *agrD* and characteristic of group IV in the other *agr* genes (Goerke *et al.*, 2005a). This disparity between the *agrD* sequence and other *agr* genes at the same locus may account for AIP IV abolishing or activating AgrC I in different studies (Goerke *et al.*, 2003).

As some AIPs inhibit heterologous Agr and subsequently virulence factors controlled by this regulator, this model of peptides capable of controlling virulence factor production was thus extensively studied to be potentially used as therapeutics against staphylococcal infections (see section on *in vivo* virulence of *agr*$^-$ mutants). The only feature common to these AIPs, with the exception of that of *Staphylococcus intermedius*, is a cysteine residue five amino acids from the C terminus (Fig. 6.1D). The first synthetic peptide was not efficient. Indeed, mass spectrometry found that this synthetic peptide was dimeric, probably due to intermolecular bonds between cysteine residues. In contrast, the native AIP was found to be monomeric, with a molecular mass 18±1 atomic mass units lower than that predicted based on amino-acid sequence. It has been suggested that the native peptide forms an intramolecular bond between the carboxy group of the peptide and the sulfhydryl group of the cysteine (Fig. 6.1D). A synthetic cyclic peptide was indeed capable of activating the *agr* locus of its own group (Ji *et al.*, 1997). The residues constituting the macrocyclic structure have been referred to as endocyclic residues, whereas those in the N-terminal tail are known as exocyclic residues (Lyon *et al.*, 2002b).

About 58% of group I AIPs purified from a special culture medium allowing AIP overproduction are methionyl sulfoxide AIPs, generated by *S*-oxidation of the methionine thioether side-chain. These derivatives can neither activate nor inhibit the *agr* locus (McDowell *et al.*, 2001). Furthermore, the native N-terminus of AIP III and AIP from *S. epidermidis* is required for their activity (Otto *et al.*, 1998; Otto *et al.*, 1999; Lyon *et al.*, 2002b), suggesting that this region may be important for AIP function.

The structure-activity relationships of AIP I and II have been investigated in detail and we now have a precise idea of their mechanism of action. The replacement, in turn, of the same amino acid by an L-alanine residue in AIP I and AIP II showed that the effect of this replacement was most pronounced for endocyclic amino acids. Such replacements reduced the activating and inhibitory properties of these mutated AIPs by a factor of up to 600 (Mayville *et al.*, 1999; McDowell *et al.*, 2001). Replacement of the thioester bond (-S-) by an ester lactone (-O-) or an amide lactam (-NH-) resulted in AIPs I and II with a very low activating capacity and a potent inhibitory effect (Mayville *et al.*, 1999; McDowell *et al.*, 2001; Lyon *et al.*, 2002a).

Deletion of the tail region converts the macrocycle AIP II into an inhibitor (Lyon *et al.*, 2000; Lyon *et al.*, 2002b) of the receptors of all groups (Lyon *et al.*, 2000). One of the key residues required for AIP II activation is present in the tail. Indeed, replacement of the exocyclic Asn-3 by an alanine results in a self-antagonist (Fig. 6.1D) (Lyon *et al.*, 2002a). The cyclic structure of AIP II is therefore crucial for activation and inhibition, whereas the N-terminal tail of this molecule is important only for activation (Mayville *et al.*, 1999).

A truncated AIP I lacking its tail acts as a partial agonist, weakly activating AgrC I at high concentrations and inhibiting its own activation at low concentrations (Lyon *et al.*, 2002b). The endocyclic residue in position 5 of AIP I has been shown to be crucial for its activation properties (Fig. 6.1D) (McDowell *et al.*, 2001; Lyon *et al.*, 2002b). Indeed, the replacement of the Asp-5 present into the cyclic structure by an alanine residue results in the production of an AIP I molecule that inhibits its own group (McDowell *et al.*, 2001; Lyon *et al.*, 2002b). AIP IV differs from AIP I by a single amino acid, in position 5, and this position is crucial for the specificity of both AIPs (Lyon *et al.*, 2002b). A chimera of AIPs in which exocylic and endocyclic regions of AIP I and II were switched cannot activate any receptor, but is nonetheless inhibitory (Lyon *et al.*, 2002b). Thus, cross-inhibitory action appears to be more permissive than activation. This may account for the AIPs of various structures inhibiting the same receptor (Lyon *et al.*, 2002a).

Three groups of AIPs have been described in *S. epidermidis* and *S. intermedius* (Otto *et al.*, 1998; Dufour *et al.*, 2002; Ji *et al.*, 2005; Sung *et al.*, 2006). *S. intermedius* is the only staphylococcal species in which the AIPs are based on a lactone rather than a thiolactone. AIPs from *S. intermedius* activate their cognate receptors and inhibit all four groups of *S. aureus*. AIP I, II and III of *S. aureus* weakly represses AgrC from *S. intermedius* (Ji *et al.*, 2005). A mutated AIP of *S. intermedius* (Ser-5-Cys) activates its cognate receptor much less efficiently than the wild-type AIP, whereas a mutant (Ser-5-Ala) has been shown to lose its activation and inhibitory properties (Ji *et al.*, 2005).

AgrC and signal transduction

This 46 kDa protein consists of an N-terminal region with five to six transmembrane domains and a cytoplasmic C-terminal domain (Fig. 6.1E) (Lina *et al.*, 1998). It has been suggested that the N-terminal region, which carries a putative first transmembrane domain, is located outside the cell because a chimera consisting of the N-terminal 33 residues of AgrC fused to PhoA is found in the extracellular medium of *E. coli* (Lina *et al.*, 1998). However, the possibility that the N-terminal region is located in the cytoplasm in *S. aureus* with the first transmembrane domain crossing the membrane cannot be ruled out (Lyon *et al.*, 2002b). The cytoplasmic membrane of an *agr*$^+$ staphylococcal strain can phosphorylate a protein of the same molecular size as AgrC in the presence of [γ-^{32}P]-ATP and early post-exponential culture supernatants. Purified native AIP also activates membrane phosphorylation. However, no phosphorylation is observed in *agr*$^-$ strains, in the absence of supernatant culture or in the presence of the supernatant culture of an *agr*$^-$ strain. Furthermore, prior treatment of the membrane with an antibody specific for AgrC inhibits phosphorylation. This strongly suggests that AgrC is activated by AIP (Lina *et al.*, 1998). An AgrC variant from Ser-176 to the stop codon and containing the last transmembrane domain and the C-terminal cytoplasmic region has similar properties to the native protein (Fig. 6.1E). In contrast, the cytoplasmic region of AgrC from Ser-199 to the stop codon displays spontaneous activation independent of the absence or presence of AIP (Lina *et al.*, 1998).

The expression of *agrCA* from one group in an *agr*$^-$ strain of another group is activated by the group of origin of the *agrCA* and inhibited by the group of origin of the *agr*$^-$ strain. Thus, AgrCA alone, rather than the background of the strain, is responsible for group specificity (Lyon *et al.*, 2000). The activation of AgrC by AIP is dose-dependent, with maximal activation generally observed at a particular concentration of AIP (Lyon *et al.*, 2000). The N-terminal domain sequences are highly divergent, whereas the C-terminal cytoplasmic domain is highly conserved (Wright *et al.*, 2004). A chimera composed of the sensor domain of one group and the cytoplasmic domain of another group is activated by an AIP from the same group as the sensor domain, suggesting that the sensor domain is responsible for group specificity (Lyon *et al.*, 2002a). Thus, AgrC consists of a sensor or input domain, principally detected in the membrane, which recognizes AIP and a transmitter domain carrying the histidine residue and present in the cytoplasm. Wright *et al.* recently constructed chimeras of the sensor domain (Wright *et al.*, 2004). A proximal region from one group (from residue 1 to 90) was associated with the distal region of another group (from residue 90 to 210) (Fig. 6.1E). Greater divergence of the parental sequences seems to be associated with poorer functioning of the chimera. The chimera is active if the distal and proximal regions display 54 to 87% sequence identity. For example, chimeras I:IV or IV:I are activated by AIP I or IV but inhibited by AIP II. Interestingly, the IV:I chimera is more activated by AIP I than the I:IV chimera. Furthermore, the I:IV chimera is more activated by AIP I than AgrC IV. This suggests that the distal domain is indeed responsible for AIP recognition, but that the proximal domain also plays a role (Wright *et al.*, 2004). Chimeras III:I and III:IV are activated by AIP I and IV, respectively but to levels two magnitudes lower than observed with their cognate receptors, AgrC I and IV. Thus, for AgrC I and IV, specificity is determined by the distal domain, whereas the proximal domain is involved in responsiveness. In contrast, 32% sequence identity results in a non-functional chimera. Chimeras I:III and IV:III are activated by all AIPs, but inhibited by none (Wright *et al.*, 2004).

Inhibition and activation by AIPs are reversible (Lyon *et al.*, 2002a). Furthermore, the binding of AIP II to its cognate receptor, AgrC, is abolished by AIP I, suggesting that the binding sites of agonist AIPs overlap with those of inhibitory AIPs, indicating that competition may occur between the agonist and antagonist AIPs (Lyon *et al.*, 2002a). AIPs increase in hydrophobicity from the N to the C terminus. The C-terminal hydrophobic region is required for interaction with AgrC. A model of binding between AIP and AgrC has been proposed (Lyon *et al.*, 2002a). According to this model, the C-terminal endocyclic region of AIP binds to a hydrophobic region present between the distal and proximal regions of AgrC, and the specific residues necessary for activation (Asp-5 for AIP I, Tyr-5 for AIP IV and Asn-3 for AIP II) activate the distal domain of AgrC. The presence of an AIP lacking these residues results in the inhibition of AgrC (Wright *et al.*, 2004). Thus, AgrC is a complex protein different from classical histidine kinases, which usually contain only two transmembrane domains. The recognition of AIPs by AgrC is a subtle process that seems to involve several domains. Furthermore, this interaction is complicated by the need to distinguish between different groups of Agr, some of which are able to inhibit AgrC.

AgrB and pheromone processing

AgrB is a 23 kDa protein located in the cytoplasmic membrane and consists of four hydrophobic transmembrane domains (Leu-43 to Leu-60, Phe-82 to Ile-99, Tyr-143 to Lys-160 and Ile-167 to Ile-184), two hydrophilic domains potentially embedded in the membrane (Thr-65 to Ala-76 and Ala-127 to Ile-138), a cytoplasmic N-terminal domain and a region between Ile-104 and Ala-124 that constitutes an external loop (Fig. 6.1F) (Zhang *et al.*, 2002). AgrB is necessary and sufficient for the processing of AgrD (Ji *et al.*, 1995; Ji *et al.*, 1997; Zhang *et al.*, 2002; Qiu *et al.*, 2005). The interaction between AgrB and AgrD is group-specific. The interesting case of the AgrB of *S. intermedius* has recently been studied (Ji *et al.*, 2005). AgrB I and AgrB II from *S. aureus* and AgrB from *S. intermedius* cannot process AgrD from the other species. Interestingly, a mutated AgrD (Ser-27-Cys) from *S. intermedius* generating an autoinducing thiolactone peptide is processed by its AgrB. In contrast, a mutated AgrD I (Cys-28-Ser) from *S. aureus* generating an autoinducing lactone peptide is not modified by AgrB I (Ji *et al.*, 2005). It has therefore been suggested that the AgrD of *S. intermedius* evolved from an AgrD with a cysteine-containing AIP and that the different Agr proteins adapt to this change (Ji *et al.*, 2005).

By constructing different AgrB chimeras with regions from group I and others from group II, Zhang *et al.* showed that the first transmembrane domain and the extracellular loop are crucial for the activity of group I AgrB. In contrast, the two hydrophilic transmembrane domains are essential for group II AgrB activity (Zhang and Ji, 2004). Furthermore, several chimeras are able to modify AgrD from both groups, suggesting that the proteolytic processing is common to both groups. Although the N-terminal region is required for AgrB, it does not contribute to the proteolytic properties of this molecule. This region may be needed for the binding of the propeptide AgrD to AgrB. His-77 and Cys-84 are essential for the proteolytic properties of AgrB. These residues are conserved in the AgrB of different groups and are located on the inner surface of the cytoplasmic membrane (Fig. 6.1F) (Qiu *et al.*, 2005). Cysteine and histidine residues are generally involved in the activity of cysteine proteases. However, the inhibition of AgrB by various inhibitors of cysteine proteases did not provide definitive proof that AgrB was indeed a cysteine protease (Qiu *et al.*, 2005). AgrB is thus an interesting protein with two functions. Its endopeptidase activity promotes two different proteolytic cleavages of AgrD to produce the AIP. AgrB probably also acts as an oligopeptide transporter secreting AIP into the extracellular domain, although this remains to be proved (Zhang and Ji, 2004).

AgrA and regulation

The presence of an active AgrA protein is required for the expression of P2 and P3. Response regulators are usually composed of an N-terminal receiver domain containing an aspartate residue phosphorylated by histidine kinase and a C-terminal region displaying DNA-binding activity (West and Stock, 2001). The DNA-binding region of response regulators may take various forms. Some are helix-turn-helix or winged helix domains (Nikolskaya and Galperin, 2002). AgrA is a 28 kDa protein and belongs to the AlgR/AgrA/LytR family of regulators (Morfeldt *et al.*, 1996a; Nikolskaya and Galperin, 2002). These regulators have a DNA binding domain that seems to consist of four β-strands, two α-helices, another β-strand and a final α-helix (Nikolskaya and Galperin, 2002). AgrA is a cytoplasmic protein (Morfeldt *et al.*, 1996b). It was assumed that AgrA, like other response regulators, bound to the promoter region of the *agr* locus. However, it has only recently been demonstrated that AgrA does indeed bind to this region. Koenig *et al.* showed that AgrA binds to direct repeats in the P3 region – ACAGTTAAG-12 bp-CTAGTTAAG – and the P2 region – ACAGTTAAG-12 bp-ACAGTTAGG (Koenig *et al.*, 2004). These sequences are very similar to the consensus sequence recognized by transcriptional regulators of the AlgR/AgrA/LytR family (Nikolskaya and Galperin, 2002). The affinity of AgrA binding to the P3 promoter is higher than that for the binding of this molecule to the P2 promoter. Mutagenesis studies have shown that an AC to CT mutation in position 1 of the second repeat increases the affinity of AgrA binding to the P3 region, whereas an A to G mutation in position 8 does not affect binding affinity (Koenig *et al.*, 2004). Interestingly, in searches for these direct repeats with 1 mismatch over the entire genome, the authors detected no other gene potentially regulated by AgrA, suggesting that the role of AgrA is, indeed, regulation of the *agr* locus and that the effector of Agr is RNAIII rather than AgrA (Koenig *et al.*, 2004). The RN4220 strain is a derivative of the NCTC8325 strain that has been cured of its three prophages and treated with nitrosoguanidine to accept *E. coli* DNA (Kreiswirth *et al.*, 1983). RN4220 presents an *agr*$^-$ phenotype, and does not produce α-toxin. Interestingly, this strain produces RNAIII, but much later and in smaller quantities than in the wild-type strain, probably resulting in an absence of haemolysin production (Traber and Novick, 2006). This phenotype appears to result from a mutation in the 3′ region of the *agrA* gene. Indeed, a stretch of A residues is found just upstream from the stop codon. The wild-type strain contains seven A residues, whereas RN4220 contains eight. This stretch of A residues probably constitutes a slippery sequence, causing a slipped-mispairing mutation. This mutation promotes a frame shift, resulting in the C-terminus of the mutated AgrA being KKNIIR rather than the KKI of the wild-type strain. It remains unclear how the addition of three amino acids at the C terminus modifies AgrA activity, but it is thought that this addition may modify binding to the P2-P3 promoter region or its stability (Traber and Novick, 2006). This mutation has also been found in a clinical strain presenting a similar phenotype. This mutation may thus occur *in vivo* and may allow bacteria to regulate Agr function (Traber and Novick, 2006).

Regulation of *agr* transcript production by other regulators

Regulation of the activation of P2 and P3 has been extensively studied. As described above, AgrA binds to this promoter region. Indeed, expression from P2 and P3 is first modulated by the activation cascade involving each Agr protein when AIP reaches a threshold concentration in the extracellular medium. The expression of the *agr* locus is also controlled by other regulators. A *sarA* deletion dramatically decreases the synthesis of RNAII and RNAIII, indicating that SarA upregulates *agr* expression (Cheung and Projan, 1994; Morfeldt *et al.*, 1996b). SarA binds to the P2-P3 promoter region (Heinrichs *et al.*, 1996; Morfeldt *et al.*, 1996b; Cheung *et al.*, 1997a; Chien and Cheung, 1998; Rechtin *et al.*, 1999). Interestingly, there are 19 to 20 bp between the –10 and –35 boxes of the P2 and P3 promoters – a particularly long spacing. The deletion of 3 bp from this region increases RNAIII synthesis in a wild-type strain, an *agr* mutant and *E. coli.* The distance between the –10 and –35 boxes of the promoter region has thus a marked effect on P3 activity, resulting in the dependence of P3 on Agr and/or SarA (Morfeldt *et al.*, 1996b; Schumacher *et al.*, 2001). SarR is also involved in regulating P2-P3 activation (see section on SarR) and SarX has been shown to bind to the

P2-P3 region (see section on SarX). The regulation of the P2-P3 region is complex, involving the binding of four regulators to this region. Furthermore, many other regulators (see the description of other regulators below) are known to modify RNAIII synthesis, probably indirectly through other regulators. For example, SigB downregulates RNAIII production (Bischoff *et al.*, 2001; Horsburgh *et al.*, 2002).

Agr evolution

Agr groups seem to be linked to clonal structure (Peacock *et al.*, 2002). It has therefore been suggested that they determine subdivisions within *S. aureus* (Wright *et al.*, 2005b). Indeed, a strong relationship between the results of genotypic methods, such as PFGE or MLST, and Agr groups has been observed, suggesting that *agr* may have evolved early in staphylococcal evolution, when nucleotide polymorphisms associated with strain typing appeared (Wright *et al.*, 2005b). However, a more recent study found that clonal complexes belong to two subspecies groups and that *agr I* to *III* were detected in both these subspecies. This has led to the suggestion that differentiation between *agr I, II* and *III* occurred after (at T2) species differentiation (occurring at T1) (Robinson *et al.*, 2005). T2 may correspond to the differentiation of two subspecies, each carrying the three *agr* groups. Differentiation between *agr I* and *IV* occurred later, at T3, to give the *agr I/IV* group (Robinson *et al.*, 2005). More recently, the analysis of 161 strains identified ten dominant lineages closed to the MLST clonal clusters. Each lineage possesses a core variable set of genes arranged throughout the chromosome. Thus, these lineages appear to have a common ancestor and to have evolved independently with fortuitous horizontal gene exchange. *agr I, II* and *III* belong to the core variable genes and are present in each lineage (Lindsay *et al.*, 2006). Thus, *agr* evolved as part of the core variable genes.

In vitro regulation of virulence factor genes by Agr

The regulation of virulence factor expression by the Agr system has been extensively studied *in vitro*. This system regulates a large number of virulence factor genes. The expression of most exoprotein genes is enhanced by the Agr system (*hla, hlb, hld, hlgCB, tst, sspABC, splABDF, aur, scp, seb, sec, sed, eta, etb, geh, plc, lip, nuc, sak, cap, fame, scn, lukPV*) (Janzon *et al.*, 1986; Recsei *et al.*, 1986; Gaskill and Khan, 1988; Mahmood and Khan, 1990; Cheung *et al.*, 1992; Sheehan *et al.*, 1992; Dassy *et al.*, 1993; Daugherty and Low, 1993; Tremaine *et al.*, 1993; Chamberlain and Imanoel, 1996; Arvidson, 2000; Bronner *et al.*, 2000; Novick, 2000; Arvidson and Tegmark, 2001; Dunman *et al.*, 2001; Reed *et al.*, 2001; Luong and Lee, 2002; Schmidt *et al.*, 2004; Rooijakkers *et al.*, 2006). In contrast, Agr decreases the expression of several cell wall-associated protein genes such as *fnbA, fnbB, spa* and *coa* (Janzon *et al.*, 1986; Recsei *et al.*, 1986; Patel *et al.*, 1992; Saravia-Otten *et al.*, 1997; Xiong *et al.*, 2004). Most of these virulence factors were found to be similarly down- or upregulated in a proteomic analysis (Ziebandt *et al.*, 2004). Some clinical strains display lower than normal levels of RNAIII expression and mutations in the *agr* locus. As Agr upregulates *cap* expression, it has been suggested that an inactive *agr* locus may result in an untypeable capsule (Cocchiaro *et al.*, 2006). The regulation of virulence factor genes by Agr depends on the strain used and, thus, on genetic background. Indeed, protease activity was found to be low only in an *agr*$^-$ mutant of strain RN6390, whereas the other six *agr*$^-$strains (laboratory or clinical) studied displayed levels of protease activity similar to those of the corresponding wild-type strain. Similarly, haemolytic activity was found to be upregulated by Agr in five strains, including RN6390, whereas it is not affected in two other strains (Blevins *et al.*, 2002). In RN6390, Agr upregulates exoproteins and downregulates cell-wall associated proteins. UAMS-1, a clinical strain isolated from a patient with osteomyelitis, shows the opposite phenotype. Thus, Agr plays a less significant role in the clinical strain UAMS-1 than in the laboratory strain RN6390, whereas SarA seems to be important in UAMS-1 (Cassat *et al.*, 2006).

The mechanism of *hla* and *spa* regulation by Agr has been elucidated (see section RNAIII). However, it remains unclear how Agr regulates the other virulence factor genes. The modulation of Rot production is one of the mechanisms used by Agr (see section on Rot). Indeed, Rot mediates the regulation of *sspA* and *aur* expression by Agr (Oscarsson *et al.*, 2006b)

In vivo role of the *agr* locus, relationship between agr groups, colonization and virulence

As a given Agr group can inhibit the other staphylococcal groups, it has been suggested that staphylococci may use this interaction to elicit the evolution of new strains *in vivo*. Several studies have indeed shown that some *agr* groups are found in a specific genetic background. Most of the *S. aureus* strains responsible for toxic-shock syndrome belong to group III (Ji *et al.*, 1997), whereas 53% of strains carrying the *eta* gene responsible for staphylococcal scalded skin syndrome belong to group IV (Jarraud *et al.*, 2000). Group I and II strains are the most frequently isolated in cases of nasal colonization and from patients presenting infections, whereas group III is present but less frequent in such cases and group IV is rare (Papakyriacou *et al.*, 2000; van Leeuwen *et al.*, 2000; Peacock *et al.*, 2002; Goerke *et al.*, 2003; Lina *et al.*, 2003; Shopsin *et al.*, 2003; Gilot and van Leeuwen, 2004). A specific distribution of *agr* groups is therefore observed *in vivo*, suggesting that some *agr* groups may be involved in colonization and/or pathogenesis. However, the role of *agr* groups in colonization remains unclear. Several studies reported no relationship between the colonization ability of *S. aureus* and *agr* groups (Kahl *et al.*, 2003; Shopsin *et al.*, 2003; Cespedes *et al.*, 2005). Furthermore, *S. aureus* strains with various *agr* groups may be present in the nose of the same subject (van Leeuwen *et al.*, 2000). AIP I from *S. epidermidis* strongly inhibits *S. aureus* AgrC I, II and III, whereas it does not inhibit AgrC IV. In contrast, AIPs from *S. aureus* are highly inefficient against *S. epidermidis* of group I, as only AIP IV slightly inhibits *S. epidermidis* (Otto *et al.*, 2001). Thus, *S. epidermidis* may take advantage of its ability to repress *S. aureus agr in vivo*. However, no correlation was found between the inhibitory action of *S. epidermidis in vitro* and the colonization of the nose by particular staphylococcal species (Lina *et al.*, 2003). It was therefore suggested that no one group can expand at the expense of the others by inhibiting other strains and affecting colonization dynamics. However, other studies have reported that some *agr* groups predominate in the nose and have concluded that *agr* groups may be involved in staphylococcal colonization (Goerke *et al.*, 2003).

No difference has been observed in the distribution of the *agr* groups encountered in healthy controls, patients with cystic fibrosis and patients with invasive infections (Peacock *et al.*, 2002; Goerke *et al.*, 2003; Kahl *et al.*, 2003). In contrast, the distribution of *agr* groups in patients with chronic wounds and intubated patients is significantly different from that in healthy subjects (Goerke *et al.*, 2003). In human clinical strains, some *agr* groups are isolated more frequently than others (Gilot and van Leeuwen, 2004). In a mouse model of mastitis, larger numbers of bacteria are recovered from mammary glands infected with a group I Agr strain than from mammary glands infected with strains of other Agr groups. Group I strains are isolated the most frequently from cows with bovine mastitis (about 88%) (Buzzola *et al.*, 2007). In the hornworm, *Manduca sexta* (see section on *in vivo* virulence of *agr*$^-$ mutants below), *agr* groups are involved in determining relative fitness *in vivo*. Indeed, strains from some *agr* groups may grow more easily than others. Interestingly, this interference is not observed in culture medium *in vitro* (Fleming *et al.*, 2006). Furthermore, in a mouse model, abscess infection by a group I strain was found to be reduced by the injection of 2× IC_{100} of AIP II (Mayville *et al.*, 1999). Thus, *agr* groups may interfere with growth *in vivo*, depending on the environment encountered by *S. aureus*.

In vivo virulence of agr$^-$ mutants

The effect of *agr* deletion on virulence has been tested in several animal models – the fruit fly *Drosophila melanogaster*, the roundworm *Caenorhabditis elegans*, the silkworm *Bombyx mori* and the tobacco hornworm *M. sexta*. *Drosophila* has been widely used to study host-pathogen interactions. It secretes antimicrobial peptides in response to infection and its innate immune system is very efficient and resembles that of humans (Lemaitre *et al.*, 1997). Indeed, the Toll receptors, which are crucial for innate immune recognition, were first identified in *Drosophila*. Flies are inoculated with *S. aureus* by pricking the dorsal thorax with a needle soaked in *S. aureus* suspension. Interestingly, the *agr*$^-$ mutation does not decrease the virulence of *S. aureus* in this model. Furthermore, the deletion of none of the main regulators involved in virulence in other animal models – SarA, SaeSR and ArlSR – has any effect on virulence in this model (Needham *et al.*, 2004). *C. elegans* is a nematode used as a model organism for studies of relationships between hosts and pathogens, such as *S. aureus* (Garsin *et al.*, 2001). Nematodes placed on agar plates are fed with different staphylococcal strains. *S. aureus* colonizes the nematode intestine, causing the death of the animal. In contrast, *Bacillus subtilis* does not kill the nematode. An *agr*$^-$ mutant was shown to result in significantly fewer deaths, suggesting that *agr* is indeed necessary for virulence in this model (Sifri *et al.*, 2003; Bae *et al.*, 2004). The silkworm (*B. mori*) is an invertebrate that has been used to study the virulence of several pathogens. It is large enough for the direct inoculation of the haemolymph with bacteria. In this model, *agr* deletion decreases the number of animal deaths (Kaito *et al.*, 2005). Unlike the other model animals studied, the hornworm, *M. sexta*, can grow at 37°C. *S. aureus* is injected into haemocoel and infected worm fail to grow and die. The deletion of *agr* significantly decreases the number of hornworm deaths (Fleming *et al.*, 2006).

Mice are the animals most frequently used for studies of staphylococcal virulence. Murine models mimicking systemic infections have been used to evaluate the ability of bacteria to become attached, to evade host defences, to disseminate and to survive in the spleen. Dissemination is not necessary for local infections. Staphylococci need to deploy defences against PMNs and to adapt to low O_2 levels and to nutrient limitation (Benton *et al.*, 2004). Indeed, several murine models have been developed for studies of the *in vivo* effect of staphylococcal virulence factors and/or regulators: skin or wound infections, eye infections, pneumonia by intranasal inoculation, mastitis, sepsis, osteomyelitis, arthritis and pyelonephritis. A model of endocarditis has also been developed in rabbit and rat (Collins and Tarkowski, 2000). Another model has been developed to study staphylococcal interactions with implants. In this model, sterile polytetrafluoroethylene tubes are implanted subcutaneously into guinea pigs or rabbits for the study of foreign body infections (Zimmerli *et al.*, 1982; Goerke *et al.*, 2001; Yarwood *et al.*, 2002).

The presence of an *agr* deletion in the murine abscess model decreases the number of staphylococci recovered in the lesions (Mayville *et al.*, 1999; Schwan *et al.*, 2003). Similarly, *agr*$^-$ mutants isolated from the liver and spleen in a systemic murine bacteraemia model are much less than the wild-type strain (Schwan *et al.*, 2003). In a rabbit model of endocarditis, the deletion of *agr* only slightly decreased the number of bacteria recovered from vegetations, whereas staphylococcal adhesion to aortic valvular vegetations was strongly attenuated in the absence of *agr* (Cheung *et al.*, 1994a). In an experimental model of endophthalmitis, *agr* was shown to be crucial for *S. aureus* virulence and the double *agr*$^-$ *sar*$^-$ mutant was almost avirulent (Booth *et al.*, 1997). In a murine model of sepsis and musculoskeletal infection, the deletion of *agr* significantly decreased the incidence of osteomyelitis and arthritis (Blevins *et al.*, 2003). In a rabbit model of acute exogenous osteomyelitis, *agr* deletion resulted in similar observations: a decrease in the incidence of osteomyelitis and in the severity of the infection. Interestingly, the *agr*$^-$ mutant was still able to colonize bones and to generate histopathological signs of osteomyelitis (Gillaspy *et al.*, 1995).

In contrast, in a murine wound model, the deletion of *agr* was found to have no effect on the number of bacteria recovered from the lesion (Schwan *et al.*, 2003). In a mouse model of pneumonia induced by intranasal inoculation, the *agr*$^-$ mutant caused slightly fewer mouse deaths than the wild type. However, it was not associated with a decrease in the number of cases of pneumonia. Pneumonia is caused by an infiltration of PMNs and macrophages, resulting in tissue damage. Indeed, the presence of an *agr* deletion has no effect on the stimulation of IL-8, a chemokine that contributes to the recruitment of PMNs. Thus, airway inflammation leading to pneumonia is not *agr*-dependent, but *agr* is instead involved in systemic infections causing death (Heyer *et al.*, 2002). In an experimental murine brain abscess model, single *sarA*$^-$ and *agr*$^-$ mutants have no significant effect on virulence and replication, whereas the double *sarA*$^-$ *agr*$^-$ mutant is associated with significantly lower levels of virulence and replication in the brain and an attenuation of the release of inflammatory mediators, such as IL-1α and β, IL-6, TNF-α, MIP-1α and β, MIP-2 (Kielian *et al.*, 2001). In murine models of abscesses and wounds, the percentage of non-haemolytic strains increases significantly during infection (from 5% at the time of injection to 30% at seven days), and these strains often harbour mutations in *agr* genes. In contrast, in systemic infections, the bacteria recovered from the liver and spleen are haemolytic and have an efficient Agr system. These results suggest that *agr* is required for systemic infections (Schwan *et al.*, 2003). Thus, the Agr system is indeed a major virulence regulator that plays a crucial role in various infections caused by *S. aureus*.

A global inhibitor of Agr has been synthesized. This truncated group II AIP consists of the thiolactone ring without the N-terminal region and it inactivates all groups *in vitro* (Lyon *et al.*, 2000). This inhibitor is also active *in vivo*, reducing the effect of the Agr system in a murine abscess model. However, a single dose of this global inhibitor is active for only 3 h (Wright *et al.*, 2005a). This compound is thus a promising drug for the treatment of staphylococcal infections involving Agr regulation.

In vivo regulation of virulence genes by Agr

RNAII appears about 3 h before RNAIII *in vitro*. In a rabbit model of endocarditis, RNAII activation is induced early in vegetations, whereas RNAIII production increases gradually. RNAIII activation in vegetations is correlated with bacterial density, does not require RNAII and depends on tissue. Indeed, plasma RNAIII concentration remains constant and low (Xiong *et al.*, 2002). In a murine model of subcutaneous abscess, RNAIII synthesis increases 3 h after inoculation. Activity then decreases from 8 to 24 h and there is another peak in activity between 24 and 72 h. The number of PMNs at the infection site is maximal at 6 h and the decrease in RNAIII activity in neutropenic mice is limited to the time period between 4 and 8 h, suggesting that the decrease in activity observed between 8 and 24 h is due to the action of PMNs (Wright *et al.*, 2005a). Interestingly, inoculation with sterile supernatants from *agr*$^{+}$ strains induces lesions similar to those obtained with live staphylococci, whereas inoculation with supernatants from *agr*$^{-}$ strains induces no lesions. This strongly suggests that the lesions observed in abscesses are caused by exoproteins regulated by Agr (Wright *et al.*, 2005a). In a guinea pig implant model, RNAIII is less strongly expressed in exudates than *in vitro* and is not correlated with bacterial density. A method for the direct detection of mRNAs in biological samples showed RNAIII levels in sputa from cystic fibrosis patients to be low. Furthermore, there is no correlation between the amount of RNAIII and bacterial density. The levels of *spa* mRNA is also low. Thirteen strains from 17 samples were found to carry an efficient *agr* locus in this study, suggesting that *agr* is not expressed and may not be involved in *spa* expression of staphylococci in patients with this disease (Goerke *et al.*, 2000).

In the rabbit model of endocarditis, the level of *cap5* expression in vegetations infected with the *agr*$^{-}$ mutant is significantly lower than that in vegetations infected with the wild-type strain. Similar observations have been reported for renal abscesses, but not for the spleen. Thus, *agr* upregulates *cap5* expression both *in vivo* and *in vitro* (van Wamel *et al.*, 2002). FnBPA is crucial for adhesion and persistence in endothelial cells and is thus a major staphylococcal virulence factor for endovascular infection. The expression of *fnbA* is similarly downregulated by Agr *in vitro* and in vegetations in the same model of endocarditis, but is also dependent on other regulators, such as SarA (Xiong *et al.*, 2004). In contrast, *hla* expression in the endocarditis model is only slightly impaired by *agr* deletion (Xiong *et al.*, 2006). In a guinea pig implant model, *hla* expression *in vivo* is high in the presence of the *agr* deletion (Goerke *et al.*, 2001). In a similar model in rabbit, RNAIII and RNAII expression is lower than that observed *in vitro*, whereas the expression of *hla* is stronger (Yarwood *et al.*, 2002). Furthermore, the expression of RNAIII is strongly repressed by serum, whereas serum increases *hla* expression (Yarwood *et al.*, 2002). These results strongly suggest that *agr* is not involved in the expression of *hla* in these models. Thus, depending on the type of infection and the virulence factor, *agr* may or may not be required for the regulation of virulence gene expression.

Fibrinogen is involved in neutrophil recruitment, the release of cytokines and chemokines and several other responses to inflammation. Fibrinogen depletion in mice increases the survival of the wild-type strain, whereas it has no effect on the survival of the *agr*⁻ mutant (Rothfork *et al.*, 2003). RNAIII production is activated by fibrinogen or serum both *in vivo* and *in vitro*. Furthermore, it has been shown that clumping concentrates the AIP secreted by bacteria, and is sufficient to induce quorum sensing. As fibrinogen induces the clumping of staphylococci, the presence of fibrinogen may also increase bacterial density, activating the synthesis of RNAII and RNAIII (Rothfork *et al.*, 2003). Interestingly, this study used a different staphylococcal strain to the study mentioned earlier and showing expression of RNAIII is repressed by serum (Yarwood *et al.*, 2002). Thus, this suggests that host components such as fibrinogen may modulate the synthesis of RNAIII *in vivo*.

Role of Agr in the relationship between staphylococci and mammalian cells

In an invasion assay, bovine mammary epithelial cells such as MAC-T are infected by *S. aureus* for 2 h and intracellular bacteria are then recovered after cell lysis. The deletion of *agr* results in an increase of the number of bacteria recovered from MAC-T with much higher survival rates at 72 h reported for the *agr*⁻ mutant than for the wild-type strain (Wesson *et al.*, 1998). Interestingly, the *agr*⁻ mutant cannot induce apoptosis, whereas the wild-type strain induces the apoptosis of endothelial and epithelial cells (Wesson *et al.*, 1998; Haslinger-Loffler *et al.*, 2005). At 30 min, 1 h and 2 h after the infection of MAC-T cells, *S. aureus* is observed in intact bound endosomes. At 3 h, *S. aureus* is detected in the cytoplasm (Bayles *et al.*, 1998; Shompole *et al.*, 2003). In contrast, the *agr*⁻ mutant is still found in the endosomes at 5 h. Furthermore, RNAIII expression peaks at 2 and 4 h for the wild-type strain. Interestingly, *hla* and *hlb* expression levels are similar to that for RNAIII, except that the peak occurs at 5 h rather than 4 h. It has therefore been suggested that the initial production of RNAIII at 2 h mediates escape from the endosomes, whereas the second peak leads to cell damage due to the production of haemolysins (Shompole *et al.*, 2003). The number of bacteria taken up by MAC-T cells is higher for group I strains than for strains of other Agr groups. Thus, the ability of group I strains to survive in cells may account for the persistence of this group in mammary glands (Buzzola *et al.*, 2007).

If the phagosomal uptake pathway is inefficient, host cells may initiate another process: autophagy. This mechanism results in the sequestering of unwanted cytoplasmic components by double membranes and their transport to lysosomes. A recent study showed that *S. aureus* may use this pathway. Staphylococci are internalized via the endophagosomal pathway. Multilamellar structures then attach to the vacuoles containing *S. aureus* to form autophagosomes. These structures are usually fused to lysosomes to obtain an autophagolysosome. However, *S. aureus* can abolish the maturation of the autophagosomes and their subsequent fusion with lysosomes. *S. aureus* may then escape into the cytoplasm, triggering cell death via a caspase-independent pathway (Schnaith *et al.*, 2007). Interestingly, *agr*⁻ mutants cannot induce autophagy and do not initiate the formation of autophagosomes. They are therefore unable to trigger cell death. Thus, *agr* regulates a factor essential for autophagy. In the presence of an activator of autophagy, *agr*⁻ mutants may induce autophagosome formation and cell death. This suggests that the formation of autophagosomes is required for *S. aureus*-induced cell death and that Agr is crucial for this process (Schnaith *et al.*, 2007). Agr is also involved in the interaction with mammalian cell surfaces. Indeed, Agr upregulates syndecan-1 ectodomain shedding (see section on cytotoxins) probably by upregulating *hla* and *hlb* (Park *et al.*, 2004). Thus, *agr* plays a vital role in the processes of internalization, intracellular localization, apoptosis induction and in the shedding of extracellular proteins from host cells.

The *sarA* locus

Structure and regulation of the *sarA* locus

The *sarA* locus (staphylococcal accessory regulatory locus) was first identified by screening a Tn*917* insertion mutant library for a decrease in exotoxin production (Cheung *et al.*, 1992). The *sarA* gene encodes the 15 kDa protein SarA (Cheung and Projan, 1994). Three promoters generate three different transcripts that terminate at the same stem-loop sequence (Fig. 6.2). The *sarA* transcript (0.6 kb) is initiated at P1, *sarB* (1.15 kb) at P2 and *sarC* (0.8 kb) at P3 (Bayer *et al.*, 1996; Manna *et al.*, 1998). These transcripts are differentially expressed during the growth phase, whereas the SarA protein is produced in similar amounts during all growth phases (Blevins *et al.*, 1999). Indeed, *sarA* and *sarB* are expressed mostly from the mid- to late exponential growth phase and *sarC* is expressed during the late-exponential and stationary phases (Bayer *et al.*, 1996; Manna *et al.*, 1998). P1 and P2 are σ^A-classical promoters, whereas P3 has a sequence typical of promoters dependent on the σ^B factor (SigB) (see section on Sigma B). P2 and P3 are weaker promoters than P1, but they contribute to the transcription of the *sarA* gene to produce SarA (Deora *et al.*, 1997; Manna *et al.*, 1998; Bischoff *et al.*, 2001; Cheung and Manna, 2005). A small ORF3 located between P1 and P3 does not seem to be required for SarA production, whereas another ORF, ORF4, located between P2 and P3 is required for full SarA production (Fig. 6.2) (Chien *et al.*, 1998; Wolz *et al.*, 2000; Cheung and Manna, 2005). Deletions in *sarA* attenuate the transcription of *sarA*, consistent with autoregulation (Chien *et al.*, 1998; Manna *et al.*, 1998). Indeed, SarA represses P1 promoter activity, but has no effect on P3 (Chakrabarti and Misra, 2000). The main regulator of SarA is SarR, which binds to the *sarA* promoter region (see section on SarR). SigB upregulates *sarC* transcript levels (Bischoff *et al.*, 2001; Nair *et al.*, 2003; Ziebandt *et al.*, 2004). However, the amounts of SarA protein are similar in the wild-type strain and the *sigB*$^-$ mutant (Horsburgh *et al.*, 2002).

SarA is a dimeric protein that can bind DNA (Rechtin *et al.*, 1999). There is some debate about the conserved A/T-rich consensus sequence recognized by SarA, which has been reported to be either 7 or 26 bp long (Chien *et al.*, 1999; Sterba *et al.*, 2003). However, it has been suggested that both consensus sequences are probably incorrect as the 26 bp sequence is not found upstream from the promoters recognized by SarA and the 7 bp consensus sequence is present more than 2,500 times in the N315 genome sequence (Roberts *et al.*, 2006). It has therefore been suggested that SarA may be like the *E. coli* proteins Fis and IHF, which regulate transcription by mediating structural changes in DNA and DNA supercoiling (Schumacher *et al.*, 2001; Roberts *et al.*, 2006). Indeed, SarA has some properties in common with these proteins. They are all small proteins (about 9 kDa) that may bind to A/T-rich regions. Furthermore, histone-like proteins have also been shown to bind to mRNA and to modulate mRNA stability or translation (Deighan *et al.*, 2000; Balandina *et al.*, 2001; Roberts *et al.*, 2006). It has been shown that 90% of *S. aureus* mRNA has a half-life of less than 5 minutes. A microarray-based comparison of the half-lives of mRNAs from a wild-type strain with that of mRNAs from a *sarA* mutant showed that most transcripts had similar half-lives in the two strains. In contrast, 138 mRNAs including those for the *cna*, *fib*, *spa* and *sspB* genes, had a significantly shorter half-life in the *sarA*$^-$ mutant. These transcripts were stable in the wild-type strain (half-life of 15 to 30 min), whereas their half-life decreased to 5 minutes in the *sarA*$^-$ mutant (Roberts *et al.*, 2006).

SarA can associate with DNA, but no classical DNA-binding motif has been identified in this protein. Several crystal structures have been reported. SarA has been reported to consist of a five-helix region and three β-strands (Liu *et al.*, 2006). However, SarA has also been reported to be composed of a four-helix core region, a β-hairpin and a carboxy-terminal loop

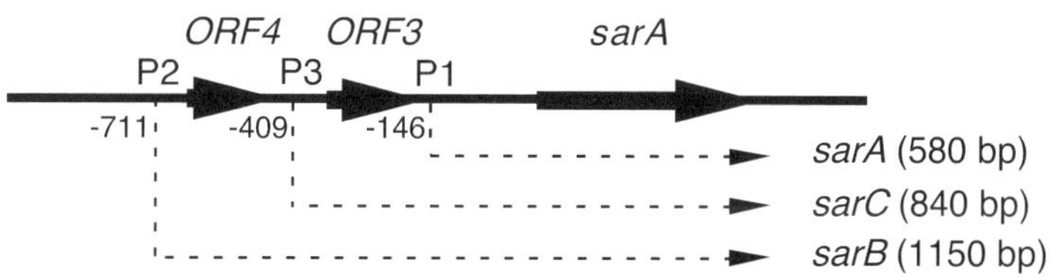

Figure 6.2 The *sarA* locus with its three transcripts, *sarA*, *sarB* and *sarC*, synthesized from three promoters, P1, P2 and P3, respectively. ORF4 and ORF3, upstream from the *sarA* gene, are also indicated, as described by Chien *et al.* (1998). Dotted lines represent the transcripts.

(Schumacher *et al.*, 2001). The reason for these differences in the reported structure of SarA is unknown. On binding to DNA, SarA undergoes conformational changes in the β-hairpin and the C-terminal loop region, whereas the core region remains stable (Schumacher *et al.*, 2001). The protein subsequently encases the DNA (Cheung and Zhang, 2001; Schumacher *et al.*, 2001; Liu *et al.*, 2006). The single cysteine residue present in SarA, Cys9, is located in the DNA-binding channel and is therefore heavily involved in the interaction with DNA. As cysteine may be oxidized, aerobic conditions may decrease SarA activity (Schumacher *et al.*, 2001). Indeed, low-oxygen conditions increase SarA activity (Chan and Foster, 1998b). The oxidation of this cysteine has also been confirmed in a SarA homologue, MgrA (see section on MgrA). The SarA residues crucial for binding to DNA are R84 and R90, whereas the residues required for repressor function are D8, C9, E11, D88 and E89 (Liu *et al.*, 2006).

In vitro regulation of virulence factor genes by *sarA*

SarA upregulates *hla, hlb, hld, seb, tst, fnbA, fnbB, hlgCB, eap, splABDF, fame, lukPV* and downregulates *aur, nuc, cna, chp, sea, sak, sspABC, scp, plc* and *spa* (Cheung *et al.*, 1992; Cheung *et al.*, 1994a; Cheung and Ying, 1994; Chamberlain and Imanoel, 1996; Cheung *et al.*, 1997b; Blevins *et al.*, 1999; Chien *et al.*, 1999; Bronner *et al.*, 2000; Chakrabarti and Misra, 2000; Wolz *et al.*, 2000; Arvidson and Tegmark, 2001; Dunman *et al.*, 2001; Ziebandt *et al.*, 2001; Bronner *et al.*, 2004; Xiong *et al.*, 2004; Trotonda *et al.*, 2005; Rooijakkers *et al.*, 2006). SarA binds to the promoter region of *hla, ssp, spa, fnbB, fnbA, sec* and *cna* (Blevins *et al.*, 1999; Chien *et al.*, 1999; Tegmark *et al.*, 2000; Sterba *et al.*, 2003; Oscarsson *et al.*, 2005; Trotonda *et al.*, 2005). SarA represses or activates *hla*, depending on the levels of SigB and SarS in the strain (Oscarsson *et al.*, 2006a). The transcription of *lip* and *geh* is not modified in a *sarA* deletion mutant, whereas the amounts of lipase (Lip) and glycerolester hydrolase (Geh) proteins are much lower in the *sarA*⁻ mutant than in the wild-type, suggesting that these proteins are degraded (Ziebandt *et al.*, 2001). SarA appears to be a crucial regulator of protease genes such as *sspA* and *aur* (Oscarsson *et al.*, 2006b). Indeed, *sarA* deletion strongly increases the production of serine protease (SspA) and cysteine protease (ScpA) and the activity of aureolysin (Aur) (Lindsay and Foster, 1999; Karlsson *et al.*, 2001; Karlsson and Arvidson, 2002). The deletion of *sarA* decreases binding to fibronectin and the amount of protein A (Cheung *et al.*, 1992; Wolz *et al.*, 2000). SspA is known to degrade FnBPs (McGavin *et al.*, 1997). It has therefore been suggested that serine protease, which is present in larger amounts, degrades FnBPs, Spa, and also secreted proteins such as alpha-haemolysin and lipases (Lindsay and Foster, 1999; Karlsson *et al.*, 2001).

SarA has been shown to be crucial for biofilm formation, as it induces expression of the *ica* operon, an important factor for biofilm formation (Beenken *et al.*, 2003; Valle *et al.*, 2003). It also regulates the expression of *bap*, which encodes a protein involved in biofilm formation (Cucarella *et al.*, 2001). The mechanisms involved in biofilm formation will not be reviewed in this chapter.

In vivo role of the *sarA* locus

The deletion of *sarA* decreases the number of bacteria recovered from aortic valvulus vegetations in a rabbit model of endocarditis (Cheung *et al.*, 1994a). This attenuation of endocarditis due to *sarA* deletion is correlated with a significant decrease in adhesion to aortic valvular vegetations and to human cells, such as HUVEC (human umbilical vein epithelial cells) (Cheung *et al.*, 1994a; Cheung *et al.*, 1994b). Interestingly, SarA upregulates *fnbA in vitro* and in a rabbit endocarditis model (Xiong *et al.*, 2004). P2 and P3 are weaker promoters than P1 *in vitro*, but a different pattern is observed in the vegetations of a rabbit endocarditis model. P2, which is a weak promoter *in vitro*, is strongly activated *in vivo*, mostly at the surface, and not in the centre of vegetations. P1 is activated at both the centre and periphery of vegetations, whereas P3 is not activated, as *in vitro* (Cheung *et al.*, 1998). In a murine model of haematogenous arthritis, the *sarA*⁻ mutant causes weaker arthritic symptoms than the wild-type strain. Furthermore, the number of circulating polymorphonuclear leukocytes and the amount of IgG synthesized in the presence of the *sarA*⁻ mutant are smaller than those induced by the wild-type strain (Nilsson *et al.*, 1996). In a murine model of sepsis and musculoskeletal infection, the absence of *sarA* significantly decreases the incidence of osteomyelitis and arthritis (Blevins *et al.*, 2003). In a murine model of renal abscess, the deletion of *sarA* does not result in the isolation of fewer bacteria from the kidney than for the wild-type strain (Kropec *et al.*, 2005). In a murine model of pneumonia induced by intranasal inoculation, the *sarA*$^-$ mutant causes fewer deaths than the wild type. However, it is not associated with a decrease in the number of cases of pneumonia because, like *agr*, it has no effect on the release of IL-8 necessary to cause pneumonia (Heyer *et al.*, 2002). Thus, SarA is an important regulator involved in systemic infections.

SarA is required for staphylococcal survival in neutrophils. *sarA*$^-$ mutants are found mostly in small phagosomes, whereas the wild-type strain is observed in large vacuoles corresponding to phagosomes promoted by macropinocytic uptake. Thus, SarA is necessary for the intraphagosomal distribution of staphylococci in neutrophils (Gresham *et al.*, 2000). The deletion of *sarA*, like that of *agr*, increases the survival of bacteria in MAC-T cells, but the deletion of *sarA* also decreases the level of *S. aureus*-induced apoptosis (Wesson *et al.*, 1998). This strongly suggests that the intracellular distribution of *S. aureus* is crucial for *S. aureus*-induced cell death and that SarA plays a vital role in this process.

The Sae system

Structure and regulation of the Sae system

The *sae* locus (staphylococcal accessory element locus) is a 'two-component' system with four elements. The functions of SaeP (or ORF4) and SaeQ (or ORF3) are unknown (Novick and Jiang, 2003; Steinhuber *et al.*, 2003). SaeR and SaeS are the response regulator and histidine kinase, respectively, of the two-component system (Fig. 6.3) (Giraudo *et al.*, 1999). Four overlapping transcripts (T1 to T4) are transcribed from three different promoters (Novick and Jiang, 2003; Steinhuber *et al.*, 2003). T1, T2 and T3 have a common 3′ end and are initiated from three different promoters, P1, P2 and P3, respectively, whereas T4 is a small monocistronic mRNA, encoded by the *saeP* gene and initiated from P1 (Fig. 6.3). Expression of the *sae* locus is regulated during the growth phase. T3 is mostly produced during the exponential growth phase, whereas P1 and P2 are activated after the exponential growth phase. Agr, Sar and Sae are required for full expression of these four transcripts (Novick and Jiang, 2003).

The Sae system upregulates the expression of *hla*, *hlb*, *hld* and *coa* independently of Agr and SarA (Giraudo *et al.*, 1994; Giraudo *et al.*, 1996; Giraudo *et al.*, 1997; Goerke *et al.*, 2005b). Furthermore, the expression of *hld*, *seb*, *efb*, *eap*, *lukF*, *lukM*, *hlgACB*, *chp*, *scn*, *fnbB* and *fnbA* is upregulated by Sae, whereas the production of Plc, SspA and Aur and the transcription of *spa* are downregulated by Sae (Giraudo *et al.*, 1996; Rampone *et al.*, 1996; Arvidson and Tegmark, 2001; Novick and Jiang, 2003; Harraghy *et al.*, 2005; Liang *et al.*, 2006; Rogasch *et al.*, 2006; Rooijakkers *et al.*, 2006; Yamazaki *et al.*, 2006; Kuroda *et al.*, 2007). The *S. aureus* strain Smith 5R usually forms rough colonies and produces γ-haemolysin, but after several subcultures in broth, this strain forms smooth colonies and no

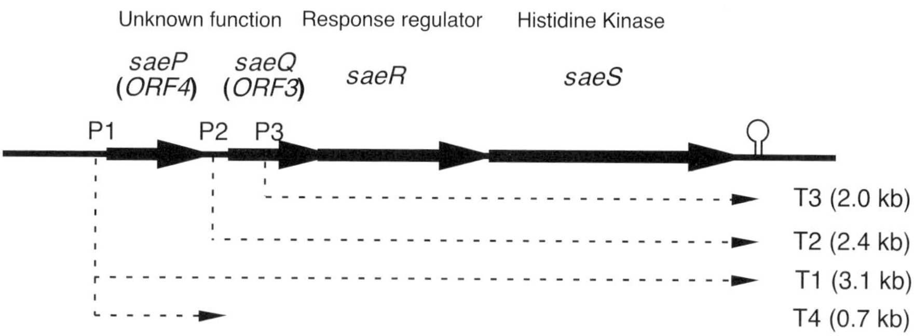

Figure 6.3 The *sae* locus, composed of four genes (T1 to T4), transcribed to give four transcripts, from three different promoters (P1 to P3) (Novick and Jiang, 2003; Steinhuber *et al.*, 2003). Dotted lines represent the transcripts.

longer produces γ-haemolysin. During subculture *in vitro*, spontaneous mutations were found to have occurred in the *sae* genes, inactivating this locus. The introduction of an active *sae* locus restored γ-haemolysin production (Yamazaki *et al.*, 2006). Similar results were obtained with *agr*: serial subcultures of a clinical strain *in vitro* led to the emergence of mutations within the *agr* locus (Somerville *et al.*, 2002a).

In vivo role of the Sae system

Sae seems to play a crucial role in the virulence of *S. aureus*. In the *C. elegans* model of staphylococcal virulence, inactivation of *sae* significantly decreases virulence (Bae *et al.*, 2004). Following the intraperitoneal injection of wild-type *S. aureus* or its *sae*$^-$ mutant in mice, lower levels of mouse mortality are observed with the *sae*$^-$ mutant than with the wild-type strain (Giraudo *et al.*, 1996; Rampone *et al.*, 1996). In a murine model of systemic infection, *sae* deletion results in the recovery of significantly fewer bacteria from the spleen than for the wild-type strain. Similarly, in a murine abscess model, fewer bacteria were isolated from lesions infected with the *sae*$^-$ mutant than from lesions infected with the wild-type strain (Benton *et al.*, 2004). In haematogenous pyelonephritis, far fewer bacteria were recovered from the kidneys of mice infected with the *sae*$^-$ mutant strain than from those of mice infected with the wild-type strain (Liang *et al.*, 2006).

Studies in which a dialysis sac was implanted in the peritoneal cavity of a sheep showed that the *sae*$^-$ mutant produced half as much protein as the wild-type strain (Rampone *et al.*, 1996). In a guinea pig model of implant infection, *sae*, *hla* and *coa* were found to be much less strongly expressed in a *sae*$^-$ mutant than in the wild-type strain, suggesting that the expression of these genes in this model requires efficient *sae* expression (Goerke *et al.*, 2001; Goerke *et al.*, 2005b). Furthermore, the level of *hla* expression observed in a *sae*$^-$ mutant in an experimental rabbit endocarditis model was found to be significantly lower than that in the wild-type strain (Xiong *et al.*, 2006). Sae is therefore essential for the regulation of haemolysin genes *in vivo* and *in vitro*.

The deletion of *sae* impaired the ability of staphylococci to adhere to lung epithelial cells and to be taken up by those cells. It also decreased the level of apoptosis induced by the staphylococci (Liang *et al.*, 2006).

The Arl system

The genes of the ArlSR (autolysis-related locus) system were first isolated by screening a Tn*917* library for genes regulating the expression of the multidrug efflux pump gene, *norA* (Fournier *et al.*, 2000). ArlSR is a classical two-component system composed of a histidine kinase, ArlS, and a response regulator, ArlR (Fournier and Hooper, 2000). The ArlSR system decreases the production of β-haemolysin, lipase and coagulase (Fournier *et al.*, 2001). ArlSR also downregulates *hla*, *sspA*, *spa*, *lukD*, *lukE*, *hlgC* and *splB* transcription, whereas it increases *hld* and *cap* transcription (Fournier *et al.*, 2001; Liang *et al.*, 2005; Luong and Lee, 2006, Meier *et al.*, 2007). It remains unclear how the Arl system regulates virulence factor gene expression. Arl has been shown to increase or decrease RNAIII synthesis, depending on the strain used, and to increase the amount of *sarA* and *mgrA* mRNAs (Fournier *et al.*, 2001; Liang *et al.*, 2005; Luong and Lee, 2006). The effect of Arl on *spa* is dependent on SarA and its effect on *cap* transcription requires MgrA (Fournier *et al.*, 2001; Luong and Lee, 2006), suggesting that the effects of the Arl system on virulence gene expression may also involve other regulators. A deletion in *arlR* has been detected in strain N315 (Kuroda *et al.*, 2001). It has been suggested that mutations inactivating regulators such as the Arl system may be responsible for untypeable capsules of several clinical strains (Cocchiaro *et al.*, 2006). An analysis of the regulation of the *arlRS* operon has shown that two main transcripts (1.5 and 2.7 kb in size) are present in strain RN6390 (Fournier *et al.*, 2001). Additional minor transcripts were observed in other strains (Meier *et al.*, 2007). The expression of *arl* increases during the post-exponential growth phase, is not autoregulated, but is stimulated by both SarA and Agr (Fournier *et al.*, 2001). MgrA upregulates *arl* expression or has no effect on *arl* transcription, depending on the strain used, but this protein binds to the promoter region of *arlRS* (Ingavale *et al.*, 2003; Luong and Lee, 2006). SigB upregulates synthesis of the *arlRS* transcript in some strains. As no σ^B promoter has been detected upstream from the operon, it has been suggested that SigB regulates *arlRS* expression indirectly (Bischoff *et al.*, 2004). The Arl system has been shown to downregulate biofilm formation by an unknown mechanism (Fournier and Hooper, 2000; Toledo-Arana *et al.*, 2005). Interestingly, the Arl system also modulates DNA supercoiling that may regulate virulence factor gene expression (see section on future trends) (Fournier and Klier, 2004).

Two studies have shown that Arl is required for full virulence in *S. aureus*. In a murine model of systemic infection and in a murine model of abscess, far fewer *arl*$^-$ mutant bacteria than wild-type bacteria were recovered from the liver and abscess, respectively (Benton *et al.*, 2004). In a murine model of haematogenous pyelonephritis, the *arlR*$^-$ mutant was found to be less virulent than the wild-type strain (Liang *et al.*, 2005). The Arl system seems to be important for staphylococcal virulence, although the mechanisms involved remain unknown.

Sar homologues

Sar homologues belong to the MarR winged-helix family of transcriptional regulators (Ellison and Miller, 2006). This family comprises two subgroups of proteins. Small proteins of 13 to 16 kDa, such as SarA, SarR, SarT and SarX, bind to DNA as dimers. Larger proteins of 29 to 30 kDa, such as Rot, SarS, SarT, SarU, SarV, MgrA and TcaR, have an N-terminal region and a C-terminal region with sequences similar to those of SarA, a central domain with a sequence bearing no resemblance to that of SarA (Liu *et al.*, 2001; Schumacher *et al.*, 2001; Li *et al.*, 2003; Chen *et al.*, 2006) and also bind to DNA.

Rot

A library of an *agr* null mutant subjected to mutagenesis with the Tn*917* transposon was screened for the restoration of protease and haemolytic activity. Inactivation of the *rot* (repressor of toxin) gene restored both protease and α-toxin activity in the *agr*$^-$ mutant. Interestingly, the deletion of *rot* from a wild-type strain has no effect on these activities (McNamara *et al.*, 2000). Rot has been shown to regulate a large number of virulence factor genes in *agr*$^-$ mutants. It decreases the transcription of *hla, hlb, splABCDEF, hlgCB, sspABC, geh, ebhAB* and enhances the expression of *spa, clfB, coa, sarS* (Said-Salim *et al.*, 2003). Indeed, Rot is an important regulator of proteases such as SspA and Aur (Oscarsson *et al.*, 2006b). Rot binds to the *spa* promoter region (Oscarsson *et al.*, 2005). It has also been shown to downregulate the expression of *seb* by binding to its promoter region. The expression of *seb* is upregulated by *agr* and the deletion of *rot* restores *seb* expression in an *agr*$^-$ mutant. This effect is observed only during the post-exponential growth phase, suggesting that *rot* expression may be modulated by *agr* (Tseng and Stewart, 2005). Indeed, Rot generally represses the expression of genes upregulated by Agr. The relationship between Rot and the Agr system is complex (Said-Salim *et al.*, 2003). In an *agr*$^-$ mutant background, *rot* deletion increases RNAIII synthesis (McNamara *et al.*, 2000). Levels of Rot protein have been found to be higher in an *agr*$^-$ mutant than in the wild type, whereas *rot* transcription is not modified or may even decrease, suggesting that *rot* regulation by *agr* does not occur at the transcriptional level (McNamara *et al.*, 2000; Said-Salim *et al.*, 2003; Geisinger *et al.*, 2006). The translation of *rot* mRNA by RNAIII is regulated in a similar manner to that of the *spa* mRNA by RNAIII. Indeed, regions in the 5′UTR of the *rot* mRNA are complementary for RNAIII loops 7, 13 and 14 (Fig. 6.1B). The 3′ domain of RNAIII binds to two complementary sequences of *rot* mRNA that contains UUGGGA. Hairpin 14 seems to preferentially bind to the complementary UUGGGA sequence carrying the SD sequence of *rot* mRNA. Base-pairing between these two RNAs may inhibit the translation of *rot* mRNA (Geisinger *et al.*, 2006; Boisset *et al.*, 2007). The presence of two loops is necessary to prevent *rot* mRNA translation. This suggests that RNAIII interacts with *rot* mRNA in two distinct regions: hairpin 14 with the SD sequence and hairpin 7 or 13 with the second UUGGGA sequence (Boisset *et al.*, 2007). RNaseIII is also involved in the inhibition of *rot* mRNA translation. In the presence of hairpin 14, *rot* mRNA is cleaved by RNaseIII in the SD sequence (Boisset *et al.*, 2007). The opposite effects of Agr and Rot on virulence gene expression may be accounted for by the negative regulation of *rot* mRNA by RNAIII. Thus, Rot may be one of the transcriptional regulators used by Agr to modulate virulence factor gene expression. Rot transcription is also downregulated by SarA, which directly binds to the promoter region of *rot* (Manna and Ray, 2007).

In an experimental rabbit model of endocarditis, *rot* deletion has no effect on the number of bacteria recovered from the cardiac vegetation and kidney. In contrast, bacterial counts were significantly lower for the *agr*$^-$ mutant than for the wild type, whereas the bacterial count for the double *agr*$^-$ *rot*$^-$ mutant was similar to that for the wild-type strain, suggesting that Rot is probably also one of the effectors used by *agr in vivo* (McNamara and Bayer, 2005).

MgrA

MgrA (multiple global regulator, also known as NorR and Rat) was identified by screening a library of Tn*917* mutants for genes regulating capsule production (Ingavale *et al.*, 2003; Luong *et al.*, 2003; Truong-Bolduc *et al.*, 2003). MgrA upregulates capsule 8 gene transcription and nuclease production, but downregulates *hla* and *spa* transcription and protease and coagulase production (Luong *et al.*, 2003). Transcriptome analysis showed that MgrA also increases the transcription of *hlgA, lukD, lukS, lukF, lukM, lip, sak, splABCDE* and decreases that of *ebh* and *sarS* (Ingavale *et al.*, 2005; Luong *et al.*, 2006). MgrA binds to the promoter regions of *sarV*, *sarX* and *spa* (Manna *et al.*, 2004; Oscarsson *et al.*, 2005). The consensus sequence of the MgrA binding site is $TGTTGGN_8ACAACG$ (Manna *et al.*, 2004; Manna and Cheung, 2006a). Like SarA, MgrA has a single cysteine residue in the N-terminal region. This region is involved in the dimerization of the protein required for DNA binding. This cysteine residue may be oxidized by several oxidants, such as hydrogen peroxide (Schumacher *et al.*, 2001; Chen *et al.*, 2006). Oxidation of the protein leads to the dissociation of the regulator from the DNA target, whereas the presence of a reducing agent restores the binding of MgrA to its target DNA (Chen *et al.*, 2006).

In a murine abscess model, mice were infected by retro-orbital injection. Far fewer bacteria were isolated from kidney and liver of mice infected with the *mgrA*$^-$ mutant (2 to 4 orders of magnitude lower) than from the kidney and liver of mice inoculated with the wild-type strain (Chen *et al.*, 2006). Thus, MgrA is a regulator required for staphylococcal virulence.

SarS

SarS (also known as SarHI) was first identified as the protein binding to the promoter region of *hla, ssp* and *spa* (Tegmark *et al.*, 2000; Cheung *et al.*, 2001). Two transcripts (1 and 1.5 kb) are produced from two promoters, P1 and P2. The origin of another 3 kb transcript remains unknown. P1 is located 150 nt upstream from the start codon and is a σ^A promoter, whereas P2 is located 800 nt upstream from the start codon and is a σ^B promoter (Fig. 6.4) (Tegmark *et al.*, 2000) (Fig. 6.4). SarS has a similar structure to SarA and SarR: five helices and three strands. However, the helix-turn-helix motifs of SarS are not as flexible as those of other proteins such as SarA and SarR. Indeed, SarS has a more compact structure. It has therefore been suggested that the MarR family of transcriptional regulators should be classified into three groups, corresponding to SarA, SarS and MarR (Li *et al.*, 2003). The expression of *sarS* is downregulated by Agr and SarA, with no effect on the expression of the *agr* and *sarA* loci and upregulated by SarT (Fig. 6.5) (see the section on SarT below) (Tegmark *et al.*, 2000; Cheung *et al.*, 2001). The *sarS* gene is located just upstream from the *spa* gene, suggesting that this regulator may regulate this gene (Tegmark *et al.*, 2000). SarS has indeed been shown to upregulate *spa* expression,

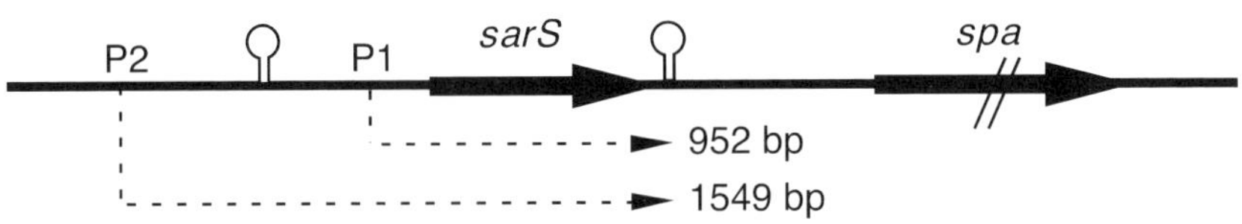

Figure 6.4 The *sarS* locus, as described by Tegmark *et al.* (2000). Two transcripts of 1 and 1.5 kb are produced from two promoters, P1 and P2. Dotted lines represent the transcripts. The double line through *spa* represents a long stretch of non-represented DNA.

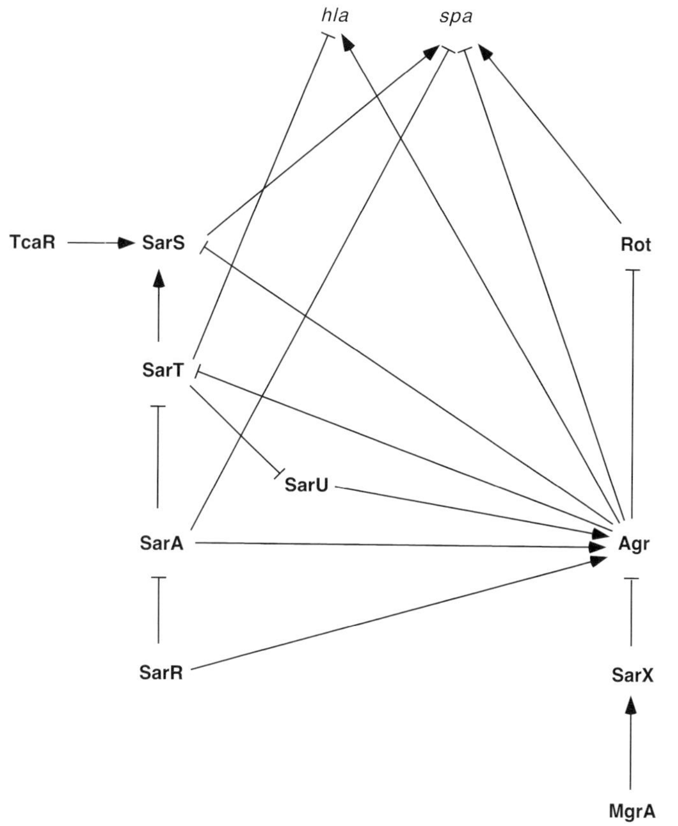

Figure 6.5 Interactions of the various members of the Sar family with virulence factor genes, as described by Schmidt *et al.* (2003). Arrows and perpendicular bars indicate positive and negative regulation, respectively.

whereas *sarS* inactivation has no effect on *hla* and *ssp* expression. SarS modifies *spa* expression independently of SarA (Tegmark *et al.*, 2000; Cheung *et al.*, 2001). SarA regulates the expression of *spa* independently of SarS, whereas Agr modulates *spa* expression through SarS and SarT (Fig. 6.5) (Cheung *et al.*, 2001). SarS and SarA binds to an overlapping binding site, as the binding of SarS affects that of SarA, and vice versa (Tegmark *et al.*, 2000; Gao and Stewart, 2004; Oscarsson *et al.*, 2005). The members of the MarR family modulate gene expression by interacting with other regulators (Ellison and Miller, 2006). SarS activates *spa*, whereas SarA represses this gene. The following model for the regulation of *spa* expression has been proposed. During the exponential growth phase, *sarS* is strongly expressed and SarS activates *spa* expression. In the post-exponential growth phase, *sarS* expression is downregulated by SarA and Agr. SarA may subsequently bind to the *spa* promoter to repress its activity.

SarR

This 14 kDa protein was first identified as the protein binding to the promoter region of *sarA*, and is known as the SarA regulator (Manna *et al.*, 1998). Its sequence is similar to that of SarA (Liu *et al.*, 2001; Manna and Cheung, 2001). Unlike SarA, SarR possesses a typical helix-turn-helix motif and displays tight interaction between the monomers (Cheung and Zhang, 2001; Liu *et al.*, 2001). The crystal structure of SarR is similar to that of SarA, with five helices, three β-strands and several loops (Liu *et al.*, 2001). Sar proteins contain a large number of Lys residues. Two winged helix motifs are found on the concave side of the protein, together with Lys and Arg residues, suggesting that this side is involved in binding to DNA. SarR may interact with DNA over a stretch of 27 bp. The interaction of this protein with DNA also seems to induce DNA bending. SarR associates with the major groove of DNA, and may also interact with the minor groove (Liu *et al.*, 2001).

SarR residues K52, K80, R82, R88 and L105 are crucial and necessary for the DNA-binding function of this protein (Manna and Cheung, 2006b). SarR downregulates the transcription of *sarA* by binding to the regions containing its three promoters. The deletion of *sarR* has no effect on the activity of *sarA* P2 and P3, whereas it increases P1 activity. SarR also upregulates the synthesis of RNAII and RNAIII, by direct binding to the P2-P3 region of the *agr* locus (Manna and Cheung, 2006b). Indeed, SarR binds to the same DNA region as SarA, but with a higher affinity than SarA. Little attenuation of RNAIII synthesis is observed in *sarA*$^-$ or *sarR*$^-$ mutants. In contrast, the decrease in RNAIII levels in the double *sarA*$^-$ *sarR*$^-$ mutant is as strong as that in the *agr*$^-$ mutant (Manna and Cheung, 2006b). This suggests that SarA and SarR bind together and are both necessary for *agr* expression. SarR preferentially recognizes an A/T-rich region containing the TAAATTAA sequence. Two consensus sequences for SarR binding are present in adjacent positions in the P2-P3 promoter region of the *agr* locus (Manna and Cheung, 2006b).

SarZ

SarZ is a 17 kDa protein and was originally identified by screening a library of multicopy genes for the restoration of haemolysin activity in a *cvfA*$^-$ mutant (see section on CvfA below). Indeed, the deletion of *cvfA* decreases haemolytic activity and, in a *cvfA*$^-$ mutant, SarZ overproduction increases haemolysin production. SarZ belongs to the MarR family and has a sequence slightly similar to that of SarA. SarZ is encoded by a monocistronic gene producing a 900 bp transcript. The expression of *sarZ* is slightly upregulated by CvfA. In *sarZ*$^-$ mutants, *hla*, *hlb* expression and RNAIII synthesis are downregulated. SarZ contains a helix-turn-helix domain and binds to DNA in the promoter region of *hla*, *asp23* and *agr*. However, this binding appears to be non-specific. Mutations affecting residues F14, K25, G67, R92 and, to a lesser extent, V43, T70 and I96 decrease DNA binding (Kaito *et al.*, 2006).

The deletion of *sarZ* decreases staphylococcal virulence in a silkworm model, but to a lesser extent than the *cvfA*$^-$ mutants. Mutations impairing DNA binding also decrease virulence in this model. Furthermore, in a murine model of systemic infection, the number of bacteria recovered from the spleen and kidney is lower for the *sarZ*$^-$ mutant than for the wild-type strain (Kaito *et al.*, 2006). Thus, SarZ, like other members of the Sar family, plays a significant role in staphylococcal virulence.

SarT, SarU, SarV, SarX

SarT, SarU, SarV and SarX were first identified as protein homologues of SarA. They have been shown to regulate numerous virulence factor genes (Fig. 6.5) but it is not known whether they also modulate staphylococcal virulence *in vivo*.

SarT (also known as SarH3) is a 16 kDa protein encoded by an 800 bp transcript. Its expression is downregulated by SarA and Agr (Fig. 6.5) (Schmidt *et al.*, 2001). SarT downregulates *hla* and *agr* expression and upregulates *sarS* and *spa* expression (Schmidt *et al.*, 2001; Schmidt *et al.*, 2003). It has been suggested that SarA acts on *hla* through SarT, whereas Agr acts independently of this pathway (Schmidt *et al.*, 2001). SarT may regulate *spa* expression through SarS (Fig. 6.5). Indeed, SarT binds to two binding sites with a consensus sequence of AAATG(A/T)CAT in the promoter region of *sarS*. Agr probably modifies *sarS* expression through SarT (Schmidt *et al.*, 2003). SarT is missing in many strains (Lindsay *et al.*, 2006).

The *sarU* gene (also known as *sarH2*) is located upstream from *sarT* and is transcribed in the opposite direction. It has one promoter and generates two transcripts of 1.4 and 1.2 kb. SarT downregulates *sarU* expression by binding to the promoter region of this gene. SarU upregulates *spa* and *hla*. As SarU also increases the production of RNAII and RNAIII, this protein is thought to act mainly through *agr* (Fig. 6.5) (Manna and Cheung, 2003). Inactivation of SarU decreases *S. aureus* virulence in the *C. elegans* model of staphylococcal virulence (Bae *et al.*, 2004).

The *sarV* gene is downregulated by MgrA and SarA, which bind to different binding sites in the *sarV* promoter region. SarV is mainly involved in the regulation of autolysis, but has also been shown to upregulate *hla*, *splA*, *aur* and *scp* expression. The deletion of *sarV* decreases *arlRS* expression (Manna *et al.*, 2004).

The *sarX* gene has only one promoter – a σ^A promoter – but generates two transcripts. The 1.5 kb transcript is produced in much smaller amounts than the 0.5 kb transcript, and both are produced mainly during the post-exponential growth phase. MgrA upregulates *sarX* expression, binding to the promoter region of this gene (see section on MgrA) and *sarX* is autoregulated. SarX downregulates RNAII and RNAIII, probably by binding to the P2-P3 region of the *agr* locus (Fig. 6.5). SarX downregulates *hla, hlb* and *sspA,* possibly indirectly, through *agr* (Manna and Cheung, 2006a).

TcaR

The *tca* (teicoplanin-associated) operon contains three genes (*tcaRAB*) and was first identified as encoding a modulator of resistance to teicoplanin and methicillin (Brandenberger *et al.*, 2000). The sequence of TcaR displays some similarity to that of SarA. This protein belongs to the MarR-like transcriptional regulator family. Transcriptome analysis has shown that TcaR regulates only three genes: *sarS, spa* and *sasF*. SasF is a cell-wall associated protein of unknown function (Roche *et al.*, 2003). The deletion of *tcaR* decreases the expression of *spa* and *sarS* but increases that of *sasF*. The double $sarS^-$ $tcaR^-$ mutant has similar levels of *spa* transcription to the $sarS^-$ mutant, suggesting that *sarS* is required for the effects of TcaR on *spa* expression (McCallum *et al.*, 2004). Interestingly, in strains derived from 8325–4, the *tcaR* gene is inactivated by a stop codon and displays weak *sarS* and *spa* expression (McCallum *et al.*, 2004). The transcription of *ica* has been shown to be increased in a *tcaR* deletion mutant, but PNAG (poly-N-acetylglucosamine surface polysaccharide) production and adhesion to polystyrene are not affected in this mutant (Jefferson *et al.*, 2004). Thus, the main function of TcaR seems to be to regulate *spa* expression through *sarS*.

Sigma factor B (σ^B)

Structure and regulation of the *sigma B* locus

RNA polymerase transcribes DNA to mRNA ready for protein production. The RNA polymerase core enzyme is composed of four subunits ($\alpha^2\beta\beta'$). This core enzyme interacts with a σ factor to form the RNA polymerase holoenzyme that recognizes specific promoter sequences, leading to the initiation of transcription. In addition to the classical σ^A factor, only two alternative sigma factors, σ^B and σ^H, have been found in *S. aureus* (Wu *et al.*, 1996; Deora *et al.*, 1997; Kuroda *et al.*, 2001). The σ^B has been studied in detail in *B. subtilis*, and is known to play an important role in the bacterial response to environmental stress and energy depletion. In *S. aureus*, the *sigB* gene encoding σ^B (or SigB) belongs to a chromosomal operon of four genes (*rsbU, rsbV, rsbW* and *sigB*) (Fig. 6.6) (Wu *et al.*, 1996). In *B. subtilis*, Rsb proteins (regulators of Sigma B) determine the availability of SigB for transcription, depending on growth conditions. RsbV, an anti-anti-sigma factor, is phosphorylated by RsbW, an anti-sigma factor and protein kinase. RsbW can also bind to SigB. RsbU, a serine/threonine phosphatase, dephosphorylates RsbV. The phosphorylation state of RsbV determines the availability of SigB. In unstressed bacteria, RsbV is phosphorylated. As a result, RsbW binds to SigB rather than RsbV. SigB is therefore unable to form the RNA polymerase holoenzyme in classical growth conditions. In stress conditions, RsbV is dephosphorylated by a phosphatase such as RsbU, and interacts with RsbW, whereas SigB, which remains free, may

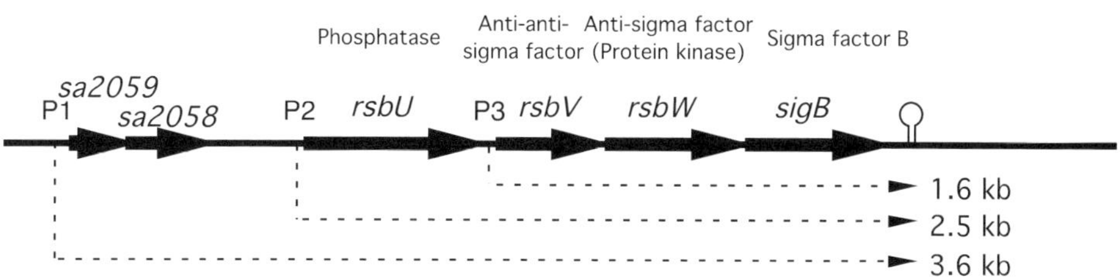

Figure 6.6 The *sigB* operon of *S. aureus* COL, composed of four genes transcribed from three different promoters, P1, P2 and P3, as described by Senn *et al.* (2005). Dotted lines represent the transcripts.

associate with the RNA polymerase core enzyme to activate transcription (Zhang *et al.*, 2005). Similar interactions have been shown to occur in *S. aureus*: RsbU associates with RsbV, RsbV with RsbW and RsbW with SigB (Miyazaki *et al.*, 1999; Senn *et al.*, 2005).

The *sigB* operon is a complex locus. Several transcripts are initiated from three different promoters. P1, a σ^A promoter, generates a 3.6 kb transcript including all four genes, P2 generates a 2.5 kb transcript and P3, a σ^B promoter, generates a 1.6 kb transcript (Fig. 6.6) (Wu *et al.*, 1996; Senn *et al.*, 2005). The 3.6 kb transcript is detected only during the early exponential phase, the transcript from P2 is detected mostly during the early exponential phase and the transcript from P3 is detected at its maximal levels between the post-exponential and stationary growth phases. Transcripts containing the *sigB* gene increase in abundance throughout the exponential growth phase, decreasing in number when bacteria enter the stationary phase. The SigB protein is produced in similar amounts in all growth phases (Senn *et al.*, 2005). The first experiments to study this locus were performed with RN6390, a well characterized strain derived from 8325–4. However, this strain presents an 11 bp deletion in the *rsbU* gene (Kullik *et al.*, 1998). RsbU is required for SigB activity in *S. aureus*. Indeed, SigB is almost inactive in RN6390 and *sigB* is expressed much more strongly in clinical strains with efficient RsbU than in RN6390 (Gertz *et al.*, 1999; Giachino *et al.*, 2001). The *sigB* operon is also autoregulated. Functional *rsbU* and *rsbV* genes are required for the full expression of *sigB* and SigB is required for the expression of *sigB*, *rsbV* and *rsbW* (Palma and Cheung, 2001; Bischoff *et al.*, 2004). In an *rsbV*$^-$ mutant and, to a lesser extent, in the *rsbU*$^-$ mutant, the P3 promoter of the *sarA* locus displays lower than normal levels of activity (Palma and Cheung, 2001).

In *S. aureus*, σ^B modulates the expression of several genes with a consensus σ^B promoter (Kullik *et al.*, 1998). It also regulates genes encoding virulence factors, upregulating *cap5*, *coa*, *clfA*, *fnbA* and downregulating *aur*, *nuc*, *seb*, *splABCDEF*, *hla*, *hlb*, *hlgACB*, *lip*, *geh*, *lukD*, *lukF*, *lukM*, *plc*, *sak*, *sspABC* and *scp* (Cheung *et al.*, 1999; Nicholas *et al.*, 1999; Ziebandt *et al.*, 2001; Horsburgh *et al.*, 2002; Bischoff *et al.*, 2004; Homerova *et al.*, 2004; Ziebandt *et al.*, 2004; Entenza *et al.*, 2005; Pane-Farre *et al.*, 2006). As SigB downregulates the *hla* and protease genes, it has been suggested that some clinical isolates producing large amounts of α-haemolysin or protease may lack *sigB*. Indeed, V8, which is known to produce large amounts of protease, has a stop codon in its *sigB* gene, and Wood 46, which produces large amounts of α-haemolysin, has an IS element in its *rsbU* gene. The introduction of a plasmid containing the *sigB* gene decreases haemolytic and proteolytic activity (Karlsson-Kanth *et al.*, 2006). The deletion of *sigB* also increases bacterial aggregation and binding to fibrinogen and fibronectin (Kullik *et al.*, 1998; Cheung *et al.*, 1999; Bischoff *et al.*, 2001; Nair *et al.*, 2003). SigB is involved in staphylococcal adhesion to platelet-fibrin clots mimicking cardiac vegetations (Entenza *et al.*, 2005). It is also required for biofilm formation and upregulates *ica* expression (Rachid *et al.*, 2000; Bateman *et al.*, 2001; Knobloch *et al.*, 2004). IS256, which integrates into the *ica* operon of *S. epidermidis*, can also integrate into the *rsbU* gene, thereby decreasing *icaA* expression and biofilm formation (Ziebuhr *et al.*, 1997; Conlon *et al.*, 2004).

In vivo role of Sigma B

In a model of murine subcutaneous abscess, similar numbers of bacteria were recovered from lesions infected with the *sigB*$^-$ mutant and lesions infected with the wild-type strain (Horsburgh *et al.*, 2002). Similar results were obtained with a murine wound infection model, a murine haematogenous pyelonephritis model and a rat osteomyelitis model (Nicholas *et al.*, 1999). In a rat model of endocarditis, the overproduction of SigB was found to increase bacterial density in the spleen and aortic vegetation at 16 h, with bacterial density subsequently decreasing, as shown by measurements at 48 h. In a murine model of sepsis and arthritis, the *rsbU*$^+$ strain induces more severe arthritis and marked weight loss, giving higher mortality rates than the *rsbU*$^-$ mutant. Furthermore, *sigB*$^-$ mutants are more persistent than the wild-type strain in the kidneys and joints (Jonsson *et al.*, 2004).

Levels of *sigB* expression in a guinea pig foreign implant model have been shown to be similar to those *in vitro* (Senn *et al.*, 2005). Salicylic acid activates *sigB* expression, decreasing the expression of *agr* and *sarA* in an experimental rabbit endocarditis model. These findings may account for the reported antibacterial properties of aspirin (Kupferwasser *et al.*, 2003).

SigB may be involved in the internalization in osteoblasts, probably by increasing the expression of MSCRAMMS, which is required for internalization (Jett and Gilmore, 2002; Nair *et al.*, 2003). *S. aureus*-induced apoptosis is attenuated in a *sigB* mutant (Haslinger-Loffler *et al.*, 2005). Thus, despite its involvement in the regulation of many virulence factors, SigB seems to play only a minor role in the pathogenicity of staphylococci. However, SigB may be active and play a role in staphylococcal virulence only in the presence of specific conditions. Indeed, *sigB* expression is activated by aspirin and other stresses (Kupferwasser *et al.*, 2003; Pane-Farre *et al.*, 2006).

The SrrAB system

The SrrAB (staphylococcal respiratory response or SrhSR system) is another two-component system homologous to ResDE in *B. subtilis*. SrrAB is involved in global energy modulation in response to changes in oxygen availability (Throup *et al.*, 2001; Yarwood *et al.*, 2001). SrrA, the response regulator, is present into the cytoplasm, whereas SrrB, the histidine kinase, is present in the membrane (Pragman *et al.*, 2004). SrrAB upregulates the expression of *tst*, *spa* and *icaR* in aerobic conditions, probably by binding directly to their promoter regions (Yarwood *et al.*, 2001; Pragman *et al.*, 2004; Pragman *et al.*, 2007). The *srrAB* operon is autoregulated. Two transcripts of 0.7 and 2.5 kb are produced from a single promoter (Yarwood *et al.*, 2001; Pragman *et al.*, 2004).

SrrAB is necessary for the survival of staphylococci in neutrophils. It increases staphylococcal resistance to non-oxidative reactions produced by neutrophils probably by enhancing the production of polysaccharide intercellular adhesin encoded by the *ica* operon (Ulrich *et al.*, 2007). In the *C. elegans* model of staphylococcal virulence, inactivation of *srrAB* results in a decreased virulence (Bae *et al.*, 2004). The overproduction of SrrAB decreases virulence in a model of endocarditis (Pragman *et al.*, 2004). Furthermore, SrrAB is required for full virulence in a murine model of sepsis (Richardson *et al.*, 2006).

Target of RNAIII-activating protein (TRAP)

RAP (RNAIII-activating protein) was first identified as a 38 kDa protein secreted by *S. aureus* that activated expression from the P3 promoter and the production of RNAIII (Balaban and Novick, 1995a). RAP appears to be an orthologue of RplB, the L2 ribosomal protein. RAP is secreted in similar amounts in all growth phases (Korem *et al.*, 2003). It is also detected in the supernatant of an *agr*$^-$ mutant culture, suggesting that RAP does not operate through the Agr pathway. Antibodies against RAP abolish RNAIII synthesis (Balaban *et al.*, 1998). The vaccination of mice with purified RAP increases the percentage of healthy mice following bacterial challenge (Balaban *et al.*, 1998).

RIP is a peptide secreted by coagulase-negative staphylococci such as *Staphylococcus xylosus*; it inhibits RNA III synthesis (Balaban and Novick, 1995a). Its sequence is YSPXTNF (X is either cysteine, a tryptophan or a modified amino acid) (Gov *et al.*, 2001). RIP has been shown to prevent infections induced by *S. aureus*, such as keratitis, osteomyelitis, mastitis (Balaban *et al.*, 2000). It also reduces biofilm formation (Balaban *et al.*, 2003).

Reports of the presence of RIP and RAP in the culture supernatants of staphylococci have remained controversial as another group was unable to replicate this finding (Novick *et al.*, 2000).

RAP induces the phosphorylation of TRAP. TRAP is a 21 kDa folded protein composed of α-helices and β-strands. The effects of RAP on RNAIII synthesis are dependent on TRAP. The *trap* gene is expressed throughout growth and is found in various staphylococcal species (Gov *et al.*, 2004). TRAP phosphorylation is activated by RAP and inhibited by RIP and AIP from strains of the same group (Balaban *et al.*, 2001). TRAP is associated with the membrane but has no transmembrane domain, suggesting that its association with the membrane may involve other proteins. Three histidine residues (His-66,

His-79 and His-154) are phosphorylated and required for its activity. No RNAIII synthesis occurs in *trap*⁻ mutants. Lower mortality rates were recorded for a *trap*⁻ mutant than for the wild type in a murine model of cellulitis (Gov *et al.*, 2004). TRAP has been shown to modulate the expression of several virulence factor genes. It upregulates the expression of *hla*, *hlb*, *hld*, *hlgB*, *cap*, *lip*, *geh*, *sspABC*, *scpA*, *plc*, *aur* and downregulates *clfB* and *spa*, consistent with a regulatory pattern similar to that for Agr (Korem *et al.*, 2003). However, other studies found that the deletion of *trapP* does not impair the formation of biofilm, the production of haemolysins, the transcription of *agr* and the staphylococcal virulence in a murine model of septic arthritis infection (Shaw *et al.*, 2007; Tsang *et al.*, 2007). Furthermore, recent data suggest that the phenotype observed with the deletion of *traP* may be due to the coexistence of a mutation in the *agrA* gene, thereby inactivating the Agr system and producing an *agr*⁻ phenotype (Adhikari *et al.*, 2007).

Other regulators

Clp proteins

Clp (caseinolytic protease) proteins consist of a proteolytic subunit – ClpP or ClpQ – associated with an ATPase specificity factor – ClpC, ClpB, ClpX, ClpL or ClpY in *S. aureus*. The proteolytic core of the complex is formed from two heptameric rings of the proteolytic subunit. The ATPase factor is attached to the proteolytic core and determines substrate specificity, allowing the substrate to enter the proteolytic core (Chatterjee *et al.*, 2005). In *E. coli*, these proteins mediate ATP-dependent protein folding, unfolding and assembly. During heat shock, proteins may unfold and aggregate. Clp proteins refold some of these proteins and degrade those that cannot be repaired (Wawrzynow *et al.*, 1995; Frees *et al.*, 2003).

These proteins have been studied in detail in *E. coli*, but much less is known about their function in *S. aureus*. Clp proteins are all involved in heat stress responses (not reviewed in this chapter). Several studies have showed that ClpX and, to a lesser extent, ClpP are required for virulence of *S. aureus*. ClpC is involved in biofilm formation (Becker *et al.*, 2001; Frees *et al.*, 2004), whereas ClpB, ClpL, ClpY and ClpQ seem to be involved only in the regulation of heat tolerance (Frees *et al.*, 2004; Frees *et al.*, 2005b). Clp proteins also seem to play various roles in virulence. Some, such as ClpX, ClpP and ClpC, seem to play crucial roles in regulating virulence gene expression, probably by interacting with global regulators such as Agr, SarA, Rot or SarS.

ClpX

ClpX regulates the expression of various virulence genes. The deletion of *clpX* results in the repression of *hla* and *sspA* transcription. Furthermore, *clpX*⁻ mutants display only low levels of protease production on plates containing skimmed milk. Interestingly, *agr* expression and AIP production are strongly attenuated in *clpX*⁻ mutants, as in *agr* deletion mutants (Frees *et al.*, 2003; Frees *et al.*, 2005a). Further studies are required, but it seems possible that ClpX regulates *hla* expression through the *agr* locus. The deletion of *rot* restores *sspA* transcription to wild-type levels in a *clpX*⁻ mutant. Thus, *rot* is required for the effects of ClpX on *sspA* (Frees *et al.*, 2005a). The deletion of *clpX* dramatically decreases *spa* transcription. Studies with double *agr*⁻ *clpX*⁻ mutant have suggested that ClpX is required for the effects of *agr* on *spa* transcription (Frees *et al.*, 2005a). ClfA and ClfB are present in two forms: 110 and 140 kDa. The 110 kDa form probably corresponds to a truncated protein cleaved by a metalloprotease. The truncated form is more abundant in the wild-type strain, whereas the 140 kDa protein predominates in the *clpX*⁻ mutant. As *clpX* deletion decreases the production of extracellular protease, it is possible that ClfA and ClfB are processed less efficiently in this mutant. The amount of ClfA is similar in the wild-type strain and the *clpX*⁻ mutant, whereas levels of ClfB production in the *clpX*⁻ mutant are significantly higher than in the wild type, due to higher levels of transcription of the corresponding gene. Furthermore, *fnbA* expression is stronger in the *clpX*⁻ mutant than in the wild-type strain (Frees *et al.*, 2005a). ClpX is required for several other functions involved in virulence (Frees *et al.*, 2004). ClpX decreases biofilm formation. However, it remains unclear how ClpX modulates adhesion to polystyrene.

The *clpX*$^-$ mutant is correctly taken up by bovine mammary epithelial cells, but cannot replicate in these cells. Agr is known to be required for escape from the endosome during replication in cells (see section on Role of Agr in the relationship between staphylococci and mammalian cells) (Shompole *et al.*, 2003). The *clpX*$^-$ mutant may therefore be unable to replicate because this deletion inhibits *agr* expression (Frees *et al.*, 2004).

A transposon signature-tagged mutagenesis study showed that *clpX* deletion decreased the virulence of *S. aureus* in a murine model of bacteraemia (Mei *et al.*, 1997). Another study confirmed these results in a murine abscess model (Frees *et al.*, 2003). ClpX is therefore clearly required for *S. aureus* virulence.

ClpP

ClpP has an effect on virulence gene expression similar to, but weaker than that of ClpX. ClpP regulates the expression of several virulence and regulator genes involved in virulence. The *clpP* deletion mutant displays repression of *hla*, *sspA*, *splBCD*, *cap*, *nuc*, *geh*, *aur* and *lip* expression and increase of *hlb* expression (Frees *et al.*, 2003; Frees *et al.*, 2005a; Michel *et al.*, 2006). ClpP does not affect *spa* transcription, but it does regulate ClfB, ClfA and FnbA production and/or transcription in a similar manner to ClpX (Frees *et al.*, 2003; Frees *et al.*, 2005a). Finally, the *clpP* deletion decreases the transcription of *arlRS*, *mgrA*, *sigB* and, to a lesser extent, *sarR*, but also slightly increases *sarT* transcription (Michel *et al.*, 2006). It also attenuates *agr* transcription and AIP production, but less so than *clpX* deletion (Frees *et al.*, 2003; Frees *et al.*, 2005a). Its effects on *sspA* expression requires Rot. The deletion of *clpP*, unlike that of *clpX*, enhances biofilm formation. As the deletion of the *agr* or *arl* locus increases biofilm formation (Fournier and Hooper, 2000; Vuong *et al.*, 2000), ClpP may modify biofilm formation through Agr and/or Arl. ClpP is also required for replication in epithelial cells (Frees *et al.*, 2004).

A *clpP* deletion mutant displays attenuated virulence in a mouse skin abscess model mimicking a human wound infection, with fewer bacteria recovered from lesions than for the wild-type strain (Frees *et al.*, 2003).

ClpC

ClpC is the last of the four genes in its operon. ClpC has been implicated in virulence chiefly in the domain of biofilm formation. Levels of *clpC* expression are much higher in staphylococcal biofilms than in bacteria with a planktonic mode of growth (Becker *et al.*, 2001). Furthermore, the deletion of *clpC* inhibits biofilm formation, suggesting that *clpC* is required for biofilm formation (Frees *et al.*, 2004).

SvrA

SvrA (staphylococcal virulence regulator) was identified in a signature-tagged mutagenesis study in a murine model of bacteraemia. This locus is required for full virulence of *S. aureus* (Mei *et al.*, 1997). It is a membrane-associated protein of 48.8 kDa, with 12 putative membrane-spanning domains (Garvis *et al.*, 2002). In *svrA* deletion mutants, levels of α-toxin, β-haemolysin and γ toxin production are significantly lower than in the wild type, due to lower levels of transcription of the corresponding genes. However, levels of Spa production are higher, due to higher rates of transcription, and the increase in transcription rate observed is greater than that for *agr*$^-$ mutants. Interestingly, the *svrA*$^-$ mutant also displays no RNAIII and RNAII expression, suggesting that SvrA may modify virulence gene expression through the *agr* locus. Following the intraperitoneal injection of wild-type *S. aureus* or its *svrA*$^-$ mutant in mice, the survival of the *svrA*$^-$ mutant, in the spleen, was lower than that of the wild type by a factor of 550, and lower than that of the *agr* mutant by a factor of 10. Survival rates were, however, similar to those of the *svrA*$^-$ *agr*$^-$ mutant. It has been proposed that SvrA may be the transmembrane protein involved in the attachment of TRAP to the membrane (Garvis *et al.*, 2002). These results remain controversial as another group did not find that the deletion of *svrA* impaired *agr* expression and haemolysin production. It has been suggested that a mutation inactivating the *agr* locus may be responsible for the phenotype observed in the absence of *svrA* (Chen and Novick, 2007).

MsrR

MsrA (methionine sulfoxide reductase) reduces the S form of methionine sulfoxide to methionine (Moskovitz *et al.*, 2002). *S. aureus* has two *msrA* genes. One is upstream from *msrR* and transcribed in the opposite direction. MsrR belongs to the LytR-CpsA-Psr family of transcriptional regulators. The N-terminal region of MsrR probably acts as a transmembrane domain, with the C-terminal region acting as an extracellular domain (Rossi *et al.*, 2003). MsrR was first identified during the screening of a Tn*551* transposon library for increased susceptibility to methicillin. Expression of *msrR* increases during the exponential growth phase, decreasing thereafter. The expression of *msrA* is downregulated by MsrR. Levels of *hla* transcription are higher in the *msrR*$^-$ mutant than in the wild type. The transcription of *spa* and RNAIII is slightly upregulated by MsrR. The three *sarBCA* transcripts of the *sarA* locus are also more abundant during the early exponential phase in *msrR* deletion mutants (Rossi *et al.*, 2003). Thus, although MsrR has a marked effect on *sarA* transcripts, its effect on virulence gene expression is minor. *In vivo* experiments in animals are required to determine whether this regulator does indeed play a significant role in staphylococcal virulence.

CvfA, CvfB

CvfA and CvfB (conserved virulence factors A and B) were first identified during a search for new virulence factor genes. About 100 genes of unknown function were inactivated and their virulence tested in the silkworm model (see section of the *in vivo* virulence of the *agr*$^-$ mutant) (Kaito *et al.*, 2005).

The function of *cvfA* is unknown. It is composed of a putative transmembrane domain in the N-terminal region and an internal region containing a KH domain involved in RNA binding and an HD domain characteristic of a family of metal-dependent phosphodiesterases (Siomi *et al.*, 1994; Aravind and Koonin, 1998; Kaito *et al.*, 2005). Mutation of the catalytic residues conserved in these two domains inhibits the activity of this protein. Deletion of *cvfA*, a monocistronic gene, increases *spa* expression and decreases the production of haemolysin and RNAII and RNAIII levels. The *cvfA*$^-$ mutant is much less virulent in silkworm than the wild-type strain (Kaito *et al.*, 2005).

CvfB, encoded by a monocistronic gene, upregulates *hla* expression and protease production. It slightly downregulates *spa* expression at the transcriptional level. CvfB also upregulates RNAIII. In the silkworm model of staphylococcal virulence, the deletion of *cvfB* results in avirulence. In the murine model of systemic infection, fewer bacteria are recovered from the spleen for the *cvfB*$^-$ mutant than for the wild-type strain (Matsumoto *et al.*, 2007). These experiments were carried out in strain RN4220, which carries a mutation in *agrA* (see section on AgrA). Experiments carried out in strain 8325–4, which has a fully functional *agr* locus, showed that the deletion of *cvfB* in an *agr*$^+$ background has no effect on haemolytic activity, whereas it decreases protease activity, as in strain RN4220. Thus, CvfB interacts with Agr in the regulation of some virulence factor genes (e.g. haemolysin genes) but not in the regulation of others (e.g. protease genes) (Matsumoto *et al.*, 2007).

Thus, CvfA and CvfB appear to play an important role in inducing staphylococcal infections, and further studies are required to determine how they modulate virulence.

CcpA

Several virulence factors are regulated by the presence of glucose in the culture medium (Duncan and Cho, 1972; Chan and Foster, 1998a). The phosphoenolpyruvate-dependent phosphotransferase system (PTS) is involved in carbohydrate uptake and is thus essential for the regulation of catabolic gene expression. Carbon catabolite repression regulates the expression of genes involved in the utilization of carbon sources. For example, the presence of glucose in the medium leads to repression of the operons involved in the catabolism of other carbon sources, such as lactose (Oskouian and Stewart, 1987). One of the mechanisms bacteria use to repress or activate gene expression is the binding of transcriptional regulators, such as CcpA (catabolite control protein A). In gram-positive bacteria, the signalling protein detecting glucose-6-phosphate generation through glycolysis is HPrK (HPr kinase). HPrK phosphorylates HPr at a serine residue. HPr(Ser-P) associates with CcpA and this complex binds to specific sequences – catabolite-response elements (or

CREs) – modifying gene expression (Warner and Lolkema, 2003). Whether the gene is activated or repressed depends on the position of these sequences in the promoter region. The deletion of *ccpA* in *S. aureus* increases the transcription of *spa* in the presence of glucose, whereas this deletion has only a slight effect in the absence of glucose. A CRE sequence has been detected upstream from the *spa* promoter, suggesting that CcpA may bind directly to the *spa* promoter in the presence of glucose. In contrast, RNAIII is produced in smaller amounts in *ccpA*$^-$ mutants than in the wild type, and this effect is probably indirect because no CRE sequences have been found in the P3 region (Seidl *et al.*, 2006). Capsule operon expression is downregulated by glucose. The deletion of *ccpA* restores *cap* mRNA production in the presence of glucose, whereas no difference in capsule operon expression is observed between the wild-type strain and the *ccpA*$^-$ mutant in the absence of glucose (Seidl *et al.*, 2006). This suggests that CcpA downregulates *cap* expression when glucose is present. This action is probably indirect, as no CRE consensus sequence has been identified in the promoter region of *cap*. CcpA is thus clearly involved in the regulation of several virulence factors in the presence of glucose. Further studies of the importance of this regulator in the regulation of virulence *in vivo* studies are required to confirm the involvement of CcpA in staphylococcal virulence.

Aconitase

S. aureus transforms glucose into pyruvate by the pentose phosphate and glycolytic pathways. In aerobic culture, this pyruvate is oxidized to acetate, which is then oxidized in the tricarboxylic acid cycle (Somerville *et al.*, 2002b). Aconitase is one of the enzymes involved in this cycle. Inactivation of the gene encoding this enzyme (*acnA*) decreases the production of Geh, Sec, serine protease, Plc, Hla and Hlb. The corresponding mutant is also unable to clump in the presence of fibrinogen. RNAIII production and *sigB* transcription are upregulated by AcnA in the stationary phase. Finally, in a soft-tissue infection model, the *acnA*$^-$ mutant takes longer than the wild type to cause lesions, but the lesions formed are no less severe (Somerville *et al.*, 2002b). These results indicate that *acnA* is required for the expression of various virulence factor genes, expression from the *agr* P3 promoter, and for full virulence *in vivo*.

Msa

The *msa* (modulator of SarA) gene was first identified by screening a Tn*551* library to detect genes involved in the regulation of SarA. The deletion of *msa* slightly decreases *sarA* expression. Msa is a putative membrane protein with a conserved domain present characteristic of the AcrB/AcrD/AcrF integral membrane protein family (Sambanthamoorthy *et al.*, 2006). Upstream from *msa* is *cspA*, which encodes a cold shock protein. The expression of the *cspA* is almost entirely abolished in an *msa* deletion mutant. Furthermore, the deletion of *msa* increases the expression of several virulence genes (*aur* and *sspA*), whereas it decreases *fnbA* transcription. It is possible that *msa* regulates these genes through *sarA*, although this has not been demonstrated. Msa also modifies the expression of other virulence genes, such as *spa* and *hla*. However, its effects differ between strains (Sambanthamoorthy *et al.*, 2006). The role and importance of this protein in the virulence regulon therefore remain unknown.

HtrA

HtrA (high temperature requirement) has been implicated in the degradation of damaged proteins in *E. coli* and may also act as a chaperone in some conditions. In *S. aureus*, two *htrA* genes, *htrA1* and *htrA2*, have been detected. The deletion of either of these genes has no significant effect on virulence gene expression in the COL and RN6390 strains. In contrast, the deletion of both *htrA1* and *htrA2* results in decreases in haemolytic activity and RNAIII production in the RN6390 strain. The production of other factors such as nuclease does not seem to be impaired by the deletion of both *htrA1* and *htrA2* in the COL strain. Similarly, in a rat model of endocarditis, the double *htrA1*$^-$ *htrA2*$^-$ mutant has a similar level of virulence to the wild-type strain COL. In contrast, fewer bacteria are recovered from the spleen and cardiac vegetations for the double *htrA1*$^-$ *htrA2*$^-$ mutant than for the wild-type strain RN6390. This further confirms that genetic background is crucial for the regulation of virulence factor genes (Rigoulay *et al.*, 2005).

Conclusions

The more we learn about *Staphylococcus*, the more complex the regulation of its virulence appears. Indeed, *S. aureus* is a fascinating bacterium that is capable of causing infections through multiple pathogenesis mechanisms in environments as diverse as the lungs, bones, blood and skin. It possesses an array of virulence factors involved in different steps of the infection process. These factors are modulated by global virulence regulators, which adapt the production of virulence factors, enabling the bacterium to survive and to induce infection in specific, often hostile environments. Many studies in the last 20 years have shown that the virulence regulation network of this bacterium is highly complex. Virulence regulators may be simple transcriptional regulators or two-component systems, or even RNA molecules modulating the production of a virulence factor by interacting with its mRNA. Furthermore, proteins with primary functions other than the regulation of virulence, such as Clp proteins, aconitase or CcpA, are also involved in virulence regulation, often via interactions with major virulence regulators.

Future trends

Several important questions remain unanswered. It has been shown that behaviour *in vitro* does not necessarily reflect behaviour *in vivo*. For example, SigB, which regulates the expression of numerous virulence factor genes *in vitro*, does not appear to be a major regulator of virulence in the various animal models tested. Furthermore, the pattern of gene expression *in vivo* is often very different from that *in vitro* (Pragman and Schlievert, 2004). Even the addition of serum to the culture medium can have a major effect on gene expression (Yarwood *et al.*, 2002). Thus, although *in vitro* experiments are a good first step in studies of a regulator or a virulence factor, only *in vivo* studies can really show us how *S. aureus* causes infections. Improved infection models, and technologies to investigate *in vivo* gene expression will be important.

The expression of virulence factor genes may differ between strains. This suggests that unknown factors and/or genetic background may also be involved in virulence regulation. Mutations in virulence regulator genes seem to occur frequently in clinical strains (e.g. mutations in the *agr* locus, the *rsbU* operon, the *sae* locus and the *arl* operon have been described) (Kuroda *et al.*, 2001; Somerville *et al.*, 2002a; Karlsson-Kanth *et al.*, 2006; Yamazaki *et al.*, 2006). Some variation is strongly associated with lineages (Lindsay *et al.*, 2006). Thus, the influence of the genetic background may be due to the presence of mutations in characterized or unknown virulence regulator and/or factor genes.

Another intriguing question concerns the co-ordination of this network for the efficient production of virulence factors, with such a large number of interactions occurring within the network. The staphylococcal virulence network is modulated by many regulators, interacting with many promoter regions. Indeed, at least four different proteins bind to the promoter region of *spa* (SarA, SarS, Rot and MgrA). Furthermore, the translation of *spa* mRNA is also inhibited by RNAIII from the *agr* locus. The gene encoding this virulence factor therefore seems to be regulated by five different pathways. At least four proteins are known to bind to the P2-P3 region of the *agr* locus (AgrA, SarA, SarR and SarX). Interestingly, most of the proteins that bind to these promoter regions belong to the Sar family. The members of the MarR family are known to modulate gene expression by interacting with other regulators (Ellison and Miller, 2006). A previous study has suggested that the members of the Sar family may also be similar to the histone-like proteins found in *E. coli* that modulate gene expression by inducing conformational changes in DNA (Roberts *et al.*, 2006). It would be interesting to determine whether the members of this family also have such functions. Indeed, the expression of several virulence factor genes (*eta* and *spa*) has been shown to be regulated by DNA supercoiling and some regulators may use this mechanism to modulate virulence factor gene expression (Sheehan *et al.*, 1992; Fournier and Klier, 2004). Whole genome technologies such as microarrays and systems biology will be beneficial (See Chapter 1).

Only the signal of the Agr two-component system has been well studied, and those of the

Sae, Arl and Srr systems remain unknown. The main function of other proteins involved in the regulation of virulence – SvrA, Msa, CvfA and CvfB – has also yet to be established.

Virulence regulators have been proposed to be alternative targets for new therapeutics against staphylococcal diseases. For example, a global inhibitor of Agr is capable of attenuating infection in a murine abscess model (Wright *et al.*, 2005a). Proteins of the Sar family appear to be susceptible to oxidation. Indeed, oxidative reagents induce the dissociation of MgrA from DNA and thereby inactivate it (Chen *et al.*, 2006). Thus, much remains to be clarified if we are to understand fully the mechanisms of staphylococcal virulence and to control infections.

References

Adhikari, R.P., Arvidson, S., and Novick, R.P. (2007). A nonsense mutation in *agrA* accounts for the defect in *agr* expression and the avirulence of *Staphylococcus aureus* 8325–4 *traP*::kan. Infect. Immun. *75*, 4534–4540.

Ahmed, S., Meghji, S., Williams, R.J., Henderson, B., Brock, J.H., and Nair, S.P. (2001). *Staphylococcus aureus* fibronectin binding proteins are essential for internalization by osteoblasts but do not account for differences in intracellular levels of bacteria. Infect. Immun. *69*, 2872–2877.

Almeida, R.A., Matthews, K.R., Cifrian, E., Guidry, A.J., and Oliver, S.P. (1996). *Staphylococcus aureus* invasion of bovine mammary epithelial cells. J. Dairy Sci. *79*, 1021–1026.

Alouf, J.E., and Muller-Alouf, H. (2003). Staphylococcal and streptococcal superantigens: molecular, biological and clinical aspects. Int. J. Med. Microbiol. *292*, 429–440.

Aravind, L., and Koonin, E.V. (1998). The HD domain defines a new superfamily of metal-dependent phosphohydrolases. Trends Biochem. Sci. *23*, 469–472.

Arvidson, S. (2000). Extracellular enzymes. In Gram-positive pathogens, V.A. Fischetti, R.P. Novick, J.J. Ferreti, D.A. Portnoy, and J.I. Rood, eds. (Washington, D.C., American Society for Microbiology), pp. 379–385.

Arvidson, S., and Tegmark, K. (2001). Regulation of virulence determinants in *Staphylococcus aureus*. Int. J. Med. Microbiol. *291*, 159–170.

Autret, N., Raynaud, C., Dubail, I., Berche, P., and Charbit, A. (2003). Identification of the *agr* locus of *Listeria monocytogenes*: role in bacterial virulence. Infect. Immun. *71*, 4463–4471.

Baddour, L.M., Tayidi, M.M., Walker, E., McDevitt, D., and Foster, T.J. (1994). Virulence of coagulase-deficient mutants of *Staphylococcus aureus* in experimental endocarditis. J. Med. Microbiol. *41*, 259–263.

Bae, T., Banger, A.K., Wallace, A., Glass, E.M., Aslund, F., Schneewind, O., and Missiakas, D.M. (2004). *Staphylococcus aureus* virulence genes identified by *bursa aurealis* mutagenesis and nematode killing. Proc. Natl. Acad. Sci. U S A *101*, 12312–12317.

Baker, M.D., and Acharya, K.R. (2004). Superantigens: structure-function relationships. Int. J. Med. Microbiol. *293*, 529–537.

Balaban, N., Collins, L.V., Cullor, J.S., Hume, E.B., Medina-Acosta, E., Vieira da Motta, O., O'Callaghan, R., Rossitto, P.V., Shirtliff, M.E., Serafim da Silveira, L., *et al.* (2000). Prevention of diseases caused by *Staphylococcus aureus* using the peptide RIP. Peptides *21*, 1301–1311.

Balaban, N., Giacometti, A., Cirioni, O., Gov, Y., Ghiselli, R., Mocchegiani, F., Viticchi, C., Del Prete, M.S., Saba, V., Scalise, G., and Dell'Acqua, G. (2003). Use of the quorum-sensing inhibitor RNAIII-inhibiting peptide to prevent biofilm formation *in vivo* by drug-resistant *Staphylococcus epidermidis*. J. Infect. Dis. *187*, 625–630.

Balaban, N., Goldkorn, T., Gov, Y., Hirshberg, M., Koyfman, N., Matthews, H.R., Nhan, R.T., Singh, B., and Uziel, O. (2001). Regulation of *Staphylococcus aureus* pathogenesis via target of RNAIII-activating Protein (TRAP). J. Biol. Chem. *276*, 2658–2667.

Balaban, N., Goldkorn, T., Nhan, R.T., Dang, L.B., Scott, S., Ridgley, R.M., Rasooly, A., Wright, S.C., Larrick, J.W., Rasooly, R., and Carlson, J.R. (1998). Autoinducer of virulence as a target for vaccine and therapy against *Staphylococcus aureus*. Science *280*, 438–440.

Balaban, N., and Novick, R.P. (1995a). Autocrine regulation of toxin synthesis by *Staphylococcus aureus*. Proc. Natl. Acad. Sci. U S A 92, 1619–1623.

Balaban, N., and Novick, R.P. (1995b). Translation of RNAIII, the *Staphylococcus aureus agr* regulatory RNA molecule, can be activated by a 3′-end deletion. FEMS Microbiol. Lett. *133*, 155–161.

Balandina, A., Claret, L., Hengge-Aronis, R., and Rouviere-Yaniv, J. (2001). The *Escherichia coli* histone-like protein HU regulates *rpoS* translation. Mol. Microbiol. 39, 1069–1079.

Bantel, H., Sinha, B., Domschke, W., Peters, G., Schulze-Osthoff, K., and Janicke, R.U. (2001). alpha-Toxin is a mediator of *Staphylococcus aureus*-induced cell death and activates caspases via the intrinsic death pathway independently of death receptor signaling. J. Cell. Biol. *155*, 637–648.

Baran, J., Guzik, K., Hryniewicz, W., Ernst, M., Flad, H.D., and Pryjma, J. (1996). Apoptosis of monocytes and prolonged survival of granulocytes as a result of phagocytosis of bacteria. Infect. Immun. *64*, 4242–4248.

Baran, J., Weglarczyk, K., Mysiak, M., Guzik, K., Ernst, M., Flad, H.D., and Pryjma, J. (2001). Fas (CD95)-Fas ligand interactions are responsible for monocyte apoptosis occurring as a result of phagocytosis and killing of *Staphylococcus aureus*. Infect. Immun. *69*, 1287–1297.

Bateman, B.T., Donegan, N.P., Jarry, T.M., Palma, M., and Cheung, A.L. (2001). Evaluation of a tetracycline-inducible promoter in *Staphylococcus aureus in vitro* and *in vivo* and its application in demonstrating the role of *sigB* in microcolony formation. Infect. Immun. *69*, 7851–7857.

Bayer, A.S., Ramos, M.D., Menzies, B.E., Yeaman, M.R., Shen, A.J., and Cheung, A.L. (1997). Hyperproduction of alpha-toxin by *Staphylococcus aureus* results in paradoxically reduced virulence in experimental endocarditis: a host defense role for platelet microbicidal proteins. Infect. Immun. *65*, 4652–4660.

Bayer, A.S., Sullam, P.M., Ramos, M., Li, C., Cheung, A.L., and Yeaman, M.R. (1995). *Staphylococcus aureus* induces platelet aggregation via a fibrinogen-dependent mechanism which is independent of principal platelet glycoprotein IIb/IIIa fibrinogen-binding domains. Infect. Immun. *63*, 3634–3641.

Bayer, M.G., Heinrichs, J.H., and Cheung, A.L. (1996). The molecular architecture of the *sar* locus in *Staphylococcus aureus*. J. Bacteriol. *178*, 4563–4570.

Bayles, K.W., and Iandolo, J.J. (1989). Genetic and molecular analyses of the gene encoding staphylococcal enterotoxin D. J. Bacteriol. *171*, 4799–4806.

Bayles, K.W., Wesson, C.A., Liou, L.E., Fox, L.K., Bohach, G.A., and Trumble, W.R. (1998). Intracellular *Staphylococcus aureus* escapes the endosome and induces apoptosis in epithelial cells. Infect. Immun. *66*, 336–342.

Becker, P., Hufnagle, W., Peters, G., and Herrmann, M. (2001). Detection of differential gene expression in biofilm-forming versus planktonic populations of *Staphylococcus aureus* using micro-representational-difference analysis. Appl. Environ. Microbiol. *67*, 2958–2965.

Beekhuizen, H., van de Gevel, J.S., Olsson, B., van Benten, I.J., and van Furth, R. (1997). Infection of human vascular endothelial cells with *Staphylococcus aureus* induces hyperadhesiveness for human monocytes and granulocytes. J. Immunol. *158*, 774–782.

Beenken, K.E., Blevins, J.S., and Smeltzer, M.S. (2003). Mutation of *sarA* in *Staphylococcus aureus* limits biofilm formation. Infect. Immun. *71*, 4206–4211.

Bekeredjian-Ding, I., Inamura, S., Giese, T., Moll, H., Endres, S., Sing, A., Zahringer, U., and Hartmann, G. (2007). *Staphylococcus aureus* protein A triggers T cell-independent B cell proliferation by sensitizing B cells for TLR2 ligands. J. Immunol. *178*, 2803–2812.

Benito, Y., Kolb, F.A., Romby, P., Lina, G., Etienne, J., and Vandenesch, F. (2000). Probing the structure of RNAIII, the *Staphylococcus aureus agr* regulatory RNA, and identification of the RNA domain involved in repression of protein A expression. Rna *6*, 668–679.

Benito, Y., Lina, G., Greenland, T., Etienne, J., and Vandenesch, F. (1998). Trans-complementation of a *Staphylococcus aureus agr* mutant by *Staphylococcus lugdunensis agr* RNAIII. J. Bacteriol. *180*, 5780–5783.

Benton, B.M., Zhang, J.P., Bond, S., Pope, C., Christian, T., Lee, L., Winterberg, K.M., Schmid, M.B., and Buysse, J.M. (2004). Large-scale identification of genes required for full virulence of *Staphylococcus aureus*. J. Bacteriol. *186*, 8478–8489.

Bernfield, M., Gotte, M., Park, P.W., Reizes, O., Fitzgerald, M.L., Lincecum, J., and Zako, M. (1999). Functions of cell surface heparan sulfate proteoglycans. Annu. Rev. Biochem. *68*, 729–777.

Betley, M.J., and Mekalanos, J.J. (1985). Staphylococcal enterotoxin A is encoded by phage. Science *229*, 185–187.

Betley, M.J., and Mekalanos, J.J. (1988). Nucleotide sequence of the type A staphylococcal enterotoxin gene. J. Bacteriol. *170*, 34–41.

Bhakdi, S., Muhly, M., Korom, S., and Hugo, F. (1989). Release of interleukin-1 beta associated with potent cytocidal action of staphylococcal alpha-toxin on human monocytes. Infect. Immun. *57*, 3512–3519.

Bhakdi, S., and Tranum-Jensen, J. (1991). Alpha-toxin of *Staphylococcus aureus*. Microbiol. Rev. *55*, 733–751.

Bischoff, M., Dunman, P., Kormanec, J., Macapagal, D., Murphy, E., Mounts, W., Berger-Bachi, B., and Projan, S. (2004). Microarray-based analysis of the *Staphylococcus aureus sigmaB* regulon. J. Bacteriol. *186*, 4085–4099.

Bischoff, M., Entenza, J.M., and Giachino, P. (2001). Influence of a functional *sigB* operon on the global regulators *sar* and *agr* in *Staphylococcus aureus*. J. Bacteriol. *183*, 5171–5179.

Blevins, J.S., Beenken, K.E., Elasri, M.O., Hurlburt, B.K., and Smeltzer, M.S. (2002). Strain-dependent differences in the regulatory roles of *sarA* and *agr* in *Staphylococcus aureus*. Infect. Immun. *70*, 470–480.

Blevins, J.S., Elasri, M.O., Allmendinger, S.D., Beenken, K.E., Skinner, R.A., Thomas, J.R., and Smeltzer, M.S. (2003). Role of *sarA* in the pathogenesis of *Staphylococcus aureus* musculoskeletal infection. Infect. Immun. *71*, 516–523.

Blevins, J.S., Gillaspy, A.F., Rechtin, T.M., Hurlburt, B.K., and Smeltzer, M.S. (1999). The staphylococcal accessory regulator (*sar*) represses transcription of the *Staphylococcus aureus* collagen adhesin gene (*cna*) in an *agr*-independent manner. Mol. Microbiol. *33*, 317–326.

Boden, M.K., and Flock, J.I. (1994). Cloning and characterization of a gene for a 19 kDa fibrinogen-binding protein from *Staphylococcus aureus*. Mol. Microbiol. *12*, 599–606.

Bohach, G.A., and Foster, T.J. (2000). *Staphylococcus aureus* exotoxins. In Gram-positive pathogens, V.A. Fischetti, R.P. Novick, J.J. Ferreti, D.A. Portnoy, and J.I. Rood, eds. (Washington, D.C., American Society for Microbiology), pp. 367–378.

Boisset, S., Geissmann, T., Huntzinger, E., Fechter, P., Bendridi, N., Possedko, M., Chevalier, C., Helfer, A.C., Benito, Y., Jacquier, A., *et al.* (2007). *Staphylococcus aureus* RNAIII coordinately represses the synthesis of virulence factors and the transcription regulator Rot by an antisense mechanism. Genes Dev *21*, 1353–1366.

Bokarewa, M.I., Jin, T., and Tarkowski, A. (2006). *Staphylococcus aureus*: Staphylokinase. Int. J. Biochem. Cell. Biol. *38*, 504–509.

Booth, M.C., Cheung, A.L., Hatter, K.L., Jett, B.D., Callegan, M.C., and Gilmore, M.S. (1997). Staphylococcal accessory regulator (*sar*) in conjunction with *agr* contributes to *Staphylococcus aureus* virulence in endophthalmitis. Infect. Immun. *65*, 1550–1556.

Braff, M.H., Jones, A.L., Skerrett, S.J., and Rubens, C.E. (2007). *Staphylococcus aureus* exploits cathelicidin antimicrobial peptides produced during early pneumonia to promote staphylokinase-dependent fibrinolysis. J. Infect. Dis. *195*, 1365–1372.

Bramley, A.J., Patel, A.H., O'Reilly, M., Foster, R., and Foster, T.J. (1989). Roles of alpha-toxin and beta-toxin in virulence of *Staphylococcus aureus* for the mouse mammary gland. Infect. Immun. *57*, 2489–2494.

Brandenberger, M., Tschierske, M., Giachino, P., Wada, A., and Berger-Bachi, B. (2000). Inactivation of a novel three-cistronic operon *tcaR-tcaA-tcaB* increases teicoplanin resistance in *Staphylococcus aureus*. Biochim. Biophys. Acta *1523*, 135–139.

Bronner, S., Monteil, H., and Prevost, G. (2004). Regulation of virulence determinants in *Staphylococcus aureus*: complexity and applications. FEMS Microbiol. Rev. *28*, 183–200.

Bronner, S., Stoessel, P., Gravet, A., Monteil, H., and Prevost, G. (2000). Variable expressions of *Staphylococcus aureus* bicomponent leucotoxins semi-quantified by competitive reverse transcription-PCR. Appl. Environ. Microbiol. *66*, 3931–3938.

Buggy, B.P., Schaberg, D.R., and Swartz, R.D. (1984). Intraleukocytic sequestration as a cause of persistent *Staphylococcus aureus* peritonitis in continuous ambulatory peritoneal dialysis. Am. J. Med. *76*, 1035–1040.

Buzzola, F.R., Alvarez, L.P., Tuchscherr, L.P., Barbagelata, M.S., Lattar, S.M., Calvinho, L., and Sordelli, D.O. (2007). Differential abilities of capsulated and non-capsulated *Staphylococcus aureus* isolates from diverse *agr* groups to invade mammary epithelial cells. Infect. Immun. *75*, 886–891.

Callegan, M.C., Engel, L.S., Hill, J.M., and O'Callaghan, R.J. (1994). Corneal virulence of *Staphylococcus aureus*: roles of alpha-toxin and protein A in pathogenesis. Infect. Immun. *62*, 2478–2482.

Cassat, J., Dunman, P.M., Murphy, E., Projan, S.J., Beenken, K.E., Palm, K.J., Yang, S.J., Rice, K.C., Bayles, K.W., and Smeltzer, M.S. (2006). Transcriptional profiling of a *Staphylococcus aureus* clinical isolate and its isogenic *agr* and *sarA* mutants reveals global differences in comparison to the laboratory strain RN6390. Microbiology *152*, 3075–3090.

Cespedes, C., Said-Salim, B., Miller, M., Lo, S.H., Kreiswirth, B.N., Gordon, R.J., Vavagiakis, P., Klein, R.S., and Lowy, F.D. (2005). The clonality of *Staphylococcus aureus* nasal carriage. J. Infect. Dis. *191*, 444–452.

Chakrabarti, S.K., and Misra, T.K. (2000). SarA represses *agr* operon expression in a purified *in vitro Staphylococcus aureus* transcription system. J. Bacteriol. *182*, 5893–5897.

Chamberlain, N.R., and Imanoel, B. (1996). Genetic regulation of fatty acid modifying enzyme from *Staphylococcus aureus*. J. Med. Microbiol. *44*, 125–129.

Chan, P.F., and Foster, S.J. (1998a). The role of environmental factors in the regulation of virulence-determinant expression in *Staphylococcus aureus* 8325-4. Microbiology *144*, 2469–2479.

Chan, P.F., and Foster, S.J. (1998b). Role of SarA in virulence determinant production and environmental signal transduction in *Staphylococcus aureus*. J. Bacteriol. *180*, 6232–6241.

Chatterjee, I., Becker, P., Grundmeier, M., Bischoff, M., Somerville, G.A., Peters, G., Sinha, B., Harraghy, N., Proctor, R.A., and Herrmann, M. (2005). *Staphylococcus aureus* ClpC is required for stress resistance, aconitase activity, growth recovery, and death. J. Bacteriol. *187*, 4488–4496.

Chavakis, T., Hussain, M., Kanse, S.M., Peters, G., Bretzel, R.G., Flock, J.I., Herrmann, M., and Preissner, K.T. (2002). *Staphylococcus aureus* extracellular adherence protein serves as anti-inflammatory factor by inhibiting the recruitment of host leukocytes. Nat. Med. *8*, 687–693.

Chen, J., and Novick, R.P. (2007). *svrA*, a multi-drug exporter, does not control *agr*. Microbiology *153*, 1604–1608.

Chen, P.R., Bae, T., Williams, W.A., Duguid, E.M., Rice, P.A., Schneewind, O., and He, C. (2006). An oxidation-sensing mechanism is used by the global regulator MgrA in *Staphylococcus aureus*. Nat. Chem. Biol. *2*, 591–595.

Cheung, A.L., Bayer, M.G., and Heinrichs, J.H. (1997a). *sar* genetic determinants necessary for transcription of RNAII and RNAIII in the *agr* locus of *Staphylococcus aureus*. J. Bacteriol. *179*, 3963–3971.

Cheung, A.L., Chien, Y.T., and Bayer, A.S. (1999). Hyperproduction of alpha-hemolysin in a *sigB* mutant is associated with elevated SarA expression in *Staphylococcus aureus*. Infect. Immun. *67*, 1331–1337.

Cheung, A.L., Eberhardt, K., and Heinrichs, J.H. (1997b). Regulation of protein A synthesis by the *sar* and *agr* loci of *Staphylococcus aureus*. Infect. Immun. *65*, 2243–2249.

Cheung, A.L., Eberhardt, K.J., Chung, E., Yeaman, M.R., Sullam, P.M., Ramos, M., and Bayer, A.S. (1994a). Diminished virulence of a *sar*$^-$/*agr*$^-$ mutant of *Staphylococcus aureus* in the rabbit model of endocarditis. J. Clin. Invest. *94*, 1815–1822.

Cheung, A.L., Koomey, J.M., Butler, C.A., Projan, S.J., and Fischetti, V.A. (1992). Regulation of exoprotein expression in *Staphylococcus aureus* by a locus (*sar*) distinct from *agr*. Proc. Natl. Acad. Sci. U S A *89*, 6462–6466.

Cheung, A.L., and Manna, A.C. (2005). Role of the distal *sarA* promoters in SarA expression in *Staphylococcus aureus*. Infect. Immun. *73*, 4391–4394.

Cheung, A.L., Nast, C.C., and Bayer, A.S. (1998). Selective activation of *sar* promoters with the use of green fluorescent protein transcriptional fusions as the detection system in the rabbit endocarditis model. Infect. Immun. *66*, 5988–5993.

Cheung, A.L., and Projan, S.J. (1994). Cloning and sequencing of *sarA* of *Staphylococcus aureus*, a gene required for the expression of *agr*. J. Bacteriol. *176*, 4168–4172.

Cheung, A.L., Schmidt, K., Bateman, B., and Manna, A.C. (2001). SarS, a SarA homolog repressible by *agr*, is an activator of protein A synthesis in *Staphylococcus aureus*. Infect. Immun. *69*, 2448–2455.

Cheung, A.L., Yeaman, M.R., Sullam, P.M., Witt, M.D., and Bayer, A.S. (1994b). Role of the *sar* locus of *Staphylococcus aureus* in induction of endocarditis in rabbits. Infect. Immun. *62*, 1719–1725.

Cheung, A.L., and Ying, P. (1994). Regulation of alpha- and beta-hemolysins by the *sar* locus of *Staphylococcus aureus*. J. Bacteriol. *176*, 580–585.

Cheung, A.L., and Zhang, G. (2001). Are the structures of SarA and SarR similar? Trends Microbiol. *9*, 570–573.

Chien, Y., and Cheung, A.L. (1998). Molecular interactions between two global regulators, *sar* and *agr*, in *Staphylococcus aureus*. J. Biol. Chem. *273*, 2645–2652.

Chien, Y., Manna, A.C., and Cheung, A.L. (1998). SarA level is a determinant of *agr* activation in *Staphylococcus aureus*. Mol. Microbiol. *30*, 991–1001.

Chien, Y., Manna, A.C., Projan, S.J., and Cheung, A.L. (1999). SarA, a global regulator of virulence determinants in *Staphylococcus aureus*, binds to a conserved motif essential for *sar*-dependent gene regulation. J. Biol. Chem. *274*, 37169–37176.

Choorit, W., Kaneko, J., Muramoto, K., and Kamio, Y. (1995). Existence of a new protein component with the same function as the LukF component of leukocidin or gamma-hemolysin and its gene in *Staphylococcus aureus* P83. FEBS Lett. *357*, 260–264.

Cocchiaro, J.L., Gomez, M.I., Risley, A., Solinga, R., Sordelli, D.O., and Lee, J.C. (2006). Molecular characterization of the capsule locus from non-typeable *Staphylococcus aureus*. Mol. Microbiol. *59*, 948–960.

Collins, L.V., and Tarkowski, A. (2000). Animal models of experimental *Staphylococcus aureus* infection. In Gram-positive pathogens, V.A. Fischetti, R.P. Novick, J.J. Ferreti, D.A. Portnoy, and J.I. Rood, eds. (Washington, D.C., American Society for Microbiology), pp. 422–430.

Conlon, K.M., Humphreys, H., and O'Gara, J.P. (2004). Inactivations of *rsbU* and *sarA* by IS256 represent novel mechanisms of biofilm phenotypic variation in *Staphylococcus epidermidis*. J. Bacteriol. *186*, 6208–6219.

Cooney, J., Kienle, Z., Foster, T.J., and O'Toole, P.W. (1993). The gamma-hemolysin locus of *Staphylococcus aureus* comprises three linked genes, two of which are identical to the genes for the F and S components of leukocidin. Infect. Immun. *61*, 768–771.

Couch, J.L., Soltis, M.T., and Betley, M.J. (1988). Cloning and nucleotide sequence of the type E staphylococcal enterotoxin gene. J. Bacteriol. *170*, 2954–2960.

Craven, N., and Anderson, J.C. (1979). The location of *Staphylococcus aureus* in experimental chronic mastitis in the mouse and the effect on the action of sodium cloxacillin. Br. J. Exp. Pathol. *60*, 453–459.

Cucarella, C., Solano, C., Valle, J., Amorena, B., Lasa, I., and Penades, J.R. (2001). Bap, a *Staphylococcus aureus* surface protein involved in biofilm formation. J. Bacteriol. *183*, 2888–2896.

Cunnion, K.M., Lee, J.C., and Frank, M.M. (2001). Capsule production and growth phase influence binding of complement to *Staphylococcus aureus*. Infect. Immun. *69*, 6796–6803.

Dassy, B., Hogan, T., Foster, T.J., and Fournier, J.M. (1993). Involvement of the accessory gene regulator (*agr*) in expression of type 5 capsular polysaccharide by *Staphylococcus aureus*. J. Gen. Microbiol. *139*, 1301–1306.

Daugherty, S., and Low, M.G. (1993). Cloning, expression, and mutagenesis of phosphatidylinositol-specific phospholipase C from *Staphylococcus aureus*: a potential staphylococcal virulence factor. Infect. Immun. *61*, 5078–5089.

de Haas, C.J., Veldkamp, K.E., Peschel, A., Weerkamp, F., Van Wamel, W.J., Heezius, E.C., Poppelier, M.J., Van Kessel, K.P., and van Strijp, J.A. (2004). Chemotaxis inhibitory protein of *Staphylococcus aureus*, a bacterial antiinflammatory agent. J. Exp. Med. *199*, 687–695.

Deighan, P., Free, A., and Dorman, C.J. (2000). A role for the *Escherichia coli* H-NS-like protein StpA in OmpF porin expression through modulation of *micF* RNA stability. Mol. Microbiol. 38, 126–139.

Deora, R., Tseng, T., and Misra, T.K. (1997). Alternative transcription factor σ^{SB} of *Staphylococcus aureus*: characterization and role in transcription of the global regulatory locus *sar*. J. Bacteriol. *179*, 6355–6359.

Dragneva, Y., Anuradha, C.D., Valeva, A., Hoffmann, A., Bhakdi, S., and Husmann, M. (2001). Subcytocidal attack by staphylococcal alpha-toxin activates NF-kappaB and induces interleukin-8 production. Infect. Immun. *69*, 2630–2635.

Drapeau, G.R. (1978). Role of metalloprotease in activation of the precursor of staphylococcal protease. J. Bacteriol. *136*, 607–613.

Dufour, P., Jarraud, S., Vandenesch, F., Greenland, T., Novick, R.P., Bes, M., Etienne, J., and Lina, G. (2002). High genetic variability of the *agr* locus in *Staphylococcus* species. J. Bacteriol. *184*, 1180–1186.

Duncan, J.L., and Cho, G.J. (1972). Production of staphylococcal alpha toxin. II. Glucose repression of toxin formation. Infect. Immun. *6*, 689–694.

Dunman, P.M., Murphy, E., Haney, S., Palacios, D., Tucker-Kellogg, G., Wu, S., Brown, E.L., Zagursky, R.J., Shlaes, D., and Projan, S.J. (2001). Transcription profiling-based identification of *Staphylococcus aureus* genes regulated by the *agr* and/or *sarA* loci. J. Bacteriol. *183*, 7341–7353.

Dziewanowska, K., Carson, A.R., Patti, J.M., Deobald, C.F., Bayles, K.W., and Bohach, G.A. (2000). Staphylococcal fibronectin binding protein interacts with heat shock protein 60 and integrins: role in internalization by epithelial cells. Infect. Immun. *68*, 6321–6328.

Dziewanowska, K., Patti, J.M., Deobald, C.F., Bayles, K.W., Trumble, W.R., and Bohach, G.A. (1999). Fibronectin binding protein and host cell tyrosine kinase are required for internalization of *Staphylococcus aureus* by epithelial cells. Infect. Immun. *67*, 4673–4678.

Ellison, D.W., and Miller, V.L. (2006). Regulation of virulence by members of the MarR/SlyA family. Curr. Opin. Microbiol. *9*, 153–159.

Entenza, J.M., Moreillon, P., Senn, M.M., Kormanec, J., Dunman, P.M., Berger-Bachi, B., Projan, S., and Bischoff, M. (2005). Role of *sigmaB* in the expression of *Staphylococcus aureus* cell wall adhesins ClfA and

FnbA and contribution to infectivity in a rat model of experimental endocarditis. Infect. Immun. *73*, 990–998.

Essmann, F., Bantel, H., Totzke, G., Engels, I.H., Sinha, B., Schulze-Osthoff, K., and Janicke, R.U. (2003). *Staphylococcus aureus* alpha-toxin-induced cell death: predominant necrosis despite apoptotic caspase activation. Cell Death Differ. *10*, 1260–1272.

Finck-Barbancon, V., Duportail, G., Meunier, O., and Colin, D.A. (1993). Pore formation by a two-component leukocidin from *Staphylococcus aureus* within the membrane of human polymorphonuclear leukocytes. Biochim. Biophys. Acta *1182*, 275–282.

Fleming, V., Feil, E., Sewell, A.K., Day, N., Buckling, A., and Massey, R.C. (2006). Agr interference between clinical *Staphylococcus aureus* strains in an insect model of virulence. J. Bacteriol. *188*, 7686–7688.

Flock, J.I., Froman, G., Jonsson, K., Guss, B., Signas, C., Nilsson, B., Raucci, G., Hook, M., Wadstrom, T., and Lindberg, M. (1987). Cloning and expression of the gene for a fibronectin-binding protein from *Staphylococcus aureus*. EMBO J. *6*, 2351–2357.

Foster, T.J. (2005). Immune evasion by staphylococci. Nat. Rev. Microbiol. *3*, 948–958.

Foster, T.J., and Hook, M. (1998). Surface protein adhesins of *Staphylococcus aureus*. Trends Microbiol. *6*, 484–488.

Foster, T.J., and McDevitt, D. (1994). Surface-associated proteins of *Staphylococcus aureus*: their possible roles in virulence. FEMS Microbiol. Lett. *118*, 199–205.

Fournier, B., Aras, R., and Hooper, D.C. (2000). Expression of the multidrug resistance transporter NorA from *Staphylococcus aureus* is modified by a two-component regulatory system. J. Bacteriol. *182*, 664–671.

Fournier, B., and Hooper, D.C. (2000). A new two-component regulatory system involved in adhesion, autolysis, and extracellular proteolytic activity of *Staphylococcus aureus*. J. Bacteriol. *182*, 3955–3964.

Fournier, B., and Klier, A. (2004). Protein A gene expression is regulated by DNA supercoiling which is modified by the ArlS-ArlR two-component system of *Staphylococcus aureus*. Microbiology *150*, 3807–3819.

Fournier, B., Klier, A., and Rapoport, G. (2001). The two-component system ArlS-ArlR is a regulator of virulence gene expression in *Staphylococcus aureus*. Mol. Microbiol. *41*, 247–261.

Frees, D., Chastanet, A., Qazi, S., Sorensen, K., Hill, P., Msadek, T., and Ingmer, H. (2004). Clp ATPases are required for stress tolerance, intracellular replication and biofilm formation in *Staphylococcus aureus*. Mol. Microbiol. *54*, 1445–1462.

Frees, D., Qazi, S.N., Hill, P.J., and Ingmer, H. (2003). Alternative roles of ClpX and ClpP in *Staphylococcus aureus* stress tolerance and virulence. Mol. Microbiol. *48*, 1565–1578.

Frees, D., Sorensen, K., and Ingmer, H. (2005a). Global virulence regulation in *Staphylococcus aureus*: pinpointing the roles of ClpP and ClpX in the *sar/agr* regulatory network. Infect. Immun. *73*, 8100–8108.

Frees, D., Thomsen, L.E., and Ingmer, H. (2005b). *Staphylococcus aureus* ClpYQ plays a minor role in stress survival. Arch. Microbiol. *183*, 286–291.

Fujita, T. (2002). Evolution of the lectin-complement pathway and its role in innate immunity. Nat. Rev. Immunol. *2*, 346–353.

Gao, J., and Stewart, G.C. (2004). Regulatory elements of the *Staphylococcus aureus* protein A (Spa) promoter. J. Bacteriol. *186*, 3738–3748.

Garsin, D.A., Sifri, C.D., Mylonakis, E., Qin, X., Singh, K.V., Murray, B.E., Calderwood, S.B., and Ausubel, F.M. (2001). A simple model host for identifying Gram-positive virulence factors. Proc. Natl. Acad. Sci. U S A *98*, 10892–10897.

Garvis, S., Mei, J.M., Ruiz-Albert, J., and Holden, D.W. (2002). *Staphylococcus aureus svrA*: a gene required for virulence and expression of the *agr* locus. Microbiology *148*, 3235–3243.

Gaskill, M.E., and Khan, S.A. (1988). Regulation of the enterotoxin B gene in *Staphylococcus aureus*. J. Biol. Chem. *263*, 6276–6280.

Geisinger, E., Adhikari, R.P., Jin, R., Ross, H.F., and Novick, R.P. (2006). Inhibition of *rot* translation by RNAIII, a key feature of *agr* function. Mol. Microbiol. *61*, 1038–1048.

Gemmell, C.G., Tree, R., Patel, A., O'Reilly, M., and Foster, T.J. (1991). Susceptibility to opsonophagocytosis of protein A, alpha-haemolysin and beta-toxin deficient mutants of *Staphylococcus aureus* isolated by allele-replacement. Zentbl. Bakteriol. *21* (Suppl.), 273–277.

Gertz, S., Engelmann, S., Schmid, R., Ohlsen, K., Hacker, J., and Hecker, M. (1999). Regulation of σ^B-dependent transcription of *sigB* and *asp23* in two different *Staphylococcus aureus* strains. Mol. Gen. Genet. *261*, 558–566.

Giachino, P., Engelmann, S., and Bischoff, M. (2001). Sigma(B) activity depends on RsbU in *Staphylococcus aureus*. J. Bacteriol. *183*, 1843–1852.

Gillaspy, A.F., Hickmon, S.G., Skinner, R.A., Thomas, J.R., Nelson, C.L., and Smeltzer, M.S. (1995). Role of the accessory gene regulator (*agr*) in pathogenesis of staphylococcal osteomyelitis. Infect. Immun. *63*, 3373–3380.

Gillaspy, A.F., Lee, C.Y., Sau, S., Cheung, A.L., and Smeltzer, M.S. (1998). Factors affecting the collagen binding capacity of *Staphylococcus aureus*. Infect. Immun. *66*, 3170–3178.

Gillaspy, A.F., Patti, J.M., and Smeltzer, M.S. (1997). Transcriptional regulation of the *Staphylococcus aureus* collagen adhesion gene, *cna*. Infect. Immun. *65*, 1536–1540.

Gillet, Y., Issartel, B., Vanhems, P., Fournet, J.C., Lina, G., Bes, M., Vandenesch, F., Piemont, Y., Brousse, N., Floret, D., and Etienne, J. (2002). Association between *Staphylococcus aureus* strains carrying gene for Panton–Valentine leukocidin and highly lethal necrotising pneumonia in young immunocompetent patients. Lancet *359*, 753–759.

Gilot, P., and van Leeuwen, W. (2004). Comparative analysis of *agr* locus diversification and overall genetic variability among bovine and human *Staphylococcus aureus* isolates. J. Clin. Microbiol. *42*, 1265–1269.

Giraudo, A.T., Calzolari, A., Cataldi, A.A., Bogni, C., and Nagel, R. (1999). The *sae* locus of *Staphylococcus aureus* encodes a two-component regulatory system. FEMS Microbiol. Lett. *177*, 15–22.

Giraudo, A.T., Cheung, A.L., and Nagel, R. (1997). The *sae* locus of *Staphylococcus aureus* controls exoprotein synthesis at the transcriptional level. Arch. Microbiol. *168*, 53–58.

Giraudo, A.T., Rampone, H., Calzolari, A., and Nagel, R. (1996). Phenotypic characterization and virulence of a *sae*⁻ *agr*⁻ mutant of *Staphylococcus aureus*. Can. J. Microbiol. *42*, 120–123.

Giraudo, A.T., Raspanti, C.G., Calzolari, A., and Nagel, R. (1994). Characterization of a Tn*551*-mutant of *Staphylococcus aureus* defective in the production of several exoproteins. Can. J. Microbiol. *40*, 677–681.

Goerke, C., Campana, S., Bayer, M.G., Doring, G., Botzenhart, K., and Wolz, C. (2000). Direct quantitative transcript analysis of the *agr* regulon of *Staphylococcus aureus* during human infection in comparison to the expression profile *in vitro*. Infect. Immun. *68*, 1304–1311.

Goerke, C., Esser, S., Kummel, M., and Wolz, C. (2005a). *Staphylococcus aureus* strain designation by *agr* and *cap* polymorphism typing and delineation of *agr* diversification by sequence analysis. Int. J. Med. Microbiol. *295*, 67–75.

Goerke, C., Fluckiger, U., Steinhuber, A., Bisanzio, V., Ulrich, M., Bischoff, M., Patti, J.M., and Wolz, C. (2005b). Role of *Staphylococcus aureus* global regulators *sae* and *sigmaB* in virulence gene expression during device-related infection. Infect. Immun. *73*, 3415–3421.

Goerke, C., Fluckiger, U., Steinhuber, A., Zimmerli, W., and Wolz, C. (2001). Impact of the regulatory loci *agr*, *sarA* and *sae* of *Staphylococcus aureus* on the induction of alpha-toxin during device-related infection resolved by direct quantitative transcript analysis. Mol. Microbiol. *40*, 1439–1447.

Goerke, C., Kummel, M., Dietz, K., and Wolz, C. (2003). Evaluation of intraspecies interference due to *agr* polymorphism in *Staphylococcus aureus* during infection and colonization. J. Infect. Dis. *188*, 250–256.

Goguen, J.D., Hoe, N.P., and Subrahmanyam, Y.V. (1995). Proteases and bacterial virulence: a view from the trenches. Infect. Agents Dis. *4*, 47–54.

Gomez, M.I., Lee, A., Reddy, B., Muir, A., Soong, G., Pitt, A., Cheung, A., and Prince, A. (2004). *Staphylococcus aureus* protein A induces airway epithelial inflammatory responses by activating TNFR1. Nat. Med. *10*, 842–848.

Goodyear, C.S., and Silverman, G.J. (2004). Staphylococcal toxin induced preferential and prolonged *in vivo* deletion of innate-like B lymphocytes. Proc. Natl. Acad. Sci. U S A *101*, 11392–11397.

Gov, Y., Bitler, A., Dell'Acqua, G., Torres, J.V., and Balaban, N. (2001). RNAIII inhibiting peptide (RIP), a global inhibitor of *Staphylococcus aureus* pathogenesis: structure and function analysis. Peptides *22*, 1609–1620.

Gov, Y., Borovok, I., Korem, M., Singh, V.K., Jayaswal, R.K., Wilkinson, B.J., Rich, S.M., and Balaban, N. (2004). Quorum sensing in Staphylococci is regulated via phosphorylation of three conserved histidine residues. J. Biol. Chem. *279*, 14665–14672.

Gray, G.S., and Kehoe, M. (1984). Primary sequence of the alpha-toxin gene from *Staphylococcus aureus* wood 46. Infect. Immun. *46*, 615–618.

Gresham, H.D., Lowrance, J.H., Caver, T.E., Wilson, B.S., Cheung, A.L., and Lindberg, F.P. (2000). Survival of *Staphylococcus aureus* inside neutrophils contributes to infection. J. Immunol. *164*, 3713–3722.

Haas, P.J., de Haas, C.J., Poppelier, M.J., van Kessel, K.P., van Strijp, J.A., Dijkstra, K., Scheek, R.M., Fan, H., Kruijtzer, J.A., Liskamp, R.M., and Kemmink, J. (2005). The structure of the C5a receptor-blocking domain of chemotaxis inhibitory protein of *Staphylococcus aureus* is related to a group of immune evasive molecules. J. Mol. Biol. 353, 859–872.

Haggar, A., Ehrnfelt, C., Holgersson, J., and Flock, J.I. (2004). The extracellular adherence protein from *Staphylococcus aureus* inhibits neutrophil binding to endothelial cells. Infect. Immun. *72*, 6164–6167.

Hamill, R.J., Vann, J.M., and Proctor, R.A. (1986). Phagocytosis of *Staphylococcus aureus* by cultured bovine aortic endothelial cells: model for postadherence events in endovascular infections. Infect. Immun. *54*, 833–836.

Hammel, M., Sfyroera, G., Ricklin, D., Magotti, P., Lambris, J.D., and Geisbrecht, B.V. (2007). A structural basis for complement inhibition by *Staphylococcus aureus*. Nat. Immunol. *8*, 430–437.

Harraghy, N., Hussain, M., Haggar, A., Chavakis, T., Sinha, B., Herrmann, M., and Flock, J.I. (2003). The adhesive and immunomodulating properties of the multifunctional *Staphylococcus aureus* protein Eap. Microbiology *149*, 2701–2707.

Harraghy, N., Kormanec, J., Wolz, C., Homerova, D., Goerke, C., Ohlsen, K., Qazi, S., Hill, P., and Herrmann, M. (2005). *sae* is essential for expression of the staphylococcal adhesins Eap and Emp. Microbiology *151*, 1789–1800.

Hartleib, J., Kohler, N., Dickinson, R.B., Chhatwal, G.S., Sixma, J.J., Hartford, O.M., Foster, T.J., Peters, G., Kehrel, B.E., and Herrmann, M. (2000). Protein A is the von Willebrand factor binding protein on *Staphylococcus aureus*. Blood 96, 2149–2156.

Haslinger, B., Strangfeld, K., Peters, G., Schulze-Osthoff, K., and Sinha, B. (2003). *Staphylococcus aureus* alpha-toxin induces apoptosis in peripheral blood mononuclear cells: role of endogenous tumour necrosis factor-alpha and the mitochondrial death pathway. Cell Microbiol. 5, 729–741.

Haslinger-Loffler, B., Kahl, B.C., Grundmeier, M., Strangfeld, K., Wagner, B., Fischer, U., Cheung, A.L., Peters, G., Schulze-Osthoff, K., and Sinha, B. (2005). Multiple virulence factors are required for *Staphylococcus aureus*-induced apoptosis in endothelial cells. Cell Microbiol. *7*, 1087–1097.

Heilmann, C., Niemann, S., Sinha, B., Herrmann, M., Kehrel, B.E., and Peters, G. (2004). *Staphylococcus aureus* fibronectin-binding protein (FnBP)-mediated adherence to platelets, and aggregation of platelets induced by FnBPA but not by FnBPB. J. Infect. Dis. *190*, 321–329.

Heinrichs, J.H., Bayer, M.G., and Cheung, A.L. (1996). Characterization of the *sar* locus and its interaction with *agr* in *Staphylococcus aureus*. J. Bacteriol. *178*, 418–423.

Hess, D.J., Henry-Stanley, M.J., Erickson, E.A., and Wells, C.L. (2003). Intracellular survival of *Staphylococcus aureus* within cultured enterocytes. J. Surg. Res. *114*, 42–49.

Heyer, G., Saba, S., Adamo, R., Rush, W., Soong, G., Cheung, A., and Prince, A. (2002). *Staphylococcus aureus agr* and *sarA* functions are required for invasive infection but not inflammatory responses in the lung. Infect. Immun. *70*, 127–133.

Hofer, M.F., Newell, K., Duke, R.C., Schlievert, P.M., Freed, J.H., and Leung, D.Y. (1996). Differential effects of staphylococcal toxic shock syndrome toxin-1 on B cell apoptosis. Proc. Natl. Acad. Sci. U S A *93*, 5425–5430.

Homerova, D., Bischoff, M., Dumolin, A., and Kormanec, J. (2004). Optimization of a two-plasmid system for the identification of promoters recognized by RNA polymerase containing *Staphylococcus aureus* alternative sigma factor *sigmaB*. FEMS Microbiol. Lett. *232*, 173–179.

Horsburgh, M.J., Aish, J.L., White, I.J., Shaw, L., Lithgow, J.K., and Foster, S.J. (2002). *sigmaB* modulates virulence determinant expression and stress resistance: characterization of a functional *rsbU* strain derived from *Staphylococcus aureus* 8325–4. J. Bacteriol. *184*, 5457–5467.

Hovde, C.J., Hackett, S.P., and Bohach, G.A. (1990). Nucleotide sequence of the staphylococcal enterotoxin C3 gene: sequence comparison of all three type C staphylococcal enterotoxins. Mol. Gen. Genet. *220*, 329–333.

Hudson, M.C., Ramp, W.K., Nicholson, N.C., Williams, A.S., and Nousiainen, M.T. (1995). Internalization of *Staphylococcus aureus* by cultured osteoblasts. Microb. Pathog. *19*, 409–419.

Huntzinger, E., Boisset, S., Saveanu, C., Benito, Y., Geissmann, T., Namane, A., Lina, G., Etienne, J., Ehresmann, B., Ehresmann, C., *et al.* (2005). *Staphylococcus aureus* RNAIII and the endoribonuclease III coordinately regulate *spa* gene expression. EMBO J. *24*, 824–835.

Hussain, M., Haggar, A., Heilmann, C., Peters, G., Flock, J.I., and Herrmann, M. (2002). Insertional inactivation of Eap in *Staphylococcus aureus* strain Newman confers reduced staphylococcal binding to fibroblasts. Infect. Immun. *70*, 2933–2940.

Ingavale, S., van Wamel, W., Luong, T.T., Lee, C.Y., and Cheung, A.L. (2005). Rat/MgrA, a regulator of autolysis, is a regulator of virulence genes in *Staphylococcus aureus*. Infect. Immun. *73*, 1423–1431.

Ingavale, S.S., Van Wamel, W., and Cheung, A.L. (2003). Characterization of RAT, an autolysis regulator in *Staphylococcus aureus*. Mol. Microbiol. *48*, 1451–1466.

Isberg, R.R., and Leong, J.M. (1990). Multiple beta 1 chain integrins are receptors for invasin, a protein that promotes bacterial penetration into mammalian cells. Cell *60*, 861–871.

Janzon, L., and Arvidson, S. (1990). The role of the delta-lysin gene (*hld*) in the regulation of virulence genes by the accessory gene regulator (*agr*) in *Staphylococcus aureus*. EMBO J. *9*, 1391–1399.

Janzon, L., Lofdahl, S., and Arvidson, S. (1986). Evidence for a coordinate transcriptional control of alpha-toxin and protein A synthesis in *Staphylococcus aureus*. FEMS Microbiol. Lett. *33*, 193–198.

Janzon, L., Lofdahl, S., and Arvidson, S. (1989). Identification and nucleotide sequence of the delta-lysin gene, *hld*, adjacent to the accessory gene regulator (*agr*) of *Staphylococcus aureus*. Mol. Gen. Genet. *219*, 480–485.

Jarraud, S., Lyon, G.J., Figueiredo, A.M., Gerard, L., Vandenesch, F., Etienne, J., Muir, T.W., and Novick, R.P. (2000). Exfoliatin-producing strains define a fourth *agr* specificity group in *Staphylococcus aureus*. J. Bacteriol. *182*, 6517–6522.

Jayasinghe, L., and Bayley, H. (2005). The leukocidin pore: evidence for an octamer with four LukF subunits and four LukS subunits alternating around a central axis. Protein Sci. *14*, 2550–2561.

Jefferson, K.K., Pier, D.B., Goldmann, D.A., and Pier, G.B. (2004). The teicoplanin-associated locus regulator (TcaR) and the intercellular adhesin locus regulator (IcaR) are transcriptional inhibitors of the *ica* locus in *Staphylococcus aureus*. J. Bacteriol. *186*, 2449–2456.

Jett, B.D., and Gilmore, M.S. (2002). Internalization of *Staphylococcus aureus* by human corneal epithelial cells: role of bacterial fibronectin-binding protein and host cell factors. Infect. Immun. *70*, 4697–4700.

Ji, G., Beavis, R., and Novick, R.P. (1997). Bacterial interference caused by autoinducing peptide variants. Science *276*, 2027–2030.

Ji, G., Beavis, R.C., and Novick, R.P. (1995). Cell density control of staphylococcal virulence mediated by an octapeptide pheromone. Proc. Natl. Acad. Sci. U S A *92*, 12055–12059.

Ji, G., Pei, W., Zhang, L., Qiu, R., Lin, J., Benito, Y., Lina, G., and Novick, R.P. (2005). *Staphylococcus intermedius* produces a functional *agr* autoinducing peptide containing a cyclic lactone. J. Bacteriol. *187*, 3139–3150.

Jin, T., Bokarewa, M., Foster, T., Mitchell, J., Higgins, J., and Tarkowski, A. (2004). *Staphylococcus aureus* resists human defensins by production of staphylokinase, a novel bacterial evasion mechanism. J. Immunol. *172*, 1169–1176.

Jonas, D., Walev, I., Berger, T., Liebetrau, M., Palmer, M., and Bhakdi, S. (1994). Novel path to apoptosis: small transmembrane pores created by staphylococcal alpha-toxin in T lymphocytes evoke internucleosomal DNA degradation. Infect. Immun. *62*, 1304–1312.

Jones, C.L., and Khan, S.A. (1986). Nucleotide sequence of the enterotoxin B gene from *Staphylococcus aureus*. J. Bacteriol. *166*, 29–33.

Jonsson, I.M., Arvidson, S., Foster, S., and Tarkowski, A. (2004). Sigma factor B and RsbU are required for virulence in *Staphylococcus aureus*-induced arthritis and sepsis. Infect. Immun. *72*, 6106–6111.

Jonsson, K., Signas, C., Muller, H.P., and Lindberg, M. (1991). Two different genes encode fibronectin binding proteins in *Staphylococcus aureus*. The complete nucleotide sequence and characterization of the second gene. Eur. J. Biochem. *202*, 1041–1048.

Joubert, O., Viero, G., Keller, D., Martinez, E., Colin, D.A., Monteil, H., Mourey, L., Dalla Serra, M., and Prevost, G. (2006). Engineered covalent leucotoxin heterodimers form functional pores: insights into S-F interactions. Biochem. J. *396*, 381–389.

Kahl, B.C., Becker, K., Friedrich, A.W., Clasen, J., Sinha, B., Von Eiff, C., and Peters, G. (2003). *agr*-dependent bacterial interference has no impact on long-term colonization of *Staphylococcus aureus* during persistent airway infection of cystic fibrosis patients. J. Clin. Microbiol. *41*, 5199–5201.

Kahl, B.C., Belling, G., Becker, P., Chatterjee, I., Wardecki, K., Hilgert, K., Cheung, A.L., Peters, G., and Herrmann, M. (2005). Thymidine-dependent *Staphylococcus aureus* small-colony variants are associated with extensive alterations in regulator and virulence gene expression profiles. Infect. Immun. *73*, 4119–4126.

Kahl, B.C., Goulian, M., van Wamel, W., Herrmann, M., Simon, S.M., Kaplan, G., Peters, G., and Cheung, A.L. (2000). *Staphylococcus aureus* RN6390 replicates and induces apoptosis in a pulmonary epithelial cell line. Infect. Immun. *68*, 5385–5392.

Kaida, S., Miyata, T., Yoshizawa, Y., Igarashi, H., and Iwanaga, S. (1989). Nucleotide and deduced amino acid sequences of staphylocoagulase gene from *Staphylococcus aureus* strain 213. Nucleic Acids Res. *17*, 8871.

Kaito, C., Kurokawa, K., Matsumoto, Y., Terao, Y., Kawabata, S., Hamada, S., and Sekimizu, K. (2005). Silkworm pathogenic bacteria infection model for identification of novel virulence genes. Mol. Microbiol. *56*, 934–944.

Kaito, C., Morishita, D., Matsumoto, Y., Kurokawa, K., and Sekimizu, K. (2006). Novel DNA binding protein SarZ contributes to virulence in *Staphylococcus aureus*. Mol. Microbiol. 62, 1601–1617.

Kalkum, M., Lyon, G.J., and Chait, B.T. (2003). Detection of secreted peptides by using hypothesis-driven multistage mass spectrometry. Proc. Natl. Acad. Sci. U S A *100*, 2795–2800.

Kamio, Y., Rahman, A., Nariya, H., Ozawa, T., and Izaki, K. (1993). The two Staphylococcal bi-component toxins, leukocidin and gamma-hemolysin, share one component in common. FEBS Lett. *321*, 15–18.

Kampen, A.H., Tollersrud, T., and Lund, A. (2005). *Staphylococcus aureus* capsular polysaccharide types 5 and 8 reduce killing by bovine neutrophils *in vitro*. Infect. Immun. *73*, 1578–1583.

Karakawa, W.W., Sutton, A., Schneerson, R., Karpas, A., and Vann, W.F. (1988). Capsular antibodies induce type-specific phagocytosis of capsulated *Staphylococcus aureus* by human polymorphonuclear leukocytes. Infect. Immun. *56*, 1090–1095.

Karlsson, A., and Arvidson, S. (2002). Variation in extracellular protease production among clinical isolates of *Staphylococcus aureus* due to different levels of expression of the protease repressor *sarA*. Infect. Immun. *70*, 4239–4246.

Karlsson, A., Saravia-Otten, P., Tegmark, K., Morfeldt, E., and Arvidson, S. (2001). Decreased amounts of cell wall-associated protein A and fibronectin-binding proteins in *Staphylococcus aureus sarA* mutants due to upregulation of extracellular proteases. Infect. Immun. *69*, 4742–4748.

Karlsson-Kanth, A., Tegmark-Wisell, K., Arvidson, S., and Oscarsson, J. (2006). Natural human isolates of *Staphylococcus aureus* selected for high production of proteases and alpha-hemolysin are *sigmaB* deficient. Int. J. Med. Microbiol. *296*, 229–236.

Kielian, T., Cheung, A., and Hickey, W.F. (2001). Diminished virulence of an alpha-toxin mutant of *Staphylococcus aureus* in experimental brain abscesses. Infect. Immun. *69*, 6902–6911.

Knobloch, J.K., Jager, S., Horstkotte, M.A., Rohde, H., and Mack, D. (2004). RsbU-dependent regulation of *Staphylococcus epidermidis* biofilm formation is mediated via the alternative sigma factor σ^B by repression of the negative regulator gene *icaR*. Infect. Immun. *72*, 3838–3848.

Koenig, R.L., Ray, J.L., Maleki, S.J., Smeltzer, M.S., and Hurlburt, B.K. (2004). *Staphylococcus aureus* AgrA binding to the RNAIII-*agr* regulatory region. J. Bacteriol. *186*, 7549–7555.

Korem, M., Sheoran, A.S., Gov, Y., Tzipori, S., Borovok, I., and Balaban, N. (2003). Characterization of RAP, a quorum sensing activator of *Staphylococcus aureus*. FEMS Microbiol. Lett. *223*, 167–175.

Kreiswirth, B.N. (1989). Genetics and expression of toxic shock syndrome toxin 1: overview. Rev. Infect. Dis. *11 Suppl 1*, S97–100.

Kreiswirth, B.N., Lofdahl, S., Betley, M.J., O'Reilly, M., Schlievert, P.M., Bergdoll, M.S., and Novick, R.P. (1983). The toxic shock syndrome exotoxin structural gene is not detectably transmitted by a prophage. Nature *305*, 709–712.

Kropec, A., Maira-Litran, T., Jefferson, K.K., Grout, M., Cramton, S.E., Gotz, F., Goldmann, D.A., and Pier, G.B. (2005). Poly-N-acetylglucosamine production in *Staphylococcus aureus* is essential for virulence in murine models of systemic infection. Infect. Immun. *73*, 6868–6876.

Kulig, P., Zabel, B.A., Dubin, G., Allen, S.J., Ohyama, T., Potempa, J., Handel, T.M., Butcher, E.C., and Cichy, J. (2007). *Staphylococcus aureus*-Derived Staphopain B, a Potent Cysteine Protease Activator of Plasma Chemerin. J. Immunol. *178*, 3713–3720.

Kullik, I., Giachino, P., and Fuchs, T. (1998). Deletion of the alternative sigma factor σ^B in *Staphylococcus aureus* reveals its function as a global regulator of virulence genes. J. Bacteriol. *180*, 4814–4820.

Kupferwasser, L.I., Yeaman, M.R., Nast, C.C., Kupferwasser, D., Xiong, Y.Q., Palma, M., Cheung, A.L., and Bayer, A.S. (2003). Salicylic acid attenuates virulence in endovascular infections by targeting global regulatory pathways in *Staphylococcus aureus*. J. Clin. Invest. *112*, 222–233.

Kuroda, H., Kuroda, M., Cui, L., and Hiramatsu, K. (2007). Subinhibitory concentrations of beta-lactam induce haemolytic activity in *Staphylococcus aureus* through the SaeRS two-component system. FEMS Microbiol. Lett. *268*, 98–105.

Kuroda, M., Ohta, T., Uchiyama, I., Baba, T., Yuzawa, H., Kobayashi, I., Cui, L., Oguchi, A., Aoki, K.,

Nagai, Y., *et al.* (2001). Whole genome sequencing of meticillin-resistant *Staphylococcus aureus*. Lancet *357*, 1225–1240.

Labandeira-Rey, M., Couzon, F., Boisset, S., Brown, E.L., Bes, M., Benito, Y., Barbu, E.M., Vazquez, V., Hook, M., Etienne, J., *et al.* (2007). *Staphylococcus aureus* Panton–Valentine leukocidin causes necrotizing pneumonia. Science *315*, 1130–1133.

Ladhani, S. (2003). Understanding the mechanism of action of the exfoliative toxins of *Staphylococcus aureus*. FEMS Immunol. Med. Microbiol. *39*, 181–189.

Lammers, A., Nuijten, P.J., and Smith, H.E. (1999). The fibronectin binding proteins of *Staphylococcus aureus* are required for adhesion to and invasion of bovine mammary gland cells. FEMS Microbiol. Lett. *180*, 103–109.

Lee, C.Y., and Iandolo, J.J. (1986). Lysogenic conversion of staphylococcal lipase is caused by insertion of the bacteriophage L54a genome into the lipase structural gene. J. Bacteriol. *166*, 385–391.

Lee, C.Y., Schmidt, J.J., Johnson-Winegar, A.D., Spero, L., and Iandolo, J.J. (1987). Sequence determination and comparison of the exfoliative toxin A and toxin B genes from *Staphylococcus aureus*. J. Bacteriol. *169*, 3904–3909.

Lee, J.C., Xu, S., Albus, A., and Livolsi, P.J. (1994). Genetic analysis of type 5 capsular polysaccharide expression by *Staphylococcus aureus*. J. Bacteriol. *176*, 4883–4889.

Lee, L.Y., Hook, M., Haviland, D., Wetsel, R.A., Yonter, E.O., Syribeys, P., Vernachio, J., and Brown, E.L. (2004a). Inhibition of complement activation by a secreted *Staphylococcus aureus* protein. J. Infect. Dis. *190*, 571–579.

Lee, L.Y., Liang, X., Hook, M., and Brown, E.L. (2004b). Identification and characterization of the C3 binding domain of the *Staphylococcus aureus* extracellular fibrinogen-binding protein (Efb). J. Biol. Chem. *279*, 50710–50716.

Lee, L.Y., Miyamoto, Y.J., McIntyre, B.W., Hook, M., McCrea, K.W., McDevitt, D., and Brown, E.L. (2002). The *Staphylococcus aureus* Map protein is an immunomodulator that interferes with T cell-mediated responses. J. Clin. Invest. *110*, 1461–1471.

Lemaitre, B., Reichhart, J.M., and Hoffmann, J.A. (1997). *Drosophila* host defense: differential induction of antimicrobial peptide genes after infection by various classes of microorganisms. Proc. Natl. Acad. Sci. U S A *94*, 14614–14619.

Lemjabbar, H., and Basbaum, C. (2002). Platelet-activating factor receptor and ADAM10 mediate responses to *Staphylococcus aureus* in epithelial cells. Nat. Med. *8*, 41–46.

Li, R., Manna, A.C., Dai, S., Cheung, A.L., and Zhang, G. (2003). Crystal structure of the SarS protein from *Staphylococcus aureus*. J. Bacteriol. *185*, 4219–4225.

Liang, X., and Ji, Y. (2006). Alpha-toxin interferes with integrin-mediated adhesion and internalization of *Staphylococcus aureus* by epithelial cells. Cell Microbiol. *8*, 1656–1668.

Liang, X., Yu, C., Sun, J., Liu, H., Landwehr, C., Holmes, D., and Ji, Y. (2006). Inactivation of a two-component signal transduction system, SaeRS, eliminates adherence and attenuates virulence of *Staphylococcus aureus*. Infect. Immun. *74*, 4655–4665.

Liang, X., Zheng, L., Landwehr, C., Lunsford, D., Holmes, D., and Ji, Y. (2005). Global regulation of gene expression by ArlRS, a two-component signal transduction regulatory system of *Staphylococcus aureus*. J. Bacteriol. *187*, 5486–5492.

Lin, W.S., Cunneen, T., and Lee, C.Y. (1994). Sequence analysis and molecular characterization of genes required for the biosynthesis of type 1 capsular polysaccharide in *Staphylococcus aureus*. J. Bacteriol. *176*, 7005–7016.

Lina, G., Boutite, F., Tristan, A., Bes, M., Etienne, J., and Vandenesch, F. (2003). Bacterial competition for human nasal cavity colonization: role of Staphylococcal *agr* alleles. Appl. Environ. Microbiol. *69*, 18–23.

Lina, G., Jarraud, S., Ji, G., Greenland, T., Pedraza, A., Etienne, J., Novick, R.P., and Vandenesch, F. (1998). Transmembrane topology and histidine protein kinase activity of AgrC, the *agr* signal receptor in *Staphylococcus aureus*. Mol. Microbiol. *28*, 655–662.

Lindsay, J.A., and Foster, S.J. (1999). Interactive regulatory pathways control virulence determinant production and stability in response to environmental conditions in *Staphylococcus aureus*. Mol. Gen. Genet. *262*, 323–331.

Lindsay, J.A., Moore, C.E., Day, N.P., Peacock, S.J., Witney, A.A., Stabler, R.A., Husain, S.E., Butcher, P.D., and Hinds, J. (2006). Microarrays reveal that each of the ten dominant lineages of *Staphylococcus aureus* has a unique combination of surface-associated and regulatory genes. J. Bacteriol. *188*, 669–676.

Lindsay, J.A., Ruzin, A., Ross, H.F., Kurepina, N., and Novick, R.P. (1998). The gene for toxic shock toxin is carried by a family of mobile pathogenicity islands in *Staphylococcus aureus*. Mol. Microbiol. *29*, 527–543.

Liu, Y., Manna, A., Li, R., Martin, W.E., Murphy, R.C., Cheung, A.L., and Zhang, G. (2001). Crystal structure of the SarR protein from *Staphylococcus aureus*. Proc. Natl. Acad. Sci. U S A *98*, 6877–6882.

Liu, Y., Manna, A.C., Pan, C.H., Kriksunov, I.A., Thiel, D.J., Cheung, A.L., and Zhang, G. (2006). Structural and function analyses of the global regulatory protein SarA from *Staphylococcus aureus*. Proc. Natl. Acad. Sci. U S A *103*, 2392–2397.

Lofdahl, S., Guss, B., Uhlen, M., Philipson, L., and Lindberg, M. (1983). Gene for staphylococcal protein A. Proc. Natl. Acad. Sci. U S A *80*, 697–701.

Lowy, F.D. (1998). *Staphylococcus aureus* infections. N. Engl. J. Med. *339*, 520–532.

Lowy, F.D. (2000a). Is *Staphylococcus aureus* an intracellular pathogen? Trends Microbiol. *8*, 341–343.

Lowy, F.D. (2000b). *Staphylococcus aureus*-eukaryotic cell interactions. In Gram-positive pathogens, V.A. Fischetti, R.P. Novick, J.J. Ferreti, D.A. Portnoy, and J.I. Rood, eds. (Washington, D.C., American Society for Microbiology), pp. 408–413.

Luong, T.T., Dunman, P.M., Murphy, E., Projan, S.J., and Lee, C.Y. (2006). Transcription Profiling of the *mgrA* Regulon in *Staphylococcus aureus*. J. Bacteriol. *188*, 1899–1910.

Luong, T.T., and Lee, C.Y. (2002). Overproduction of type 8 capsular polysaccharide augments *Staphylococcus aureus* virulence. Infect. Immun. *70*, 3389–3395.

Luong, T.T., and Lee, C.Y. (2006). The *arl* locus positively regulates *Staphylococcus aureus* type 5 capsule via an *mgrA*-dependent pathway. Microbiology *152*, 3123–3131.

Luong, T.T., Newell, S.W., and Lee, C.Y. (2003). Mgr, a novel global regulator in *Staphylococcus aureus*. J. Bacteriol. *185*, 3703–3710.

Lyon, G.J., Mayville, P., Muir, T.W., and Novick, R.P. (2000). Rational design of a global inhibitor of the virulence response in *Staphylococcus aureus*, based in part on localization of the site of inhibition to the receptor-histidine kinase, AgrC. Proc. Natl. Acad. Sci. U S A *97*, 13330–13335.

Lyon, G.J., Wright, J.S., Christopoulos, A., Novick, R.P., and Muir, T.W. (2002a). Reversible and specific extracellular antagonism of receptor-histidine kinase signaling. J. Biol. Chem. *277*, 6247–6253.

Lyon, G.J., Wright, J.S., Muir, T.W., and Novick, R.P. (2002b). Key determinants of receptor activation in the *agr* autoinducing peptides of *Staphylococcus aureus*. Biochemistry *41*, 10095–10104.

Mahmood, R., and Khan, S.A. (1990). Role of upstream sequences in the expression of the staphylococcal enterotoxin B gene. J. Biol. Chem. *265*, 4652–4656.

Mallonee, D.H., Glatz, B.A., and Pattee, P.A. (1982). Chromosomal mapping of a gene affecting enterotoxin A production in *Staphylococcus aureus*. Appl. Environ. Microbiol. *43*, 397–402.

Manna, A., and Cheung, A.L. (2001). Characterization of *sarR*, a modulator of *sar* expression in *Staphylococcus aureus*. Infect. Immun. *69*, 885–896.

Manna, A.C., Bayer, M.G., and Cheung, A.L. (1998). Transcriptional analysis of different promoters in the *sar* locus in *Staphylococcus aureus*. J. Bacteriol. *180*, 3828–3836.

Manna, A.C., and Cheung, A.L. (2003). *sarU*, a *sarA* homolog, is repressed by SarT and regulates virulence genes in *Staphylococcus aureus*. Infect. Immun. *71*, 343–353.

Manna, A.C., and Cheung, A.L. (2006a). Expression of SarX, a negative regulator of *agr* and exoprotein synthesis, is activated by MgrA in *Staphylococcus aureus*. J. Bacteriol. *188*, 4288–4299.

Manna, A.C., and Cheung, A.L. (2006b). Transcriptional regulation of the *agr* locus and the identification of DNA binding residues of the global regulatory protein SarR in *Staphylococcus aureus*. Mol. Microbiol. *60*, 1289–1301.

Manna, A.C., Ingavale, S.S., Maloney, M., van Wamel, W., and Cheung, A.L. (2004). Identification of *sarV* (SA2062), a new transcriptional regulator, is repressed by SarA and MgrA (SA0641) and involved in the regulation of autolysis in *Staphylococcus aureus*. J. Bacteriol. *186*, 5267–5280.

Manna, A.C., and Ray, B. (2007). Regulation and characterization of *rot* transcription in *Staphylococcus aureus*. Microbiology *153*, 1538–1545.

Marques, M.B., Weller, P.F., Parsonnet, J., Ransil, B.J., and Nicholson-Weller, A. (1989). Phosphatidylinositol-specific phospholipase C, a possible virulence factor of *Staphylococcus aureus*. J. Clin. Microbiol. *27*, 2451–2454.

Massey, R.C., Kantzanou, M.N., Fowler, T., Day, N.P., Schofield, K., Wann, E.R., Berendt, A.R., Hook, M., and Peacock, S.J. (2001). Fibronectin-binding protein A of *Staphylococcus aureus* has multiple, substituting, binding regions that mediate adherence to fibronectin and invasion of endothelial cells. Cell Microbiol. *3*, 839–851.

Massimi, I., Park, E., Rice, K., Muller-Esterl, W., Sauder, D., and McGavin, M.J. (2002). Identification of a novel maturation mechanism and restricted substrate specificity for the SspB cysteine protease of *Staphylococcus aureus*. J. Biol. Chem. *277*, 41770–41777.

Matsumoto, Y., Kaito, C., Morishita, D., Kurokawa, K., and Sekimizu, K. (2007). Regulation of Exoprotein Gene Expression by the *Staphylococcus aureus cvfB* Gene. Infect. Immun. *75*, 1964–1972.

Mayville, P., Ji, G., Beavis, R., Yang, H., Goger, M., Novick, R.P., and Muir, T.W. (1999). Structure-activity analysis of synthetic autoinducing thiolactone peptides from *Staphylococcus aureus* responsible for virulence. Proc. Natl. Acad. Sci. U S A *96*, 1218–1223.

McAleese, F.M., Walsh, E.J., Sieprawska, M., Potempa, J., and Foster, T.J. (2001). Loss of clumping factor B fibrinogen binding activity by *Staphylococcus aureus* involves cessation of transcription, shedding and cleavage by metalloprotease. J. Biol. Chem. *276*, 29969–29978.

McCallum, N., Bischoff, M., Maki, H., Wada, A., and Berger-Bachi, B. (2004). TcaR, a putative MarR-like regulator of *sarS* expression. J. Bacteriol. *186*, 2966–2972.

McDevitt, D., Francois, P., Vaudaux, P., and Foster, T.J. (1994). Molecular characterization of the clumping factor (fibrinogen receptor) of *Staphylococcus aureus*. Mol. Microbiol. *11*, 237–248.

McDowell, P., Affas, Z., Reynolds, C., Holden, M.T., Wood, S.J., Saint, S., Cockayne, A., Hill, P.J., Dodd, C.E., Bycroft, B.W., *et al.* (2001). Structure, activity and evolution of the group I thiolactone peptide quorum-sensing system of *Staphylococcus aureus*. Mol. Microbiol. *41*, 503–512.

McGavin, M.H., Krajewska-Pietrasik, D., Ryden, C., and Hook, M. (1993). Identification of a *Staphylococcus aureus* extracellular matrix-binding protein with broad specificity. Infect. Immun. *61*, 2479–2485.

McGavin, M.J., Zahradka, C., Rice, K., and Scott, J.E. (1997). Modification of the *Staphylococcus aureus* fibronectin binding phenotype by V8 protease. Infect. Immun. *65*, 2621–2628.

McNamara, P.J., and Bayer, A.S. (2005). A rot mutation restores parental virulence to an *agr*-null *Staphylococcus aureus* strain in a rabbit model of endocarditis. Infect. Immun. *73*, 3806–3809.

McNamara, P.J., Milligan-Monroe, K.C., Khalili, S., and Proctor, R.A. (2000). Identification, cloning, and initial characterization of *rot*, a locus encoding a regulator of virulence factor expression in *Staphylococcus aureus*. J. Bacteriol. *182*, 3197–3203.

Mei, J.M., Nourbakhsh, F., Ford, C.W., and Holden, D.W. (1997). Identification of *Staphylococcus aureus* virulence genes in a murine model of bacteraemia using signature-tagged mutagenesis. Mol. Microbiol. *26*, 399–407.

Meier, S., Goerke, C., Wolz, C., Seidl, K., Homerova, D., Schulthess, B., Kormanec, J., Berger-Bachi, B., and Bischoff, M. (2007). σ^B and the σ^B-Dependent *arlRS* and *yabJ-spoVG* Loci Affect Capsule Formation in *Staphylococcus aureus*. Infect. Immun. *75*, 4562–4571.

Menzies, B.E., and Kourteva, I. (1998). Internalization of *Staphylococcus aureus* by endothelial cells induces apoptosis. Infect. Immun. *66*, 5994–5998.

Michel, A., Agerer, F., Hauck, C.R., Herrmann, M., Ullrich, J., Hacker, J., and Ohlsen, K. (2006). Global regulatory impact of ClpP protease of *Staphylococcus aureus* on regulons involved in virulence, oxidative stress response, autolysis, and DNA repair. J. Bacteriol. *188*, 5783–5796.

Miyazaki, E., Chen, J.M., Ko, C., and Bishai, W.R. (1999). The *Staphylococcus aureus rsbW* (*orf159*) gene encodes an anti-sigma factor of SigB. J. Bacteriol. *181*, 2846–2851.

Moreillon, P., Entenza, J.M., Francioli, P., McDevitt, D., Foster, T.J., Francois, P., and Vaudaux, P. (1995). Role of *Staphylococcus aureus* coagulase and clumping factor in pathogenesis of experimental endocarditis. Infect. Immun. *63*, 4738–4743.

Morfeldt, E., Panova-Sapundjieva, I., Gustafsson, B., and Arvidson, S. (1996a). Detection of the response regulator AgrA in the cytosolic fraction of *Staphylococcus aureus* by monoclonal antibodies. FEMS Microbiol. Lett. *143*, 195–201.

Morfeldt, E., Taylor, D., von Gabain, A., and Arvidson, S. (1995). Activation of alpha-toxin translation in *Staphylococcus aureus* by the trans-encoded antisense RNA, RNAIII. EMBO J. *14*, 4569–4577.

Morfeldt, E., Tegmark, K., and Arvidson, S. (1996b). Transcriptional control of the *agr*-dependent virulence gene regulator, RNAIII, in *Staphylococcus aureus*. Mol. Microbiol. *21*, 1227–1237.

Moskovitz, J., Singh, V.K., Requena, J., Wilkinson, B.J., Jayaswal, R.K., and Stadtman, E.R. (2002). Purification and characterization of methionine sulfoxide reductases from mouse and *Staphylococcus aureus* and their substrate stereospecificity. Biochem. Biophys. Res. Commun *290*, 62–65.

Nair, S.P., Bischoff, M., Senn, M.M., and Berger-Bachi, B. (2003). The *sigma B* regulon influences internalization of *Staphylococcus aureus* by osteoblasts. Infect. Immun. *71*, 4167–4170.

Nakayama, J., Chen, S., Oyama, N., Nishiguchi, K., Azab, E.A., Tanaka, E., Kariyama, R., and Sonomoto, K. (2006). Revised model for *Enterococcus faecalis fsr* quorum-sensing system: the small open reading frame *fsrD* encodes the gelatinase biosynthesis-activating pheromone propeptide corresponding to staphylococcal AgrD. J. Bacteriol. *188*, 8321–8326.

Needham, A.J., Kibart, M., Crossley, H., Ingham, P.W., and Foster, S.J. (2004). *Drosophila melanogaster* as a model host for *Staphylococcus aureus* infection. Microbiology *150*, 2347–2355.

Ni Eidhin, D., Perkins, S., Francois, P., Vaudaux, P., Hook, M., and Foster, T.J. (1998). Clumping factor B (ClfB), a new surface-located fibrinogen-binding adhesin of *Staphylococcus aureus*. Mol. Microbiol. *30*, 245–257.

Nicholas, R.O., Li, T., McDevitt, D., Marra, A., Sucoloski, S., Demarsh, P.L., and Gentry, D.R. (1999). Isolation and characterization of a *sigB* deletion mutant of *Staphylococcus aureus*. Infect. Immun. *67*, 3667–3669.

Nikolskaya, A.N., and Galperin, M.Y. (2002). A novel type of conserved DNA-binding domain in the transcriptional regulators of the AlgR/AgrA/LytR family. Nucleic Acids Res. *30*, 2453–2459.

Nilsson, I.M., Bremell, T., Ryden, C., Cheung, A.L., and Tarkowski, A. (1996). Role of the staphylococcal accessory gene regulator (*sar*) in septic arthritis. Infect. Immun. *64*, 4438–4443.

Nilsson, I.M., Lee, J.C., Bremell, T., Ryden, C., and Tarkowski, A. (1997). The role of staphylococcal polysaccharide microcapsule expression in septicemia and septic arthritis. Infect. Immun. 65, 4216–4221.

Novick, R.P. (2000). Pathogenicity factors and their regulation. In Gram-positive pathogens, V.A. Fischetti, R.P. Novick, J.J. Ferreti, D.A. Portnoy, and J.I. Rood, eds. (Washington, D.C., American Society for Microbiology), pp. 392–407.

Novick, R.P., and Jiang, D. (2003). The staphylococcal *saeRS* system coordinates environmental signals with *agr* quorum sensing. Microbiology *149*, 2709–2717.

Novick, R.P., Projan, S.J., Kornblum, J., Ross, H.F., Ji, G., Kreiswirth, B., Vandenesch, F., and Moghazeh, S. (1995). The *agr P2* operon: an autocatalytic sensory transduction system in *Staphylococcus aureus*. Mol. Gen. Genet. *248*, 446–458.

Novick, R.P., Ross, H.F., Figueiredo, A.M.S., Abramochkin, G., Muir, T., Balaban, N., Singh, B., Goldkorn, T., Rasooly, A., Torres, J.V., and Uziel, O. (2000). Activation and inhibition of the staphylococcal *AGR* system. Science *287*, 391a.

Novick, R.P., Ross, H.F., Projan, S.J., Kornblum, J., Kreiswirth, B., and Moghazeh, S. (1993). Synthesis of staphylococcal virulence factors is controlled by a regulatory RNA molecule. EMBO J. *12*, 3967–3975.

O'Brien, L., Kerrigan, S.W., Kaw, G., Hogan, M., Penades, J., Litt, D., Fitzgerald, D.J., Foster, T.J., and Cox, D. (2002a). Multiple mechanisms for the activation of human platelet aggregation by *Staphylococcus aureus*: roles for the clumping factors ClfA and ClfB, the serine-aspartate repeat protein SdrE and protein A. Mol. Microbiol. *44*, 1033–1044.

O'Brien, L.M., Walsh, E.J., Massey, R.C., Peacock, S.J., and Foster, T.J. (2002b). *Staphylococcus aureus* clumping factor B (ClfB) promotes adherence to human type I cytokeratin 10: implications for nasal colonization. Cell Microbiol. *4*, 759–770.

O'Riordan, K., and Lee, J.C. (2004). *Staphylococcus aureus* capsular polysaccharides. Clin. Microbiol. Rev. *17*, 218–234.

Oscarsson, J., Harlos, C., and Arvidson, S. (2005). Regulatory role of proteins binding to the *spa* (protein A) and *sarS* (staphylococcal accessory regulator) promoter regions in *Staphylococcus aureus* NTCC 8325–4. Int. J. Med. Microbiol. *295*, 253–266.

Oscarsson, J., Kanth, A., Tegmark-Wisell, K., and Arvidson, S. (2006a). SarA is a repressor of *hla* (alpha-hemolysin) transcription in *Staphylococcus aureus*: its apparent role as an activator of *hla* in the prototype strain NCTC 8325 depends on reduced expression of *sarS*. J. Bacteriol. *188*, 8526–8533.

Oscarsson, J., Tegmark-Wisell, K., and Arvidson, S. (2006b). Coordinated and differential control of aureolysin (*aur*) and serine protease (*sspA*) transcription in *Staphylococcus aureus* by *sarA*, *rot* and *agr* (RNAIII). Int. J. Med. Microbiol. *296*, 365–380.

Oskouian, B., and Stewart, G.C. (1987). Cloning and characterization of the repressor gene of the *Staphylococcus aureus* lactose operon. J. Bacteriol. *169*, 5459–5465.

Otto, M., Echner, H., Voelter, W., and Gotz, F. (2001). Pheromone cross-inhibition between *Staphylococcus aureus* and *Staphylococcus epidermidis*. Infect. Immun. *69*, 1957–1960.

Otto, M., Sussmuth, R., Jung, G., and Gotz, F. (1998). Structure of the pheromone peptide of the *Staphylococcus epidermidis agr* system. FEBS Lett. *424*, 89–94.

Otto, M., Sussmuth, R., Vuong, C., Jung, G., and Gotz, F. (1999). Inhibition of virulence factor expression in *Staphylococcus aureus* by the *Staphylococcus epidermidis agr* pheromone and derivatives. FEBS Lett. *450*, 257–262.

Ouyang, S., and Lee, C.Y. (1997). Transcriptional analysis of type 1 capsule genes in *Staphylococcus aureus*. Mol. Microbiol. *23*, 473–482.

Palma, M., and Cheung, A.L. (2001). σ^B activity in *Staphylococcus aureus* is controlled by RsbU and an additional factor(s) during bacterial growth. Infect. Immun. *69*, 7858–7865.

Palma, M., Haggar, A., and Flock, J.I. (1999). Adherence of *Staphylococcus aureus* is enhanced by an endogenous secreted protein with broad binding activity. J. Bacteriol. *181*, 2840–2845.

Palma, M., Nozohoor, S., Schennings, T., Heimdahl, A., and Flock, J.I. (1996). Lack of the extracellular 19-kilodalton fibrinogen-binding protein from *Staphylococcus aureus* decreases virulence in experimental wound infection. Infect. Immun. *64*, 5284–5289.

Palma, M., Shannon, O., Quezada, H.C., Berg, A., and Flock, J.I. (2001). Extracellular fibrinogen-binding protein, Efb, from *Staphylococcus aureus* blocks platelet aggregation due to its binding to the alpha-chain. J. Biol. Chem. *276*, 31691–31697.

Palmqvist, N., Patti, J.M., Tarkowski, A., and Josefsson, E. (2004). Expression of staphylococcal clumping factor A impedes macrophage phagocytosis. Microbes Infect. *6*, 188–195.

Pane-Farre, J., Jonas, B., Forstner, K., Engelmann, S., and Hecker, M. (2006). The *sigmaB* regulon in *Staphylococcus aureus* and its regulation. Int. J. Med. Microbiol. *296*, 237–258.

Panizzi, P., Friedrich, R., Fuentes-Prior, P., Bode, W., and Bock, P.E. (2004). The staphylocoagulase family of zymogen activator and adhesion proteins. Cell Mol. Life Sci. *61*, 2793–2798.

Papakyriacou, H., Vaz, D., Simor, A., Louie, M., and McGavin, M.J. (2000). Molecular analysis of the accessory gene regulator (*agr*) locus and balance of virulence factor expression in epidemic methicillin-resistant *Staphylococcus aureus*. J. Infect. Dis. *181*, 990–1000.

Park, P.W., Foster, T.J., Nishi, E., Duncan, S.J., Klagsbrun, M., and Chen, Y. (2004). Activation of syndecan-1 ectodomain shedding by *Staphylococcus aureus* alpha-toxin and beta-toxin. J. Biol. Chem. *279*, 251–258.

Patel, A.H., Kornblum, J., Kreiswirth, B., Novick, R., and Foster, T.J. (1992). Regulation of the protein A-encoding gene in *Staphylococcus aureus*. Gene *114*, 25–34.

Patel, A.H., Nowlan, P., Weavers, E.D., and Foster, T. (1987). Virulence of protein A-deficient and alpha-toxin-deficient mutants of *Staphylococcus aureus* isolated by allele replacement. Infect. Immun. *55*, 3103–3110.

Patti, J.M., Jonsson, H., Guss, B., Switalski, L.M., Wiberg, K., Lindberg, M., and Hook, M. (1992). Molecular characterization and expression of a gene encoding a *Staphylococcus aureus* collagen adhesin. J. Biol. Chem. *267*, 4766–4772.

Peacock, S.J., Moore, C.E., Justice, A., Kantzanou, M., Story, L., Mackie, K., O'Neill, G., and Day, N.P. (2002). Virulent combinations of adhesin and toxin genes in natural populations of *Staphylococcus aureus*. Infect. Immun. *70*, 4987–4996.

Peng, H.L., Novick, R.P., Kreiswirth, B., Kornblum, J., and Schlievert, P. (1988). Cloning, characterization, and sequencing of an accessory gene regulator (*agr*) in *Staphylococcus aureus*. J. Bacteriol. *170*, 4365–4372.

Peterson, P.K., Verhoef, J., Sabath, L.D., and Quie, P.G. (1977). Effect of protein A on staphylococcal opsonization. Infect. Immun. *15*, 760–764.

Phonimdaeng, P., O'Reilly, M., Nowlan, P., Bramley, A.J., and Foster, T.J. (1990). The coagulase of *Staphylococcus aureus* 8325–4. Sequence analysis and virulence of site-specific coagulase-deficient mutants. Mol. Microbiol. *4*, 393–404.

Potempa, J., Dubin, A., Korzus, G., and Travis, J. (1988). Degradation of elastin by a cysteine proteinase from *Staphylococcus aureus*. J. Biol. Chem. *263*, 2664–2667.

Potempa, J., Fedak, D., Dubin, A., Mast, A., and Travis, J. (1991). Proteolytic inactivation of alpha-1-anti-chymotrypsin. Sites of cleavage and generation of chemotactic activity. J. Biol. Chem. *266*, 21482–21487.

Potempa, J., Watorek, W., and Travis, J. (1986). The inactivation of human plasma alpha 1-proteinase inhibitor by proteinases from *Staphylococcus aureus*. J. Biol. Chem. *261*, 14330–14334.

Pragman, A.A., Ji, Y., and Schlievert, P.M. (2007). Repression of *Staphylococcus aureus* SrrAB using inducible antisense *srrA* alters growth and virulence factor transcript levels. Biochemistry *46*, 314–321.

Pragman, A.A., and Schlievert, P.M. (2004). Virulence regulation in *Staphylococcus aureus*: the need for *in vivo* analysis of virulence factor regulation. FEMS Immunol. Med. Microbiol. *42*, 147–154.

Pragman, A.A., Yarwood, J.M., Tripp, T.J., and Schlievert, P.M. (2004). Characterization of virulence factor regulation by SrrAB, a two-component system in *Staphylococcus aureus*. J. Bacteriol. *186*, 2430–2438.

Prevost, G., Couppie, P., Prevost, P., Gayet, S., Petiau, P., Cribier, B., Monteil, H., and Piemont, Y. (1995a). Epidemiological data on *Staphylococcus aureus* strains producing synergohymenotropic toxins. J. Med. Microbiol. *42*, 237–245.

Prevost, G., Cribier, B., Couppie, P., Petiau, P., Supersac, G., Finck-Barbancon, V., Monteil, H., and Piemont, Y. (1995b). Panton–Valentine leucocidin and gamma-hemolysin from *Staphylococcus aureus* ATCC 49775 are encoded by distinct genetic loci and have different biological activities. Infect. Immun. *63*, 4121–4129.

Projan, S.J., Kornblum, J., Kreiswirth, B., Moghazeh, S.L., Eisner, W., and Novick, R.P. (1989). Nucleotide sequence: the beta-hemolysin gene of *Staphylococcus aureus*. Nucleic Acids Res. *17*, 3305.

Projan, S.J., and Novick, R.P. (1997). The molecular basis of pathogenicity. In The staphylococci in human disease, K.B. Crossley, and G.L. Archer, eds. (New-York, N.Y., Churchill Livingstone), pp. 55–81.

Qiu, R., Pei, W., Zhang, L., Lin, J., and Ji, G. (2005). Identification of the putative staphylococcal AgrB catalytic residues involving the proteolytic cleavage of AgrD to generate autoinducing peptide. J. Biol. Chem. *280*, 16695–16704.

Rachid, S., Ohlsen, K., Wallner, U., Hacker, J., Hecker, M., and Ziebuhr, W. (2000). Alternative transcription factor σ^B is involved in regulation of biofilm expression in a *Staphylococcus aureus* mucosal isolate. J. Bacteriol. *182*, 6824–6826.

Rahman, A., Nariya, H., Izaki, K., Kato, I., and Kamio, Y. (1992). Molecular cloning and nucleotide sequence of leukocidin F-component gene (*lukF*) from methicillin resistant *Staphylococcus aureus*. Biochem. Biophys. Res. Commun *184*, 640–646.

Rampone, H., Martinez, G.L., Giraudo, A.T., Calzolari, A., and Nagel, R. (1996). *In vivo* expression of exoprotein synthesis with a Sae mutant of *Staphylococcus aureus*. Can. J. Vet. Res. *60*, 237–240.

Rechtin, T.M., Gillaspy, A.F., Schumacher, M.A., Brennan, R.G., Smeltzer, M.S., and Hurlburt, B.K. (1999). Characterization of the SarA virulence gene regulator of *Staphylococcus aureus*. Mol. Microbiol. *33*, 307–316.

Recsei, P., Kreiswirth, B., O'Reilly, M., Schlievert, P., Gruss, A., and Novick, R.P. (1986). Regulation of exoprotein gene expression in *Staphylococcus aureus* by *agr*. Mol. Gen. Genet. *202*, 58–61.

Reed, S.B., Wesson, C.A., Liou, L.E., Trumble, W.R., Schlievert, P.M., Bohach, G.A., and Bayles, K.W. (2001). Molecular characterization of a novel *Staphylococcus aureus* serine protease operon. Infect. Immun. *69*, 1521–1527.

Rice, K., Peralta, R., Bast, D., de Azavedo, J., and McGavin, M.J. (2001). Description of staphylococcus serine protease (*ssp*) operon in *Staphylococcus aureus* and nonpolar inactivation of *sspA*-encoded serine protease. Infect. Immun. *69*, 159–169.

Richardson, A.R., Dunman, P.M., and Fang, F.C. (2006). The nitrosative stress response of *Staphylococcus aureus* is required for resistance to innate immunity. Mol. Microbiol. *61*, 927–939.

Rigoulay, C., Entenza, J.M., Halpern, D., Widmer, E., Moreillon, P., Poquet, I., and Gruss, A. (2005). Comparative analysis of the roles of HtrA-like surface proteases in two virulent *Staphylococcus aureus* strains. Infect. Immun. *73*, 563–572.

Roberts, C., Anderson, K.L., Murphy, E., Projan, S.J., Mounts, W., Hurlburt, B., Smeltzer, M., Overbeek, R., Disz, T., and Dunman, P.M. (2006). Characterizing the effect of the *Staphylococcus aureus* virulence factor regulator, SarA, on log-phase mRNA half-lives. J. Bacteriol. *188*, 2593–2603.

Robinson, D.A., Monk, A.B., Cooper, J.E., Feil, E.J., and Enright, M.C. (2005). Evolutionary genetics of the accessory gene regulator (*agr*) locus in *Staphylococcus aureus*. J. Bacteriol. *187*, 8312–8321.

Roche, F.M., Downer, R., Keane, F., Speziale, P., Park, P.W., and Foster, T.J. (2004). The N-terminal A domain of fibronectin-binding proteins A and B promotes adhesion of *Staphylococcus aureus* to elastin. J. Biol. Chem. *279*, 38433–38440.

Roche, F.M., Massey, R., Peacock, S.J., Day, N.P., Visai, L., Speziale, P., Lam, A., Pallen, M., and Foster, T.J. (2003). Characterization of novel LPXTG-containing proteins of *Staphylococcus aureus* identified from genome sequences. Microbiology *149*, 643–654.

Rogasch, K., Ruhmling, V., Pane-Farre, J., Hoper, D., Weinberg, C., Fuchs, S., Schmudde, M., Broker, B.M., Wolz, C., Hecker, M., and Engelmann, S. (2006). Influence of the two-component system SaeRS on global gene expression in two different *Staphylococcus aureus* strains. J. Bacteriol. *188*, 7742–7758.

Rollof, J., Braconier, J.H., Soderstrom, C., and Nilsson-Ehle, P. (1988). Interference of *Staphylococcus aureus* lipase with human granulocyte function. Eur. J. Clin. Microbiol. Infect. Dis. *7*, 505–510.

Rooijakkers, S.H., Ruyken, M., Roos, A., Daha, M.R., Presanis, J.S., Sim, R.B., van Wamel, W.J., van Kessel, K.P., and van Strijp, J.A. (2005a). Immune evasion by a staphylococcal complement inhibitor that acts on C3 convertases. Nat. Immunol. *6*, 920–927.

Rooijakkers, S.H., Ruyken, M., van Roon, J., van Kessel, K.P., van Strijp, J.A., and van Wamel, W.J. (2006). Early expression of SCIN and CHIPS drives instant immune evasion by *Staphylococcus aureus*. Cell Microbiol. *8*, 1282–1293.

Rooijakkers, S.H., van Kessel, K.P., and van Strijp, J.A. (2005b). Staphylococcal innate immune evasion. Trends Microbiol. *13*, 596–601.

Rooijakkers, S.H., van Wamel, W.J., Ruyken, M., van Kessel, K.P., and van Strijp, J.A. (2005c). Anti-opsonic properties of staphylokinase. Microbes Infect. *21*, 21.

Rossi, J., Bischoff, M., Wada, A., and Berger-Bachi, B. (2003). MsrR, a putative cell envelope-associated element involved in *Staphylococcus aureus sarA* attenuation. Antimicrob. Agents Chemother. *47*, 2558–2564.

Rothfork, J.M., Dessus-Babus, S., Van Wamel, W.J., Cheung, A.L., and Gresham, H.D. (2003). Fibrinogen depletion attenuates *Staphyloccocus aureus* infection by preventing density-dependent virulence gene upregulation. J. Immunol. *171*, 5389–5395.

Rzychon, M., Sabat, A., Kosowska, K., Potempa, J., and Dubin, A. (2003). Staphostatins: an expanding new group of proteinase inhibitors with a unique specificity for the regulation of staphopains, *Staphylococcus* spp. cysteine proteinases. Mol. Microbiol. *49*, 1051–1066.

Said-Salim, B., Dunman, P.M., McAleese, F.M., Macapagal, D., Murphy, E., McNamara, P.J., Arvidson, S., Foster, T.J., Projan, S.J., and Kreiswirth, B.N. (2003). Global regulation of *Staphylococcus aureus* genes by Rot. J. Bacteriol. *185*, 610–619.

Sako, T., and Tsuchida, N. (1983). Nucleotide sequence of the staphylokinase gene from *Staphylococcus aureus*. Nucleic Acids Res. *11*, 7679–7693.

Sambanthamoorthy, K., Smeltzer, M.S., and Elasri, M.O. (2006). Identification and characterization of *msa* (SA1233), a gene involved in expression of SarA and several virulence factors in *Staphylococcus aureus*. Microbiology *152*, 2559–2572.

Saravia-Otten, P., Muller, H.P., and Arvidson, S. (1997). Transcription of *Staphylococcus aureus* fibronectin binding protein genes is negatively regulated by *agr* and an *agr*-independent mechanism. J. Bacteriol. *179*, 5259–5263.

Sau, S., and Lee, C.Y. (1996). Cloning of type 8 capsule genes and analysis of gene clusters for the production of different capsular polysaccharides in *Staphylococcus aureus*. J. Bacteriol. *178*, 2118–2126.

Sawai, T., Tomono, K., Yanagihara, K., Yamamoto, Y., Kaku, M., Hirakata, Y., Koga, H., Tashiro, T., and Kohno, S. (1997). Role of coagulase in a murine model of hematogenous pulmonary infection induced by intravenous injection of *Staphylococcus aureus* enmeshed in agar beads. Infect. Immun. *65*, 466–471.

Schmidt, K.A., Donegan, N.P., Kwan, W.A., Jr., and Cheung, A. (2004). Influences of *sigmaB* and *agr* on expression of staphylococcal enterotoxin B (*seb*) in *Staphylococcus aureus*. Can. J. Microbiol. *50*, 351–360.

Schmidt, K.A., Manna, A.C., and Cheung, A.L. (2003). SarT influences *sarS* expression in *Staphylococcus aureus*. Infect. Immun. *71*, 5139–5148.

Schmidt, K.A., Manna, A.C., Gill, S., and Cheung, A.L. (2001). SarT, a repressor of alpha-hemolysin in *Staphylococcus aureus*. Infect. Immun. *69*, 4749–4758.

Schnaith, A., Kashkar, H., Leggio, S.A., Addicks, K., Kronke, M., and Krut, O. (2007). *Staphylococcus aureus* subvert autophagy for induction of caspase-independent host cell death. J. Biol. Chem. *282*, 2695–2706.

Schumacher, M.A., Hurlburt, B.K., and Brennan, R.G. (2001). Crystal structures of SarA, a pleiotropic regulator of virulence genes in *S. aureus*. Nature *409*, 215–219.

Schwan, W.R., Langhorne, M.H., Ritchie, H.D., and Stover, C.K. (2003). Loss of hemolysin expression in *Staphylococcus aureus agr* mutants correlates with selective survival during mixed infections in murine abscesses and wounds. FEMS Immunol. Med. Microbiol. *38*, 23–28.

Schwarz-Linek, U., Werner, J.M., Pickford, A.R., Gurusiddappa, S., Kim, J.H., Pilka, E.S., Briggs, J.A., Gough, T.S., Hook, M., Campbell, I.D., and Potts, J.R. (2003). Pathogenic bacteria attach to human fibronectin through a tandem beta-zipper. Nature *423*, 177–181.

Seidl, K., Stucki, M., Ruegg, M., Goerke, C., Wolz, C., Harris, L., Berger-Bachi, B., and Bischoff, M. (2006). *Staphylococcus aureus* CcpA affects virulence determinant production and antibiotic resistance. Antimicrob. Agents Chemother. *50*, 1183–1194.

Senn, M.M., Giachino, P., Homerova, D., Steinhuber, A., Strassner, J., Kormanec, J., Fluckiger, U., Berger-Bachi, B., and Bischoff, M. (2005). Molecular analysis and organization of the *sigmaB* operon in *Staphylococcus aureus*. J. Bacteriol. *187*, 8006–8019.

Shaw, L., Golonka, E., Potempa, J., and Foster, S.J. (2004). The role and regulation of the extracellular proteases of *Staphylococcus aureus*. Microbiology *150*, 217–228.

Shaw, L.N., Jonsson, I.M., Singh, V.K., Tarkowski, A., and Stewart, G.C. (2007). Inactivation of *traP* has no effect on the agr quorum-sensing system or virulence of *Staphylococcus aureus*. Infect. Immun. *75*, 4519–4527.

Sheehan, B.J., Foster, T.J., Dorman, C.J., Park, S., and Stewart, G.S. (1992). Osmotic and growth-phase dependent regulation of the *eta* gene of *Staphylococcus aureus*: a role for DNA supercoiling. Mol. Gen. Genet. *232*, 49–57.

Shompole, S., Henon, K.T., Liou, L.E., Dziewanowska, K., Bohach, G.A., and Bayles, K.W. (2003). Biphasic intracellular expression of *Staphylococcus aureus* virulence factors and evidence for Agr-mediated diffusion sensing. Mol. Microbiol. *49*, 919–927.

Shopsin, B., Mathema, B., Alcabes, P., Said-Salim, B., Lina, G., Matsuka, A., Martinez, J., and Kreiswirth, B.N. (2003). Prevalence of *agr* specificity groups among *Staphylococcus aureus* strains colonizing children and their guardians. J. Clin. Microbiol. *41*, 456–459.

Sieprawska-Lupa, M., Mydel, P., Krawczyk, K., Wojcik, K., Puklo, M., Lupa, B., Suder, P., Silberring, J., Reed, M., Pohl, J., *et al.* (2004). Degradation of human antimicrobial peptide LL-37 by *Staphylococcus aureus*-derived proteinases. Antimicrob. Agents Chemother. *48*, 4673–4679.

Sifri, C.D., Begun, J., Ausubel, F.M., and Calderwood, S.B. (2003). *Caenorhabditis elegans* as a model host for *Staphylococcus aureus* pathogenesis. Infect. Immun. *71*, 2208–2217.

Signas, C., Raucci, G., Jonsson, K., Lindgren, P.E., Anantharamaiah, G.M., Hook, M., and Lindberg, M. (1989). Nucleotide sequence of the gene for a fibronectin-binding protein from *Staphylococcus aureus*: use of this peptide sequence in the synthesis of biologically active peptides. Proc. Natl. Acad. Sci. U S A *86*, 699–703.

Sinha, B., Francois, P., Que, Y.A., Hussain, M., Heilmann, C., Moreillon, P., Lew, D., Krause, K.H., Peters, G., and Herrmann, M. (2000). Heterologously expressed *Staphylococcus aureus* fibronectin-binding proteins are sufficient for invasion of host cells. Infect. Immun. *68*, 6871–6878.

Sinha, B., Francois, P.P., Nusse, O., Foti, M., Hartford, O.M., Vaudaux, P., Foster, T.J., Lew, D.P., Herrmann, M., and Krause, K.H. (1999). Fibronectin-binding protein acts as *Staphylococcus aureus* invasin via fibronectin bridging to integrin alpha5beta1. Cell Microbiol. *1*, 101–117.

Siomi, H., Choi, M., Siomi, M.C., Nussbaum, R.L., and Dreyfuss, G. (1994). Essential role for KH domains in RNA binding: impaired RNA binding by a mutation in the KH domain of FMR1 that causes fragile X syndrome. Cell *77*, 33–39.

Somerville, G.A., Beres, S.B., Fitzgerald, J.R., DeLeo, F.R., Cole, R.L., Hoff, J.S., and Musser, J.M. (2002a). *In vitro* serial passage of *Staphylococcus aureus*: changes in physiology, virulence factor production, and *agr* nucleotide sequence. J. Bacteriol. *184*, 1430–1437.

Somerville, G.A., Chaussee, M.S., Morgan, C.I., Fitzgerald, J.R., Dorward, D.W., Reitzer, L.J., and Musser, J.M. (2002b). *Staphylococcus aureus* aconitase inactivation unexpectedly inhibits post-exponential-phase growth and enhances stationary-phase survival. Infect. Immun. *70*, 6373–6382.

Song, L., Hobaugh, M.R., Shustak, C., Cheley, S., Bayley, H., and Gouaux, J.E. (1996). Structure of staphylococcal alpha-hemolysin, a heptameric transmembrane pore. Science *274*, 1859–1866.

Steinhuber, A., Goerke, C., Bayer, M.G., Doring, G., and Wolz, C. (2003). Molecular architecture of the regulatory locus *sae* of *Staphylococcus aureus* and its impact on expression of virulence factors. J. Bacteriol. *185*, 6278–6286.

Sterba, K.M., Mackintosh, S.G., Blevins, J.S., Hurlburt, B.K., and Smeltzer, M.S. (2003). Characterization of *Staphylococcus aureus* SarA binding sites. J. Bacteriol. *185*, 4410–4417.

Su, Y.C., and Wong, A.C. (1995). Identification and purification of a new staphylococcal enterotoxin, H. Appl. Environ. Microbiol. *61*, 1438–1443.

Sung, J.M., Chantler, P.D., and Lloyd, D.H. (2006). Accessory gene regulator locus of *Staphylococcus intermedius*. Infect. Immun. *74*, 2947–2956.

Supersac, G., Prevost, G., and Piemont, Y. (1993). Sequencing of leucocidin R from *Staphylococcus aureus* P83 suggests that staphylococcal leucocidins and gamma-hemolysin are members of a single, two-component family of toxins. Infect. Immun. *61*, 580–587.

Tegmark, K., Karlsson, A., and Arvidson, S. (2000). Identification and characterization of SarH1, a new global regulator of virulence gene expression in *Staphylococcus aureus*. Mol. Microbiol. *37*, 398–409.

Tegmark, K., Morfeldt, E., and Arvidson, S. (1998). Regulation of *agr*-dependent virulence genes in *Staphylococcus aureus* by RNAIII from coagulase-negative staphylococci. J. Bacteriol. *180*, 3181–3186.

Tenover, F.C., and Gaynes, R.P. (2000). The epidemiology of *Staphylococcus* infections. In Gram-positive pathogens, V.A. Fischetti, R.P. Novick, J.J. Ferreti, D.A. Portnoy, and J.I. Rood, eds. (Washington, D.C., American Society for Microbiology), pp. 414–421.

Thakker, M., Park, J.S., Carey, V., and Lee, J.C. (1998). *Staphylococcus aureus* serotype 5 capsular polysaccharide is antiphagocytic and enhances bacterial virulence in a murine bacteremia model. Infect. Immun. *66*, 5183–5189.

Throup, J.P., Zappacosta, F., Lunsford, R.D., Annan, R.S., Carr, S.A., Lonsdale, J.T., Bryant, A.P., McDevitt, D., Rosenberg, M., and Burnham, M.K. (2001). The *srhSR* gene pair from *Staphylococcus aureus*: genomic and proteomic approaches to the identification and characterization of gene function. Biochemistry *40*, 10392–10401.

Toledo-Arana, A., Merino, N., Vergara-Irigaray, M., Debarbouille, M., Penades, J.R., and Lasa, I. (2005). *Staphylococcus aureus* develops an alternative, *ica*-independent biofilm in the absence of the *arlRS* two-component system. J. Bacteriol. *187*, 5318–5329.

Traber, K., and Novick, R. (2006). A slipped-mispairing mutation in AgrA of laboratory strains and clinical isolates results in delayed activation of *agr* and failure to translate delta- and alpha-haemolysins. Mol. Microbiol. *59*, 1519–1530.

Tremaine, M.T., Brockman, D.K., and Betley, M.J. (1993). Staphylococcal enterotoxin A gene (*sea*) expression is not affected by the accessory gene regulator (*agr*). Infect. Immun. *61*, 356–359.

Trotonda, M.P., Manna, A.C., Cheung, A.L., Lasa, I., and Penades, J.R. (2005). SarA positively controls *bap*-dependent biofilm formation in *Staphylococcus aureus*. J. Bacteriol. *187*, 5790–5798.

Truong-Bolduc, Q.C., Zhang, X., and Hooper, D.C. (2003). Characterization of NorR protein, a multifunctional regulator of *norA* expression in *Staphylococcus aureus*. J. Bacteriol. *185*, 3127–3138.

Tsang, L.H., Daily, S.T., Weiss, E.C., and Smeltzer, M.S. (2007). Mutation of *traP* in *Staphylococcus aureus* has no impact on expression of *agr* or biofilm formation. Infect. Immun. *75*, 4528–4533.

Tseng, C.W., and Stewart, G.C. (2005). Rot repression of enterotoxin B expression in *Staphylococcus aureus*. J. Bacteriol. *187*, 5301–5309.

Tzianabos, A.O., Wang, J.Y., and Lee, J.C. (2001). Structural rationale for the modulation of abscess formation by *Staphylococcus aureus* capsular polysaccharides. Proc. Natl. Acad. Sci. U S A *98*, 9365–9370.

Ulrich, M., Bastian, M., Cramton, S.E., Ziegler, K., Pragman, A.A., Bragonzi, A., Memmi, G., Wolz, C., Schlievert, P.M., Cheung, A., and Doring, G. (2007). The staphylococcal respiratory response regulator SrrAB induces *ica* gene transcription and polysaccharide intercellular adhesin expression, protecting *Staphylococcus aureus* from neutrophil killing under anaerobic growth conditions. Mol. Microbiol. *65*, 1276–1287.

Ulvatne, H., Haukland, H.H., Samuelsen, O., Kramer, M., and Vorland, L.H. (2002). Proteases in *Escherichia coli* and *Staphylococcus aureus* confer reduced susceptibility to lactoferricin B. J. Antimicrob. Chemother. *50*, 461–467.

Valle, J., Toledo-Arana, A., Berasain, C., Ghigo, J.M., Amorena, B., Penades, J.R., and Lasa, I. (2003). SarA and not sigmaB is essential for biofilm development by *Staphylococcus aureus*. Mol. Microbiol. *48*, 1075–1087.

van Leeuwen, W., van Nieuwenhuizen, W., Gijzen, C., Verbrugh, H., and van Belkum, A. (2000).

Population studies of methicillin-resistant and -sensitive *Staphylococcus aureus* strains reveal a lack of variability in the *agrD* gene, encoding a staphylococcal autoinducer peptide. J. Bacteriol. *182*, 5721–5729.

van Wamel, W., Xiong, Y.Q., Bayer, A.S., Yeaman, M.R., Nast, C.C., and Cheung, A.L. (2002). Regulation of *Staphylococcus aureus* type 5 capsular polysaccharides by *agr* and *sarA in vitro* and in an experimental endocarditis model. Microb. Pathog. *33*, 73–79.

van Wamel, W.J., Rooijakkers, S.H., Ruyken, M., van Kessel, K.P., and van Strijp, J.A. (2006). The innate immune modulators staphylococcal complement inhibitor and chemotaxis inhibitory protein of *Staphylococcus aureus* are located on beta-hemolysin-converting bacteriophages. J. Bacteriol. *188*, 1310–1315.

Vandenesch, F., Kornblum, J., and Novick, R.P. (1991). A temporal signal, independent of *agr*, is required for *hla* but not *spa* transcription in *Staphylococcus aureus*. J. Bacteriol. *173*, 6313–6320.

Vandenesch, F., Projan, S.J., Kreiswirth, B., Etienne, J., and Novick, R.P. (1993). Agr-related sequences in *Staphylococcus lugdunensis*. FEMS Microbiol. Lett. *111*, 115–122.

Vann, J.M., and Proctor, R.A. (1987). Ingestion of *Staphylococcus aureus* by bovine endothelial cells results in time- and inoculum-dependent damage to endothelial cell monolayers. Infect. Immun. *55*, 2155–2163.

Vann, J.M., and Proctor, R.A. (1988). Cytotoxic effects of ingested *Staphylococcus aureus* on bovine endothelial cells: role of *Staphylococcus aureus* alpha-hemolysin. Microb. Pathog. *4*, 443–453.

Veldkamp, K.E., Heezius, H.C., Verhoef, J., van Strijp, J.A., and van Kessel, K.P. (2000). Modulation of neutrophil chemokine receptors by *Staphylococcus aureus* supernate. Infect. Immun. *68*, 5908–5913.

Vesga, O., Groeschel, M.C., Otten, M.F., Brar, D.W., Vann, J.M., and Proctor, R.A. (1996). *Staphylococcus aureus* small colony variants are induced by the endothelial cell intracellular milieu. J. Infect. Dis. *173*, 739–742.

Vieira, O.V., Botelho, R.J., and Grinstein, S. (2002). Phagosome maturation: aging gracefully. Biochem. J. *366*, 689–704.

Voyich, J.M., Braughton, K.R., Sturdevant, D.E., Whitney, A.R., Said-Salim, B., Porcella, S.F., Long, R.D., Dorward, D.W., Gardner, D.J., Kreiswirth, B.N., *et al.* (2005). Insights into mechanisms used by *Staphylococcus aureus* to avoid destruction by human neutrophils. J. Immunol. *175*, 3907–3919.

Vuong, C., Saenz, H.L., Gotz, F., and Otto, M. (2000). Impact of the *agr* quorum-sensing system on adherence to polystyrene in *Staphylococcus aureus*. J. Infect. Dis. *182*, 1688–1693.

Walsh, E.J., O'Brien, L.M., Liang, X., Hook, M., and Foster, T.J. (2004). Clumping factor B, a fibrinogen-binding MSCRAMM (microbial surface components recognizing adhesive matrix molecules) adhesin of *Staphylococcus aureus*, also binds to the tail region of type I cytokeratin 10. J. Biol. Chem. *279*, 50691–50699.

Warner, J.B., and Lolkema, J.S. (2003). CcpA-dependent carbon catabolite repression in bacteria. Microbiol. Mol. Biol. Rev. *67*, 475–490.

Watts, A., Ke, D., Wang, Q., Pillay, A., Nicholson-Weller, A., and Lee, J.C. (2005). *Staphylococcus aureus* strains that express serotype 5 or serotype 8 capsular polysaccharides differ in virulence. Infect. Immun. *73*, 3502–3511.

Wawrzynow, A., Wojtkowiak, D., Marszalek, J., Banecki, B., Jonsen, M., Graves, B., Georgopoulos, C., and Zylicz, M. (1995). The ClpX heat-shock protein of *Escherichia coli*, the ATP-dependent substrate specificity component of the ClpP-ClpX protease, is a novel molecular chaperone. EMBO J. *14*, 1867–1877.

Wesson, C.A., Liou, L.E., Todd, K.M., Bohach, G.A., Trumble, W.R., and Bayles, K.W. (1998). *Staphylococcus aureus* Agr and Sar global regulators influence internalization and induction of apoptosis. Infect. Immun. *66*, 5238–5243.

West, A.H., and Stock, A.M. (2001). Histidine kinases and response regulator proteins in two-component signaling systems. Trends Biochem. Sci. *26*, 369–376.

Wolz, C., Pohlmann-Dietze, P., Steinhuber, A., Chien, Y.T., Manna, A., van Wamel, W., and Cheung, A. (2000). Agr-independent regulation of fibronectin-binding protein(s) by the regulatory locus *sar* in *Staphylococcus aureus*. Mol. Microbiol. 36, 230–243.

Wright, A.J., Higginbottom, A., Philippe, D., Upadhyay, A., Bagby, S., Read, R.C., Monk, P.N., and Partridge, L.J. (2007). Characterisation of receptor binding by the chemotaxis inhibitory protein of *Staphylococcus aureus* and the effects of the host immune response. Mol. Immunol. *44*, 2507–2517.

Wright, J.S., 3rd, Jin, R., and Novick, R.P. (2005a). Transient interference with staphylococcal quorum sensing blocks abscess formation. Proc. Natl. Acad. Sci. U S A *102*, 1691–1696.

Wright, J.S., 3rd, Lyon, G.J., George, E.A., Muir, T.W., and Novick, R.P. (2004). Hydrophobic interactions drive ligand-receptor recognition for activation and inhibition of staphylococcal quorum sensing. Proc. Natl. Acad. Sci. U S A *101*, 16168–16173.

Wright, J.S., 3rd, Traber, K.E., Corrigan, R., Benson, S.A., Musser, J.M., and Novick, R.P. (2005b). The *agr* radiation: an early event in the evolution of staphylococci. J. Bacteriol. *187*, 5585–5594.

Wu, S., de Lencastre, H., and Tomasz, A. (1996). Sigma-B, a putative operon encoding alternate sigma factor of *Staphylococcus aureus* RNA polymerase: molecular cloning and DNA sequencing. J. Bacteriol. *178*, 6036–6042.

Xiong, Y.Q., Bayer, A.S., Yeaman, M.R., Van Wamel, W., Manna, A.C., and Cheung, A.L. (2004). Impacts of *sarA* and *agr* in *Staphylococcus aureus* strain Newman on fibronectin-binding protein A gene expression and fibronectin adherence capacity *in vitro* and in experimental infective endocarditis. Infect. Immun. *72*, 1832–1836.

Xiong, Y.Q., Van Wamel, W., Nast, C.C., Yeaman, M.R., Cheung, A.L., and Bayer, A.S. (2002). Activation and transcriptional interaction between *agr* RNAII and RNAIII in *Staphylococcus aureus in vitro* and in an experimental endocarditis model. J. Infect. Dis. *186*, 668–677.

Xiong, Y.Q., Willard, J., Yeaman, M.R., Cheung, A.L., and Bayer, A.S. (2006). Regulation of *Staphylococcus aureus* alpha-toxin gene (*hla*) expression by *agr*, *sarA*, and *sae in vitro* and in experimental infective endocarditis. J. Infect. Dis. *194*, 1267–1275.

Yamaguchi, T., Hayashi, T., Takami, H., Ohnishi, M., Murata, T., Nakayama, K., Asakawa, K., Ohara, M., Komatsuzawa, H., and Sugai, M. (2001). Complete nucleotide sequence of a *Staphylococcus aureus* exfoliative toxin B plasmid and identification of a novel ADP-ribosyltransferase, EDIN-C. Infect. Immun. *69*, 7760–7771.

Yamazaki, K., Kato, F., Kamio, Y., and Kaneko, J. (2006). Expression of gamma-hemolysin regulated by *sae* in *Staphylococcus aureus* strain Smith 5R. FEMS Microbiol. Lett. *259*, 174–180.

Yarwood, J.M., McCormick, J.K., Paustian, M.L., Kapur, V., and Schlievert, P.M. (2002). Repression of the *Staphylococcus aureus* accessory gene regulator in serum and *in vivo*. J. Bacteriol. *184*, 1095–1101.

Yarwood, J.M., McCormick, J.K., and Schlievert, P.M. (2001). Identification of a novel two-component regulatory system that acts in global regulation of virulence factors of *Staphylococcus aureus*. J. Bacteriol. *183*, 1113–1123.

Zhang, L., Gray, L., Novick, R.P., and Ji, G. (2002). Transmembrane topology of AgrB, the protein involved in the post-translational modification of AgrD in *Staphylococcus aureus*. J. Biol. Chem. *277*, 34736–34742.

Zhang, L., and Ji, G. (2004). Identification of a staphylococcal AgrB segment(s) responsible for group-specific processing of AgrD by gene swapping. J. Bacteriol. *186*, 6706–6713.

Zhang, L., Lin, J., and Ji, G. (2004). Membrane anchoring of the AgrD N-terminal amphipathic region is required for its processing to produce a quorum-sensing pheromone in *Staphylococcus aureus*. J. Biol. Chem. *279*, 19448–19456.

Zhang, S., Iandolo, J.J., and Stewart, G.C. (1998). The enterotoxin D plasmid of *Staphylococcus aureus* encodes a second enterotoxin determinant (*sej*). FEMS Microbiol. Lett. *168*, 227–233.

Zhang, S., Reeves, A., Woodbury, R.L., and Haldenwang, W.G. (2005). Coexpression patterns of σ^B regulators in *Bacillus subtilis* affect σ^B inducibility. J. Bacteriol. *187*, 8520–8525.

Ziebandt, A.K., Becher, D., Ohlsen, K., Hacker, J., Hecker, M., and Engelmann, S. (2004). The influence of *agr* and *sigmaB* in growth phase dependent regulation of virulence factors in *Staphylococcus aureus*. Proteomics 4, 3034–3047.

Ziebandt, A.K., Weber, H., Rudolph, J., Schmid, R., Hoper, D., Engelmann, S., and Hecker, M. (2001). Extracellular proteins of *Staphylococcus aureus* and the role of SarA and sigma B. Proteomics *1*, 480–493.

Ziebuhr, W., Heilmann, C., Gotz, F., Meyer, P., Wilms, K., Straube, E., and Hacker, J. (1997). Detection of the intercellular adhesion gene cluster (*ica*) and phase variation in *Staphylococcus epidermidis* blood culture strains and mucosal isolates. Infect. Immun. *65*, 890–896.

Zimmerli, W., Waldvogel, F.A., Vaudaux, P., and Nydegger, U.E. (1982). Pathogenesis of foreign body infection: description and characteristics of an animal model. J. Infect. Dis. *146*, 487–497.

The Response of *S. aureus* to Environmental Stimuli

7

Malcolm J. Horsburgh

Abstract

The success of *S. aureus* as an opportunistic human pathogen is typically ascribed to its versatility via an inherent capacity to respond to changes in its environment. The range of stimuli to which it responds are not completely characterized, however significant advances have been made in our understanding of how this bacterium modulates gene expression in response to anaerobiosis, metal ion limitation, nitrosative and oxidative stress, temperature and pH change, starvation and antibiotic-directed cell wall stress.

Introduction

The success of *S. aureus* as an opportunistic pathogen is typically ascribed to its versatility via an inherent capacity to modulate behaviour in response to changes in its environment. The primary niche for the bacterium in humans is the squamous epithelium of the anterior nares. To initiate colonization at this site the bacterium might typically have been mobilized from the nose to the skin of the hand of one individual, then onto a surface before being transferred via the hand of a second individual to their nose (Fig. 7.1). Thus for transmission to be successful the bacterium must survive in, and adapt to, a range of different environments as it moves from host to host. *S. aureus* has thus evolved to have a high level of adaptability for environmental survival during its mobilization between niches. Moreover it has been argued that the more complex nature of the transmission cycle of *S. aureus* in comparison to *S. epidermidis*, which is readily transmitted from skin to skin with no known barriers for transmission, is a factor responsible

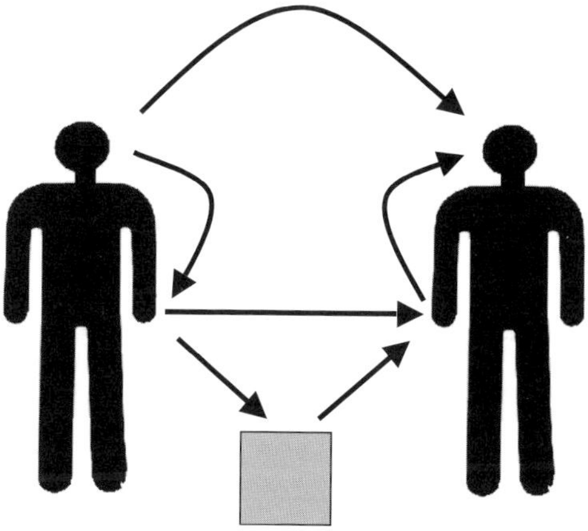

Figure 7.1 The transmission cycle of *S. aureus*. The arrows show potential routes for transmission, which can be direct via contact or indirect via survival on fomites.

for the impressive arsenal of host subversive components that play a role in the pathogenesis of *S. aureus* disease (Massey *et al.*, 2006).

It can also be presupposed that *S. aureus* has evolved to possess a suitably diverse set of survival mechanisms to facilitate efficient cycling between contrasting environments that impose physiological stress upon the cell (Clements and Foster, 1999). Environmental changes will impose altered conditions upon the bacterium to which it must respond rapidly and effectively to thrive. During transmission, via the route described above, *S. aureus* will be exposed to stress imposed via a wide variety of factors including desiccation, osmolarity changes, nutrient limitation, pH changes, temperature changes, exposure to interference from other bacteria in the nose and on the skin, and the antimicrobial action of skin lipids. During an infection *S. aureus*, potentially, has access to a wide variety of tissues and thus unique niches in which it is capable of manifesting pathologic processes. Infection will expose the cell to yet further new environmental stimuli to which it must now respond. The challenge for the bacterium is thus multifactorial, and for the researcher who is attempting to dissect the responses of the cell these responses will be observed as complex inter-connecting response networks.

This chapter will focus on several key responses made by *S. aureus* to the gamut of environmental conditions encountered during its life cycle. The range of environmental conditions discussed are those pertaining to adaptation relevant to *in vivo* conditions e.g. reduction in oxygen concentration, metal ion limitation, oxidative and nitrosative stress, stress response networks mediated by the accessory sigma factor σ^B, cell wall stress and antibiotics, reduced pH, temperature shift and starvation. Of course, adaptation to the environment includes quorum sensing but this is covered in Chapter 6, which describes the regulation of virulence by Agr, so discussion will be very limited here. Besides the well-studied Agr locus it is surprising just how little is known about the signal transduction associated with the two-component regulatory systems in *S. aureus*. Even for increasingly well-studied systems such as SrrAB, VraSR and SaeRS, the actual environmental or cellular-derived information that activates the system is unknown. Identification of the stimuli and how they activate receptors could potentially lead to the design of novel therapeutics to interfere with gene expression pathways necessary for virulence. Accordingly over the next few years there is likely to be an explosion of activity in staphylococcal biology that will greatly extend our knowledge of cell signalling in this organism and how this relates to adaptation and pathogenesis.

Reduced oxygen tension

Response

S. aureus is a facultative anaerobe that survives in an aerobic environment during transmission and on the skin. Little is known of the oxygen concentration during replication in the anterior nares although this environment would be expected to be aerobic. The typical pathology associated with *S. aureus* infections is an abscess and it is in this environment that reduced oxygen will routinely be encountered (Park *et al.*, 1992).

In *S. aureus* a reduction in the environmental concentration of molecular oxygen is sensed via the activity of SrrAB (Throup *et al.*, 2001; Yarwood *et al.*, 2001). This two-component sensor-regulatory pair controls the expression of components required to redirect metabolism from a respiratory pathway, in the presence of oxygen, to a fermentative pathway in its absence. SrrAB is a homologue of the ResDE system, described in detail for *Bacillus subtilis*, which regulates aerobic or anaerobic respiration in a direct and an indirect manner (Nakano *et al.*, 1996; Sun *et al.*, 1996). In *B. subtilis*, the ResDE signal transduction acts in concert with the transcriptional regulator Fnr to achieve effective control of gene expression (Geng *et al.*, 2007). So far there has been no identification of an Fnr-like regulator in *S. aureus*; however, this role might be fulfilled by the cytoplasmic NreAB two-component system (Kamps *et al.*, 2004) described later in this chapter.

An allelic replacement mutant of SrrAB displays a pronounced growth defect when cultured anaerobically on rich medium (tryptose soy broth), but aerobic growth is unaffected (Throup *et al.*, 2001). Analysis of the expression changes of the SrrAB mutant and its isogenic parent were resolved using 2D gel electrophoresis to determine regulatory function. This demonstrated that SrrAB responds to oxygen availability to regulate the expression of enzymes that modulate energy production. This study was conducted in a medium with no alternative electron acceptors, so the major change to metabolism observed in the SrrAB mutant during respiration in the absence of oxygen was its reduced ability to switch to fermentation (Throup *et al.*, 2001). Consequently the amount of lactate dehydrogenase and alcohol dehydrogenase enzymes was downregulated. These enzymes are central to anaerobic fermentation since, via reoxidation of NADH, these catalyse the formation of lactate or ethanol where lactate is the major product of fermentation produced in *S. aureus*.

The anaerobic growth defect of the *srrAB* mutant was proposed to be a consequence of the repression of F0/F1 ATPase, concomitant with a failure to reduce expression of NADH dehydrogenase under fermentative conditions. This antagonises the generation of a proton gradient across the membrane causing abortive initiation of the respiratory electron transport chain (Throup *et al.*, 2001). In a *srrAB* mutant the enzymes fumarase, aconitase and succinyl-CoA synthetase were derepressed. Typically under anaerobic conditions TCA cycle activity is minimized and without this the production of NADH continues. Since NADH must be reoxidized by fermentative enzymes it was proposed that the failure to induce lactate dehydrogenase in the mutant accentuates the loss of equilibrium in the cell (Throup *et al.*, 2001).

A more complete analysis of the response made by *S. aureus* to a shift from aerobic to anaerobic growth conditions was recently undertaken using transcriptomics and proteomics (Fuchs *et al.*, 2007). This study confirmed that glycolytic enzymes and fermentative enzymes were upregulated and TCA cycle enzymes were downregulated under these conditions. Moreover, they identified an increased expression of genes encoding formate and lactate transporters, nitrate respiration and nitrate reduction. In addition to these metabolic components, the anaerobic stimulon was found to include proteins belonging to the Clp machinery (ClpL, ClpP) and several virulence-associated loci (Fuchs *et al.*, 2007). In addition to the SrrAB two-component system, which senses an unknown signal associated with a reduction in environmental oxygen levels, the *S. aureus* genome also encodes a homologue of the *S. carnosus* NreAB (nitrogen regulation) two-component system (Kamps *et al.*, 2004). This has been proposed to function as an intracellular oxygen sensor that might be a staphylococcal equivalent of the FNR sensor that is widely dispersed among other bacteria. In *S. carnosus* it was determined that NreB showed reversible activation properties consistent with an FeS protein of the FNR type (Kamps *et al.*, 2004).

Oxygen concentration and its effect on virulence

The switch between aerobic and anaerobic environments was observed to modulate the expression of several virulence-associated genes, including *pls*, *hlY*, *splCD*, *epiG* and *isaB* (Fuchs *et al.*, 2007). This extended a link between oxygen concentration and virulence that was previously demonstrated with respect to the expression of TSST-1 in different clinical strains (Ross and Onderdonk, 2000; Yarwood and Schlievert, 2000) and altered SarA regulation in response to reduced oxygen (Chan and Foster, 1998; Lindsay and Foster, 1999). Several putative regulator proteins were found by Fuchs *et al.*, (2007) to be induced under oxygen-limiting conditions suggestive of a wider network.

SrrAB contributes to cellular fitness by conferring adaptability to changes in environmental oxygen tension. In a murine pyelonephritis model of infection the *SrrAB* mutant was attenuated, with a 3-log drop in survival compared to the parent strain (Throup *et al.*, 2001). The reason for the observed attenuation is unknown but is likely to be multifactorial. Several studies from the Schlievert laboratory have examined the role of the SrrAB two-component system for its contribution to the expression of virulence determinants (Yarwood *et al.*, 2001; Pragman *et al.*, 2007). A link with virulence was proposed as a result of SrrAB-dependent alterations in transcription of *agr*RNAIII and the genes encoding Spa and TSST-1. However this study was performed in the *agr* mutant strain, RN4220, thereby affecting interpretation of regulatory data relating to cell density-dependent signalling. Irrespective of this, the study proposed that SrrAB might act as a mechanistic link between aspects of metabolism, environmental signal transduction and virulence regulation. SrrAB-dependent transcription differences existed between microaerobic and aerobic culture conditions (Yarwood *et al.*, 2001). Anaerobic conditions were not tested. SrrAB positively regulated expression of TSST-1 and this induction was most strong in microaerobic versus aerobic conditions. Expression of Spa was negatively regulated by SrrAB in microaerobic conditions and positively regulated in aerobic conditions.

Analysis of the SrrAB regulon via proteomics provided little insight into other phenotypic changes in the behaviour of *S. aureus* with respect to oxygen availability. Cramton *et al.* (2001) determined that anaerobic culture increased expression of the *icaADBC* operon, which encodes proteins responsible for the synthesis of β-1,6-linked N-acetylglucosaminoglycan polysaccharide. This polymer constitutes the polysaccharide intercellular adhesin (PIA) that functions as the major mediator of biofilm formation in *S. aureus* (McKenney *et al.*, 1998; Cramton *et al.*, 1999). Increased expression of the *ica* operon was due to increased transcription and was observed in several separate strains indicating there is a common response (Cramton *et al.*, 2001). A more recent analysis has confirmed that SrrAB is the mediator responsible for significantly increasing production of PIA, via enhanced *ica* transcription, under anaerobic growth conditions (Ulrich *et al.*, 2007). Moreover, the production of PIA under these conditions was determined to contribute to survival from phagocytosis, whereby PIA production was protective against non-oxidative killing mechanisms. In contrast, the authors demonstrated that PIA does not protect *S. aureus* from neutrophil killing in aerobic conditions since PIA did not limit reactive oxygen species mediated killing (Ulrich *et al.*, 2007).

Metal ion limitation

Response

Metal ions are essential micronutrients for all life forms and bacteria have numerous metal transport and regulation mechanisms. In particular, magnesium, iron and manganese are essential for growth, being important co-factors in many enzymes, and the cell has specific uptake pathways. *S. aureus* has evolved to sense the intracellular concentration of several metal ions using metalloregulator proteins belonging to the ferric uptake regulator (Fur) and diphtheria toxin regulator (DtxR) families. The *S. aureus* genome encodes three Fur homologues, which have been ascribed roles in the regulation of iron uptake (Fur, SA1329) (Horsburgh *et al.*, 2001b), regulation of iron storage and antioxidant proteins (PerR, SA1678) (Horsburgh *et al.*, 2001a) and Zinc homeostasis (Zur, SA1383) (Lindsay and Foster, 2001). It also encodes one DtxR homologue, MntR, which controls manganese homeostasis (Horsburgh *et al.*, 2002b).

Iron homeostasis

There is a near universal requirement for iron by pathogenic bacteria and this stems from its redox potential and thus its excellent versatility in a vast array of catalytic reactions (Miller and Britigan, 1997), in particular it is an essential co-factor in DNA replication enzymes necessary for reproduction. Iron is plentiful in the human host, but is sequestered away into host proteins and complexes. The challenge for *S. aureus* is to obtain sufficient quantities of iron from the host environment where it is very effectively maintained at growth-limiting concentrations. To accentuate the problem of availability for an invading pathogen, the concentration of the most easily available iron is decreased further by hypoferraemia, the host's innate response to infection. The availability of free iron is limited in plasma by transferrin and within cells by ferritin, which the host uses for the management of iron. The potential availability of iron in the host is, however, significant with erythrocytes having large quantities of haem-iron complexes (Maresso and Schneewind, 2006). Typically *S. aureus* is a pathogen of the skin, yet its aggressive behaviour which damages tissue facilitates its dissemination systemically. Entry into the venous network could provide a seemingly endless supply of iron. However, iron uptake must be carefully managed since iron will react with endogenously generated reactive oxygen species from metabolism such as peroxide or superoxide to produce very harmful hydroxyl radicals (Miller and Britigan, 1997). Thus there is a concomitant requirement for oxidative stress defence and there are deleterious implications for *S. aureus* if they acquire a high intracellular iron store (Repine *et al.*, 1981). Therefore, the benefits of iron for cellular function must be carefully balanced with antioxidants.

Iron transport

The evolution of *S. aureus* has endowed it with an impressive array of iron uptake mechanisms (Fig. 7.2), a feature that is likely to add to its success as a commensal and pathogen of humans. A major mechanism for iron acquisition is via the secretion of siderophores, which are low molecular weight chelators of ferric iron (Fig. 7.2). The biosynthesis, role and regulation of the siderophore staphylobactin were recently described (Dale *et al.*, 2004a) and other, possibly distinct, siderophores (staphyloferrin A and B, aureochelin) have been described from different strains of *S. aureus* (Meiwes *et al.*, 1990). Heterologous hydroxamate siderophores can also be utilized and this might be advantageous for the ecological fitness of the bacterium in mixed populations. The importance of iron acquisition via siderophore-directed uptake was supported by the profound attenuation observed in a staphylobactin biosynthesis mutant. In a murine kidney abscess model of infection the mutant was effectively cleared from the mouse after six days in contrast to the parent control, which reduced only 1-log during this time (Dale *et al.*, 2004a). A very sharp drop in viability was observed between day five and day six suggesting that during the infection the iron could not be transported into the cell by other means efficiently enough to maintain survival in the host. Since bacteria were grown in rich medium prior to inoculation of the mouse, they would have initially contained a reservoir of iron in the storage proteins ferritin and MrgA, and this could have become critically depleted by day 6.

In *S. aureus* siderophore-mediated iron acquisition is central for the use of iron complexed by transferrin (Park *et al.*, 2005). Several proteins have been demonstrated to possess transferrin binding capabilities, glyceraldehyde 3-phosphate dehydrogenase (GAPDH) (Modun *et al.*, 1994; Modun *et al.*, 1998) and the cell wall binding protein IsdA (Taylor and Heinrichs, 2002), however they are likely to have only an ancillary role in the utilization of transferrin-iron (Park *et al.*, 2005).

Two further, distinct iron uptake mechanisms are encoded in the *S. aureus* genome (Fig. 7.2). Two membrane proteins with high levels of identity to GTP driven ferrous iron (Fe^{2+}) transporters are encoded in the genome, *feoB* (SA2337, SA2369), indicating that ferrous iron transport appears highly probable, although the activity of these transporters remains to be confirmed experimentally. *S. aureus* encodes a more complex, multicomponent pathway for the acquisition of iron from haem. The avail-

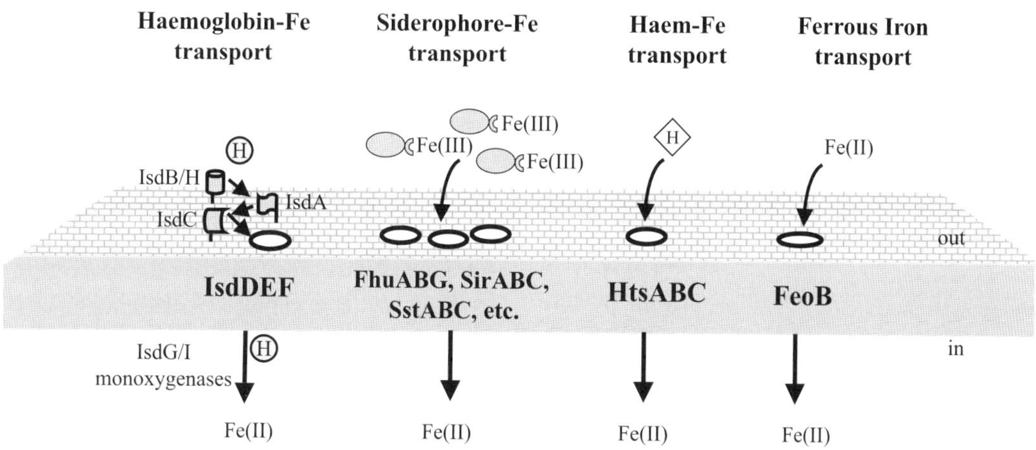

Figure 7.2 The Fur-regulated iron transport mechanisms of *S. aureus*. The four major mechanisms of Fur-dependent iron sequestration and transport are illustrated. Siderophore-iron mediated transport is achieved via several separate transport systems for staphylobactin and heterologous hydroxamate-iron complexes (see Table 7.1).

ability of haem-chelated iron is potentially very large and represents an obvious solution to the iron supply problem for an invading pathogen. Sequestration of this haem iron is achieved via the activity of the multicomponent Isd family of proteins (Maresso and Schneewind, 2006). This uptake pathway comprises four separate cell wall anchored proteins (Fig. 7.2), IsdB/H, IsdA, IsdC, that are proposed to shuttle the haem that is liberated from haemoglobin or haptoglobin through the cell wall to a membrane-localized permease complex, IsdDEF, which transfers the haem-iron complex to the cytoplasm (Wu *et al.*, 2005; Torres *et al.*, 2006). Here, via the action of two monoxygenases, IsdG/I, the tetrapyrrole ring is cleaved to liberate the iron (Skaar *et al.*, 2004a).

In a competition assay *S. aureus* displayed no obvious preference for whether the cytoplasmic supply of iron was derived from either transferrin (siderophore-acquired) or haem (Isd complex-acquired) (Skaar *et al.*, 2004b). A greater accumulation of haem-derived iron compared to transferrin-derived iron was observed at the cell membrane/cell wall and this was interpreted to reflect an iron source preference. Moreover, it was suggested that the porphyrin complexed iron in haem might be utilized directly at the membrane as a cofactor for redox active proteins (Skaar *et al.*, 2004b). The direct use of the porphyrin-iron would also serve to reduce synthesis by the bacterium and might reduce the toxicity of the iron by virtue of its chelation away from harmful reactive oxygen species.

A haemin-iron efflux mechanism was recently identified from analysis of the *S. aureus* proteome when cultured with haemin as the sole iron source (Friedman *et al.*, 2006). Haemin is the Fe^{3+} oxidation product of haem. The Hrt transport system (SA2149–SA2150) was strongly upregulated (>45-fold) under these conditions and mutants in the operon were defective for growth in haemin. The regulation of this transport mechanism is not Fur-regulated (see below) but is instead controlled via the activity of the HssRS two-component signal transduction system (SA2151–SA2152) (Stauff *et al.*, 2007). Although haem is an important source of iron for the cell, haem is also toxic at elevated concentrations and consequently the intracellular concentration of haem is controlled via expression of the HrtAB encoded efflux pump (Stauff *et al.*, 2007).

Iron uptake regulation by Fur

Fur (SA1329) typically functions as a transcriptional repressor that utilizes iron as a corepressor to regulate the operons within its regulon; in *S. aureus* these operons encode the different iron uptake mechanisms described in the previous section. The Fur protein–iron complex characteristically binds to an inverted repeat motif (Fur box) encoded within the promoter region of a regulated gene, repressing transcription at micromolar iron concentrations. Using the Fur motif identified in the promoter region of known regulated genes to search the *S. aureus* N315 database a total of 11 confirmed operons and 11 putatively Fur-regulated operons can be identified (Table 7.1) (Horsburgh *et al.*, 2001b; M.J. Horsburgh, unpublished). The regulon demonstrates that Fur functions primarily as an iron uptake regulator.

Inactivation of *fur* leads to the iron-independent overexpression of the Fur regulon (Horsburgh *et al.*, 2001b) (Table 7.1). A *fur* mutant displays a profound growth defect (Horsburgh *et al.*, 2001b; Richardson *et al.*, 2006) that is a consequence of unregulated and probably excessive iron transport into the cell coupled with reduced expression of catalase; *katA* is positively regulated by Fur (Horsburgh *et al.*, 2001b). The coupling of iron uptake with the concomitant expression of antioxidants like catalase or peroxidase is observed in many bacteria (Grifantini *et al.*, 2004; Sabri *et al.*, 2006) and is a preventative measure to reduce formation of the deleterious hydroxyl radical formed by reaction of peroxide with iron (Fenton reaction) (Miller and Britigan, 1997). The *fur* mutant strain is outgrown by more rapid growing suppressors, upon continued subculture, which have near normal levels of catalase expression (M.J. Horsburgh, unpublished). The mechanism behind the Fur-dependent positive regulation of catalase expression remains unexplored, however, positive regulation by Fur has been observed in other bacteria (Masse and Gottesman, 2002).

A recent proteomic analysis of iron regulation indicated that an array of proteins have modu-

Table 7.1 Fur repressed operons in *S. aureus* identified from observed transcriptional repression studies (Horsburgh *et al.*, 2001b, Dale *et al.*, 2004a, Dale *et al.*, 2004b, Allard 2006), proteomic studies (Friedman *et al.*, 2006) or from *in silico* analysis using search pattern facility in Aureolist (http://genolist.pasteur.fr/AureoList) (M.J. Horsburgh, unpublished) for the *S. aureus* N315 genome. A revised consensus sequence is shown.

Operon	*Names and Function*	*Fur box motif*	*Motif Location*	*Confirmed*
SA0111 - SA0109	*siderophore transport (sirABC)*	tgataatgattctca	*-48*	*Yes*
SA0112 - SA0120	*staphylobactin synthesis*	tgatagtgagaatca	*-66*	*Yes*
SA0257	*methyltransferase*	tgataataattatca	*-34*	
SA0331 - SA0333	*iron permease and peroxidase*	tgataattattatca	*-155*	
SA0335	*putative sec-independent secretion component*	tgataatcattatcg	*-65*	
SA0602 - SA0604	*fhuCBG ferrichrome transporter*	tgataatcattatca	*-100*	*Yes*
SA0688 - SA0691	*ferrichrome uptake (sstABCD)*	tgataatgattatca	*-41*	*Yes*
SA0806	*putative NADH dehydrogenase*	tgataataataatca	*-57*	
SA0976	*isdB*	tgataatgattatca	*-50*	*Yes*
SA0977	*isdA*	taataatgattatca	*-77*	*Yes*
SA0978 - SA0983	*isdCDEF srtB isdG*	tgataatcattatta	*-148*	*Yes*
SA1184	*aconitate hydratase (citB)*	tgataattattctca	*-196*	
SA1552	*isdH*	tgataattattatca	*-48*	*Yes*
SA1979 - SA1977	*haem transport system (htsABC)*	tgataattgttatca	*-57*	*Yes*
SA1982 - SA1980	*putative siderophore transporter*	tgataatgattctta	*-43*	*Yes*
SA1983	*putative siderophore transporter*	tgataataagaatca	*-79*	
SA2001	*oxidoreductase*	tgataataagaatca	*-115*	
SA2079	*ferrichrome uptake (fhuD)*	tgataatcatttca	*-28*	*Yes*
SA2162	*thioredoxin reductase*	tgataatgattctta	*-108*	
SA2217 - SA2216	*ABC transporter*	tgataatcgttatca	*-149*	
SA2304	*fructose bisphosphate aldolase*	tgataatcattgtaa	*-68*	*Yes*
SA2337 - SA2338	*ferrous iron transporter (feoB1)*	tgataatgattatta	*-39*	
	Consensus Fur box sequence	**tgataat-attatca**		

lated expression in response to iron limitation some of which were either directly or indirectly Fur-regulated (Friedman *et al.*, 2006). This study also demonstrated that when iron is limiting the respiratory pathway is redirected to upregulate glycolysis, with a concomitant downregulation of the TCA cycle, resulting in fermentative metabolism which generates lactate and other acidic fermentative products. This accumulation of acidic components reduces the pH of the immediate environment and the authors proposed that this pH decrease will effectively enhance iron solubility. This mechanism might also occur *in vivo*, where the bacteria grow in a more effectively buffered environment, since a tissue cage model of infection (Allard *et al.*, 2006) using a small microarray of 460 genes revealed some of the same metabolic enzyme changes observed by Friedman *et al.* (2006). Unfortunately the work of Allard *et al.* (2006) compared iron limitation vs. iron overload vs. tissue cage in their study making interpretation difficult and greatly limiting its usefulness due to the pleiotropic effect that iron overload has on the cell (see the later section on PerR).

Role of Fur in virulence

A strain of 8325-4 with an allelic exchange mutation of the *fur* gene showed reduced virulence in a murine abscess model of infection when compared to its isogenic parent strain (Horsburgh *et al.*, 2001b). This indicates the importance of iron metabolism regulated by Fur in *S. aureus*. The reason for the reduced virulence is not clear but might be multifactorial due to the unregulated uptake of iron into the cell combined with reduced catalase expression. Certainly the *fur* mutant displays severely reduced resistance to many different imposed stresses suggesting that the overall fitness of the mutant is significantly impaired (Horsburgh *et al.*, 2001a; Horsburgh *et al.*, 2001b).

PerR, an iron storage and antioxidant regulator

The cellular function of a second Fur homologue in *S. aureus* was characterized by allelic replacement of its cognate gene, *perR* (SA1678). Physiological analysis determined that the mutant was hyper-resistant to hydrogen peroxide but possessed wild-type levels of resistance to cumene hydroperoxide (Horsburgh *et al.*, 2001a; Horsburgh *et al.*, 2001b). Consequently it was determined that the function of this protein was similar to that of the *Bacillus subtilis* PerR protein, which was known to function as a transcriptional repressor of genes encoding a peroxide resistance regulon. Studies in *B. subtilis* demonstrated that PerR was a manganese-responsive repressor (Herbig and Helmann, 2001; Lee and Helmann, 2007) and analysis of *S. aureus* similarly demonstrated that addition of micromolar levels of Mn in the growth medium strongly repressed the expression of catalase (Horsburgh *et al.*, 2001a). A *S. aureus perR* strain of 8325–4 showed normal growth in rich medium but had a slower growth rate in metal depleted medium indicating it had a metal ion concentration dependent phenotype (Horsburgh *et al.*, 2001a).

Putative PerR binding sites were identified by searching the *S. aureus* 8325 genome database for potential regulatory motifs similar to an inverted repeat located between the −35 and −10 elements of the *katA* gene promoter. This search revealed many potentially regulated operons encoding proteins with apparent roles in cellular resistance to reactive oxygen species or metal ion regulation (Horsburgh *et al.*, 2001a). A PerR-dependent regulon was confirmed via the construction of *lacZ* fusions to eight separate operons and expression studies of the *perR* mutant and its isogenic 8325-4 parent. In addition, analysis of the cytoplasmic protein fraction of the *perR* inactivated strain revealed that it overexpressed several different proteins. N-terminal sequencing of these proteins identified them as alkylhydroperoxide reductase C, and MrgA and ferritin, two putative iron storage proteins (Horsburgh *et al.*, 2001a). The PerR-dependent regulon that is known at present is shown in Table 7.2.

The PerR-dependent regulon is likely to encompass a wider set of genes than is shown in Table 7.2. The PerR binding motif is AT rich making *in silico* determination of the regulon more difficult, however several oxidoreductases, peroxidases and stress related proteins have putative PerR binding motifs upstream of their coding regions (M.J. Horsburgh, unpublished).

Expression analysis of individual PerR regulon genes demonstrated that all were repressed transcriptionally in the presence of micromolar quantities of Mn(II), when assayed in a chemically defined medium (Horsburgh *et al.*, 2001a). Conversely the addition of excess levels of iron markedly increased transcription of the regulon. In some cases such as with ferritin and catalase, their genes were transcribed at a level that was similar or greater to that seen in a *perR* mutant. This was suggestive of complex regulation at these loci; transcription of *katA* was positively regulated by Fur (Horsburgh *et al.*, 2001b). In

Table 7.2 PerR-dependent regulon of *S. aureus* (Horsburgh *et al.*, 2001a).

Operon	*Names and Function*	*PerR box motif*
SA1170	*Catalase (katA)*	AATTATAATTATTATAAAT
SA0366-SA0365	*Akyl hydroperoxide reductase (ahpCF)*	GATTAGAATTATTATAATT
SA1941	*MrgA (ferritin-like) (mrgA)*	GATTAGAATTATTATAATA
SA1709	*Ferritin (ftn)*	AATTATAATTATTATTATT
SA0719	*Thioredoxin reductase (trxB)*	GCATATAATTATTATTATT
SA1680-SA1679	*Oxidoreductase, 3-phosphoglycerate dehydrogenase (bcp pdh)*	TCATAAAATTATTATAATG
SA1329	*Fur (fur)*	TTTTTTAATTATTAGTAGG
SA1678	*PerR (perR)*	AATAATAATTATTATATAA
	Consensus PerR box sequence	--TTAtAATTATTATaA--

the case of transcription of the gene encoding ferritin (*ftn*) there is also the possibility of positive regulation mediated by Fur or independent regulation (Horsburgh *et al.*, 2001a; Horsburgh *et al.*, 2001b); a separate study of *ftn* transcription failed to resolve this definitively, but repeated the observations of iron- and manganese-dependent regulation described above (Morrissey *et al.*, 2004).

Studies of PerR-mediated regulation have demonstrated that PerR functions as a transcriptional repressor that uses Mn(II) as its corepressor. PerR is a redox sensor that senses H_2O_2 by metal-catalysed histidine oxidation, and in *B. subtilis* it regulates a similar regulon (Herbig and Helmann, 2001; Moore *et al.*, 2005; Lee and Helmann, 2006). In *S. aureus*, however, PerR differs in its ability to regulate transcription via the relative amounts of Mn(II) and Fe(II). In the presence of Mn(II) the transcription of the regulon is repressed but in the presence of Fe(II) the regulon is derepressed (Horsburgh *et al.*, 2001a). As Fe(II) is titrated into a Mn(II) containing medium the repression is also lifted showing that the presence of iron has a dominant function (Horsburgh *et al.*, 2001b; Morrissey *et al.*, 2004) (Fig. 7.3). This has a very simple explanation and produces a working model for redox sensing in *S. aureus*. The presence of Fe(II) in the cell is deleterious if it is not chelated to prevent the Fenton reaction with endogenously generated peroxides. Thus *S. aureus* uses PerR to sense peroxide levels and via the concentrations of iron and manganese in the cell it modulates the expression of antioxidants and iron storage proteins accordingly. In this manner the potential toxicity of iron is reduced since the cell actively reduces the amount of peroxide available to react with iron by inducing peroxidases and simultaneously chelates the iron into the storage proteins induced to reduce the intracellular iron pool (Horsburgh *et al.*, 2001b; Horsburgh *et al.*, 2002c). In *B. subtilis* PerR also functions as a peroxide sensor to increase PerR regulon expression when peroxide is added (Herbig and Helmann, 2001). This observation was repeated with *S. aureus* using transcriptional fusions in the presence of very high concentrations of exogenously added peroxide (Horsburgh *et al.*, 2001a). However, two separate studies have since failed to observe peroxide mediated induction of the PerR regulon, questioning whether the primary function of PerR in *S. aureus* is that of a metal ion sensor/endogenous peroxide sensor rather than as an exogenous peroxide sensor (Weber *et al.*, 2004; Chang *et al.*, 2006). There is evidence that PerR senses endogenous peroxide since deletion of several different members of the PerR regulon results in increased expression of the whole PerR regulon (Horsburgh *et al.*, 2001a; Horsburgh *et al.*, 2001b; Cosgrove *et al.*, 2007). This behaviour was first described in *B. subtilis* and was found to be a consequence of increased peroxide stress within the cell (Bsat *et al.*, 1996). The responsiveness of PerR to the concentration of Mn(II) within the cell raises questions about its cellular role and why peroxide resistance/iron uptake should be regulated by this metal ion. It was subsequently shown that Mn(II) has the ability to quench reactive oxygen species

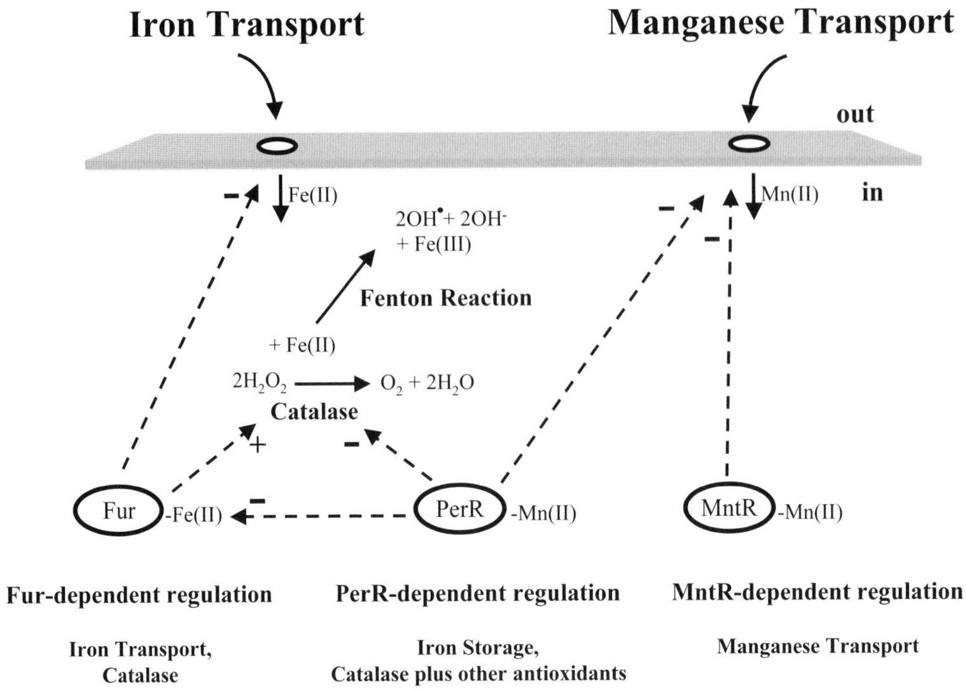

Figure 7.3 The interactive network regulating homeostasis of iron and manganese and expression of antioxidants and iron storage proteins. Regulation of expression by Fur, PerR and MntR proteins is shown with dashed arrows and positive (+) or negative (–) regulation is indicated.

(Archibald and Fridovich, 1981; Al-Maghrebi *et al.*, 2002) and this is discussed more fully in the later section on manganese homeostasis.

The transcription of *perR* is autoregulated by virtue of a PerR-binding motif in its own promoter (Horsburgh *et al.*, 2001a). In this manner the cell maintains an effective level of the protein. PerR also regulates the expression of Fur and in doing so it perfectly coordinates the uptake and supply of iron with endogenous peroxide levels and demonstrates the importance of cooperativity between the Fur and PerR regulons (Horsburgh *et al.*, 2002b; Horsburgh *et al.*, 2002c) (Fig. 7.3).

Role of PerR in virulence

When a PerR mutant strain of 8325-4 was tested for its relative virulence compared to its isogenic parent it was found to have a greater than one log reduced survival in a murine abscess model of infection (Horsburgh *et al.*, 2001a). This was the first indication that PerR proteins were important in bacterial pathogenesis and subsequent studies have observed a similar involvement of PerR in the virulence of *Streptococcus pyogenes* (Ricci *et al.*, 2002), *Listeria monocytogenes* (Rea *et al.*, 2005) and *Enterococcus faecalis* (Verneuil *et al.*, 2005). There is no clear explanation for the reduced virulence of a PerR mutant, given that it is hyper-resistant to peroxides and might be expected to have increased *in vivo* survival, particularly in neutrophils. Recent studies with *Streptococcus pyogenes* have clearly shown that PerR has a pleiotropic effect in this bacterium, over and above its own regulon (Brenot *et al.*, 2007). Similarly, uncharacterized indirect effects might explain the survival defect in *S. aureus*.

The PerR regulon and hypochlorous acid resistance

Derepression of the PerR regulon was also observed after exposure of *S. aureus* to hypochlorous acid (HOCl), which is known to generate reactive oxygen species. The addition of HOCl resulted in a simultaneous decrease in superoxide dismutase (SOD) enzyme activity (Maalej *et al.*, 2006). *S. aureus* contains two SODs, SodA and SodM, which reduce superoxide radicals to hydrogen peroxide (Clements *et al.*, 1999a; Valderas and Hart, 2001; Karavolos *et al.*, 2003). A mechanistic explanation for the coordination between the decrease in SOD levels and the

PerR regulon was not proposed in this study, other than to suggest it was due to an increase in superoxide anions (Maalej *et al.*, 2006). Possible explanations for the observed derepression of the PerR regulon are that reduced SOD activity causes an increased superoxide anion concentration which in turn liberates iron in the cell; alternatively, superoxide anion could directly affect the PerR:Fe form. Increases in the cellular Fe concentration would need to be rapidly controlled and induction of the PerR regulon which coordinates antioxidant defence and iron storage would help to restore balance to the system.

Manganese homeostasis

Mn is a growth requirement for many different pathogens and has been ascribed specific cellular roles as a cofactor in enzymes for metabolism, catabolism and signal transduction (Yocum and Pecoraro, 1999; Jakubovics and Jenkinson, 2001). Enzymatic and non-enzymatic roles have been described for Mn(II) with respect to protecting the cell from oxidative stress. Moreover, many bacteria, including *S. aureus* (Clements and Foster, 1999; Valderas and Hart, 2001), contain a Mn-superoxide dismutase (SOD) that catalyses the reduction of superoxide radicals. In humans the concentration range of manganese is appreciable and ranges from low concentrations on the skin (0.05 μM), to 30 fold higher in blood (1.5–2.4 μM) and yet higher still in the central nervous system (160–180 μM) (Posey *et al.*, 1999) indicating that Mn(II) is more freely available compared to the more sequestered iron.

Manganese transport

The transport of manganese is important for several key pathogens and two distinct transport systems have been described. ATP-binding cassette (ABC) Mn transporters (Claverys, 2001) contribute to disease caused by *E. faecalis* (Singh *et al.*, 1998), *Streptococcus pneumoniae* and *Strep. parasanguinis* (Burnette-Curley *et al.*, 1995; Berry and Paton, 1996), and *Yersinia pestis* (Bearden and Perry, 1999). The *mntABC* operon (SA0589-SA0587) of *S. aureus* encodes an ABC-type transporter that is required for the uptake of Mn (Horsburgh *et al.*, 2002b); it has homology to Mn transporters from *Strep. gordonii* (Kolenbrander *et al.*, 1998) and *B. subtilis* (Que and Helmann, 2000) and a transporter of unknown specificity from *S. epidermidis* (Cockayne *et al.*, 1998). The *mntABC* operon is likely to encode the active Mn(II) transport system in *S. aureus* membrane vesicles described by Perry and Silver (1982), which transports Mn(II) and Cd(II). A *mntABC* mutant displayed an increased MIC to Cd(II) (Horsburgh *et al.*, 2002b).

Bacterial Nramp homologues, called MntH, are selective Mn transporters that are implicated in defence from reactive oxygen species and have a role in pathogenesis (Kehres *et al.*, 2000; Makui *et al.*, 2000). *S. aureus* encodes a MntH homologue (SA0956) which is implicated in manganese transport (Horsburgh *et al.*, 2002b) but its metal ion specificity has not been determined experimentally.

Regulation of manganese transport

A sole DtxR-like protein called MntR is encoded within the *S. aureus* genome and its central role for the transport of manganese was determined by allelic replacement (Horsburgh *et al.*, 2002b). DtxR-like proteins from other bacteria, including ScaR from *Strep. gordonii* (Jakubovics *et al.*, 2000), TroR from *Treponema pallidum* (Posey *et al.*, 1999) and MntR from *B. subtilis* (Que and Helmann, 2000) and *E. coli* (Patzer and Hantke, 2001) are transcriptional repressors of Mn(II) uptake systems including Mn(II) ABC transporters and MntH (an Nramp homologue). In *S. aureus* the transcription of *mntABC* was repressed by Mn(II) and this regulation was MntR dependent (Fig. 7.3). In contrast, transcription of the *mntH* gene was not repressed by high Mn(II) concentrations, and inactivation of MntR reduced *mntH* transcript levels (Horsburgh *et al.*, 2002b). MntR-dependent repression of *mntABC* occurred above 0.1 μM Mn(II) and within the physiologically relevant concentration range that *S. aureus* will encounter during infection in different regions of the body.

The *mntR* and *mntABC* operons are transcribed divergently and a putative MntR binding motif was found to be situated over the identified transcriptional start site for *mntABC*. A very similar MntR binding motif was identified in the −35 region of the *mntH* promoter (Horsburgh *et al.*, 2002b).

The physiological role of manganese in S. aureus

In *S. aureus* manganese functions as a corepressor to transcriptionally regulate the PerR and the MntR regulons. The Mn(II) bound PerR repressor also functions to repress the transcription of Fur. A potential functional link between the MntR, PerR and Fur regulons was investigated to determine whether Mn(II) availability within the cell, regulated by MntR, modulated expression of the Fur regulon (Horsburgh *et al.*, 2002b). Hierarchical control was clearly observed and transcription of *ftn* and *katA* was demonstrated to be PerR dependent via MntR directed regulation of Mn(II) concentration (Horsburgh *et al.*, 2002b). Thus in *S. aureus*, PerR is an important regulator of iron storage via ferritin and the ferritin-like MrgA protein and it modulates Fur expression in response to the intracellular manganese concentration. The interplay between the MntR and PerR regulatory pathway was reinforced with evidence that *mntABC* transcription was also repressed by PerR and this was Mn-dependent (Horsburgh *et al.*, 2002b; Horsburgh *et al.*, 2002c). The complex interactions between the Fur, PerR and MntR regulons is shown in Fig. 7.3.

Peroxide resistance is inextricably linked to cellular iron concentration because, deleteriously for the cell, H_2O_2 reacts with Fe(II) to produce OH• (Miller and Britigan, 1997). Consequently, to avoid the harmful effects of the Fenton reaction the bacterium coordinates iron accumulation with the endogenous production of peroxide and the supply of Mn. The repression of peroxide defence by Mn might appear counterproductive given that Mn treated cells display greater sensitivity to external peroxide (Horsburgh *et al.*, 2002b). The scavenging activity of Mn(II) occurs with free Mn (II) or when the metal ion is either coordinated to lactate or orthophosphate (Horsburgh *et al.*, 2002b). It is apparent therefore that *S. aureus* can utilize Mn(II) ions to reduce peroxide stress in the cell which in turn will reduce demand for PerR regulon expression. The coordinated expression of Mn and Fe supply enables a rapid response if extra iron or extra peroxide alters the equilibrium. Thus, overall the PerR regulon might act to increase cellular fitness.

Contribution of manganese transport to S. aureus virulence

A murine abscess model of infection was used to examine the relative virulence of allelic replacement mutants that were constructed in the *mntR*, *mntA* and *mntH* genes (Horsburgh *et al.*, 2002b). Individually none of these loci contributed to infection in the studied model. A double *mnt* mutant did reveal a statistically reduced survival in the abscess model suggesting that manganese transport positively contributes to pathogenesis.

Zinc transport

A third Fur-like protein encoded in the genome of *S. aureus* was found to have homology to the zinc uptake regulator of *B. subtilis*. To confirm this putative function in *S. aureus* an allelic replacement of *zur* was generated (Lindsay and Foster, 2001). *zur* is situated at the end of a three gene operon, downstream from two putative membrane proteins with homology to zinc and other metal ion transporters. Confirmation that Zur was a regulator of the putative metal transporters was achieved by the construction and assay of transcriptional fusions to the transporter genes while maintaining an intact copy of *zur* under P*spac* control (Lindsay and Foster, 2001).

Cellular response to nitrosative stress

The innate immune response utilizes nitric oxide (NO•) as an essential antimicrobial metabolite and as an immunomodulatory effector. It occupies a central role in the elimination of pathogens such as *S. aureus* via local high concentrations that are generated by activated neutrophils during tissue infection. The intrinsic chemical nature of NO• leads to its rapid oxidation to generate reactive nitrogen species that react with the bacteria to cause nitrosylation of complexed metal centres, most importantly iron, the nitrosylation of protein thiol groups and at higher concentrations damage occurs due to deamination of DNA bases, lipid peroxidation and nitration of tyrosine residues in proteins (Schapiro *et al.*, 2003).

The cellular response to NO• has been studied in several key bacteria and was recently examined in *S. aureus* via physiological and transcriptomic analyses (Richardson *et al.*, 2006). This revealed that the SrrAB and Fur regulons, described above, were essential for resistance to NO•. The Fur regulon was derepressed during the nitrosative stress and this is likely to be a consequence of the reaction between NO and the ferrous iron bound to Fur (Richardson *et al.*, 2006). This reaction generates dinitrosyl-iron complexes, which are known to inactivate the Fur repressor. The reduced survival of a *fur* mutant and the essential nature of the Fur regulon can be attributed, in all likeliness, to the derepressed uptake of iron into the cell in the Fur mutant increasing the available reactive iron pool. Involvement of the SrrAB regulon was proposed to be a result of impaired respiratory function (Richardson *et al.*, 2006). Moreover, SrrAB activated expression of the *hmp*-encoded flavohaemoglobin and this protected the cell from the effects of NO• *in vitro*. Flavohaemoglobin is required for the *in vivo* survival of *S. aureus* since an *hmp* mutant was avirulent in a murine infection model (Richardson *et al.*, 2006).

The regulation of the nitrosative stress response involves the Fur and SrrAB regulons; the overarching mechanisms responsible for the induction of the SrrAB response and Hmp expression remain unclear.

Stress and sigma B

The alternative sigma factor, σ^B, occupies a central role in the cellular physiology and virulence-associated gene transcription of *S. aureus* (Chan *et al.*, 1998; Kullik *et al.*, 1998; Nicholas *et al.*, 1999; Giachino *et al.*, 2001; Horsburgh *et al.*, 2002a). These two areas of cell biology are tightly interwoven in a complex and overlapping regulatory network and accordingly it is difficult to have a discussion of physiology without mention of virulence gene expression. The role of σ^B in the overall network of virulence gene regulation is dealt with in Chapter 6; consequently this chapter will focus on its primary cellular role, which is the control of stress responses subsequent to the perception of an environmental change.

In *S. aureus*, the *rsbUrsbVrsbWsigB* operon is transcribed by σ^A and *rsbVrsbWsigB* transcription is autoregulated via a σ^B promoter upstream of *rsbV* (Kullik and Giachino, 1997; Senn *et al.*, 2005);. RsbW functions as an anti-sigma factor (Miyazaki *et al.*, 1999; Senn *et al.*, 2005) and RsbV is proposed to be an anti-anti-sigma factor, similar to its known activity in *Bacillus subtilis* (Palma and Cheung, 2001; Senn *et al.*, 2005). During growth under non-stress activating conditions σ^B activity is controlled via binding to RsbW limiting its ability to transcribe from its cognate promoters. Since *S. aureus* lacks many of the known σ^B hierarchical control systems present in *B. subtilis* (including RsbR, RsbS and RsbT), it appears that RsbU must function as a key stress reception module (Kullik and Giachino, 1997). When stress is perceived RsbV is dephosphorylated by RsbU, producing an activated form that binds to and antagonizes RsbW releasing σ^B to binding to the RNAP complex (Fig. 7.4).

Environmental sensing via σ^B

S. aureus perceives a range of different environmental changes and transduces this information to activate σ^B, with stressors including elevated temperature, salt (Kullik and Giachino, 1997; Pane-Farre *et al.*, 2006), and alkaline pH (Pane-Farre *et al.*, 2006). This demonstrates that there is an overall general stress response network that is controlled by the direct activation of σ^B in the cell (Fig. 7.4). Although RsbV and RsbW function in a similar manner to their *B. subtilis* counterparts, the absence of many control proteins means the overall regulation of the partner-switch mechanism controlling σ^B activity in *S. aureus* differs markedly from that in *B. subtilis*. Moreover, overexpression of *rsbU* was sufficient to activate σ^B (Senn *et al.*, 2005). Further study demonstrated that cytosolic proteins could interfere with the function of RsbU altering its regulatory capacity thereby modulating activation of the sigma B factor. This latter property is beginning to be realized in a wider cellular context. Shaw *et al.*, (2006) identified many genes that, when inactivated, altered the expression of alpha-haemolysin and protease activity in a σ^B-dependent manner. The identified genes were located very close to each other in the genome highlighting a 'hotspot' region where individual transposon insertions altered σ^B activity. Insertions in this region resulted in changes to expression of Asp23, α-haemolysin and proteases, at least in the SH1000 strain used for the study (Shaw *et al.*, 2006). The pathway(s)

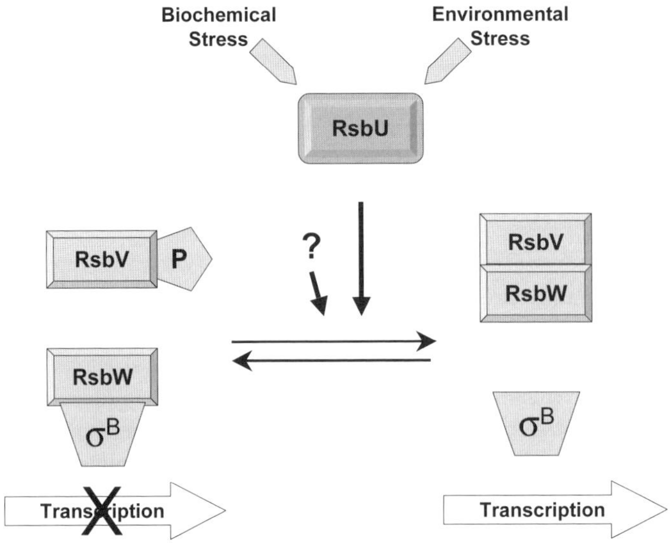

Figure 7.4 Regulatory cascade for the accessory sigma factor σ^B. Environmental stresses (temperature, pH, salt) are sensed and transduced via changes in the dephosphorylating activity of RsbU on RsbV (Kullik and Giachino, 1997); (Pane-Farre *et al.*, 2006). This dephosphorylation causes a partner switch of RsbW/σ^B to RsbV/RsbW releasing σ^B for transcription from its cognate promoters. Biochemical stresses within endogenous metabolic pathways are also proposed to act via RsbU which is a positive regulator of σ^B activation (Shaw *et al.*, 2006). The activity marked (?) refers to an additional regulatory factor highlighted by Palma and Cheung (2001). Other unlabelled putative effectors have been described (Senn *et al.*, 2005).

whereby these insertions modulate the activity of σ^B was not determined; however, the authors speculated that the action is via RsbU directed alterations to the regulatory phosphorylation cascade (Shaw *et al.*, 2006) (Fig. 7.4). It was previously hypothesized by Senn *et al.*, (2005) that the regulatory cascade contained input points whereby nutritional stress, physical stress and growth phase information could enter the pathway to modulate the activity of σ^B.

A full appreciation of the pleiotropic involvement of σ^B in many different cellular processes was realized fully only after the *S. aureus* 8325 strain lineage used most commonly for genetic studies was revealed to have an 11 bp deletion within *rsbU* that rendered the encoded protein defective (Giachino *et al.*, 2001). To circumvent this issue, an unmarked, functional *rsbU* replacement version of the commonly used 8325-4 strain was generated (Horsburgh *et al.*, 2002a). This study, together with that by Giachino *et al.*, (2001) which used a marked cis-complementation of *rsbU*, lead to a greater appreciation of the central involvement of σ^B in the cell. It was clearly shown that expression of σ^B reduced the expression of the *agr* locus via an unknown mechanism leading to a reduction in the expression of a range of virulence loci regulated directly by *agr*.

Several phenotypic changes in these repaired *rsbU* strains were observed to result in altered stress resistance (Giachino *et al.*, 2001; Horsburgh *et al.*, 2002a). Increased *katA* transcription was observed during growth of SH1000 (8325-4 *rsbU*+) compared to an isogenic *sigB* mutant. Mapping of the promoter identified a σ^B transcriptional start site revealing that catalase is regulated by this accessory sigma factor, in addition to PerR and Fur (Horsburgh *et al.*, 2002b). Increased catalase activity was observed in σ^B-replete strains (Giachino *et al.*, 2001; Horsburgh *et al.*, 2002a) and this could be the factor responsible for enhancing the starvation survival of SH1000 above that of 8325-4 (Horsburgh *et al.*, 2002a).

Phenotypically the most obvious change associated with the restoration of σ^B activity is the increased production of the orange pigment staphyloxanthin. This triterpenoid carotenoid is located in the cell membrane where it is known to function as an antioxidant protecting the cell from peroxide, superoxide and hydroxyl radicals and hypochloride (Clauditz *et al.*, 2006)). A *crtM* mutant was shown to be highly attenuated in a murine abscess model of infection compared to its isogenic parent strain (Liu *et al.*, 2005). This is surprising since no attenuation was observed in similar models of infection for strains with mutations in *katA*, or the double mutants *sodA sodM* and *katA ahpC* (Clements *et al.*, 1999a; Horsburgh *et al.*, 2001a; Karavolos *et al.*, 2003). The work of (Liu *et al.*, 2005) would suggest, therefore, that staphyloxanthin plays a more important *in vivo* antioxidant defence role than these components or that it has an undiscovered central role for *in vivo* survival. σ^B-dependent expression of pigment also contributes to UV resistance, with σ^B-replete strains having increased UV tolerance, mostly in the near-UV range (Giachino *et al.*, 2001). This is most likely to result from carotenoid production, since these pigments have been shown to be protective in other genera (Tuveson *et al.*, 1988).

Recent studies have determined the extent of the σ^B regulon via the use of microarrays to identify differential gene transcription. Over 250 genes were affected transcriptionally by the presence/absence of σ^B activity and of 198 genes that were upregulated most had a clearly identified σ^B promoter-like element upstream (Bischoff *et al.*, 2004). The largest proportion of the regulated genes were found to be involved in a variety of central cellular pathways involved in intermediary metabolism, cell wall and capsule formation and membrane transport systems (Bischoff *et al.*, 2004; Pane-Farre *et al.*, 2006). This is somewhat different to that of the well-studied *B. subtilis* system making it a poor comparison and suggesting σ^B in *S. aureus* serves an altogether different role. The transcription analysis studies using microarray also determined the large extent to which genes encoding virulence-associated loci were regulated by σ^B. Previous studies had identified that alpha- haemolysin (Giachino *et al.*, 2001; Ziebandt *et al.*, 2001; Horsburgh *et al.*, 2002a), clumping factor (Nair *et al.*, 2003), coagulase (Miyazaki *et al.*, 1999; Nair *et al.*, 2003), fibronectin binding protein A (Nair *et al.*, 2003), lipases (Kullik *et al.*, 1998; Ziebandt *et al.*, 2001) and proteases (Horsburgh *et al.*, 2002a) were regulated by σ^B and the array study added to the regulon genes for capsule, γ-haemolysin, leukocidin and several putative virulence factors (Bischoff *et al.*, 2004).

The link between cellular stress and transcriptional modulation of virulence-associated loci appears to be complex and at this point involves several unidentified components. Regardless of this, the accumulated data suggest a sensitive system that is incredibly well poised for responding to environmental stimuli. Moreover, this complex multi-tiered response network is likely to explain many aspects of the incredible versatility of *S. aureus* as a commensal and a pathogen.

Antibiotics and cell wall stress

Much of the notoriety afforded to *S. aureus* strains, such as MRSA, is a consequence of its formidable acquisition of determinants conferring resistance to antibiotics. The primary target of a broad range of antibiotics is against the cell wall and thus peptidoglycan synthesis (Chapter 8). In the environment bacteria have to survive a diverse array of cell wall or cell membrane active compounds released by the host or competitor species and have necessarily evolved a compensatory response network to manage and repair the consequent, potentially life-threatening assault. Recently our knowledge of the cellular response of *S. aureus* to antibiotics and the stress this imposes on the cell was greatly increased. Proteomic and transcriptomic studies (Singh *et al.*, 2001; Utaida *et al.*, 2003) revealed the existence of a cell wall stress stimulon that was induced in response to the action of antibiotics, including oxacillin, D-cycloserine, bacitracin (Singh *et al.*, 2001; Utaida *et al.*, 2003) and vancomycin (Gardete *et al.*, 2006; McAleese *et al.*, 2006; McCallum *et al.*, 2006). These antibiotics are active against different steps of peptidoglycan biosynthesis so it was of interest that they affected gene expression to generate a common response to the damage in addition to antibiotic-specific changes in transcription. The existence of a core, global cell wall stress stimulon demonstrated that the cell has a strategic response that relies on a central framework around which it embroiders more specific antibiotic-dependent changes. The core stimulon was found to consist of fifteen genes (Singh *et al.*, 2001; Utaida *et al.*, 2003; McAleese *et al.*, 2006) although, additionally, it was also shown that there was a pronounced strain dependence to this core stimulon (McCallum *et al.*, 2006) in response to vancomycin. This was due in part to differential sensitivities of the strains to the antibiotic.

The mechanism supporting the induction of the core stimulon is dependent on the two-component signal transduction operon VraSR, which itself forms part of the cell wall stimulon (Gardete *et al.*, 2006; Yin *et al.*, 2006). It was clearly shown that VraSR was essential for sensing inhibition of several stages of cell wall synthesis. Moreover, the signal that was perceived was specific to the cell wall and no cross-stimulation was observed from other environmental perturbations that were tested, including temperature, high osmolarity or pH (Kuroda *et al.*, 2003). VraSR has a role in the activation of cell wall synthesis via steps in biosynthesis (MurZ) and polymerization (PBP2 and SgtB) of peptidoglyan and its disruption results with concomitant increases in sensitivity to cell wall inhibitory antibiotics including glycopeptides, fosfomycin and β-lactams. VraSR thus appears to be key to the antibiotic-induced response in *S. aureus* due to the observation that its inactivation ablates the transcriptional response to cell wall inhibitors (Gardete *et al.*, 2006; Yin *et al.*, 2006). It remains to be unambiguously determined whether VraSR is the moiety with overall hierarchical control of the response or whether it is indirectly activated. Certainly the data presented so far demonstrate that VraSR is acutely responsive to alterations in the amount of PBP2 and it was elegantly determined that decreases in PBP2 regulated by *spac* promoter control were matched with an increase in *vraSR* transcription (Gardete *et al.*, 2006). VraSR represents a key environmental sensor-regulator in *S. aureus* that perceives disruption to cell wall synthesis and modulates expression of its regulon of cell wall synthesis/repair enzymes and the requisite stress response to relieve growth retardation.

In addition to the cellular response to cell wall active antibiotics, studies have examined the effects on gene transcription resulting from exposure to quinolones such as ciprofloxacin and trimethoprim (Cirz *et al.*, 2007). These antibiotics induce double-stranded breaks in the chromosome and stalled replication forks, which produce single-stranded DNA. ssDNA directly binds RecA which in turn binds to LexA catalysing its autoproteolysis thereby inducing the SOS response within the cell. This effect is not limited to quinolones and similar SOS induction is observed for β-lactams (Miller *et al.*, 2004) and rifamycins (Cirz *et al.*, 2005). Several studies have demonstrated that the antibiotic-induced SOS response via exposure to subinhibitory concentrations of ciprofloxacin, trimethoprim or β-lactams induced bacteriophage synthesis and horizontal gene transfer of virulence factors (Goerke *et al.*, 2006; Maiques *et al.*, 2006)). Moreover, it leads to the expression of bacteriophage encoded virulence factors.

Acid pH

During its lifecycle, *S. aureus* will be exposed to environmental fluctuations in pH. The skin is typically 1.5–2.5 pH units lower (Ehlers *et al.*, 2001) than that found in an abscess (Bessman *et al.*, 1989) or the tightly buffered conditions in blood (Robinson, 1975). A complex cellular response was observed for *S. aureus* cultured at a reduced pH. The transcription of *sae*, the virulence-associated, two-component signal transduction locus was shown to be significantly different under these conditions; a temporal transition in the transcription pattern for this locus did not occur at pH 5.5 but was observed to occur at pH 7.5 (Weinrick *et al.*, 2004). The excreted protein profiles were also substantially different for cells grown in media at pH 5 demonstrating a pleiotropic effect upon protein expression due to reduced pH. The finding that the classical *agr*-dependent postexponential induction of genes encoding virulence factors that occurs at pH 7.5 did not occur at pH 5.5 in the studied growth medium, which has potential implications for the *in vivo* lifestyle of *S. aureus*. In an attempt to rationalize the transcriptional response of the cell to a mild acid environment, the mild acid stimulon (MAS) was identified transcriptomically via microarray (Weinrick *et al.*, 2004). This large-scale analysis confirmed the altered regulation of genes encoding extracellular virulence factors. Moreover, the transcription of genes involved in transport of sugars and peptides, intermediary metabolism and pH homeostasis were identified to be modulated under these conditions. The analysis of the MAS by Weinrick *et al.* (2004) documented the effects of a continuous exposure to acid. This approach is distinct to the sudden imposition of an acid shock, whereby the cell must rapidly adapt to low pH. A recent microarray study investigated the response of *S. aureus* to acid shock with rapid acidification to pH 4.5 (Bore *et al.*, 2007). This revealed there was much overlap between the response to acid shock and the MAS. During acid shock there were observed to be specific changes that were not part of the MAS suggesting different responses in acid adapted and acid shocked cells. These distinct changes include, for shock, increased transcription of genes encoding enzymes in the pentose phosphate pathway which will enable NADH regeneration and increased transcription of oxidative stress genes *katA* and *sodA* (Bore *et al.*, 2007).

Temperature shift

The transcriptional response made by *S. aureus* to rapid temperature change was identified by microarray analysis (Anderson *et al.*, 2006). Distinct but overlapping changes in gene transcription were observed between the two temperature shifts (10°C cold shock or 42°C heat shock) when cultures were moved from 37°C. The study identified that a major factor affecting the response to this environmental change was an alteration in mRNA half-life (Anderson *et al.*, 2006). This was not accompanied by a thorough physiological investigation of changes concomitant with the temperature shift and downstream effects of stability on protein accumulation were not determined, making the net physiological consequence rather unclear.

The physiological contribution of several members of the Clp ATPases family to heat stress has been determined. Mutants of the *clpC* and *clpB* genes had an inability to grow at high temperature and an inability to induce thermotolerance, respectively. This identified that these Clp proteins provide protective functions to the cell at increased temperature and they were further shown to be negatively regulated by CtsR, the heat shock transcriptional regulator (Frees *et al.*, 2004). ClpC was proposed to target heat-denatured proteins for degradation by the ClpP peptidase. The ClpP proteolytic complex also serves to support growth at increased temperature (Frees *et al.*, 2004).

Starvation survival

S. aureus responds to nutrient limitation by development of a stable starvation state that is activated following glucose (carbon) or multiple nutrient starvation, but this state is not induced by amino acid or phosphate starvation (Watson *et al.*, 1998b). As the cells enter into the starvation state their size was observed to decrease (Fig. 7.5), they had reduced numbers of division septa, became more electron dense and had increased resistance to acid, heat and oxidizing agents. The maintenance of viability in this altered, starved state requires cell wall and protein synthesis since the addition of penicillin G or chloramphenicol, respectively, prevented culture of the organisms. Further analysis of protein expression using [^{35}S]methionine incorporation hinted at differential protein expression in cells entering and maintaining the starvation state compared to post-exponential phase grown cells (Watson *et al.*, 1998b). Moreover, there is evidence for the existence of long-lived transcripts maintained during starvation for more efficient recovery when nutrients become available again (Clements and Foster, 1998). The differentially expressed proteins and the regulatory processes

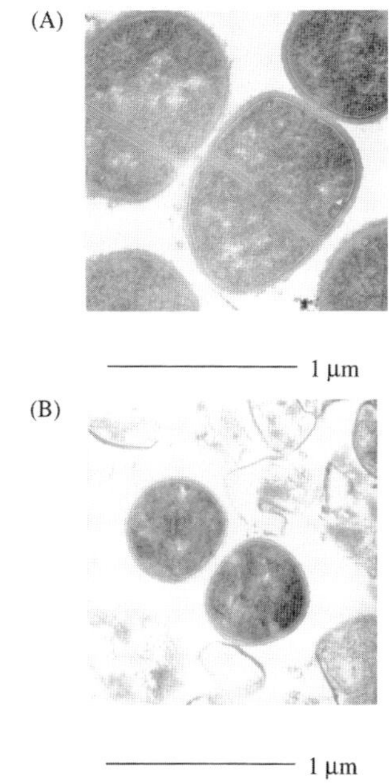

Figure 7.5 Cells of *S. aureus* growing without nutrient limitation (A) compared with those 25 days after glucose starvation (B) (Watson *et al.*, 1998b). Scale bars = 1 μm.

controlling the entry of cells into the starvation state have not been identified.

A screen for components required for starvation identified mutants with reduced recovery. Many of the components required for recovery from the starvation state were found to be associated with oxidative stress resistance and included superoxide dismutase A (Clements *et al.*, 1999a), catalase (Horsburgh *et al.*, 2001a), and cysteine synthase (Lithgow *et al.*, 2004). Further components required for starvation survival included the transcription factor RpoE, HprT for purine recycling and haem A synthase (Watson *et al.*, 1998a; Clements *et al.*, 1999b). Most of these starvation survival determinants were found to exert their effect in the recovery phase upon exit from the starved state, a time when the impact of endogenously generated oxidative stress from the resumption of metabolism will have a detrimental effect if the cell has inadequate resistance (Watson *et al.*, 1998b).

Future direction

A flurry of activity in the post-genome era of staphylococcal biology has dramatically increased our knowledge of environmental responses and survival mechanisms. An increasing use of microarray technology and the imminent take-up and development of applications using high-throughput sequencing technology is likely to mean that our understanding of gene transcription will develop faster in the future. In concert with genomics many new developments in proteomic technology will help elucidate response networks in the pathogen.

Several key areas demand better understanding for this pathogen. The range of factors to which the pathogen responds is still poorly understood. Indeed many two-component signal transduction systems, of which *S. aureus* has 16–17 dependent on the strain, have no known stimulus. Moreover, the study of response networks is performed *in vitro* in most cases so the analysis of *in vivo* gene expression remains an important goal for future. This is especially pertinent for designing inhibitors to response networks and thereby developing new treatments for this increasingly formidable pathogen.

References

Al-Maghrebi, M., Fridovich, I., and Benov, L. (2002). Manganese supplementation relieves the phenotypic deficits seen in superoxide-dismutase-null Escherichia coli. Arch. Biochem. Biophys. *402*, 104–109.

Allard, M., Moisan, H., Brouillette, E., Gervais, A. L., Jacques, M., Lacasse, P., Diarra, M. S., and Malouin, F. (2006). Transcriptional modulation of some *Staphylococcus aureus* iron-regulated genes during growth *in vitro* and in a tissue cage model *in vivo*. Microbes Infect. *8*, 1679–1690.

Anderson, K. L., Roberts, C., Disz, T., Vonstein, V., Hwang, K., Overbeek, R., Olson, P. D., Projan, S. J., and Dunman, P. M. (2006). Characterization of the *Staphylococcus aureus* heat shock, cold shock, stringent, and SOS responses and their effects on log-phase mRNA turnover. J. Bacteriol. *188*, 6739–6756.

Archibald, F. S., and Fridovich, I. (1981). Manganese, superoxide dismutase, and oxygen tolerance in some lactic acid bacteria. J. Bacteriol. *146*, 928–936.

Bearden, S. W., and Perry, R. D. (1999). The Yfe system of Yersinia pestis transports iron and manganese and is required for full virulence of plague. Mol. Microbiol. *32*, 403–414.

Berry, A. M., and Paton, J. C. (1996). Sequence heterogeneity of PsaA, a 37-kilodalton putative adhesin essential for virulence of Streptococcus pneumoniae. Infect. Immun. *64*, 5255–5262.

Bessman, A. N., Page, J., and Thomas, L. J. (1989). *In vivo* pH of induced soft-tissue abscesses in diabetic and nondiabetic mice. Diabetes *38*, 659–662.

Bischoff, M., Dunman, P., Kormanec, J., Macapagal, D., Murphy, E., Mounts, W., Berger-Bachi, B., and Projan, S. (2004). Microarray-based analysis of the *Staphylococcus aureus* sigmaB regulon. J. Bacteriol. *186*, 4085–4099.

Bore, E., Langsrud, S., Langsrud, O., Rode, T. M., and Holck, A. (2007). Acid-shock responses in *Staphylococcus aureus* investigated by global gene expression analysis. Microbiol. *153*, 2289–2303.

Brenot, A., Weston, B. F., and Caparon, M. G. (2007). A PerR-regulated metal transporter (PmtA) is an interface between oxidative stress and metal homeostasis in Streptococcus pyogenes. Mol. Microbiol. *63*, 1185–1196.

Bsat, N., Chen, L., and Helmann, J. D. (1996). Mutation of the *Bacillus subtilis* alkyl hydroperoxide reductase (ahpCF) operon reveals compensatory interactions among hydrogen peroxide stress genes. J. Bacteriol. *178*, 6579–6586.

Burnette-Curley, D., Wells, V., Viscount, H., Munro, C. L., Fenno, J. C., Fives-Taylor, P., and Macrina, F. L. (1995). FimA, a major virulence factor associated with Streptococcus parasanguis endocarditis. Infect. Immun. *63*, 4669–4674.

Chan, P. F., and Foster, S. J. (1998). Role of SarA in virulence determinant production and environmental signal transduction in *Staphylococcus aureus*. J. Bacteriol. *180*, 6232–6241.

Chan, P. F., Foster, S. J., Ingham, E., and Clements, M. O. (1998). The *Staphylococcus aureus* alternative sigma factor sigmaB controls the environmental stress response but not starvation survival or pathogenicity in a mouse abscess model. J. Bacteriol. *180*, 6082–6089.

Cirz, R. T., Chin, J. K., Andes, D. R., de Crecy-Lagard, V., Craig, W. A., and Romesberg, F. E. (2005). Inhibition of mutation and combating the evolution of antibiotic resistance. PLoS Biol. 3, e176.

Cirz, R. T., Jones, M. B., Gingles, N. A., Minogue, T. D., Jarrahi, B., Peterson, S. N., and Romesberg, F. E. (2007). Complete and SOS-mediated response of *Staphylococcus aureus* to the antibiotic ciprofloxacin. J. Bacteriol. *189*, 531–539.

Clauditz, A., Resch, A., Wieland, K. P., Peschel, A., and Gotz, F. (2006). Staphyloxanthin plays a role in the fitness of *Staphylococcus aureus* and its ability to cope with oxidative stress. Infect. Immun. *74*, 4950–4953.

Claverys, J. P. (2001). A new family of high-affinity ABC manganese and zinc permeases. Res. Microbiol. *152*, 231–243.

Clements, M. O., and Foster, S. J. (1998). Starvation recovery of *Staphylococcus aureus* 8325–4. Microbiol. *144 (Pt 7)*, 1755–1763.

Clements, M. O., and Foster, S. J. (1999). Stress resistance in *Staphylococcus aureus*. Trends Microbiol. *7*, 458–462.

Clements, M. O., Watson, S. P., and Foster, S. J. (1999a). Characterization of the major superoxide dismutase of *Staphylococcus aureus* and its role in starvation survival, stress resistance, and pathogenicity. J. Bacteriol. *181*, 3898–3903.

Clements, M. O., Watson, S. P., Poole, R. K., and Foster, S. J. (1999b). CtaA of *Staphylococcus aureus* is required for starvation survival, recovery, and cytochrome biosynthesis. J. Bacteriol. *181*, 501–507.

Cockayne, A., Hill, P. J., Powell, N. B., Bishop, K., Sims, C., and Williams, P. (1998). Molecular cloning of a 32-kilodalton lipoprotein component of a novel iron-regulated Staphylococcus epidermidis ABC transporter. Infect. Immun. *66*, 3767–3774.

Cosgrove, K., Coutts, G., Jonsson, I. M., Tarkowski, A., Kokai-Kun, J. F., Mond, J. J., and Foster, S. J. (2007). Catalase (KatA) and alkyl hydroperoxide reductase (AhpC) have compensatory roles in peroxide stress resistance and are required for survival, persistence, and nasal colonization in *Staphylococcus aureus*. J. Bacteriol. *189*, 1025–1035.

Cramton, S. E., Gerke, C., Schnell, N. F., Nichols, W. W., and Gotz, F. (1999). The intercellular adhesion (ica) locus is present in *Staphylococcus aureus* and is required for biofilm formation. Infect. Immun. *67*, 5427–5433.

Cramton, S. E., Ulrich, M., Gotz, F., and Doring, G. (2001). Anaerobic conditions induce expression of polysaccharide intercellular adhesin in *Staphylococcus aureus* and *Staphylococcus epidermidis*. Infect. Immun. *69*, 4079–4085.

Dale, S. E., Doherty-Kirby, A., Lajoie, G., and Heinrichs, D. E. (2004a). Role of siderophore biosynthesis in virulence of *Staphylococcus aureus*: identification and characterization of genes involved in production of a siderophore. Infect. Immun. *72*, 29–37.

Dale, S. E., Sebulsky, M. T., and Heinrichs, D. E. (2004b). Involvement of SirABC in iron-siderophore import in *Staphylococcus aureus*. J. Bacteriol. *186*, 8356–8362.

Ehlers, C., Ivens, U. I., Moller, M. L., Senderovitz, T., and Serup, J. (2001). Females have lower skin surface pH than men. A study on the surface of gender, forearm site variation, right/left difference and time of the day on the skin surface pH. Skin Res. Technol. *7*, 90–94.

Frees, D., Chastanet, A., Qazi, S., Sorensen, K., Hill, P., Msadek, T., and Ingmer, H. (2004). Clp ATPases are required for stress tolerance, intracellular replication and biofilm formation in *Staphylococcus aureus*. Mol. Microbiol. *54*, 1445–1462.

Friedman, D. B., Stauff, D. L., Pishchany, G., Whitwell, C. W., Torres, V. J., and Skaar, E. P. (2006). *Staphylococcus aureus* redirects central metabolism to increase iron availability. PLoS Pathog. *2*, e87.

Fuchs, S., Pane-Farre, J., Kohler, C., Hecker, M., and Engelmann, S. (2007). Anaerobic gene expression in *Staphylococcus aureus*. J. Bacteriol..

Gardete, S., Wu, S. W., Gill, S., and Tomasz, A. (2006). Role of VraSR in antibiotic resistance and antibiotic-induced stress response in *Staphylococcus aureus*. Antimicrob. Agents Chemother. *50*, 3424–3434.

Geng, H., Zhu, Y., Mullen, K., Zuber, C. S., and Nakano, M. M. (2007). Characterization of ResDE-dependent fnr transcription in *Bacillus subtilis*. J. Bacteriol. *189*, 1745–1755.

Giachino, P., Engelmann, S., and Bischoff, M. (2001). Sigma(B) activity depends on RsbU in *Staphylococcus aureus*. J. Bacteriol. *183*, 1843–1852.

Goerke, C., Koller, J., and Wolz, C. (2006). Ciprofloxacin and trimethoprim cause phage induction and virulence modulation in *Staphylococcus aureus*. Antimicrob. Agents Chemother. *50*, 171–177.

Grifantini, R., Frigimelica, E., Delany, I., Bartolini, E., Giovinazzi, S., Balloni, S., Agarwal, S., Galli, G., Genco, C., and Grandi, G. (2004). Characterization of a novel Neisseria meningitidis Fur and iron-regulated operon required for protection from oxidative stress: utility of DNA microarray in the assignment of the biological role of hypothetical genes. Mol. Microbiol. *54*, 962–979.

Herbig, A. F., and Helmann, J. D. (2001). Roles of metal ions and hydrogen peroxide in modulating the interaction of the *Bacillus subtilis* PerR peroxide regulon repressor with operator DNA. Mol. Microbiol. *41*, 849–859.

Horsburgh, M. J., Aish, J. L., White, I. J., Shaw, L., Lithgow, J. K., and Foster, S. J. (2002a). sigmaB modulates virulence determinant expression and stress resistance: characterization of a functional rsbU strain derived from *Staphylococcus aureus* 8325–4. J. Bacteriol. *184*, 5457–5467.

Horsburgh, M. J., Clements, M. O., Crossley, H., Ingham, E., and Foster, S. J. (2001a). PerR controls oxidative stress resistance and iron storage proteins and is required for virulence in *Staphylococcus aureus*. Infect. Immun. *69*, 3744–3754.

Horsburgh, M. J., Ingham, E., and Foster, S. J. (2001b). In *Staphylococcus aureus*, fur is an interactive regulator with PerR, contributes to virulence, and Is necessary for oxidative stress resistance through positive regulation of catalase and iron homeostasis. J. Bacteriol. *183*, 468–475.

Horsburgh, M. J., Wharton, S. J., Cox, A. G., Ingham, E., Peacock, S., and Foster, S. J. (2002b). MntR modulates expression of the PerR regulon and superoxide resistance in *Staphylococcus aureus* through control of manganese uptake. Mol. Microbiol. *44*, 1269–1286.

Horsburgh, M. J., Wharton, S. J., Karavolos, M., and Foster, S. J. (2002c). Manganese: elemental defence for a life with oxygen. Trends Microbiol. *10*, 496–501.

Jakubovics, N. S., and Jenkinson, H. F. (2001). Out of the iron age: new insights into the critical role of manganese homeostasis in bacteria. Microbiol. *147*, 1709–1718.

Jakubovics, N. S., Smith, A. W., and Jenkinson, H. F. (2000). Expression of the virulence-related Sca (Mn2+) permease in Streptococcus gordonii is regulated by a diphtheria toxin metallorepressor-like protein ScaR. Mol. Microbiol. *38*, 140–153.

Kamps, A., Achebach, S., Fedtke, I., Unden, G., and Gotz, F. (2004). Staphylococcal NreB: an O(2)-sensing histidine protein kinase with an O(2)-labile iron-sulphur cluster of the FNR type. Mol. Microbiol. *52*, 713–723.

Karavolos, M. H., Horsburgh, M. J., Ingham, E., and Foster, S. J. (2003). Role and regulation of the superoxide dismutases of *Staphylococcus aureus*. Microbiol. *149*, 2749–2758.

Kehres, D. G., Zaharik, M. L., Finlay, B. B., and Maguire, M. E. (2000). The NRAMP proteins of Salmonella typhimurium and Escherichia coli are selective manganese transporters involved in the response to reactive oxygen. Mol. Microbiol. *36*, 1085–1100.

Kolenbrander, P. E., Andersen, R. N., Baker, R. A., and Jenkinson, H. F. (1998). The adhesion-associated sca operon in Streptococcus gordonii encodes an inducible high-affinity ABC transporter for Mn2+ uptake. J. Bacteriol. *180*, 290–295.

Kullik, I., Giachino, P., and Fuchs, T. (1998). Deletion of the alternative sigma factor sigmaB in *Staphylococcus aureus* reveals its function as a global regulator of virulence genes. J. Bacteriol. *180*, 4814–4820.

Kullik, I. I., and Giachino, P. (1997). The alternative sigma factor sigmaB in *Staphylococcus aureus*: regulation of the sigB operon in response to growth phase and heat shock. Arch. Microbiol. *167*, 151–159.

Kuroda, M., Kuroda, H., Oshima, T., Takeuchi, F., Mori, H., and Hiramatsu, K. (2003). Two-component system VraSR positively modulates the regulation of cell-wall biosynthesis pathway in *Staphylococcus aureus*. Mol. Microbiol. *49*, 807–821.

Lee, J. W., and Helmann, J. D. (2006). The PerR transcription factor senses H2O2 by metal-catalysed histidine oxidation. Nature *440*, 363–367.

Lee, J. W., and Helmann, J. D. (2007). Functional specialization within the Fur family of metalloregulators. Biometals *20*, 485–499.

Lindsay, J. A., and Foster, S. J. (1999). Interactive regulatory pathways control virulence determinant production and stability in response to environmental conditions in *Staphylococcus aureus*. Mol. Gen. Genet. *262*, 323–331.

Lindsay, J. A., and Foster, S. J. (2001). zur: a Zn(2+)-responsive regulatory element of *Staphylococcus aureus*. Microbiol. *147*, 1259–1266.

Lithgow, J. K., Hayhurst, E. J., Cohen, G., Aharonowitz, Y., and Foster, S. J. (2004). Role of a cysteine synthase in *Staphylococcus aureus*. J. Bacteriol. *186*, 1579–1590.

Liu, G. Y., Essex, A., Buchanan, J. T., Datta, V., Hoffman, H. M., Bastian, J. F., Fierer, J., and Nizet, V. (2005). *Staphylococcus aureus* golden pigment impairs neutrophil killing and promotes virulence through its antioxidant activity. J. Exp. Med. *202*, 209–215.

Maalej, S., Dammak, I., and Dukan, S. (2006). The impairment of superoxide dismutase coordinates the derepression of the PerR regulon in the response of *Staphylococcus aureus* to HOCl stress. Microbiol. *152*, 855–861.

Maiques, E., Ubeda, C., Campoy, S., Salvador, N., Lasa, I., Novick, R. P., Barbe, J., and Penades, J. R. (2006). beta-lactam antibiotics induce the SOS response and horizontal transfer of virulence factors in *Staphylococcus aureus*. J. Bacteriol. *188*, 2726–2729.

Makui, H., Roig, E., Cole, S. T., Helmann, J. D., Gros, P., and Cellier, M. F. (2000). Identification of the Escherichia coli K-12 Nramp orthologue (MntH) as a selective divalent metal ion transporter. Mol. Microbiol. *35*, 1065–1078.

Maresso, A. W., and Schneewind, O. (2006). Iron acquisition and transport in *Staphylococcus aureus*. Biometals *19*, 193–203.

Masse, E., and Gottesman, S. (2002). A small RNA regulates the expression of genes involved in iron metabolism in Escherichia coli. Proc. Natl Acad. Sci. USA *99*, 4620–4625.

Massey, R. C., Horsburgh, M. J., Lina, G., Hook, M., and Recker, M. (2006). The evolution and maintenance of virulence in *Staphylococcus aureus*: a role for host-to-host transmission? Nat. Rev. Microbiol. *4*, 953–958.

McAleese, F., Wu, S. W., Sieradzki, K., Dunman, P., Murphy, E., Projan, S., and Tomasz, A. (2006). Overexpression of genes of the cell wall stimulon in clinical isolates of *Staphylococcus aureus* exhibiting vancomycin-intermediate- *S. aureus*-type resistance to vancomycin. J. Bacteriol. *188*, 1120–1133.

McCallum, N., Spehar, G., Bischoff, M., and Berger-Bachi, B. (2006). Strain dependence of the cell wall-damage induced stimulon in *Staphylococcus aureus*. Biochim. Biophys. Acta *1760*, 1475–1481.

McKenney, D., Hubner, J., Muller, E., Wang, Y., Goldmann, D. A., and Pier, G. B. (1998). The ica locus of Staphylococcus epidermidis encodes production of the capsular polysaccharide/adhesin. Infect. Immun. *66*, 4711–4720.

Meiwes, J., Fiedler, H. P., Haag, H., Zahner, H., Konetschny-Rapp, S., and Jung, G. (1990). Isolation and characterization of staphyloferrin A, a compound with siderophore activity from Staphylococcus hyicus DSM 20459. FEMS Microbiol. Lett. *55*, 201–205.

Miller, C., Thomsen, L. E., Gaggero, C., Mosseri, R., Ingmer, H., and Cohen, S. N. (2004). SOS response induction by beta-lactams and bacterial defense against antibiotic lethality. Science *305*, 1629–1631.

Miller, R. A., and Britigan, B. E. (1997). Role of oxidants in microbial pathophysiology. Clin. Microbiol. Rev. *10*, 1–18.

Miyazaki, E., Chen, J. M., Ko, C., and Bishai, W. R. (1999). The *Staphylococcus aureus* rsbW (orf159) gene encodes an anti-sigma factor of SigB. J. Bacteriol. *181*, 2846–2851.

Modun, B., Evans, R. W., Joannou, C. L., and Williams, P. (1998). Receptor-mediated recognition and uptake of iron from human transferrin by *Staphylococcus aureus* and *Staphylococcus epidermidis*. Infect. Immun. *66*, 3591–3596.

Modun, B., Kendall, D., and Williams, P. (1994). Staphylococci express a receptor for human transferrin: identification of a 42-kilodalton cell wall transferrin-binding protein. Infect. Immun. *62*, 3850–3858.

Moore, C. M., Gaballa, A., Hui, M., Ye, R. W., and Helmann, J. D. (2005). Genetic and physiological responses of *Bacillus subtilis* to metal ion stress. Mol. Microbiol. *57*, 27–40.

Morrissey, J. A., Cockayne, A., Brummell, K., and Williams, P. (2004). The staphylococcal ferritins are differentially regulated in response to iron and manganese and via PerR and Fur. Infect. Immun. *72*, 972–979.

Nair, S. P., Bischoff, M., Senn, M. M., and Berger-Bachi, B. (2003). The sigma B regulon influences internalization of *Staphylococcus aureus* by osteoblasts. Infect. Immun. *71*, 4167–4170.

Nakano, M. M., Zuber, P., Glaser, P., Danchin, A., and Hulett, F. M. (1996). Two-component regulatory proteins ResD-ResE are required for transcriptional activation of fnr upon oxygen limitation in *Bacillus subtilis*. J. Bacteriol. *178*, 3796–3802.

Nicholas, R. O., Li, T., McDevitt, D., Marra, A., Sucoloski, S., Demarsh, P. L., and Gentry, D. R. (1999). Isolation and characterization of a sigB deletion mutant of *Staphylococcus aureus*. Infect. Immun. *67*, 3667–3669.

Palma, M., and Cheung, A. L. (2001). sigma(B) activity in *Staphylococcus aureus* is controlled by RsbU and an additional factor(s) during bacterial growth. Infect. Immun. *69*, 7858–7865.

Pane-Farre, J., Jonas, B., Forstner, K., Engelmann, S., and Hecker, M. (2006). The sigmaB regulon in *Staphylococcus aureus* and its regulation. Int. J. Med. Microbiol. *296*, 237–258.

Park, M. K., Myers, R. A., and Marzella, L. (1992). Oxygen tensions and infections: modulation of microbial growth, activity of antimicrobial agents, and immunologic responses. Clin. Infect. Dis. *14*, 720–740.

Park, R. Y., Sun, H. Y., Choi, M. H., Bai, Y. H., and Shin, S. H. (2005). *Staphylococcus aureus* siderophore-mediated iron-acquisition system plays a dominant and essential role in the utilization of transferrin-bound iron. J. Microbiol. *43*, 183–190.

Patzer, S. I., and Hantke, K. (2001). Dual repression by Fe(2+)-Fur and Mn(2+)-MntR of the mntH gene, encoding an NRAMP-like Mn(2+) transporter in Escherichia coli. J. Bacteriol. *183*, 4806–4813.

Posey, J. E., Hardham, J. M., Norris, S. J., and Gherardini, F. C. (1999). Characterization of a manganese-dependent regulatory protein, TroR, from Treponema pallidum. Proc. Natl Acad. Sci. USA *96*, 10887–10892.

Pragman, A. A., Ji, Y., and Schlievert, P. M. (2007). Repression of *Staphylococcus aureus* SrrAB using inducible antisense srrA alters growth and virulence factor transcript levels. Biochemistry *46*, 314–321.

Que, Q., and Helmann, J. D. (2000). Manganese homeostasis in *Bacillus subtilis* is regulated by MntR, a bifunctional regulator related to the diphtheria toxin repressor family of proteins. Mol. Microbiol. 35, 1454–1468.

Rea, R., Hill, C., and Gahan, C. G. (2005). Listeria monocytogenes PerR mutants display a small-colony phenotype, increased sensitivity to hydrogen peroxide, and significantly reduced murine virulence. Appl. Environ. Microbiol. *71*, 8314–8322.

Repine, J. E., Fox, R. B., and Berger, E. M. (1981). Hydrogen peroxide kills *Staphylococcus aureus* by reacting with staphylococcal iron to form hydroxyl radical. J. Biol. Chem. *256*, 7094–7096.

Ricci, S., Janulczyk, R., and Bjorck, L. (2002). The regulator PerR is involved in oxidative stress response and iron homeostasis and is necessary for full virulence of Streptococcus pyogenes. Infect. Immun. *70*, 4968–4976.

Richardson, A. R., Dunman, P. M., and Fang, F. C. (2006). The nitrosative stress response of *Staphylococcus aureus* is required for resistance to innate immunity. Mol. Microbiol. *61*, 927–939.

Robinson, J. R. (1975). Fundamentals of acid-base regulation, 5th edn (London, England.: Blackwell Scientific).

Ross, R. A., and Onderdonk, A. B. (2000). Production of toxic shock syndrome toxin 1 by *Staphylococcus aureus* requires both oxygen and carbon dioxide. Infect. Immun. *68*, 5205–5209.

Sabri, M., Leveille, S., and Dozois, C. M. (2006). A SitABCD homologue from an avian pathogenic Escherichia coli strain mediates transport of iron and manganese and resistance to hydrogen peroxide. Microbiol. *152*, 745–758.

Schapiro, J. M., Libby, S. J., and Fang, F. C. (2003). Inhibition of bacterial DNA replication by zinc mobilization during nitrosative stress. Proc. Natl Acad. Sci. USA *100*, 8496–8501.

Sebulsky, M. T., Shilton, B. H., Speziali, C. D., and Heinrichs, D. E. (2003). The role of FhuD2 in iron(III)-hydroxamate transport in *Staphylococcus aureus*. Demonstration that FhuD2 binds iron(III)-hydroxamates but with minimal conformational change and implication of mutations on transport. J. Biol. Chem. *278*, 49890–49900.

Senn, M. M., Giachino, P., Homerova, D., Steinhuber, A., Strassner, J., Kormanec, J., Fluckiger, U., Berger-Bachi, B., and Bischoff, M. (2005). Molecular analysis and organization of the sigmaB operon in *Staphylococcus aureus*. J. Bacteriol. *187*, 8006–8019.

Shaw, L. N., Aish, J., Davenport, J. E., Brown, M. C., Lithgow, J. K., Simmonite, K., Crossley, H., Travis, J., Potempa, J., and Foster, S. J. (2006). Investigations into sigmaB-modulated regulatory pathways governing extracellular virulence determinant production in *Staphylococcus aureus*. J. Bacteriol. *188*, 6070–6080.

Singh, K. V., Coque, T. M., Weinstock, G. M., and Murray, B. E. (1998). *In vivo* testing of an *Enterococcus faecalis* efaA mutant and use of efaA homologs for species identification. FEMS Immunol. Med. Microbiol. *21*, 323–331.

Singh, V. K., Jayaswal, R. K., and Wilkinson, B. J. (2001). Cell wall-active antibiotic induced proteins of *Staphylococcus aureus* identified using a proteomic approach. FEMS Microbiol. Lett. *199*, 79–84.

Skaar, E. P., Gaspar, A. H., and Schneewind, O. (2004a). IsdG and IsdI, heme-degrading enzymes in the cytoplasm of *Staphylococcus aureus*. J. Biol. Chem. *279*, 436–443.

Skaar, E. P., Humayun, M., Bae, T., DeBord, K. L., and Schneewind, O. (2004b). Iron-source preference of *Staphylococcus aureus* infections. Science *305*, 1626–1628.

Stauff, D. L., Torres, V. J., and Skaar, E. P. (2007). Signaling and DNA binding activities of the *staphylococcus aureus* HssR-HssS two-component system required for heme sensing. J. Biol. Chem. *282*, 26111–26121.

Sun, G., Sharkova, E., Chesnut, R., Birkey, S., Duggan, M. F., Sorokin, A., Pujic, P., Ehrlich, S. D., and Hulett, F. M. (1996). Regulators of aerobic and anaerobic respiration in *Bacillus subtilis*. J. Bacteriol. *178*, 1374–1385.

Taylor, J. M., and Heinrichs, D. E. (2002). Transferrin binding in *Staphylococcus aureus*: involvement of a cell wall-anchored protein. Mol. Microbiol. *43*, 1603–1614.

Throup, J. P., Zappacosta, F., Lunsford, R. D., Annan, R. S., Carr, S. A., Lonsdale, J. T., Bryant, A. P., McDevitt, D., Rosenberg, M., and Burnham, M. K. (2001). The srhSR gene pair from *Staphylococcus aureus*: genomic and proteomic approaches to the identification and characterization of gene function. Biochemistry *40*, 10392–10401.

Torres, V. J., Pishchany, G., Humayun, M., Schneewind, O., and Skaar, E. P. (2006). *Staphylococcus aureus* IsdB is a hemoglobin receptor required for heme iron utilization. J. Bacteriol. *188*, 8421–8429.

Tuveson, R. W., Larson, R. A., and Kagan, J. (1988). Role of cloned carotenoid genes expressed in Escherichia coli in protecting against inactivation by near-UV light and specific phototoxic molecules. J. Bacteriol. *170*, 4675–4680.

Ulrich, M., Bastian, M., Cramton, S. E., Ziegler, K., Pragman, A. A., Bragonzi, A., Memmi, G., Wolz, C., Schlievert, P. M., Cheung, A., and Doring, G. (2007). The staphylococcal respiratory response regulator SrrAB induces ica gene transcription and polysaccharide intercellular adhesin expression, protecting *Staphylococcus aureus* from neutrophil killing under anaerobic growth conditions. Mol. Microbiol. *65*, 1276–1287.

Utaida, S., Dunman, P. M., Macapagal, D., Murphy, E., Projan, S. J., Singh, V. K., Jayaswal, R. K., and Wilkinson, B. J. (2003). Genome-wide transcriptional profiling of the response of *Staphylococcus aureus* to cell-wall-active antibiotics reveals a cell-wall-stress stimulon. Microbiol. *149*, 2719–2732.

Valderas, M. W., and Hart, M. E. (2001). Identification and characterization of a second superoxide dismutase gene (sodM) from *Staphylococcus aureus*. J. Bacteriol. *183*, 3399–3407.

Verneuil, N., Rince, A., Sanguinetti, M., Posteraro, B., Fadda, G., Auffray, Y., Hartke, A., and Giard, J. C. (2005). Contribution of a PerR-like regulator to the oxidative-stress response and virulence of *Enterococcus faecalis*. Microbiol. *151*, 3997–4004.

Watson, S. P., Antonio, M., and Foster, S. J. (1998a). Isolation and characterization of *Staphylococcus aureus* starvation-induced, stationary-phase mutants defective in survival or recovery. Microbiology *144*, 3159–3169.

Watson, S. P., Clements, M. O., and Foster, S. J. (1998b). Characterization of the starvation-survival response of *Staphylococcus aureus*. J. Bacteriol. *180*, 1750–1758.

Weinrick, B., Dunman, P. M., McAleese, F., Murphy, E., Projan, S. J., Fang, Y., and Novick, R. P. (2004). Effect of mild acid on gene expression in *Staphylococcus aureus*. J. Bacteriol. *186*, 8407–8423.

Wu, R., Skaar, E. P., Zhang, R., Joachimiak, G., Gornicki, P., Schneewind, O., and Joachimiak, A. (2005). *Staphylococcus aureus* IsdG and IsdI, heme-degrading enzymes with structural similarity to monooxygenases. J. Biol. Chem. *280*, 2840–2846.

Yarwood, J. M., McCormick, J. K., and Schlievert, P. M. (2001). Identification of a novel two-component regulatory system that acts in global regulation of virulence factors of *Staphylococcus aureus*. J. Bacteriol. *183*, 1113–1123.

Yarwood, J. M., and Schlievert, P. M. (2000). Oxygen and carbon dioxide regulation of toxic shock syndrome toxin 1 production by *Staphylococcus aureus* MN8. J. Clin. Microbiol. *38*, 1797–1803.

Yin, S., Daum, R. S., and Boyle-Vavra, S. (2006). VraSR two-component regulatory system and its role in induction of pbp2 and vraSR expression by cell wall antimicrobials in *Staphylococcus aureus*. Antimicrob. Agents Chemother. *50*, 336–343.

Yocum, C. F., and Pecoraro, V. L. (1999). Recent advances in the understanding of the biological chemistry of manganese. Curr. Opin. Chem. Biol. *3*, 182–187.

Ziebandt, A. K., Weber, H., Rudolph, J., Schmid, R., Hoper, D., Engelmann, S., and Hecker, M. (2001). Extracellular proteins of *Staphylococcus aureus* and the role of SarA and sigma B. Proteomics *1*, 480–493.

Mechanisms of β-Lactam and Glycopeptide Resistance in *Staphylococcus aureus*

8

Mariana G. Pinho

Abstract

Worldwide spread of multidrug resistant *Staphylococcus aureus* poses serious challenges to chemotherapy, which were recently emphasized by the appearance of vancomycin resistant *S. aureus* (VRSA) strains. Vancomycin has become the standard therapeutic agent against methicillin-resistant *S. aureus* (MRSA) strains, for which the choice of treatment is limited by the accumulation of a number of other antibiotic resistance markers acquired during recent evolution of the staphylococcal genome. This chapter summarizes specific resistance mechanisms against β-lactam and glycopeptide antibiotics, as well as the mechanisms for synergism between these two classes of antibiotics, focusing on the molecular aspects as well as on the whole-cell response to the presence of antibiotics that target cell wall synthesis.

Introduction

Staphylococcus aureus is a major human pathogen, capable of causing a range of illnesses from minor skin infections to life-threatening infections such as toxic shock, endocarditis, severe pneumonia or septicaemia. It is an extraordinarily versatile bacteria in terms of its number of virulence factors as well its ability to adapt to rapidly changing environments, especially to the ever-increasing antibiotic selective pressure. The importance of antimicrobial chemotherapy to treat *S. aureus* infections can not be overstated since, if left untreated, some staphylococcal infections have mortality rates of approximately 90% (Smith and Vickers, 1960)

The first battle against *S. aureus* infections, as well as most other bacterial pathogens, was won with the introduction of penicillin into clinical practice in the early 1940s, which greatly reduced mortality due to bacterial infections. However, within a short time, resistant bacteria that failed to respond to penicillin treatment were identified. The first report on extracts from bacteria that could destroy penicillin was actually published in 1940, even prior to wide use of this antibiotic as a therapeutic agent (Abraham and Chain, 1940). Penicillinase (later called β-lactamase) producing strains of *S. aureus*, able to hydrolyse penicillin, quickly became endemic in many hospitals, and later found their way into the community. Penicillin therefore became ineffective for the treatment of most staphylococcal infections, prompting the development of penicillinase-resistant β-lactams, such as methicillin. By the late 1960s, more than 80% of both hospital and community-acquired *S. aureus* isolates were resistant to penicillin, and that number has since increased to over 90% of *S. aureus* isolates (Projan, 2000; Lowy, 2003).

Methicillin was first introduced in Europe in 1959 as an antimicrobial agent but, once again, it was not long before the first staphylococci isolates exhibiting resistance to methicillin were reported (Barber, 1961; Jevons, 1961). This type of resistance was effective against all β-lactam antibiotics and was termed 'intrinsic' because it did not involve the degradation of the antibiotic by a β-lactamase (Seligman, 1966).

Since then, MRSA clones have spread across borders in waves of clonal dissemination and started to cause serious hospitals infections worldwide. *S. aureus* is now one of the leading causes of nosocomial infections. The incidence of MRSA strains varies depending on the country and hospital, but it can correspond to 60–80% of all *S. aureus* isolates in some hospitals (Lowy, 2003; Oliveira *et al.*, 2002). The problem of MRSA dissemination has been a critical one for clinicians, as the therapeutic outcome from MRSA infections is worse than from methicillin-sensitive *S. aureus* (MSSA) infections (Cosgrove *et al.*, 2003). One of the reasons for an increase in mortality associated with MRSA bacteraemia may be the requirement for use of less effective drugs to treat these infections, rather than enhanced virulence of MRSA strains (Lowy, 2003).

The pattern of MRSA spread is reminiscent of penicillin resistance a few decades earlier: although initially confined to hospitals, since the late 1990s, MRSA strains have also emerged in the community (CA-MRSA), leading to serious and sometimes lethal infections in previously healthy individuals, without known risk factors or health-care contact (Chambers, 2001).

The scenario of antibiotic resistance in staphylococci is aggravated by the fact that the genetic background of the MRSA clones spread worldwide, seems to be particularly suited for the accumulation of further resistance markers. In fact *S. aureus* could be considered to be a paradigm among bacteria for the quote 'Given sufficient time and drug use, antibiotic resistance will emerge' (Levy, 1998). Multidrug resistance is now the norm for clinical isolates of this pathogen, which has developed resistance mechanisms against virtually all classes of available antibiotics. The glycopeptide antibiotic vancomycin is one of the few drugs that have remained effective against MRSA, becoming the preferred agent of treatment for severe MRSA infections. However, in 1997, the first isolates with decreased susceptibility to vancomycin were identified in Japan (Hiramatsu *et al.*, 1997) and have been detected in different continents since then (Gemmell, 2004). Although the minimum inhibitory concentrations (MICs) of these vancomycin intermediate resistant *S. aureus* (VISA) strains is limited to the range 8–16 μg/ml, it is nevertheless sufficient to cause complications in therapy including treatment failure (Sieradzki *et al.*, 1999b; Hiramatsu, 2001). More recently, in 2002, the first highly vancomycin resistant *S. aureus* (VRSA) strains were isolated in two hospitals in the United States, with MICs over 32μg/ml (CDC, 2002; Sievert *et al.*, 2002).

MRSA and VRSA strains currently constitute the major challenges for successful chemotherapy of *S. aureus* infections and this chapter will address the current knowledge of the genetic basis of resistance mechanisms to methicillin and vancomycin.

Mode of action of β-lactam and glycopeptide antibiotics

The structural integrity and shape of bacterial cells is maintained by the cell wall, a large mesh-like polymer which surrounds the cell and is essential for cell viability. The main constituent of the cell wall is the peptidoglycan, also called murein, a polymer made up of long glycan chains which are cross-linked via flexible peptide bridges to form a strong but elastic structure.

Muropeptides, the basic units of peptidoglycan, are synthesized in the cytoplasm as pentapeptide chains linked to disaccharides composed of *N*-acetylglucosamine (GlcNAc) and *N*-acetylmuramic acid (MurNAc) (see insert in Fig. 8.1). This precursor is linked to the lipid-carrier bactoprenol (C55 in Fig. 8.1), on the inside of the cellular membrane. In *S. aureus* a five glycine cross-bridge, used for cross-linking different peptides, is attached to the third amino acid – lysine – on the pentapeptide chain, by FemXAB proteins. The precursor is then translocated to outside of the cell, presumably with the aid of an unidentified flipase, and incorporated into the growing network of peptidoglycan by the penicillin-binding proteins (PBPs). PBPs are enzymes that catalyse the transglycosylation and transpeptidation reactions responsible for the formation of the glycosidic and peptide bonds of the peptidoglycan, respectively. These reactions occur mainly, if not exclusively, at the division septum of *S. aureus* (Pinho and Errington, 2003).

In the transglycosylation reaction, which leads to the formation of the glycan strands, the reducing end of the MurNAc of the nascent lipid-linked peptidogycan strand is linked to the

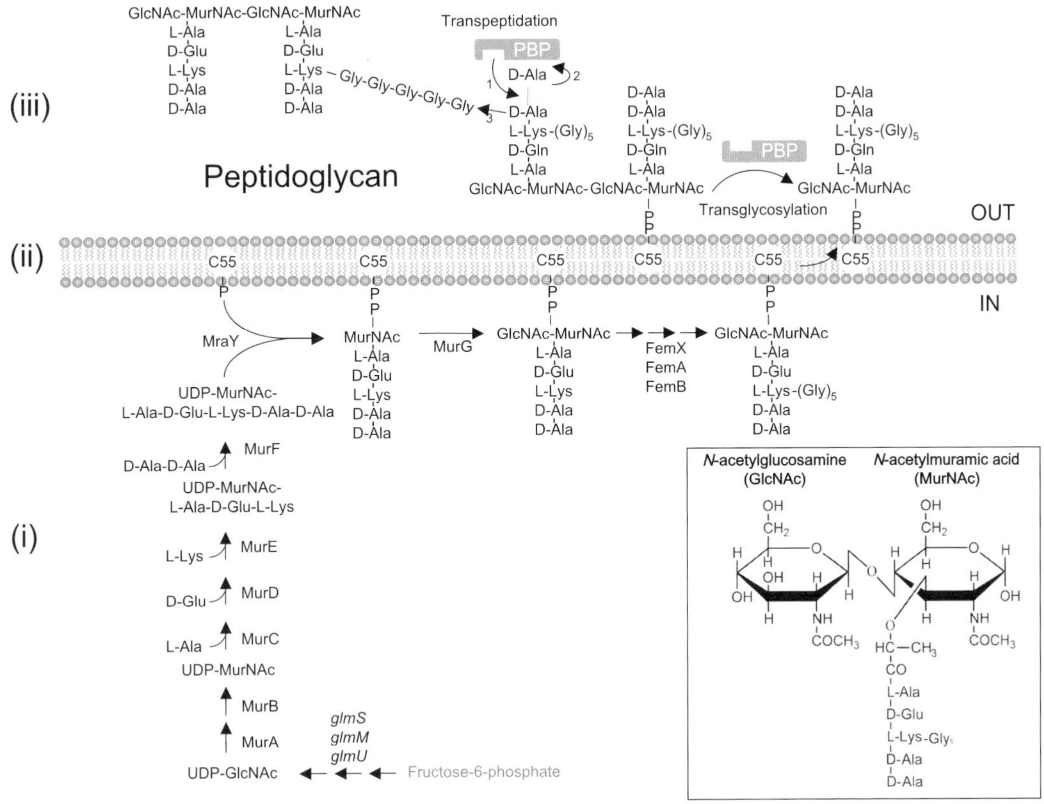

Figure 8.1 Peptidoglycan biosynthesis. The biosynthesis of peptidoglycan can be divided in three different stages: (i) the formation of nucleotide sugar linked precursors UDP-GlcNAc and UDP-N-acetylmuramyl-pentapeptide, that takes place in the cytoplasm, (ii) the transfer of phospho-N-acetylmuramyl-pentapeptide and UDP-GlcNAc to a lipid carrier, undecaprenyl phosphate or bactoprenol (C55-P), to yield disaccharide-(pentapeptide)-pyrophosphate-undecaprenol, that takes place in the cytoplasmic membrane and (iii) the incorporation of this completed subunit into the growing peptidoglycan by transglycosylation and transpeptidation, that takes place at the extracellular side of the cytoplasmic membrane. In the transglycosylation reaction for the formation of glycan strands, the nascent lipid-linked peptidoglycan strand is transferred onto the C-4 carbon of the glucosamine residue of the lipid-linked peptidoglycan precursor, with the release of undecaprenyl-pyrophosphate. During transpeptidation, the reaction inhibited by β-lactam antibiotics, the D-Ala-D-Ala bond is first cleaved, providing the energy necessary for the reaction to occur, and an enzyme-substrate intermediate is formed (1), with the concomitant release of the terminal D-Ala (2). A second step involves the transfer of the peptidyl moiety to an acceptor which is the last glycine of the cross-bridge in the case of *S. aureus* (3).

glucosamine residue of the lipid-linked peptidoglycan precursor, with concomitant release of the lipid carrier (Fig. 8.1). In the transpeptidation reaction, which results in cross-linking of the peptidoglycan, the D-Ala-D-Ala bond of one stem peptide is first cleaved and an enzyme-substrate intermediate is formed, releasing the terminal D-Ala. A subsequent step involves the transfer of the peptidyl moiety to the final glycine of the pentaglycine cross-bridge. This reaction results in the formation of a peptide bond between the D-$Ala_{(4)}$ of a donor peptide and the amino group of the glycine of an acceptor peptide (Fig. 8.1).

β-Lactam antibiotics, of which penicillin and oxacillin are examples (Fig. 8.2), inhibit the transpeptidation reaction by acting as suicide substrate homologues that bind almost irreversibly to the transpeptidase active site of PBPs, resulting in the formation of a stable ester-linked acyl enzyme which virtually lacks catalytic activity (Ghuysen, 1991). Therefore, when grown in the presence of β-lactams, *S. aureus* cells are unable to form pentaglycine cross-links and the cell wall loses its integrity. The β-lactam antibiotic methicillin is no longer commercially available and was replaced in clinical practice and in *in vitro* studies

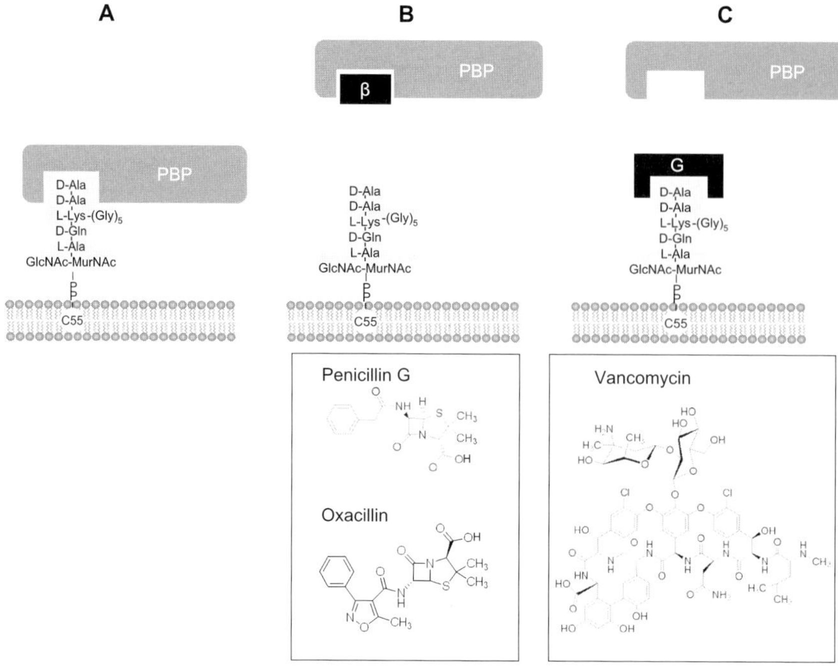

Figure 8.2 Schematic representation of the mode of action of β-lactams and glycopeptides and their chemical structures. Penicillin-binding proteins (PBPs), enzymes that catalyse the last steps of bacterial cell wall biosynthesis, bind to the terminal D-Alanyl-D-alanine residues of peptidoglycan precursors to carry out transpeptidation (A). β-lactam antibiotics (β) are structurally analogous to D-Ala-D-Ala. PBPs can react with these antibiotics by cleaving the β-lactam bond and forming a stable penicilloyl-enzyme intermediate that does not react further. Acylation of PBPs by β-lactams therefore inactivates the enzyme (B). Glycopeptides (G) inhibit cell wall synthesis by a completely different mechanism. They bind the terminal D-Ala-D-Ala residues of the peptidoglycan precursor preventing access of PBPs to their natural substrate (C). The chemical structures of β-lactam antibiotics penicillin G and oxacillin and the glycopeptide antibiotic vancomycin are also shown.

by oxacillin, which will be mentioned in place of methicillin throughout most of this text, as it is identical to methicillin regarding the mode of action and the bacterial mechanisms of resistance against it.

Clinically used glycopeptide antibiotics, of which vancomycin is the most relevant example, also inhibit cell wall synthesis but by a completely different mechanism: they bind to the C-terminal D-Ala-D-Ala residues of the externally oriented lipid-linked peptidoglycan precursor. Interaction between D-Ala-D-Ala residues and glycopeptide antibiotics involves the formation of five hydrogen bonds and results in the formation of a non-covalent but stable complex, that effectively sequesters the D-Ala-D-Ala dipeptide (Pootoolal *et al.*, 2002). This complex physically prevents the use of the peptidoglycan precursor by transpeptidases, i.e. it blocks the access of PBPs to their natural substrates (Fig. 8.2). Glycopeptide antibiotics may also inhibit transglycosylation, an effect that is independent of binding to the D-Ala-D-Ala residues and which may be dependent on direct binding to transglycosylase proteins (Ge *et al.*, 1999).

Mechanism of resistance to β-lactam antibiotics

The *mecA* gene

S. aureus has two primary β-lactam resistance mechanisms: the expression of β-lactamase enzymes, encoded by the *blaZ* gene, which hydrolyse β-lactams such as penicillin; and expression of PBP2A (also called PBP2′), encoded by the *mecA* gene, which is responsible for higher level β-lactam resistance, including against penicillinase-resistant antibiotics such as oxacillin. It is this second mechanism that will be the focus of this chapter, as it is responsible for the blanket resistance against all β-lactam antibiotics in MRSA strains, and therefore that causes a challenge to chemotherapy of staphylococcal infections. The intrinsic resistance to all β-lactam antibiotics delivered by PBP2A, results from the fact that this protein, in contrast to the four native PBPs of *S. aureus*, has very low affinity for β-lactam antibiotics and therefore its transpeptidase domain remains active in the presence of otherwise inhibitory concentrations of β-lactams.

PBP2A is a high molecular weight class B PBP, with three domains: an N-terminal membrane anchor, a central non-penicillin binding domain of unknown function and a C-terminal transpeptidase domain containing the three conserved motifs characteristic of transpeptidases – SXXK (where S is the active site serine), (S/Y)X(N/C) and (K/H)(T/S) G (Goffin and Ghuysen, 1998). Comparison of the kinetic parameters of PBP2A with β-lactam sensitive PBPs indicates that the β-lactam resistance of PBP2A is mainly due to inefficient formation of the acyl-PBP intermediate, and not due to a more rapid breakdown of the acyl-PBP intermediate (Lu *et al.*, 1999). The structure of the soluble domain of PBP2A revealed that although the C-terminal domain has a folding pattern typical of PBP transpeptidases, it has some unique structural deviations that may explain its low affinity for β-lactams. The main difference is that the active site motif containing the nucleophilic serine (Ser^{403}) is located on an α-helix sequestered within an extended narrow groove. The poor acylation rate of PBP2A is due to this distorted active site which must undergo a conformational change for acylation to occur (Lim and Strynadka, 2002).

The origin of PBP2A is unknown. It is not native to *S. aureus* species, as there is no allelic equivalent of *mecA* in methicillin susceptible *S. aureus* strains. Clues to the extraspecies origin of this gene came from the identification of a *mecA* homologue, present in each one of a large number of independent isolates of *Staphylococcus sciuri*, which was suggested to be an evolutionary ancestor of the *S. aureus mecA* (Couto *et al.*, 1996; Wu *et al.*, 1996b). The *mecA* gene from *S. sciuri* and *S. aureus* show 88% amino acid similarity and this value increases to 96% within the transpeptidase domains (Wu *et al.*, 1996b). Despite this similarity, strains of *S. sciuri* carrying the *mecA* homologue showed no appreciable degree of methicillin resistance, so presumably this gene, native of *S. sciuri*, has a physiological function that is not related to antibiotic resistance (Couto *et al.*, 1996). However *S. sciuri mecA* can be recruited to become an antibiotic resistance determinant under selective pressure: *S. sciuri* mutants selected in the presence of increasing concentrations of antibiotic had a drastic increase in the transcription rate of the *mecA* homologue, due to a point mutation in the promoter. This 'activated' gene was able to confer some degree of resistance when introduced into a *S. aureus* strain (Wu *et al.*, 2001; Couto *et al.*, 2003), constituting further evidence in favour of the relationship between the two *mecA* homologues. It should be noted however, that the difference at the sequence level is high enough to indicate that *S. aureus* did not acquire *mecA* directly from *S. sciuri* strains.

The *S. aureus mecA* gene is located on a mobile genetic element, the staphylococcal cassette chromosome *mec* (SCC*mec*), which is horizontally transferable among staphylococcal species (Katayama *et al.*, 2000). Up until now, six different types of SCC*mec* cassettes have been described (Ito *et al.*, 1999; Ito *et al.*, 2001; Ma *et al.*, 2002; Ito *et al.*, 2004; Hanssen and Ericson Sollid, 2006; Oliveira *et al.*, 2006), which differ in size (from 21 to 67 kb), in genetic composition and may include the presence of additional drug-resistance genes. Common features to SCC*mec* cassettes include the presence of the *mecA* gene, with a functional or non-functional *mecI-mecR1* regulatory region and a set of unique recombinase genes, *ccrA* and *ccrB*, which are required for both its excision from and integration into the *S. aureus* chromosome (Katayama *et al.*, 2000). SCC*mec* type V is an exception as it carries a single recombinase, *ccrC* (Ito *et al.*, 2004). Despite the potential mobility, the presence of *mecA* is restricted to a few major MRSA clones, within five groups of related genotypes (Enright *et al.*, 2002). It appears that the *mec* element has been imported into *S. aureus* on more than one, but in a limited number of occasions, followed by horizontal transfer of the SCC*mec* into different lineages (Enright *et al.*, 2002; Oliveira *et al.*, 2002). However, there is increasing evidence of recombination, reorganization and transfer of SCC*mec* elements in recent years (Chapter 3).

Interestingly, MSSA clinical isolates that belong to major MRSA lineages are more transformable with a plasmid containing the *mecA* gene and better able to maintain the plasmid and express resistance than MSSA strains from other lineages, suggesting that some backgrounds are more easily adaptable to whatever fitness cost might be imposed by acquisition of SCC*mec* or *mecA* (Katayama *et al.*, 2005). However the factors that contribute to an enhanced epidemicity – defined as potential for colonization, invasion and spread – of a few MRSA clones are still not fully understood.

Regulation of the *mecA* gene

The *mecA* gene can be regulated by two different regulatory systems, encoding similar sensor-transducer and repressor proteins. One is the regulatory element *mecR1-mecI*, which is located upstream of *mecA* in the SCC*mec*, and the other is the *blaR1-blaI* element, located upstream of the structural gene *blaZ* which encodes a β-lactamase. In both cases the regulatory genes are divergently transcribed from the structural genes. MecI and BlaI are repressors that can each block the transcription of both *mecA* and *blaZ*, by binding as homodimers to palindromic sites within the promoter-operator regions of these genes (Hackbarth and Chambers, 1993; Gregory *et al.*, 1997; Sharma *et al.*, 1998; Lewis and Dyke, 2000). MecR1 and BlaR1 are the sensor-transducers which bind β-lactam antibiotics, generating a transmembrane signal that results in the removal of the repressor from DNA binding sites, inducing transcription of both structural and regulatory genes (Gregory *et al.*, 1997; Hardt *et al.*, 1997). This signalling mechanism has been studied in more detail with the *bla* genes (Zhang *et al.*, 2001), but due to the high degree of homology between *mec* and *bla* regulatory systems (Garcia-Castellanos *et al.*, 2003; Garcia-Castellanos *et al.*, 2004; Marrero *et al.*, 2006) it is thought that they both work in a very similar manner and therefore the *bla* regulation system is used as a working model for both *blaZ* and *mecA* regulation.

Induction of the *bla* system initiates with the binding of β-lactam antibiotics to the extracellular penicillin-binding domain of the sensor protein BlaR1 (Fig. 8.3). The structure of the sensor domain of BlaR1 shares the greatest similarity with β-lactam hydrolyzing class D β-lactamases (Wilke *et al.*, 2004), but in BlaR1 the acylation of active site serine is made essentially irreversible by subsequent *N*-decarboxylation of the active site Lys^{392} (carboxylation of this residue has been proposed to be a prerequisite for acylation by β-lactams to occur) (Golemi-Kotra *et al.*, 2003; Birck *et al.*, 2004; Thumanu *et al.*, 2006; Cha and Mobashery, 2007). Circular dichroism and stopped-flow IR spectroscopy indicate that the binding of penicillin to BlaR1 results in a significant conformational change in the sensor domain (Golemi-Kotra *et al.*, 2003; Thumanu *et al.*, 2006). However, currently available X-ray structures for the apo- and penicillin-acylated BlaR1 sensor domain do not support this idea, suggesting instead that additional extracellular segments of the protein are involved in signal transmission (Wilke *et al.*, 2004). Therefore the exact molecular mechanism for transmission of the extracellular signal through the membrane is still unknown. We do know however that β-lactam binding to BlaR1 promotes rapid autocatalytic cleavage of this protein, which is a prometalloprotease. Once cleaved, the cytoplasmic transducer domain becomes an activated metalloprotease which then, directly or together with another (currently unidentified) factor, cleaves BlaI. BlaI can only bind to its target DNA as intact homodimers. Upon cleavage, BlaI becomes incapable of dimerizing and therefore of binding to the DNA. The C-terminal proteolytic cleavage site of BlaI is more accessible when the repressor is bound to DNA then when it is in solution, suggesting that the induction cascade involves cleavage of bound rather than of free repressor (Safo *et al.*, 2005). Release of BlaI from the intergenic operator sites allows transcription of both *blaZ* and *blaR1-blaI* genes with the concomitant expression of β-lactamase as well as expression of the regulatory proteins. Autoregulation of the regulatory proteins is important because BlaR1, once cleaved, can no longer transmit a signal and therefore intact BlaR1 must be continuously synthesized for the sensing of antibiotic molecules in the environment. Once the extracellular concentration of antibiotic decreases, BlaR1 is

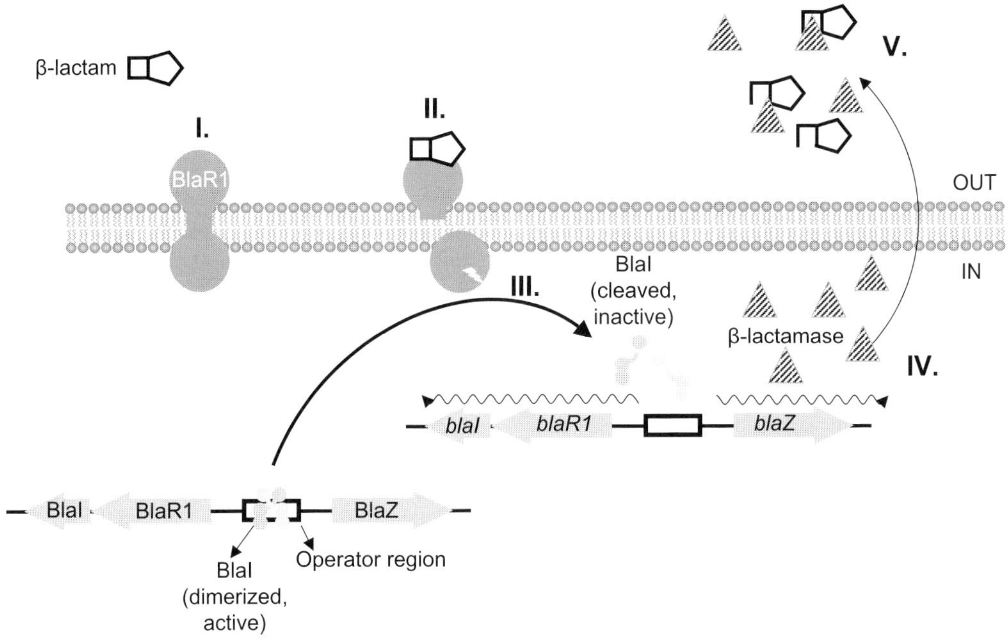

Figure 8.3 Induction of β-lactamase production by β-lactam antibiotics. BlaR1 is the sensor protein of a two-component system which detects extracellular β-lactam antibiotics (I). Upon binding of the β-lactam to the sensor domain of BlaRI, the protein is autocatalytically cleaved, activating the cytoplasmic zinc metalloprotease domain (II). Once cleaved, the activated cytoplasmic transducer domain of BlaRI cleaves BlaI, either directly or together with another (unidentified) factor (III). BlaI is a repressor protein that binds as dimers to the intergenic region between *blaZ* (which encodes the β-lactamase) and *blaR1-blaI*, preventing transcription of both the structural gene and of its cognate regulatory proteins. Cleaved BlaI is unable to dimerize and therefore to bind DNA. Release of BlaI from the operator sites allows transcription of *blaZ, blaR1* and *blaI* with the concomitant expression of β-lactamase (IV), which is then exported to the extracellular medium where it can hydrolyse penicillinase susceptible β-lactam antibiotics (V). Regulation of *mecA* gene by MecR1 and MecI is thought to occur in a similar manner, leading to the production of PBP2A.

no longer autoactivated and it no longer cleaves BlaI. Thus the intracellular concentration of uncleaved BlaI increases, it dimerizes, binds its DNA recognition sites and represses expression of the *bla* genes (Zhang *et al.*, 2001).

Despite the strong similarities between the repressors MecI and BlaI, and the interchangeability in regulation of either *mecA* or *blaZ*, there are important differences regarding the sensor-transducing proteins MecR1 and BlaR1: each sensor transducer works only with its cognate repressor and the kinetics of signal transduction are very different as BlaR1 leads to induction of *blaZ/mecA* transcription within minutes, while MecR1 induction of *mecA* transcription takes hours (Archer and Bosilevac, 2001; Marrero *et al.*, 2006).

Besides *mecA* regulation by the *mec* and *bla* regulatory systems, other factors seem to affect *mecA* transcription, although in these cases the mechanism is not yet understood. In MRSA strain COL, construction of a conditional *pbp2* or *pbpB* mutant (the gene encoding for native PBP2) revealed a link between the transcription of *pbp2* and *mecA* genes, where suppression of *pbp2* expression results in a gradual decline of the *mecA* transcript (Gardete *et al.*, 2006a). Also in strain COL, the *mecA* transcription rate appears to decrease with decreased *murE* or *murF* expression (Gardete *et al.*, 2004; Sobral *et al.*, 2006). MurE and MurF are ligases that catalyse cytoplasmic steps of the synthesis of the peptidoglycan precursor (see Fig. 8.1). However, decreased expression of both *murE* and *murF* genes is associated with decreased expression of *pbp2* (Gardete *et al.*, 2004; Sobral *et al.*, 2006), and it is therefore possible that *mecA* expression in *murE* or *murF* conditional mutants is actually dependent on *pbp2* expression and not directly on the expression of the two ligases. Strain COL constitutively produces PBP2A, as it does not contain functional *mecI/mecR1* and lacks *blaI/BlaR1* (Oliveira *et al.*, 2001). It is therefore possible that transcription of the acquired *mecA* gene, when lacking dedicated controlling elements, is somehow influenced by the regulatory network that controls the transcription of native *pbp2* (Gardete *et al.*, 2006a).

Native genes required for expression of β-lactam resistance

The *mecA* gene is the principal determinant of oxacillin resistance in MRSA strains. However, several additional native genes of *S. aureus*, also present in MSSA strains, have been identified as being essential for the full expression of oxacillin resistance. Most of these genes, called *aux* (auxiliary) or *fem* (factor essential for methicillin resistance) genes, were identified by screening transposon libraries generated in an oxacillin resistant background, for loss of resistance (Kornblum *et al.*, 1986; Berger-Bachi *et al.*, 1992; de Lencastre and Tomasz, 1994; de Lencastre *et al.*, 1999). Inactivation of *fem* genes in the presence of an intact *mecA* usually results in strains with a heterogeneous profile of oxacillin resistance, i.e. the majority of cells have a low MIC, while a subpopulation, which appears at variable frequencies depending on the strain (from 10^{-7} to 10^{-3}), is highly resistant. This resistant subpopulation results from mostly unidentified mutations in the staphylococcal genome (but not in the *mec* gene) and, with few exceptions, it maintains high levels of resistance even in the absence of selective pressure (Berger-Bachi and Rohrer, 2002; Ryffel *et al.*, 1994). Although *fem* genes may not be essential for bacterial viability, they need to be intact for oxacillin resistance to be fully expressed and therefore they can be considered good candidate targets for the development of drugs which would act synergistically with oxacillin, and therefore could resensitize MRSA strains to β-lactams.

The approximately 30 *fem* genes that have been identified so far are mostly housekeeping genes that encode, among others, proteins of unknown function, proteins with putative sensory/regulatory activities (protein kinases or ABC transporters), alternative transcription factors (*sigB*) or proteins that have a direct or indirect role in peptidoglycan metabolism/structure (de Lencastre *et al.*, 1999; Berger-Bachi and Rohrer, 2002). This last group includes the best studied *fem* genes and Table 8.1 summarizes their function in peptidoglycan metabolism (see also Fig. 8.1).

Although the biochemical function of many *fem* genes involved in peptidoglycan biosynthesis is known, the exact reason for why their alteration reduces oxacillin resistance remains undetermined. Mutations in some of these genes lead to the production of structurally abnormal peptidoglycan precursors which may be unable to compete effectively with the antibiotic for the active site of PBP2A (de Lencastre *et al.*, 1994; de Lencastre *et al.*, 1999). Abnormal muropeptides, when integrated into the peptidoglycan network, could also create localized structural defects or could trigger the unregulated action of cell wall hydrolases (de Lencastre *et al.*, 1999; Tomasz, 2000). Addition of β-lactams could aggravate the structural defects either by increasing the proportion of abnormal precursor in the peptidoglycan, which would jeopardize the structural stability of the cell, or by inducing a suicidal activation of enzymes that catalyse cell wall turnover. Even though the mechanistic details remain to be better elucidated, it is clear that interfering with the biosynthesis of the peptidoglycan precursor in different reactions, either by gene mutation (Berger-Bachi and Tschierske, 1998; Ludovice *et al.*, 1998; Rohrer *et al.*, 1999; Sobral *et al.*, 2003) or by using inhibitors of early steps in peptidoglycan biosynthesis such as fosfomycin, β-chloro-D-alanine or D-cycloserine (Sieradzki and Tomasz, 1997), leads to a drastic reduction in expression of oxacillin resistance. Therefore it seems that a non-limiting rate of production of normal peptidoglycan precursors is essential for proper functioning of PBP2A. Interestingly the amount of PBP2A itself does not seem to be determinant. Although production of PBP2A is probably required to be above a certain threshold concentration, there is no correlation between the total amount of PBP2A in clinical isolates and levels of drug resistance (Hartman and Tomasz, 1986).

Functional cooperation between acquired PBP2A and native PBP2

One of the *fem* genes, for which a molecular explanation for decreased oxacillin resistance upon

Table 8.1 *fem* genes known to be involved in cell wall metabolism/structure

Gene	Function or effect on cell wall composition/metabolism	References
fmhB (femX)	Synthesis of pentaglycin cross-links: addition of 1st glycine	(Rohrer *et al.*, 1999)
femA	Synthesis of pentaglycin cross-links: addition of 2nd and 3rd glycine	(Berger-Bachi *et al.*, 1992; Stranden *et al.*, 1997)
femB	Synthesis of pentaglycin cross-links: addition of 4th and 5th glycine	(Berger-Bachi *et al.*, 1992; Henze *et al.*, 1993; Stranden *et al.*, 1997)
femC (glnR)	Glutamine synthetase repressor: involved in the amidation of the stem peptide glutamate residues	(Berger-Bachi *et al.*, 1992; Gustafson *et al.*, 1994; Ornelas-Soares *et al.*, 1993)
femD (glmM)	Phosphoglucosamine mutase: catalyses the synthesis of glucosamine-1-phosphate a cytoplasmic peptidoglycan precursor	(Berger-Bachi *et al.*, 1992; Jolly *et al.*, 1997; Wu *et al.*, 1996a)
femF (murE)	UDP-N-acetylmuramyl tripeptide synthetase: catalyses the incorporation of lysine into the peptidoglycan stem peptide	(Ludovice *et al.*, 1998; Ornelas-Soares *et al.*, 1994)
murF	UDP-N-acetylmuramoyl-tripeptide-D-Alanyl-D-Alanine ligase: Incorporates the D-Ala-D-Ala dipeptide into the peptidoglycan stem peptide	(Sobral *et al.*, 2006; Sobral *et al.*, 2003)
pbp2	Penicillin binding protein 2, catalyses the last steps of peptidoglycan synthesis, namely transpeptidation and transglycosylation	(Pinho *et al.*, 2001a; Pinho *et al.*, 1997)
fmtA	Membrane protein whose inactivation decreases cross-linking and amidation of peptidoglycan	(Komatsuzawa *et al.*, 1999)
fmtB (mrp)	Cell surface protein whose inactivation reduces the amount of pentaglycyl-substituted monomer and increases the amount of unsubstituted pentapeptide	(Komatsuzawa *et al.*, 2000; Wu and De Lencastre, 1999)
llm	Putative membrane protein whose inactivation increases Triton X100 induced autolysis	(Maki *et al.*, 1994)
vraSR	Sensor and response regulator of a two-component system essential for the full expression of the cell wall stimulon	(Gardete *et al.*, 2006b; Kuroda *et al.*, 2003)

inactivation was determined, is *pbp2*. This gene encodes penicillin-binding protein (PBP) 2, one of the four native PBPs from *S. aureus* and the only bifunctional PBP, i.e. the only PBP which is capable of both transpeptidation and transglycosylation of the peptidoglycan in this organism. PBP2 is essential in MSSA strains but not in MRSA strains which contain the extra PBP2A, indicating that PBP2A is capable of substituting the transpeptidase domain of PBP2 (Pinho *et al.*, 2001b). PBP2A was long considered to take over the role of the native PBPs in cell wall synthesis, in the presence of β-lactam antibiotics at concentrations which inactivate the later. Therefore, it was surprising to find that PBP2 was actually required for optimal expression of oxacillin resistance (Pinho *et al.*, 1997). Mutagenesis of *pbp2* demonstrated that it is the transglycosylase domain of PBP2 that is necessary for resistance (Pinho *et al.*, 2001a). In the presence of high concentrations of β-lactam antibiotics, the low affinity transpeptidase domain of PBP2A remains active and cross-links the peptidoglycan, albeit to a lower degree than the one obtained by the action of native PBPs in the absence of β-lactams. Under these conditions, the penicillin insensitive transglycosylase domain of PBP2 becomes essential, indicating that acquired PBP2A and native PBP2 are likely to cooperate functionally in the synthesis of peptidoglycan, perhaps within a multienzymatic complex (Pinho *et al.*, 2001a).

Further evidence for the interaction between these two proteins came from cellular localization studies of PBP2. The main location for cell wall synthesis in *S. aureus* is the division septum (Pinho and Errington, 2003). Usually, in MSSA strains, PBP2 localizes to the septum presumably because it is recruited to the division site by binding to its substrate, which is localized at that place (Pinho and Errington, 2005). Yet, in the presence of oxacillin, the specific septal localization of PBP2 is lost and the protein becomes dispersed over the entire surface of the cell (Fig. 8.4), because β-lactam binding to the active site abolishes the ability of PBP2 to recognize and bind its substrate and therefore to localize properly (Pinho and Errington, 2005). However, in MRSA strains, PBP2A is able to prevent delocalization of PBP2 from the septum upon exposure to β-lactams, raising the hypothesis that PBP2A, which is insensitive to β-lactams, recognizes the substrate, localizes to the division septum and recruits PBP2 by a direct or an indirect protein-protein interaction, so that the transglycosylase domain of PBP2 can work together with the transpeptidase domain of PBP2A (Pinho and Errington, 2005).

Mechanism of resistance to vancomycin

Resistance in VISA strains

The first *S. aureus* strains with decreased susceptibility to vancomycin were isolated in Japan in 1997. Since then these strains, usually referred to as vancomycin intermediate resistant *S. aureus* (VISA), have been described in different European, North and South American and Asian countries (Gemmell, 2004). Contrary to *mecA* and *vanA* (see below) mediated resistance, VISA strains have not acquired extra genes responsible for the resistance mechanism (McAleese *et al.*, 2006). Instead they are thought to have accumulated a series of point mutations, during a gradual adaptive process, which led to changes in peptidoglycan biosynthesis. Several genes have been associated with vancomycin intermediate resistance, for example global regulators such as *agr* (Sakoulas *et al.*, 2005) or the *vraSR* two-component regulatory system (Kuroda *et al.*, 2003); genes involved in cell wall synthesis such as *pbp2* (Boyle-Vavra *et al.*, 2003; Sieradzki and Tomasz, 1999) and *pbp4* (Sieradzki *et al.*, 1999a) or genes involved in membrane biosynthesis such as *fmtC* (Nishi *et al.*, 2004). However, these correspond only to a small fraction of the genes whose expression is altered in VISA strains as shown, for example, by transcriptional analysis a VISA isolate and its vancomycin susceptible parental strain (McAleese *et al.*, 2006).

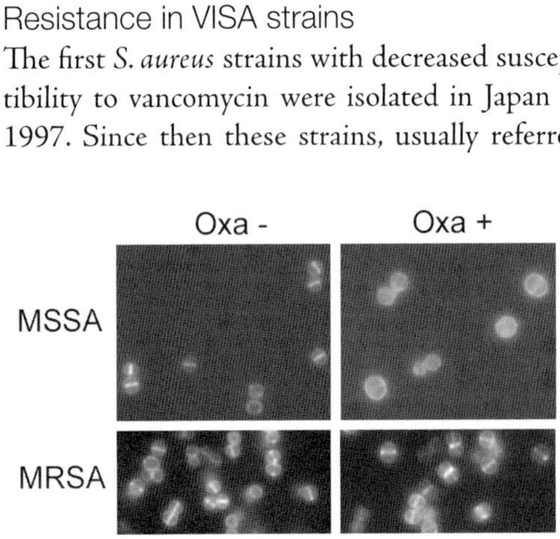

Figure 8.4 Cellular localization of PBP2. In MSSA strain RN4220, a GFP-PBP2 fusion localizes to the division septum (upper left panel). However upon addition of oxacillin (4µg/ml), the characteristic pattern of localization is lost and the protein becomes dispersed over the entire surface of the cell (upper right panel). In MRSA strain COL the septal localization is maintained even in the presence oxacillin (4µg/ml, lower right panel), indicating that PBP2A is able to maintain proper localization of PBP2 in the presence of the antibiotic.

Despite the fact that the exact nature of the mutations that are essential for the phenotypic expression of vancomycin intermediate resistance remains undetermined, a common phenotype can be found in virtually all clinical VISA strains studied. This consists of a thickened cell wall which could result from an increase in peptidoglycan synthesis and/or a decrease in cell wall turnover or autolysis (Hiramatsu, 2001; Cui *et al.*, 2003; Sieradzki and Tomasz, 2003). Additionally, VISA strains may have a lower degree of cross-linking of peptidoglycan strands, which results in a higher number of free D-Ala-D-Ala residues in the cell wall (Sieradzki *et al.*, 1999a; Sieradzki and Tomasz, 1999; Sieradzki and Tomasz, 2003). As mentioned earlier, the lethal targets of vancomycin are the D-Ala-D-Ala residues of the lipid-linked peptidoglycan precursors, which are located on the extracellular side of the cell membrane. However, vancomycin also binds D-Ala-D-Ala residues present in the mature peptidoglycan of *S. aureus*, although this binding does not lead to inhibition of peptidoglycan biosynthesis. By increasing the number of D-Ala-D-Ala residues in the mature peptidoglycan,

VISA strains increase the number of non-lethal or 'decoy' targets that can trap vancomycin in the cell wall. The proposed resistance mechanism for VISA strains states that binding of the large vancomycin molecules to a high number of non-lethal targets in the mesh-like cell wall, causes steric hindrance that prevents further penetration of antibiotic molecules (Fig. 8.5). This precludes vancomycin molecules from reaching their lethal target – the lipid-linked peptidoglycan precursor- found below the cell wall, at the cell membrane (Hiramatsu, 2001; Sieradzki *et al.*, 1999a). This has often been referred to as 'drug capture' (Sieradzki *et al.*, 1999a) or 'clogging' phenomenon (Hiramatsu, 2001), which results in a lower diffusion rate of vancomycin through the cell wall of VISA strains when compared to susceptible strains (Cui *et al.*, 2006a; Pereira *et al.*, 2007).

It is important to recognize that the lethal target of vancomycin is found at the site of cell wall synthesis. In *S. aureus*, cell wall synthesis takes place mainly, and possibly exclusively, at the division septum and not around the entire surface of the cellular membrane (Pinho and Errington, 2003). Therefore the path of vancomycin molecules to their lethal targets is not merely through approximately 20–30 nm of lateral cell wall, but it requires travelling to the tip of the division septum, which may correspond to a distance of around 500 nm when the septum is close to completion (Pereira *et al.*, 2007). Therefore, it is possible that the action of vancomycin (and other antibiotics) varies in different phases of the bacterial cell cycle (Fig. 8.5).

For many years vancomycin was the only effective treatment for many MRSA infections. Recently, new antibiotics with anti-MRSA activity were approved (quinupristin-dalfopristin, linezolid, tigecycline and daptomycin). However, VISA strains isolated from patients who did not receive daptomycin treatment, were found to have developed lower susceptibility to this antibiotic (Cui *et al.*, 2006b; Patel *et al.*, 2006;

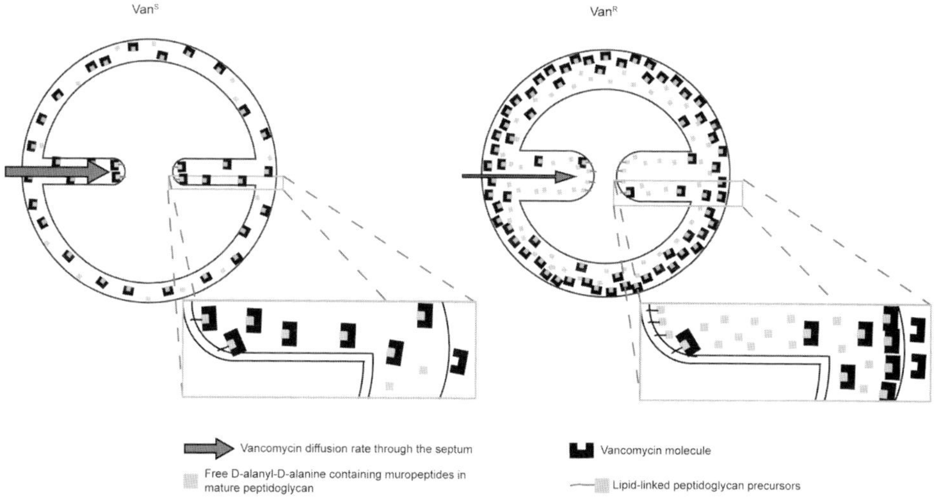

Figure 8.5 Model for vancomycin resistance in VISA strains. VISA strains (Van^R) have a thicker cell wall which contains a larger number of non-lethal D-Ala-D-Ala targets for vancomycin. Binding of the large vancomycin molecules to these 'decoy' targets hinders the progress of further antibiotic molecules to their lethal target – the lipid-linked peptidoglycan precursor – at the site of cell wall synthesis, causing a decrease in the diffusion rate of vancomycin through the cell wall of VISA strains when compared to susceptible strains. Cell wall synthesis in *S. aureus* takes place mainly, if not only, at the septum and a complete septum is synthesized before separation of the two daughter cells is initiated. Therefore the path of vancomycin to its lethal target is through the division septum and only composition changes that occur in the septum should contribute significantly to the mechanism of resistance. In resistant cells, the lower diffusion rate of vancomycin molecules to the septal tip, decreases the effective concentration of antibiotic that reaches lipid-linked precursor at the site of cell wall synthesis, per unit time. This could permit a sufficient proportion of lipid-linked precursor to escape binding by vancomycin allowing cell wall synthesis to continue. This model implies that the efficiency of vancomycin may vary during the cell cycle, as the path to the lethal targets at the septal tip is shorter in the beginning of septum formation and longer as it approaches completion. Adapted with permission from Pereira, P.M. *et al.* (2007) Antimicrob. Agents Chemother. 51, 3627–3633.

Mwangi *et al.*, 2007), indicating that, despite the different mechanisms of action, there may be common features between the resistance mechanisms to vancomycin and daptomycin, and that co-selection of resistance may occur.

Resistance in VRSA strains

High level vancomycin resistance was first reported in 1988, in clinical isolates of *Enterococcus* species (Leclercq *et al.*, 1988; Uttley *et al.*, 1988) and can be associated to one of several gene clusters. The *vanA* cluster, characterized by inducible, high-level resistance to vancomycin and teicoplanin (another glycopeptide antibiotic), is encoded by the Tn*1546* transposon. Tn*1546* contains a set of genes (see below) that provide an alternative pathway for peptidoglycan precursor biosynthesis, which is able to function in the presence of glycopeptides. More specifically, the enzymes encoded in Tn*1546* replace the carboxy terminal D-Ala-D-Ala residues of the peptidoglycan precursor with the depsipeptide D-alanyl-D-lactate (D-Ala-D-Lac). Binding of vancomycin to the normal peptidoglycan precursor involves a set of hydrogen bonds between the peptide portion of the antibiotic molecule and the D-Ala-D-Ala dipeptide. The structural change to D-Ala-D-Lac in the peptidoglycan, results in the loss of a critical hydrogen bond between the binding pocket of vancomycin and the peptide component of the substrate. The affinity of vancomycin for D-Ala-D-Lac containing cell wall precursors is therefore 1000 times lower than for the normal D-Ala-D-Ala containing precursors (Bugg *et al.*, 1991).

Noble and coworkers demonstrated that transfer of high-level vancomycin resistance from *Enterococcus faecalis* to *S. aureus* could occur by conjugation, both in an *in vitro* and in an *in vivo* model (Noble *et al.*, 1992). This was obviously a cause of immediate concern, but it took nature another decade to repeat the laboratory experiment. In 2002, the first VRSA strain was isolated from a dialysis patient in Michigan, USA, with an MIC of 1024 μg/ml (Sievert *et al.*, 2002; Clark *et al.*, 2005). Genetic analysis of this strain showed that it harbours a 57.9 kb plasmid, pLW1043, which includes a complete Tn*1546* transposon, most likely acquired from a vancomycin-resistant *E. faecalis* co-isolate (Weigel *et al.*, 2003). Tn*1546* contains nine genes (reviewed in Boneca and Chiosis, 2003; Courvalin, 2006): ORF1 and ORF2 are the transposase and resolvase enzymes required for mobilization of the transposon; *vanR* and *vanS* encode an inducible two-component regulatory system. The *vanS* gene encodes a transmembrane protein that senses vancomycin, directly or indirectly, and upon autophosphorylation transmits a signal to the response regulator protein VanR, leading to transcriptional activation of the other resistance genes. *vanH* encodes a dehydrogenase that converts pyruvate into D-lactate and *vanA* encodes the D-Ala-D-Lac ligase responsible for the synthesis of the depsipeptide (see Fig. 8.6). *vanX* encodes a D,D-dipeptidase that selectively cleaves vancomycin susceptible D-Ala-D-Ala dipeptides, removing them from the cell wall precursor pool; *vanY* encodes a membrane-bound D,D-carboxypeptidase which removes the carboxy terminal D-Ala from the normal cell wall precursor lipid-linked pentapetides. VanX and VanY therefore work together to eliminate from the precursors pool, the normal vancomycin susceptible D-Ala-D-Ala containing peptidoglycan precursors, which continue to be synthesized due to the activity of the native DdlA ligase. The exact function of *vanZ* is not clear, although its expression has been associated with teicoplanin resistance (Arthur *et al.*, 1995). Native staphylococcal PBPs, namely PBP2 (Severin *et al.*, 2004b), are able to use the modified, D-Ala-D-Lac containing precursor for cell wall synthesis, which can therefore continue even in the presence of vancomycin.

Since the first report of a VRSA strain, other VRSA strains have been isolated (Miller *et al.*, 2002; Kacica and McDonald, 2004). The second VRSA strain, isolated from a chronic foot ulcer of a patient in Pennsylvania, also carried Tn*1546* but in a plasmid different from the one in the Michigan strain (Miller *et al.*, 2002; Tenover *et al.*, 2004). Detailed analysis revealed the presence of two insertion sequences in Tn*1546* from Pennsylvania strain, absent from the Michigan strain, indicating that the two first VRSA isolates arose from independent genetic events (Clark *et al.*, 2005). Although spread of VRSA isolates to other patients has been successfully contained so far, the fact that the ge-

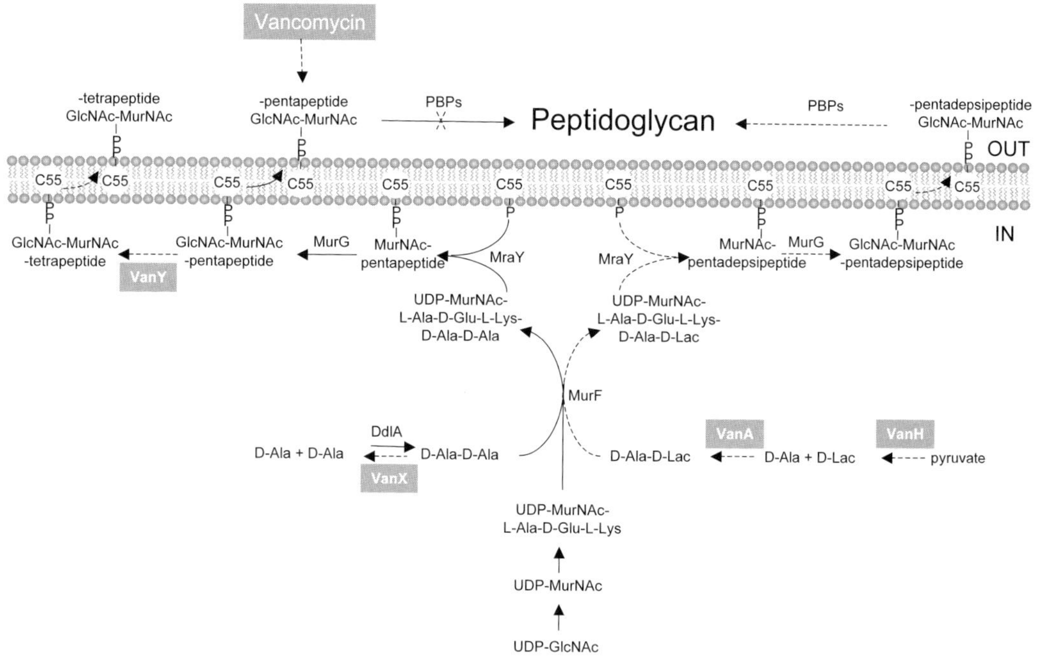

Figure 8.6 Peptidoglycan biosynthetic pathway in VRSA strains. VRSA strains have acquired a set of genes that encode an alternative pathway for peptidoglycan precursor biosynthesis (dashed arrows). Normal synthesis of peptidoglycan (solid arrows) leads to the production of a D-Ala-D-Ala terminating precursor which is translocated to the outer leaflet of the cytoplasmic membrane and incorporated into the peptidoglycan network by the action of the PBPs. Vancomycin inhibits peptidoglycan biosynthesis by binding to the terminal D-Ala-D-Ala residues of the peptidoglycan precursor and therefore preventing access of PBPs to their natural substrate. The set of *van* genes (see text for details) acquired by VRSA strains enables the bacteria to synthesize a D-Ala-D-Lac terminating peptidoglycan precursor, for which vancomycin has a very low affinity. As PBPs can still use these pentadepsipeptide precursors for cell wall synthesis, this process can continue even in the presence of the antibiotic. Adapted with permission from Boneca, I. G., and Chiosis, G. (2003) Expert Opin. Ther. Targets 7: 311–328.

netic background of these strains is identical the USA100 lineage (also known as the New York/ Japan clone, CC5) (Tenover *et al.*, 2004; Weigel *et al.*, 2003), the most common pulsed-field type in US hospitals (McDougal *et al.*, 2003), makes it likely that it is merely a question of time until VRSA strains become more common in the clinical setting.

Synergism between β-lactams and glycopeptides

The highly vancomycin resistant VRSA strains isolated in 2002 contain not only the plasmid-borne *vanA* gene complex, but also a chromosomally located *mecA* gene (Miller *et al.*, 2002; Sievert *et al.*, 2002). These strains are therefore able to express both high vancomycin and high oxacillin resistance, which is of obvious and grave concern. To better study the coexistence of *mecA* and *vanA* mediated resistance mechanisms in the same background, Severin and colleagues introduced the *vanA* encoding plasmid pLW1043 into the well characterized and sequenced strain COL, resulting in a strain that was called COLVA (Severin *et al.*, 2004a). Using this strain, they showed that although COLVA is capable of expressing high level resistance to both oxacillin and vancomycin, it can not grow when the two antibiotics are present simultaneously. The reason for this is that the pentadepsipeptide cell wall precursors, produced in the presence of vancomycin, can be used by PBP2 but are not appropriate substrates for PBP2A (Severin *et al.*, 2004b). In the presence of high concentrations of oxacillin, native PBPs, including PBP2, become inactivated and the only transpeptidase that remains functional is PBP2A which can not utilize the pentadepsipeptide and therefore cell wall synthesis can not continue (Severin *et al.*, 2004b).

Synergy between β-lactam and glycopeptide antibiotics was confirmed in four of the recent USA clinical *vanA* type MRSA strains, all of which show a greater than 100-fold reduction in vancomycin resistance in the presence of a sub-MIC concentration of oxacillin and, conversely, a decrease in oxacillin resistance of more than 100 times, in the presence of a sub-MIC concentration of vancomycin (Perichon and Courvalin, 2006).

However, a combination of vancomycin with β-lactam antibiotics may not be a good long-term therapeutic option, as strain COLVA grown in the presence of sub-MIC concentrations of oxacillin (40 μg/ml) produced an heterogeneous vancomycin resistant subpopulation, capable of growing simultaneously in high (100–500 μg/ml) concentration of vancomycin (Severin *et al.*, 2004a). The appearance of COLVA colonies capable of growing in the presence of elevated concentrations of both oxacillin and vancomycin indicates that the two resistance mechanisms can coexist, even though this is not yet the case in clinical isolates.

Stress-like response of the staphylococcal cell to cell wall synthesis inhibitors

Although β-lactams and glycopeptides inhibit the synthesis of peptidoglycan by very different mechanisms, there is a common set of genes that are upregulated in response to both classes of antibiotics. Three independent studies on genome-wide transcriptional profiling of the response of different *S. aureus* strains (N315, RN450, JH1) to the presence the various cell wall active antibiotics (vancomycin, oxacillin, D-cycloserine and bacitracin) have led to the identification of a 'cell wall stimulon', which comprises the entire set of genes that respond together to cell wall damage (Kuroda *et al.*, 2003; Utaida *et al.*, 2003; McAleese *et al.*, 2006). Each of these three initial studies identified up to 140 genes which are induced in the presence of cell wall active antibiotics. A group of 15 genes were common to all studies, despite the use of different strains, different antibiotics and different antibiotic exposure times, indicating that they may be key components of the cell wall stress stimulon (McAleese *et al.*, 2006). These 15 genes encode proteins involved in cell wall synthesis (*sgtB* encoding a monofunctional glycosyltransferase; *murZ*, also called *murA*, encoding UDP-*N*-acetylglucosamine 1-carboxylvinyl transferase 2, which catalyses one of the first steps in the synthesis of the peptidoglycan precursor); in teicoplanin resistance (*tcaA*); in posttranslational modification (*psrA*, encoding peptidyl-prolyl *cis/trans* isomerase homologue, which probably facilitates envelope protein folding; and *htrA* encoding a heatshock protein homologue); in autolysis and oxacillin resistance (*fmt*); in regulation (*vraS* encoding a two-component response regulator associated with vancomycin resistance), as well as eight hypothetical proteins with unknown functions (Utaida *et al.*, 2003; McAleese *et al.*, 2006). The gene *pbp2*, encoding penicillin-binding protein PBP2, a protein involved in both oxacillin and vancomycin resistance, is also considered to be part of the core cell wall stimulon (Kuroda *et al.*, 2003; Utaida *et al.*, 2003). Only two genes, encoding autolysins (*atl* and a hypothetical protein similar to aulolysins), were downregulated in all three studies (McAleese *et al.*, 2006).

Interesting observations regarding the cell wall stimulon came from studies using an isogenic pair of strains, isolated from one patient undergoing vancomycin therapy, composed of JH1 (the vancomycin susceptible parental strain, isolated prior to the beginning of chemotherapy) and JH9 (VISA strain derived from JH1, with decreased susceptibility of vancomycin, MIC=8μg/ml, isolated from the same patient approximately 3 months later) (Sieradzki *et al.*, 2003; McAleese *et al.*, 2006). Ten out of the fifteen genes that constitute the core cell wall stimulon, and 27 out of 72 genes which are transiently overexpressed in JH1 during vancomycin treatment, are permanently overexpressed in strain JH9 grown in the absence of antibiotics (McAleese *et al.*, 2006). It seems that, in the process of becoming resistant, JH9 cells adopted a permanently stressed state, even in the absence of antibiotics (McAleese *et al.*, 2006).

Recent sequencing of the entire genomes of these two strains shows that 35 point mutations in 31 loci were acquired during evolution of JH9 from JH1. However analysis of intermediate isolates from the same patient suggest that just four or five of those mutations (which include the *vraSR* operon) may be sufficient to achieve an MIC of 6μg/ml, suggesting that a low number of point mutations may be responsible for the complex phenotypes associated with VISA strains (Mwangi *et al.*, 2007). The function of some of the affected genes remains to be determined.

In N315, induction of one third of the genes overexpressed in the presence of vancomycin is dependent on the two-component system VraSR (Kuroda *et al.*, 2003). More importantly, induction of the entire set of 15 genes which constitute the conserved cell wall stimulon is abolished in a *vraSR* null mutant, suggesting that the sensor kinase VraS responds either to damage of the cell wall structure or to inhibition of cell wall synthesis (Kuroda *et al.*, 2003). The *vraSR* operon is also induced when cell wall synthesis is perturbed not by cell wall active antibiotics, but by depleting the enzymatic machinery of cell wall synthesis of one of its main components, PBP2, indicating that VraS is able to sense alterations that result from lack of the activity of this enzyme (Gardete *et al.*, 2006b). The exact nature of the signal which is sensed directly by VraS remains unknown, but the response regulated by this protein is obviously a complex one which seems to result in, among other phenotypes, an increase in the rate of peptidoglycan biosynthesis and a decrease in autolysis rate and which is important for both oxacillin and vancomycin resistance (Kuroda *et al.*, 2003; Boyle-Vavra *et al.*, 2006; Gardete *et al.*, 2006b).

Besides inducing a specific response to cell wall damage, β-lactam antibiotics (and not antibiotics that target protein synthesis, for example) also induce a *bona fide* SOS response, even at sub-MIC concentrations, which is dependent on the presence of RecA and results in the induction of *lexA* (Maiques *et al.*, 2006). In *Escherichia coli*, inactivation of PBP3, either by β-lactams or genetic mutation, induces the SOS response through the DpiBA two-component signal transduction system, transiently inhibiting cell division and enabling bacteria to survive an otherwise lethal antibiotic exposure (Miller *et al.*, 2004). Whether a similar system is involved in SOS induction by β-lactams in *S. aureus*, remains to be determined.

Conclusions

This chapter has focused on the resistance mechanisms of *S. aureus* to β-lactam and glycopeptide antibiotics, due to the clinical relevance of MRSA and VISA/VRSA strains. But this extremely versatile human pathogen has developed resistance mechanisms against virtually all classes of used antibiotics. Furthermore MRSA strains, previously limited to the hospital environment, have spread to the community. Therefore, the need to understand resistance mechanisms, to develop new strategies to treat staphylococcal (and other) infections, as well as to promote the rational and prudent use of antibiotics, continues to be of crucial importance.

Future trends

Evolution of antibiotic resistance relies upon two fundamental mechanisms, both of clinical relevance: generation of point mutations and acquisition of resistance genes. While the later can be easily traceable, the former are usually more elusive. The recent availability of less expensive sequencing technologies will certainly enable researchers to identify nucleotide changes responsible for increases in either antibiotic resistance or virulence, allowing a faster identification of changes in newly emerging pathogenic strains.

Most laboratory studies on antibiotic resistance use isogenic populations of cells. However, two cells of an isogenic population can behave differently at any time point, not only due to cells being at different stages of the cell cycle, but also due to epigenetic factors that lead to different gene expression in genetically identical cells. Single cell technologies may allow new interpretations of the mode of action and mechanism of resistance to antibiotics.

The overall response of bacteria to cell wall active antibiotics is obviously a multifactorial one. Many of the individual players involved in this response have been currently identified, but we will need not only to integrate the available information, but also to study in more detail each of the relevant proteins (including the large number of hypothetical proteins that are part of the core cell wall stimulon), in order to fully understand how the bacterial cell responds to antibiotic exposure. We still do not know, for example, which alterations induced upon antibiotic exposure are beneficial to the cell and are part of its 'defence' mechanism to cope with cell wall synthesis impairment and which alterations are deleterious consequences which the cell can not avoid.

Detailed genetic, biochemical and structural information about antibiotic targets and the whole cell response to the presence of antibiotics will hopefully facilitate the development of rationally designed drugs active against multi drug resistant strains. Valid alternatives to currently existing antibiotics may be compounds that do not inhibit bacterial growth on their own but either act synergistically with a primary antibiotic (by impairing expression of resistance) or decrease the virulence of bacterial cells, facilitating their elimination by the host immune system. Because they would not be lethal to the bacterial cell, selective pressure for resistance to arise could be lower, which would result in a longer clinical life time of these therapeutic agents.

Acknowledgements

I would like to thank Sérgio Filipe, Duarte Oliveira and James Yates for helpful comments on the manuscript. I also thank Pedro Pereira for assistance in preparing Fig. 8.5 and Ivo Boneca for allowing the adaptation of his illustration for Fig. 8.6. Research in the author's laboratory is currently supported by FCT grant POCI/BIA-BCM/56493/2004.

References

Abraham, E.P., and Chain, E. (1940). An enzyme from bacteria able to destroy penicillin. Nature *146*, 837.

Archer, G.L., and Bosilevac, J.M. (2001). Signaling antibiotic resistance in staphylococci. Science *291*, 1915–1916.

Arthur, M., Depardieu, F., Molinas, C., Reynolds, P., and Courvalin, P. (1995). The *vanZ* gene of Tn1546 from *Enterococcus faecium* BM4147 confers resistance to teicoplanin. Gene *154*, 87–92.

Barber, M. (1961). Methicillin-resistant staphylococci. J. Clin. Pathol. *14*, 385–393.

Berger-Bachi, B., and Rohrer, S. (2002). Factors influencing methicillin resistance in staphylococci. Arch. Microbiol. *178*, 165–171.

Berger-Bachi, B., Strassle, A., Gustafson, J.E., and Kayser, F.H. (1992). Mapping and characterization of multiple chromosomal factors involved in methicillin resistance in *Staphylococcus aureus*. Antimicrob. Agents Chemother. *36*, 1367–1373.

Berger-Bachi, B., and Tschierske, M. (1998). Role of fem factors in methicillin resistance. Drug Resist. Updat. *1*, 325–335.

Birck, C., Cha, J.Y., Cross, J., Schulze-Briese, C., Meroueh, S.O., Schlegel, H.B., Mobashery, S., and Samama, J.P. (2004). X-ray crystal structure of the acylated beta-lactam sensor domain of BlaR1 from *Staphylococcus aureus* and the mechanism of receptor activation for signal transduction. J. Am. Chem. Soc. *126*, 13945–13947.

Boneca, I.G., and Chiosis, G. (2003). Vancomycin resistance: occurrence, mechanisms and strategies to combat it. Expert Opin. Ther. Targets *7*, 311–328.

Boyle-Vavra, S., Yin, S., Challapalli, M., and Daum, R.S. (2003). Transcriptional induction of the penicillin-binding protein 2 gene in *Staphylococcus aureus* by cell wall-active antibiotics oxacillin and vancomycin. Antimicrob. Agents Chemother. *47*, 1028–1036.

Boyle-Vavra, S., Yin, S., and Daum, R.S. (2006). The VraS/VraR two-component regulatory system required for oxacillin resistance in community-acquired methicillin-resistant *Staphylococcus aureus*. FEMS Microbiol. Lett. *262*, 163–171.

Bugg, T.D., Wright, G.D., Dutka-Malen, S., Arthur, M., Courvalin, P., and Walsh, C.T. (1991). Molecular basis for vancomycin resistance in *Enterococcus faecium* BM4147: biosynthesis of a depsipeptide peptidoglycan precursor by vancomycin resistance proteins VanH and VanA. Biochemistry *30*, 10408–10415.

CDC (2002). Public Health Dispatch: Vancomycin-Resistant *Staphylococcus aureus* – Pennsylvania, 2002. MMWR *51*, 902.

Cha, J., and Mobashery, S. (2007). Lysine N(zeta)-decarboxylation in the BlaR1 protein from *Staphylococcus aureus* at the root of its function as an antibiotic sensor. J. Am. Chem. Soc. *129*, 3834–3835.

Chambers, H.F. (2001). The changing epidemiology of *Staphylococcus aureus*? Emerg. Infect. Dis. *7*, 178–182.

Clark, N.C., Weigel, L.M., Patel, J.B., and Tenover, F.C. (2005). Comparison of Tn*1546*-like elements in vancomycin-resistant *Staphylococcus aureus* isolates from Michigan and Pennsylvania. Antimicrob. Agents Chemother. *49*, 470–472.

Cosgrove, S.E., Sakoulas, G., Perencevich, E.N., Schwaber, M.J., Karchmer, A.W., and Carmeli, Y. (2003). Comparison of mortality associated with methicillin-resistant and methicillin-susceptible *Staphylococcus aureus* bacteremia: a meta-analysis. Clin. Infect. Dis. *36*, 53–59.

Courvalin, P. (2006). Vancomycin resistance in gram-positive cocci. Clin. Infect. Dis. *42 Suppl 1*, S25–34.

Couto, I., de Lencastre, H., Severina, E., Kloos, W., Webster, J.A., Hubner, R.J., Sanches, I.S., and Tomasz, A. (1996). Ubiquitous presence of a *mecA* homologue in natural isolates of *Staphylococcus sciuri*. Microb. Drug Resist. *2*, 377–391.

Couto, I., Wu, S.W., Tomasz, A., and de Lencastre, H. (2003). Development of methicillin resistance in clinical isolates of *Staphylococcus sciuri* by transcriptional activation of the *mecA* homologue native to the species. J. Bacteriol. *185*, 645–653.

Cui, L., Iwamoto, A., Lian, J.Q., Neoh, H.M., Maruyama, T., Horikawa, Y., and Hiramatsu, K. (2006a). Novel mechanism of antibiotic resistance originating in vancomycin-intermediate *Staphylococcus aureus*. Antimicrob. Agents Chemother. *50*, 428–438.

Cui, L., Ma, X., Sato, K., Okuma, K., Tenover, F.C., Mamizuka, E.M., Gemmell, C.G., Kim, M.N., Ploy, M.C., El-Solh, N., *et al.* (2003). Cell wall thickening is a common feature of vancomycin resistance in *Staphylococcus aureus*. J. Clin. Microbiol. *41*, 5–14.

Cui, L., Tominaga, E., Neoh, H.M., and Hiramatsu, K. (2006b). Correlation between reduced daptomycin susceptibility and vancomycin resistance in Vancomycin-Intermediate *Staphylococcus aureus*. Antimicrob. Agents Chemother. *50*, 1079–1082.

de Lencastre, H., de Jonge, B.L., Matthews, P.R., and Tomasz, A. (1994). Molecular aspects of methicillin resistance in *Staphylococcus aureus*. J. Antimicrob. Chemother. *33*, 7–24.

de Lencastre, H., and Tomasz, A. (1994). Reassessment of the number of auxiliary genes essential for expression of high-level methicillin resistance in *Staphylococcus aureus*. Antimicrob. Agents Chemother. *38*, 2590–2598.

de Lencastre, H., Wu, S.W., Pinho, M.G., Ludovice, A.M., Filipe, S., Gardete, S., Sobral, R., Gill, S., Chung, M., and Tomasz, A. (1999). Antibiotic resistance as a stress response: complete sequencing of a large number of chromosomal loci in *Staphylococcus aureus* strain COL that impact on the expression of resistance to methicillin. Microb. Drug Resist. *5*, 163–175.

Enright, M.C., Robinson, D.A., Randle, G., Feil, E.J., Grundmann, H., and Spratt, B.G. (2002). The evolutionary history of methicillin-resistant *Staphylococcus aureus* (MRSA). Proc. Natl Acad. Sci. USA *99*, 7687–7692.

Garcia-Castellanos, R., Mallorqui-Fernandez, G., Marrero, A., Potempa, J., Coll, M., and Gomis-Ruth, F.X. (2004). On the transcriptional regulation of methicillin resistance: MecI repressor in complex with its operator. J. Biol. Chem. *279*, 17888–17896.

Garcia-Castellanos, R., Marrero, A., Mallorqui-Fernandez, G., Potempa, J., Coll, M., and Gomis-Ruth, F.X. (2003). Three-dimensional structure of MecI. Molecular basis for transcriptional regulation of staphylococcal methicillin resistance. J. Biol. Chem. *278*, 39897–39905.

Gardete, S., de Lencastre, H., and Tomasz, A. (2006a). A link in transcription between the native *pbpB* and the acquired *mecA* gene in a strain of *Staphylococcus aureus*. Microbiol. *152*, 2549–2558.

Gardete, S., Ludovice, A.M., Sobral, R.G., Filipe, S.R., de Lencastre, H., and Tomasz, A. (2004). Role of *murE* in the expression of beta-lactam antibiotic resistance in *Staphylococcus aureus*. J. Bacteriol. *186*, 1705–1713.

Gardete, S., Wu, S.W., Gill, S., and Tomasz, A. (2006b). Role of VraSR in antibiotic resistance and antibiotic-induced stress response in *Staphylococcus aureus*. Antimicrob. Agents Chemother. *50*, 3424–3434.

Ge, M., Chen, Z., Onishi, H.R., Kohler, J., Silver, L.L., Kerns, R., Fukuzawa, S., Thompson, C., and Kahne, D. (1999). Vancomycin derivatives that inhibit peptidoglycan biosynthesis without binding D-Ala-D-Ala. Science *284*, 507–511.

Gemmell, C.G. (2004). Glycopeptide resistance in *Staphylococcus aureus*: is it a real threat? J. Infect. Chemother. *10*, 69–75.

Ghuysen, J.-M. (1991). Serine β-lactamases and penicillin-binding proteins. ARM *45*, 37–67.

Goffin, C., and Ghuysen, J.M. (1998). Multimodular penicillin-binding proteins: an enigmatic family of orthologs and paralogs. Microbiol. Mol. Biol. Rev. *62*, 1079–1093.

Golemi-Kotra, D., Cha, J.Y., Meroueh, S.O., Vakulenko, S.B., and Mobashery, S. (2003). Resistance to beta-lactam antibiotics and its mediation by the sensor domain of the transmembrane BlaR signaling pathway in *Staphylococcus aureus*. J. Biol. Chem. *278*, 18419–18425.

Gregory, P.D., Lewis, R.A., Curnock, S.P., and Dyke, K.G. (1997). Studies of the repressor (BlaI) of beta-lactamase synthesis in *Staphylococcus aureus*. Mol. Microbiol. *24*, 1025–1037.

Gustafson, J., Strassle, A., Hachler, H., Kayser, F.H., and Berger-Bachi, B. (1994). The *femC* locus of *Staphylococcus aureus* required for methicillin resistance includes the glutamine synthetase operon. J. Bacteriol. *176*, 1460–1467.

Hackbarth, C.J., and Chambers, H.F. (1993). blaI and blaR1 regulate beta-lactamase and PBP 2a production in methicillin-resistant *Staphylococcus aureus*. Antimicrob. Agents Chemother. *37*, 1144–1149.

Hanssen, A.M., and Ericson Sollid, J.U. (2006). SCC*mec* in staphylococci: genes on the move. FEMS Immunol. Med. Microbiol. *46*, 8–20.

Hardt, K., Joris, B., Lepage, S., Brasseur, R., Lampen, J.O., Frere, J.M., Fink, A.L., and Ghuysen, J.M. (1997). The penicillin sensory transducer, BlaR, involved in the inducibility of beta-lactamase synthesis in *Bacillus licheniformis* is embedded in the plasma membrane via a four-alpha-helix bundle. Mol. Microbiol. *23*, 935–944.

Hartman, B.J., and Tomasz, A. (1986). Expression of methicillin resistance in heterogeneous strains of *Staphylococcus aureus*. Antimicrob. Agents Chemother. *29*, 85–92.

Henze, U., Sidow, T., Wecke, J., Labischinski, H., and Berger-Bachi, B. (1993). Influence of *femB* on methicillin resistance and peptidoglycan metabolism in *Staphylococcus aureus*. J. Bacteriol. *175*, 1612–1620.

Hiramatsu, K. (2001). Vancomycin-resistant *Staphylococcus aureus*: a new model of antibiotic resistance. Lancet Infect. Dis. *1*, 147–155.

Hiramatsu, K., Hanaki, H., Ino, T., Yabuta, K., Oguri, T., and Tenover, F.C. (1997). Methicillin-resistant *Staphylococcus aureus* clinical strain with reduced vancomycin susceptibility. J. Antimicrob. Chemother. *40*, 135–136.

Ito, T., Katayama, Y., Asada, K., Mori, N., Tsutsumimoto, K., Tiensasitorn, C., and Hiramatsu, K. (2001). Structural comparison of three types of staphylococcal cassette chromosome *mec* integrated in the chromosome in methicillin-resistant *Staphylococcus aureus*. Antimicrob. Agents Chemother. *45*, 1323–1336.

Ito, T., Katayama, Y., and Hiramatsu, K. (1999). Cloning and nucleotide sequence determination of the entire *mec* DNA of pre-methicillin-resistant *Staphylococcus aureus* N315. Antimicrob. Agents Chemother. *43*, 1449–1458.

Ito, T., Ma, X.X., Takeuchi, F., Okuma, K., Yuzawa, H., and Hiramatsu, K. (2004). Novel type V staphylococcal cassette chromosome *mec* driven by a novel cassette chromosome recombinase, *ccrC*. Antimicrob. Agents Chemother. *48*, 2637–2651.

Jevons, M.P. (1961). 'Celbenin'-resistant staphylococci. Br. Med. J. *1*, 124–125.

Jolly, L., Wu, S., van Heijenoort, J., de Lencastre, H., Mengin-Lecreulx, D., and Tomasz, A. (1997). The *femR315* gene from *Staphylococcus aureus*, the interruption of which results in reduced methicillin resistance, encodes a phosphoglucosamine mutase. J. Bacteriol. *179*, 5321–5325.

Kacica, M., and McDonald, L.C. (2004). Vancomycin-resistant *Staphylococcus aureus*-New York, 2004. MMWR Morb. Mortal. Wkly Rep. 53, 322–323.

Katayama, Y., Ito, T., and Hiramatsu, K. (2000). A new class of genetic element, staphylococcus cassette chromosome *mec*, encodes methicillin resistance in *Staphylococcus aureus*. Antimicrob. Agents Chemother. *44*, 1549–1555.

Katayama, Y., Robinson, D.A., Enright, M.C., and Chambers, H.F. (2005). Genetic background affects stability of *mecA* in *Staphylococcus aureus*. J. Clin. Microbiol. *43*, 2380–2383.

Komatsuzawa, H., Ohta, K., Labischinski, H., Sugai, M., and Suginaka, H. (1999). Characterization of *fmtA*, a gene that modulates the expression of methicillin resistance in *Staphylococcus aureus*. Antimicrob. Agents Chemother. *43*, 2121–2125.

Komatsuzawa, H., Ohta, K., Sugai, M., Fujiwara, T., Glanzmann, P., Berger Bachi, B., and Suginaka, H. (2000). Tn*551*-mediated insertional inactivation of the *fmtB* gene encoding a cell wall-associated protein abolishes methicillin resistance in *Staphylococcus aureus*. J. Antimicrob. Chemother. *45*, 421–431.

Kornblum, J., Hartman, B.J., Novick, R.P., and Tomasz, A. (1986). Conversion of a homogeneously methicillin-resistant strain of *Staphylococcus aureus* to heterogeneous resistance by Tn*551*-mediated insertional inactivation. Eur. J. Clin. Microbiol. 5, 714–718.

Kuroda, M., Kuroda, H., Oshima, T., Takeuchi, F., Mori, H., and Hiramatsu, K. (2003). Two-component system VraSR positively modulates the regulation of cell-wall biosynthesis pathway in *Staphylococcus aureus*. Mol. Microbiol. *49*, 807–821.

Leclercq, R., Derlot, E., Duval, J., and Courvalin, P. (1988). Plasmid-mediated resistance to vancomycin and teicoplanin in *Enterococcus faecium*. N. Engl. J. Med. *319*, 157–161.

Levy, S.B. (1998). Multidrug resistance-a sign of the times. N. Engl. J. Med. *338*, 1376–1378.

Lewis, R.A., and Dyke, K.G. (2000). MecI represses synthesis from the beta-lactamase operon of *Staphylococcus aureus*. J. Antimicrob. Chemother. *45*, 139–144.

Lim, D., and Strynadka, N.C. (2002). Structural basis for the beta lactam resistance of PBP2a from methicillin-resistant *Staphylococcus aureus*. Nat. Struct. Biol. 9, 870–876.

Lowy, F.D. (2003). Antimicrobial resistance: the example of *Staphylococcus aureus*. J. Clin. Invest. *111*, 1265–1273.

Lu, W.P., Sun, Y., Bauer, M.D., Paule, S., Koenigs, P.M., and Kraft, W.G. (1999). Penicillin-binding protein 2a from methicillin-resistant *Staphylococcus aureus*: kinetic characterization of its interactions with beta-lactams using electrospray mass spectrometry. Biochemistry 38, 6537–6546.

Ludovice, A.M., Wu, S.W., and de Lencastre, H. (1998). Molecular cloning and DNA sequencing of the *Staphylococcus aureus* UDP-N-acetylmuramyl tripeptide synthetase (*murE*) gene, essential for the optimal expression of methicillin resistance. Microb. Drug Resist. 4, 85–90.

Ma, X.X., Ito, T., Tiensasitorn, C., Jamklang, M., Chongtrakool, P., Boyle-Vavra, S., Daum, R.S., and Hiramatsu, K. (2002). Novel type of staphylococcal cassette chromosome mec identified in community-acquired methicillin-resistant *Staphylococcus aureus* strains. Antimicrob. Agents Chemother. *46*, 1147–1152.

Maiques, E., Ubeda, C., Campoy, S., Salvador, N., Lasa, I., Novick, R.P., Barbe, J., and Penades, J.R. (2006). Beta-lactam antibiotics induce the SOS response and horizontal transfer of virulence factors in *Staphylococcus aureus*. J. Bacteriol. *188*, 2726–2729.

Maki, H., Yamaguchi, T., and Murakami, K. (1994). Cloning and characterization of a gene affecting the methicillin resistance level and the autolysis rate in *Staphylococcus aureus*. J. Bacteriol. *176*, 4993–5000.

Marrero, A., Mallorqui-Fernandez, G., Guevara, T., Garcia-Castellanos, R., and Gomis-Ruth, F.X. (2006). Unbound and acylated structures of the MecR1 extracellular antibiotic-sensor domain provide insights into the signal-transduction system that triggers methicillin resistance. J. Mol. Biol. *361*, 506–521.

McAleese, F., Wu, S.W., Sieradzki, K., Dunman, P., Murphy, E., Projan, S., and Tomasz, A. (2006). Overexpression of genes of the cell wall stimulon in clinical isolates of *Staphylococcus aureus* exhibiting vancomycin-intermediate- *S. aureus*-type resistance to vancomycin. J. Bacteriol. *188*, 1120–1133.

McDougal, L.K., Steward, C.D., Killgore, G.E., Chaitram, J.M., McAllister, S.K., and Tenover, F.C. (2003). Pulsed-field gel electrophoresis typing of oxacillin-resistant *Staphylococcus aureus* isolates from the United States: establishing a national database. J. Clin. Microbiol. *41*, 5113–5120.

Miller, C., Thomsen, L.E., Gaggero, C., Mosseri, R., Ingmer, H., and Cohen, S.N. (2004). SOS response induction by beta-lactams and bacterial defense against antibiotic lethality. Science 305, 1629–1631.

Miller, D., Urdaneta, V., Weltman, A., and Park, S. (2002). Public Health Dispatch: Vancomycin-Resistant *Staphylococcus aureus* – Pennsylvania, 2002. MMWR Morb. Mortal. Wkly Rep. *51*, 902.

Mwangi, M.M., Wu, S.W., Zhou, Y., Sieradzki, K., de Lencastre, H., Richardson, P., Bruce, D., Rubin, E., Myers, E., Siggia, E.D., and Tomasz, A. (2007). Tracking the *in vivo* evolution of multidrug resistance in *Staphylococcus aureus* by whole-genome sequencing. Proc. Natl Acad. Sci. USA *104*, 9451–9456.

Nishi, H., Komatsuzawa, H., Fujiwara, T., McCallum, N., and Sugai, M. (2004). Reduced content of lysyl-phosphatidylglycerol in the cytoplasmic membrane affects susceptibility to moenomycin, as well as vancomycin, gentamicin, and antimicrobial peptides, in *Staphylococcus aureus*. Antimicrob. Agents Chemother. *48*, 4800–4807.

Noble, W.C., Virani, Z., and Cree, R.G. (1992). Co-transfer of vancomycin and other resistance genes from *Enterococcus faecalis* NCTC 12201 to *Staphylococcus aureus*. FEMS Microbiol. Lett. *72*, 195–198.

Oliveira, D.C., Milheirico, C., and de Lencastre, H. (2006). Redefining a structural variant of staphylococcal cassette chromosome mec, SCCmec type VI. Antimicrob. Agents Chemother. *50*, 3457–3459.

Oliveira, D.C., Tomasz, A., and de Lencastre, H. (2001). The evolution of pandemic clones of methicillin-resistant *Staphylococcus aureus*: identification of two ancestral genetic backgrounds and the associated mec elements. Microb. Drug Resist. *7*, 349–361.

Oliveira, D.C., Tomasz, A., and de Lencastre, H. (2002). Secrets of success of a human pathogen: molecular evolution of pandemic clones of meticillin-resistant *Staphylococcus aureus*. Lancet Infect. Dis. *2*, 180–189.

Ornelas-Soares, A., de Lencastre, H., de Jonge, B., Gage, D., Chang, Y.S., and Tomasz, A. (1993). The peptidoglycan composition of a *Staphylococcus aureus* mutant selected for reduced methicillin resistance. J. Biol. Chem. *268*, 26268–26272.

Ornelas-Soares, A., de Lencastre, H., de Jonge, B.L., and Tomasz, A. (1994). Reduced methicillin resistance in a new *Staphylococcus aureus* transposon mutant that incorporates muramyl dipeptides into the cell wall peptidoglycan. J. Biol. Chem. *269*, 27246–27250.

Patel, J.B., Jevitt, L.A., Hageman, J., McDonald, L.C., and Tenover, F.C. (2006). An association between reduced susceptibility to daptomycin and reduced susceptibility to vancomycin in *Staphylococcus aureus*. Clin. Infect. Dis. *42*, 1652–1653.

Pereira, P.M., Filipe, S.R., Tomasz, A., and Pinho, M.G. (2007). Fluorescence ratio imaging microscopy shows decreased access of vancomycin to cell wall synthetic sites in Vancomycin-Resistant *Staphylococcus aureus*. Antimicrob. Agents Chemother. *51*, 3627–3633.

Perichon, B., and Courvalin, P. (2006). Synergism between beta-lactams and glycopeptides against VanA-type methicillin-resistant *Staphylococcus aureus* and heterologous expression of the *vanA* operon. Antimicrob. Agents Chemother. *50*, 3622–3630.

Pinho, M.G., de Lencastre, H., and Tomasz, A. (2001a). An acquired and a native penicillin-binding protein cooperate in building the cell wall of drug-resistant staphylococci. Proc. Natl Acad. Sci. USA *98*, 10886–10891.

Pinho, M.G., and Errington, J. (2003). Dispersed mode of *Staphylococcus aureus* cell wall synthesis in the absence of the division machinery. Mol Microbiol *50*, 871–881.

Pinho, M.G., and Errington, J. (2005). Recruitment of penicillin-binding protein PBP2 to the division site of *Staphylococcus aureus* is dependent on its transpeptidation substrates. Mol Microbiol 55, 799–807.

Pinho, M.G., Filipe, S.R., de Lencastre, H., and Tomasz, A. (2001b). Complementation of the essential peptidoglycan transpeptidase function of penicillin-binding protein 2 (PBP2) by the drug resistance protein PBP2A in *Staphylococcus aureus*. J. Bacteriol. *183*, 6525–6531.

Pinho, M.G., Ludovice, A.M., Wu, S., and De Lencastre, H. (1997). Massive reduction in methicillin resistance by transposon inactivation of the normal PBP2 in a methicillin-resistant strain of *Staphylococcus aureus*. Microb. Drug Resist. 3, 409–413.

Pootoolal, J., Neu, J., and Wright, G.D. (2002). Glycopeptide antibiotic resistance. Annu. Rev. Pharmacol. Toxicol. *42*, 381–408.

Projan, S. (2000). Antibiotic resistance in the staphylococci. In Gram-Positive Pathogens, V. A. Fischetti, R. P. Novick, J. J. Ferretti, D. A. Portnoy, and J. I. Rood, eds. (Washington, D.C., American Society for Microbiology).

Rohrer, S., Ehlert, K., Tschierske, M., Labischinski, H., and Berger-Bachi, B. (1999). The essential *Staphylococcus aureus* gene *fmhB* is involved in the first step of peptidoglycan pentaglycine interpeptide formation. Proc. Natl Acad. Sci. USA *96*, 9351–9356.

Ryffel, C., Strassle, A., Kayser, F.H., and Berger-Bachi, B. (1994). Mechanisms of heteroresistance in methicillin-resistant *Staphylococcus aureus*. Antimicrob. Agents Chemother. *38*, 724–728.

Safo, M.K., Zhao, Q., Ko, T.P., Musayev, F.N., Robinson, H., Scarsdale, N., Wang, A.H., and Archer, G.L. (2005). Crystal structures of the BlaI repressor from *Staphylococcus aureus* and its complex with DNA: insights into transcriptional regulation of the *bla* and *mec* operons. J. Bacteriol. *187*, 1833–1844.

Sakoulas, G., Eliopoulos, G.M., Fowler, V.G., Jr., Moellering, R.C., Jr., Novick, R.P., Lucindo, N., Yeaman, M.R., and Bayer, A.S. (2005). Reduced susceptibility of *Staphylococcus aureus* to vancomycin and platelet microbicidal protein correlates with defective autolysis and loss of accessory gene regulator (agr) function. Antimicrob. Agents Chemother. *49*, 2687–2692.

Seligman, S.J. (1966). Penicillinase-negative variants of methicillin-resistant *Staphylococcus aureus*. Nature *209*, 994–996.

Severin, A., Tabei, K., Tenover, F., Chung, M., Clarke, N., and Tomasz, A. (2004a). High level oxacillin and vancomycin resistance and altered cell wall composition in *Staphylococcus aureus* carrying the staphylococcal *mecA* and the enterococcal *vanA* gene complex. J. Biol. Chem. *279*, 3398–3407.

Severin, A., Wu, S.W., Tabei, K., and Tomasz, A. (2004b). Penicillin-binding protein 2 is essential for expression of high-level vancomycin resistance and cell wall synthesis in vancomycin-resistant *Staphylococcus aureus* carrying the enterococcal *vanA* gene complex. Antimicrob. Agents Chemother. *48*, 4566–4573.

Sharma, V.K., Hackbarth, C.J., Dickinson, T.M., and Archer, G.L. (1998). Interaction of native and mutant MecI repressors with sequences that regulate *mecA*, the gene encoding penicillin binding protein 2a in methicillin-resistant staphylococci. J. Bacteriol. *180*, 2160–2166.

Sieradzki, K., Leski, T., Dick, J., Borio, L., and Tomasz, A. (2003). Evolution of a vancomycin-intermediate *Staphylococcus aureus* strain *in vivo*: multiple changes in the antibiotic resistance phenotypes of a single lineage of methicillin-resistant *S. aureus* under the impact of antibiotics administered for chemotherapy. J. Clin. Microbiol. *41*, 1687–1693.

Sieradzki, K., Pinho, M.G., and Tomasz, A. (1999a). Inactivated *pbp4* in highly glycopeptide-resistant laboratory mutants of *Staphylococcus aureus*. J. Biol. Chem. *274*, 18942–18946.

Sieradzki, K., Roberts, R.B., Haber, S.W., and Tomasz, A. (1999b). The development of vancomycin resistance in a patient with methicillin-resistant *Staphylococcus aureus* infection. N. Engl. J. Med. *340*, 517–523.

Sieradzki, K., and Tomasz, A. (1997). Suppression of beta-lactam antibiotic resistance in a methicillin-resistant *Staphylococcus aureus* through synergic action of early cell wall inhibitors and some other antibiotics. J. Antimicrob. Chemother. *39 Suppl A*, 47–51.

Sieradzki, K., and Tomasz, A. (1999). Gradual alterations in cell wall structure and metabolism in vancomycin-resistant mutants of *Staphylococcus aureus*. J. Bacteriol. *181*, 7566–7570.

Sieradzki, K., and Tomasz, A. (2003). Alterations of cell wall structure and metabolism accompany reduced susceptibility to vancomycin in an isogenic series of clinical isolates of *Staphylococcus aureus*. J. Bacteriol. *185*, 7103–7110.

Sievert, D.M., Boulton, M.L., Stoltman, G., Johnson, D., Stobierski, M.G., Downes, F.P., Somsel, P.A., Rudrick, J.T., Brown, W., Hafeez, W., *et al.* (2002). *Staphylococcus aureus* Resistant to Vancomycin — United States, 2002. MMWR Morb. Mortal. Wkly Rep. *51*, 565–567.

Smith, I.M., and Vickers, A.B. (1960). Natural history of 338 treated and untreated patients with staphylococcal septicaemia (1936–1955). Lancet *1*, 1318–1322.

Sobral, R.G., Ludovice, A.M., de Lencastre, H., and Tomasz, A. (2006). Role of *murF* in cell wall biosynthesis: isolation and characterization of a *murF* conditional mutant of *Staphylococcus aureus*. J. Bacteriol. *188*, 2543–2553.

Sobral, R.G., Ludovice, A.M., Gardete, S., Tabei, K., de Lencastre, H., and Tomasz, A. (2003). Normally functioning *murF* is essential for the optimal expression of methicillin resistance in *Staphylococcus aureus*. Microb. Drug Resist. *9*, 231–241.

Stranden, A.M., Ehlert, K., Labischinski, H., and Berger-Bachi, B. (1997). Cell wall monoglycine cross-bridges and methicillin hypersusceptibility in a *femAB* null mutant of methicillin-resistant *Staphylococcus aureus*. J. Bacteriol. *179*, 9–16.

Tenover, F.C., Weigel, L.M., Appelbaum, P.C., McDougal, L.K., Chaitram, J., McAllister, S., Clark, N., Killgore, G., O'Hara, C.M., Jevitt, L., *et al.* (2004). Vancomycin-resistant *Staphylococcus aureus* isolate from a patient in Pennsylvania. Antimicrob. Agents Chemother. *48*, 275–280.

Thumanu, K., Cha, J., Fisher, J.F., Perrins, R., Mobashery, S., and Wharton, C. (2006). Discrete steps in sensing of beta-lactam antibiotics by the BlaR1 protein of the methicillin-resistant *Staphylococcus aureus* bacterium. Proc. Natl Acad. Sci. USA *103*, 10630–10635.

Tomasz, A. (2000). The staphylococcal cell wall. In Gram-Positive Pathogens, V. A. Fischetti, R. P. Novick, J. J. Ferretti, D. A. Portnoy, and J. I. Rood, eds. (Washington, American Society for Microbiology), pp. 351–360.

Utaida, S., Dunman, P.M., Macapagal, D., Murphy, E., Projan, S.J., Singh, V.K., Jayaswal, R.K., and Wilkinson, B.J. (2003). Genome-wide transcriptional profiling of the response of *Staphylococcus aureus* to cell-wall-active antibiotics reveals a cell-wall-stress stimulon. Microbiol. *149*, 2719–2732.

Uttley, A.H., Collins, C.H., Naidoo, J., and George, R.C. (1988). Vancomycin-resistant enterococci. Lancet *1*, 57–58.

Weigel, L.M., Clewell, D.B., Gill, S.R., Clark, N.C., McDougal, L.K., Flannagan, S.E., Kolonay, J.F., Shetty, J., Killgore, G.E., and Tenover, F.C. (2003). Genetic analysis of a high-level vancomycin-resistant isolate of *Staphylococcus aureus*. Science *302*, 1569–1571.

Wilke, M.S., Hills, T.L., Zhang, H.Z., Chambers, H.F., and Strynadka, N.C. (2004). Crystal structures of the Apo and penicillin-acylated forms of the BlaR1 beta-lactam sensor of *Staphylococcus aureus*. J. Biol. Chem. *279*, 47278–47287.

Wu, S., de Lencastre, H., Sali, A., and Tomasz, A. (1996a). A phosphoglucomutase-like gene essential for the optimal expression of methicillin resistance in *Staphylococcus aureus*: molecular cloning and DNA sequencing. Microb. Drug Resist. *2*, 277–286.

Wu, S., Piscitelli, C., de Lencastre, H., and Tomasz, A. (1996b). Tracking the evolutionary origin of the methicillin resistance gene: cloning and sequencing of a homologue of *mecA* from a methicillin susceptible strain of *Staphylococcus sciuri*. Microb. Drug Resist. *2*, 435–441.

Wu, S.W., and De Lencastre, H. (1999). Mrp-a new auxiliary gene essential for optimal expression of methicillin resistance in *Staphylococcus aureus*. Microb. Drug Resist. *5*, 9–18.

Wu, S.W., de Lencastre, H., and Tomasz, A. (2001). Recruitment of the *mecA* gene homologue of *Staphylococcus sciuri* into a resistance determinant and expression of the resistant phenotype in *Staphylococcus aureus*. J. Bacteriol. *183*, 2417–2424.

Zhang, H.Z., Hackbarth, C.J., Chansky, K.M., and Chambers, H.F. (2001). A proteolytic transmembrane signaling pathway and resistance to beta-lactams in staphylococci. Science *291*, 1962–1965.

Staphylococcus epidermidis and other Coagulase-Negative Staphylococci

9

Shu Yeong Queck and Michael Otto

Abstract

Over the last two decades, coagulase-negative staphylococci with the most important species *Staphylococcus epidermidis* have been recognized as important opportunistic pathogens. These abundant commensal organisms of the human skin and mucous membranes may cause serious infections, predominantly as biofilm-associated infections on indwelling medical devices, and are now the most frequent cause of hospital-acquired infections. More recently, the elucidation of the genomes of *S. epidermidis* and other coagulase-negative staphylococci, and a more pronounced interest in the molecular biology of especially *S. epidermidis* and its interaction with human host defences, have provided more detailed insight into how these bacteria cause human disease. We have learned that, although definitely more limited, the repertoire and also the regulation of virulence factors in *S. epidermidis* may differ significantly from *S. aureus*. Furthermore, as a result of the increasing volume and depth of research on this pathogen, some findings have been obtained in *S. epidermidis* that have paradigmatic character for many staphylococci and Gram-positive pathogens.

Staphylococci – introduction

Staphylococci are Gram-positive, AT-rich cocci, which are often arranged in grape-like clusters. The genus is divided into two groups, based on the ability of a species to produce the blood clotting enzyme coagulase. *Staphylococcus aureus* and *S. epidermidis* represent the most commonly found and studied coagulase-positive and coagulase-negative staphylococcal species in humans, respectively (Kloos and Schleifer, 1986).

Generally, species of the genus *Staphylococcus* are normal and abundant colonizers of the human skin and mucous membranes. Often, a predominant staphylococcal species can be found occupying specific niches of the human body (Kloos and Schleifer, 1986). The coagulase-negative staphylococci (CNS) maintain a benign relationship with their host and are only considered pathogenic in certain situations, for example, when the natural barrier such as the skin is damaged via trauma, inoculation or implantation of medical devices. In contrast, the coagulase-positive *S. aureus* is a well-established pathogen in humans. Unquestionably, there is a wider array of information pertaining to the physiology and virulence of *S. aureus* compared to the CNS. In recent years however medical interest in CNS has increased due to the escalation in the rate of CNS infections in hospitals. The two main reasons for the increasing rate of CNS infections are the spreading of antibiotic resistance among clinical CNS and the frequent use of medical devices (Raad *et al.*, 1998). According to the National Nosocomial Infections Surveillance System Report 2004 (NNIS, Centers for Disease Control and Prevention, Atlanta, GA), the rate of methicillin resistance among clinical CNS isolates was in the average of 89%, which was much higher than its pathogenic counterpart *S. aureus* (~40–50%), compared over a 6 year period from 1998 to 2003. Furthermore, CNS species are the predominant cause of nosocomial bloodstream infections, cardiovascular infections and infections of the eye, ear, nose and throat (NNIS, 1998). Many of these infections caused by CNS are the results of biofilm formation on medical devices. Biofilm-associated infections are difficult to eradicate by conventional antibiotic treatment, as the formation of biofilms dramatically increases the resistance to antimicrobial agents and defence mechanisms of host immunity.

While the pathogenesis of *S. aureus* can be attributed to the production of a number of virulence factors such as toxins and degradative exoenzymes, these factors are rarely associated in infections caused by CNS. Instead, the success of CNS to mount an infection likely relies on less aggressive factors such as those involved in adhesion to host surfaces and evasion of the host's immune system, resulting in infections that are more silent compared to those with *S. aureus*. To date, the question of what triggers transformation of CNS from the commensal flora to pathogens in humans remains unresolved. In addition, it is often difficult to distinguish if CNS isolates are the causative agents of an infection or simply contaminants during isolation since CNS are abundant flora in humans. However, with the recent advancements of genome research and data gathered from epidemiological studies, we are beginning to identify subtle differences between the commensal and pathogenic CNS populations that may provide a better schematic for distinguishing these CNS populations in the clinical settings.

Among the CNS, *S. epidermidis* is the most studied species, as it is believed to account for most of the device-related infections, whereas *S. saprophyticus* is the leading cause of uncomplicated urinary tract infections (UTI), next to the Enterobacteriaceae. The other CNS species (*S. haemolyticus, S. capitis, S. hominis, S. saccharolyticus, S. warneri, S. lugdunensis, S. cohnii*) have also occasionally been implicated in human infections. In this review, we will examine the virulence factors of CNS, with a specific focus on the predominant species in human infections, *S. epidermidis*.

Staphylococcus epidermidis – the most common CNS of device-related infections

Population and genetic diversity

Clinically, *S. epidermidis* infects mostly immunocompromised individuals such as AIDS patients, drug abusers and newborns or long-termed hospitalized and critically ill patients (Vuong and Otto, 2002; Venkatesh *et al.*, 2006). Most of the infections caused by *S. epidermidis* are associated with the formation of stress-resistant biofilms on medical devices such as indwelling catheters, or implanted device or contact lenses. In addition, the prevalence of genetically diverse methicillin-resistant *S. epidermidis* (MRSE) in the clinical setting has prevented the effective eradication of such infections with traditional antibiotic regimes (Raad *et al.*, 1998). Comparatively, a greater amount of information on the population structures of the methicillin-resistant *S. aureus* (MRSA) clones exists relative to those of MRSE. There is also a better consensus pertaining to the clonality of epidemic MRSA strains. Nevertheless, it can be surmised from the data available on infection-associated *S. epidermidis* strains that there is considerable genetic diversity, regardless of geographic or clinical origins (Dominguez *et al.*, 1996; Galdbart *et al.*, 1999; Wang *et al.*, 2003; Milisavljevic *et al.*, 2005; Miragaia *et al.*, 2007; Ueta *et al.*, 2007;). For instance, the staphylococcal cassette chromosome *mec* (SCC*mec*) element is a good illustration of the genetic diversity in *S. epidermidis*. There is a greater diversity in the SCC*mec* elements in MRSE strains, compared to those of MRSA. While only five major SCC*mec* types have been identified in MRSA strains (Katayama *et al.*, 2000; Ito *et al.*, 2001; Ma *et al.*, 2002), MRSE strains harbour at least five other uncharacterized SCC*mec* types (Miragaia *et al.*, 2007). In both MRSA and MRSE strains, methicillin resistance is mediated by *mecA*, which encodes for a penicillin binding protein (pbp) with reduced binding affinity for the β-lactam antibiotics. The *mecA* gene and its regulatory elements are located on large chromosomal fragments termed SCC*mec* elements, which can also carry, among other resistance determinants, a set of recombinases and a variety of mobile genetic elements such as transposons and insertion sequences. Similar to MRSA, SCC*mec* elements in MRSE strains are preferentially integrated at the *orfX* site, which appears to be a frequent site for genetic recombination in *S. epidermidis* (Miragaia *et al.*, 2007). In addition, certain mobile genetic elements such as IS256, a composite transposon of *Tn*4001, are often found in *S. epidermidis* strains associated with human infections (Kozitskaya *et al.*, 2004; Gu *et al.*, 2005; Yao *et al.*, 2005b). Nosocomial *S. epidermidis* isolates that harbour multiple copies of IS256 have been linked to biofilm formation and resistance to antimicrobial agents such as the aminoglycosides. There is also evidence to suggest that IS256 may result in chromosomal rearrangements in *S. epidermidis* (Ziebuhr *et al.*, 1999; Ziebuhr *et al.*, 2000). Taken together, frequent recombination events mediated by specific mobile genetic elements in the genome may play an important role in generating the genetic diversity observed in infection-associated *S. epidermidis* strains.

In recent years, there is an increased interest in using multilocus sequence typing (MLST) to characterize the population structures of epidemic MRSE strains (Wang *et al.*, 2003; Wisplinghoff *et al.*, 2003; Thomas *et al.*, 2007). Essentially, MLST characterizes the genetic population within bacterial strains by tracking changes in seven housekeeping genes. This approach is useful for elucidating the relationships between strains and to identify ancestral genotypes based on the sequence divergence of these housekeeping genes, which evolve at a relatively slow rate (Maiden *et al.*, 1998). However, there are at least two different MLST schemes proposed in *S. epidermidis* (Wang *et al.*, 2003; Wisplinghoff *et al.*, 2003), which makes it difficult to infer adequately on the genetic relatedness of strains. The main critic of the two published MLST schemes is that they were largely based on the scheme for *S. aureus* (Thomas *et al.*, 2007). Hence, an improved MLST scheme (incorporating part of the previous MLST schemes) for *S. epidermidis* was recently proposed. From a collection of *S. epidermidis* isolates of diverse geographical settings over a period of 6 years (1996–2001), a total of 74 sequence types (STs) from 217 isolates (nosocomial and commensal) was identified with the new MLST scheme, corroborating with previous studies that the *S. epidermidis* population is genetically diverse (Miragaia *et al.*, 2007). However, a single clonal lineage appears to encompass the majority of the STs (74% of the isolates) in *S. epidermidis*. STs of this lineage were found to be widely disseminated worldwide (South America, Europe, Africa and Asia) regardless of the origin of isolation, i.e. infection or colonization sites. Of particular interest, the ST2 type was the most frequently represented among the isolates, regardless of geographic or clinical origins and appears to be the ancestral genotype for the other prevalent STs. Almost all STs in this clonal lineage are methicillin resistant and carry predominantly the SCC*mec* IV element. Interestingly, the SCC*mec* IV element was also prevalent in MRSE isolates in other studies (Wisplinghoff *et al.*, 2003; Miragaia *et al.*, 2005) and importantly, this element is often associated with the highly virulent community associated-MRSA (Daum *et al.*, 2002). However, the connection between the SCC*mec* IV element and pathogenicity of staphylococci remains unclear. Nevertheless, it is apparent that one would need to analyse more bacterial strains from diverse sources with a uniform MLST scheme to better characterize the epidemiology of *S. epidermidis*.

Lessons from genome sequencing

To date, the genomes of two *S. epidermidis* strains, ATCC12228 (reference strain, non-biofilm forming) and RP62A (MRSE, biofilm forming) are completely sequenced and annotated (Zhang *et al.*, 2003; Gill *et al.*, 2005; Wei *et al.*, 2006) (http://www.ncbi.nlm.nih.gov/genomes/lproks.cgi). Such genome data have provided useful insights into the physiology of *S. epidermidis*. For example, while most of the attention has been focused on the virulence of staphylococci, little is known about their metabolism. However, from the genome of *S. epidermidis*, several metabolic pathways such as those involved in the utilization of a variety of simple and complex sugars, the phosphate pathway and mevalonate pathway were identified. This new array of information is likely to greatly accelerate our understanding on the biology of *S. epidermidis*. In addition, specific features in the genome reflect the adaptability of *S. epidermidis* to persist under specific conditions, for example, in high salt environments. A total of eight predicted sodium ion/proton exchangers and six transport systems for proline, glycine betaine or probable osmoprotectants are present in RP62A. Additionally, the mechanosensitive ion channels (MscL and MscS) and the constitutive Trk potassium ion channels, which have been shown to protect cells from osmotic shock, are present. The production of adhesion factors in *S. epidermidis* can be viewed as another adaptation to a habitat, as the bacteria need them for efficient colonization and persistence in the human host. An important feature in the pathogenesis of *S. epidermidis* in device-related infections is the formation of stress-resistant biofilms, which enables the organism to resist antimicrobial agents and defence mechanisms from innate host defence. Biofilm formation proceeds through a series of events namely, adhesion to surfaces, surface colonization and detachment of biofilms culminating in bacterial dissemination. Therefore, adhesion also constitutes the first step of the infection process for *S. epidermidis*. From the genome of *S. epidermidis*, an abundance of adhesins to abiotic surfaces, such as polystyrene, and biotic surfaces, such as vitronectin, fibrinogen and collagen can be found. This indicates that the organism is well-equipped to colonize a variety of surfaces. Notably, the autolysin E (AltE) and the fibrinogen binding proteins (Fbe/SdrG) are considered important adhesins of *S. epidermidis*.

As most infections caused by *S. epidermidis* are chronic and 'silent', for example, biofilm-related infections (Costerton *et al.*, 1999), it is not surprising that many of the established virulence factors associated with *S. aureus* infections such as enterotoxins, exotoxins and leukotoxins are absent. However, more recent findings show that *S. epidermidis* and other CNS can also produce factors that are capable of inducing an inflammatory response, notably the phenol soluble modulins (Mehlin *et al.*, 1999) (PSMs). Interestingly, it has been demonstrated in *S. epidermidis* that the levels of PSM production were found to be significantly lower in biofilms (Yao *et al.*, 2005a), indicating that *S. epidermidis* biofilm-associated infections are generally less aggressive. Additionally, novel virulence determinants have been identified in the *S. epidermidis* genomes, for example, the *cap* operon responsible for the production of polyglutamate capsule (PGA) (Kocianova *et al.*, 2005), which is a major virulence factor of *Bacillus anthracis*. Overall, it can be surmised from the genome of *S. epidermidis* that the organism is more adapted for subacute or chronic infections compared to its more virulent counterpart, *S. aureus*. Recently, it was proposed that the 'silent' nature of *S. epidermidis* infections can be attributed to the ease of its transmission from host to host, since the organism is abundant on the skin (Massey *et al.*, 2006). Conversely, acute infections by *S. aureus* may facilitate an otherwise more complex pathway of transmission, through enhancing host to host contacts, for example, frequent contacts between healthcare takers and *S. aureus*-infected individuals.

Undoubtedly, an array of information pertaining to adaptation and virulence was gathered from the sequenced genomes. However, it remains a challenge to distinguish among commensal and pathogenic *S. epidermidis* strains. One approach to resolve this issue is to search for genetic differences among the two populations of *S. epidermidis*. A comparison between the genome of ATCC12228 and RP62A revealed a total of 10297 single nucleotide polymorphisms (SNPs) with a considerable number of SNPs resulting in a non-synonymous (NS) change in amino acid sequence (Gill *et al.*, 2005; Wei *et al.*, 2006). NS SNPs are relatively common in genes encoding for cell envelope functions. Interestingly, a number of cell-wall related proteins in RP62A were also found to be more prevalent in infection-associated strains and to be immunogenic, i.e. expressed during human infection (Bowden *et al.*, 2005). Several of the NS SNPs gene products are involved in stress adaptations such as oxidative and osmotic shock. Indeed, RP62A exhibited better survival than ATCC12228 when both strains were subjected to oxidative stress (Wei *et al.*, 2006). In addition, a large number of the NS SNPs occur in genes encoding for hypothetical proteins of unknown functions. Taken together, the subtle genetic differences within the genomes of RP62A and ATCC12228 may reflect specific adaptations by pathogenic *S. epidermidis* to cope with changes in the host environment during infection, especially when NS SNPs appear to evolve more rapidly in RP62A than ATCC12228.

Another approach taken to differentiate between infection-associated *S. epidermidis* and the commensal strains is to identify specific infection-associated genetic markers. One potential marker is the *ica* locus, which encodes for the production of the polysaccharide intracellular adhesin (PIA). The production of PIA has previously been linked to mediate biofilm formation and protection against host innate immunity in *S. epidermidis*. Notably, the *ica* operon is more prevalent in biofilm forming strains such as RP62A and absent in a number of non-biofilm forming isolates, including ATCC12228. While the *ica* locus appears to be an ideal genetic marker for the identification of infection-associated *S. epidermidis* strains, there were also several studies showing that neither the *ica* locus nor biofilm formation are necessary in all *S. epidermidis* and other CNS-related infections (Chokr *et al.*, 2006; Kogan *et al.*, 2006; Chokr *et al.*, 2007; Nilsdotter-Augustinsson *et al.*, 2007; Qin *et al.*, 2007; Rohde *et al.*, 2007). A recent comparative genomic study between *S. epidermidis* isolated from prostheses infections and the skin of healthy individuals have revealed several other potential targets, in addition to previous markers such as *ica* and IS256, for distinguishing the commensal strains from their pathogenic counterparts (Yao *et al.*, 2005b). Of particular interest, a gene encoding a 190-kDa cell surface protein with high homology to a streptococcal haemagglutinin binding protein is more prevalent in infection-associated strains. The function(s) of this surface protein in *S. epidermidis* however, remains unclear.

In conclusion, analysis of the sequenced genomes revealed high genetic variability of *S. epidermidis* as a species. It is clear that such genetic diversity would render the search for markers of invasiveness in pathogenic strains of *S. epidermidis* a greater challenge. Furthermore, many of the newly identified features found in the genomes (e.g. for stress adaptations or virulence) would require experimental data to validate their role in pathogenesis of *S. epidermidis*.

Virulence determinants

While several novel putative virulence factors have been identified in the genomes of *S. epidermidis* RP62A and ATCC12228, it should be kept in mind that for many of them we do not know yet if these factors are produced or play a role in *S. epidermidis* infection. Therefore, this section will mostly focus on virulence factors, for which evidence is available from biochemical or molecular biology experiments, or from epidemiological data (Table 9.1).

In order to investigate virulence factors in *S. epidermidis*, genetic technologies are used that are largely similar to those used in *S. aureus*. Transposon mutagenesis has mostly been performed using Tn917 in *S. epidermidis* (Mack *et al.*, 1994; Heilmann *et al.*, 1996a), but mariner-based transposon mutagenesis, which commonly has a less pronounced bias for integrating into specific sequences in the genome (Bae *et al.*, 2004), should be possible. We have also extensively used microarrays mostly for transcriptional profiling (Yao *et al.*, 2005a; Yao *et al.*, 2005b; Yao *et al.*, 2006; Lai *et al.*, 2007; Li *et al.*, 2007). These arrays first were 70-mer synthetic oligonucleotide arrays based on the *S. epidermidis* RP62A sequence, which we now have replaced with Affymetrix arrays (RMLchip3) that are based on both the RP62A and ATCC12228 genomes (Zhang *et al.*, 2003; Gill *et al.*, 2005). However, molecular investigation of *S. epidermidis* has been difficult due to inefficient transformation procedures. *S. epidermidis* is usually transformed by electroporation (Augustin and Gotz, 1990), although for strain 1457 a transducing phage may be used as an alternative (Nedelmann *et al.*, 1998). Electroporation efficiency is often extremely low and for some strains may not work at all. For those strains, protoplast transformation (Gotz *et al.*, 1981) has been used with more success (C. Vuong, M. Otto, unpublished). Ligation reactions are best first transformed into *S. aureus* RN4220 and from there, the final plasmid products may be isolated and electroporated into *S. epidermidis*. Probably, a similar method may work for most CNS. Unfortunately, due to the difficulties with transformation, *S. epidermidis* can usually not be transformed with large plasmids. In our group, the largest plasmid that was successfully brought into *S. epidermidis* using protoplast transformation was ~ 18kb (C. Vuong, M. Otto, unpublished). The transformed strain was used to establish a bioluminescent *S. epidermidis* strain, Xen43, that is based on strain 1457, has the *lux* genes integrated into the chromosome, and is available from Caliper Life Sciences.

Table 9.1 Virulence factors of *S. epidermidis*

Name	Genetic locus identified (reference)	Protein/factor identified/characterized (reference)	Function(s)
PIA	*ica* (Heilmann *et al.*, 1996b)	IcaA, IcaB, IcaC, IcaD, IcaR (Gerke *et al.*, 1998; Heilmann *et al.*, 1996b; Vuong *et al.*, 2004b)	Biofilm exopolysaccharide: cell-cell adhesion, haemagglutination, inflammatory effects, evasion of host immune system
AAP	(Zhang *et al.*, 2003)	(Hussain *et al.*, 1997)	Biofilm accumulation
SSP1, SSP2 (= AAP?)		(Veenstra *et al.*, 1996)	Biofilm initiation: Attachment to polystyrene
Bhp (Bap)	(Zhang *et al.*, 2003)	Bhp	Biofilm formation
PGA	(Zhang *et al.*, 2003)	Poly-γ-D,L-glutamic acid	Immune evasion, osmoprotection
AltE	(Heilmann *et al.*, 1997)	(Heilmann *et al.*, 1997)	Autolysin/adhesin: biofilm attachment to polystyrene, vitronectin binding
Fbe/SdrG	(Nilsson *et al.*, 1998; Davis *et al.*, 2001)	(Davis *et al.*, 2001; Nilsson *et al.*, 1998)	Fibrinogen binding, inhibition of phagocytosis
Embp	(Williams *et al.*, 2002)	(Williams *et al.*, 2002)	Fibronectin binding
Lipases	*gehC* (Farrell *et al.*, 1993)	GehC (Farrell *et al.*, 1993)	Persistence in fatty secretions
	gehD (Longshaw *et al.*, 2000)	GehD (Bowden *et al.*, 2002; Longshaw *et al.*, 2000)	Persistence in fatty secretions, binding to collagen
Cysteine protease	–	(Dubin *et al.*, 2001)	Unknown, possibly tissue damage
Metalloproteases	(Teufel and Gotz, 1993)	SepA (Lai *et al.*, 2007; Teufel and Gotz, 1993)	Lipase maturation, resistance to antimicrobial peptides
Serine protease	(Ohara-Nemoto *et al.*, 2002)	GluSE (Ohara-Nemoto *et al.*, 2002)	Biofilm formation (?), degradation of fibrinogen and complement factor C5
Fatty acid-modifying enzyme (FAME)	–	(Chamberlain, 1999)	Detoxification of host-produced bactericidal fatty acids
Staphylococcus epidermidis surface (Ses) proteins	(Bowden *et al.*, 2005)	SesA, SesC, SesE, SesG, SesH, SesI, Ebh (Bowden *et al.*, 2005)	Unknown, most are possibly involved in pathogenesis
Phenol soluble modulins (PSMs)	(Mehlin *et al.*, 1999; Vuong *et al.*, 2004a; Yao *et al.*, 2005a)	(Liles *et al.*, 2001; Mehlin *et al.*, 1999; Vuong *et al.*, 2004a; Yao *et al.*, 2005a)	Biofilm detachment, several proinflammatory activities
δ-toxin (= PSMγ)	(Otto *et al.*, 1998b)	(McKevitt *et al.*, 1990)	Several proinflammatory effects
Staphyloferrin A and B	–	(Lindsay *et al.*, 1994)	Siderophores: iron uptake
SitABC	(Cockayne *et al.*, 1998)	(Cockayne *et al.*, 1998)	Possibly involved in iron uptake
Lantibiotics (epidermin, Pep5)	(Kaletta *et al.*, 1989; Schnell *et al.*, 1988)	Reviewed in (Jack and Jung, 2000)	Bacterial interference (?)

Factors involved in biofilm formation

Biofilm formation is generally considered the most important virulence trait of *S. epidermidis* and several other CNS. For this reason, research on biofilm factors in *S. epidermidis* has started relatively early and results achieved in *S. epidermidis* have had a model character for other staphylococci, including *S. aureus*, and even for other species.

The first attempts to address the molecular basis of *S. epidermidis* biofilm formation were performed in the mid 90s, using transposon mutagenesis approaches with Tn917 on a genetic level (Heilmann *et al.*, 1996a), and biochemical characterization of the exopolysaccharide substance of *S. epidermidis* on a biochemical level (Mack *et al.*, 1996). The main result of these investigations was the discovery of the biosynthetic genetic locus and the characterization of the main exopolysaccharide of *S. epidermidis*, polysaccharide intercellular adhesin, or PIA. The substance is a linear β-1,6-linked homopolymer of N-acetylglucosamine, in which a varying percentage of N-acetyl groups are removed. In the meantime, the same substance has been found in several bacteria, including important pathogens such as *S. aureus*, *Yersinia pestis*, and pathogenic *E. coli*, to name but a few (Cramton *et al.*, 1999; Darby *et al.*, 2002; Wang *et al.*, 2004). Unfortunately, the polymer has been given many different names (PNAG, PGA), although chemical analyses have found that it consists predominantly of the same substance, polymeric partially deacetylated N-acetylglucosamine.

The genetic locus driving PIA synthesis is called *ica* and comprises the *icaA*, *icaD*, *icaB*, and *icaC* genes, in this order seen from the *icaA* promoter that controls expression of the entire operon (Fig. 9.1). In addition, there is a regulatory gene called *icaR* encoded upstream and transcribed in the other direction. The *icaR* and *icaA* promoters are subject to a multitude of regulatory effects linking *ica* gene expression to virulence regulators such as *sarA* and *sigB*, and to environmental influences such as anaerobiosis, which significantly increases PIA production, or salt concentration (Cramton *et al.*, 2001; Knobloch *et al.*, 2001; Knobloch *et al.*, 2002;

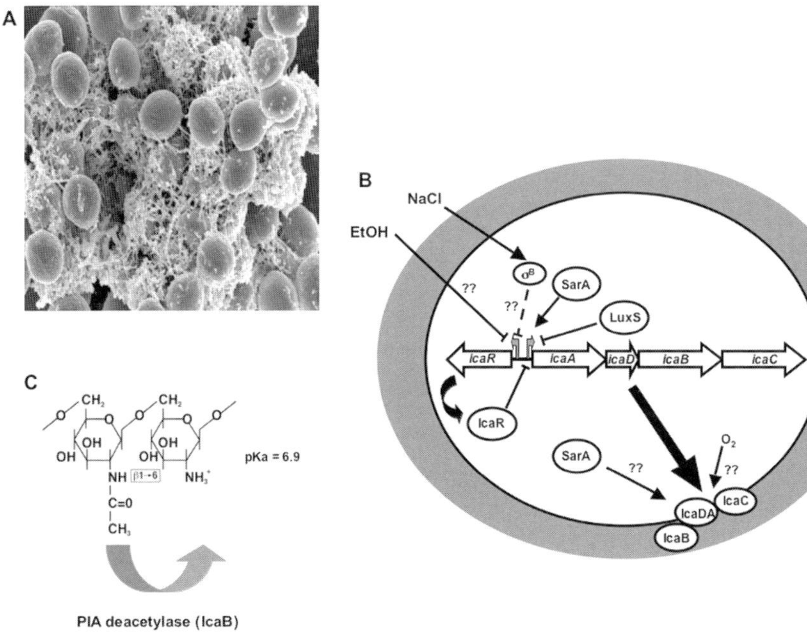

Figure 9.1 PIA (Polysaccharide intercellular adhesin). (A) Scanning electron microscopic picture of *S. epidermidis* cells embedded in extracellular matrix. Most of this matrix is made of PIA (Vuong *et al.*, 2004d). (B) Genetic organization of the *ica* gene locus encoding the PIA biosynthesis genes and the negative regulator IcaR. Regulatory influences on the *icaA* and *icaR* promoters and PIA production are shown. (C) Chemical structure of PIA and function of the PIA deacetylase IcaB. IcaB is a surface-located enzyme that removes a portion of the acetyl moieties in the N-acetylglucosamine polymer, yielding free amino groups with positive charges at acid to neutral pH.

Knobloch *et al.*, 2004; Tormo *et al.*, 2005b). The function of the *icaA* and *icaD* genes has been investigated in a groundbreaking manuscript by Gerke *et al.* demonstrating that the two gene products form the N-acetylglucosmaine transferase (Gerke *et al.*, 1998), while it is still not clear why there are two gene products needed for this function. Both IcaA and IcaD are transmembrane proteins. IcaC is another membrane protein with a putative function in the export of the growing sugar chain, an assumption based on its location and the fact that in an *icaC* mutant, sugar chains only reach a length of ~15 residues, which may be due, hypothetically, to the possibility that efficient polymerization must be linked to export (Gerke *et al.*, 1998). IcaB is the deacetylase removing N-acetyl residues in a certain percentage (~10–20%) after export of the sugar chain on the outside surface of the cell (Vuong *et al.*, 2004b). Deacetylation is critical for the function of PIA in biofilm formation and in pathogenesis. In an *icaB* mutant lacking deacetylated residues, PIA is released from the cell surface, indicating that the positive charge of the polymer is crucial for cell surface location (Vuong *et al.*, 2004b). An anchor structure that links the PIA polymer to the surface in a covalent linkage, like for many other surface polymers such as teichoic acids and polysaccharide capsules, is not known for PIA. As the results with the *icaB* mutant indicate, the electrostatic interaction with the negatively charged cell surface may be sufficient to keep PIA molecules stuck on the cell.

A further mutant discovered in the initial transposon mutagenesis approach was one in which the locus encoding the main autolysin of *S. epidermidis*, AtlE, was disrupted (Heilmann *et al.*, 1996a). The mutant was impaired in the initial attachment to the plastic surface. This effect is most likely due to the decreased surface hydrophobicity of the mutant, which again is probably due to the fact that AtlE as one of the most abundant surface proteins of *S. epidermidis* is missing in the mutant (Heilmann *et al.*, 1997). While direct attachment to plastic may occur *in vivo* during the colonization of catheters, the interaction of specific surface binding proteins with host matrix proteins that soon cover the implanted material, is most likely much more important for initial *in vivo* biofilm attachment and discussed below. Furthermore, many laboratories including ours have observed that attachment to plastic is extremely dependent on the material, and even the use of different batches of microtitre plates of the same type may lead to different results. However, there is agreement as to the fact that surface physico-chemical parameters such as hydrophobicity are the main determinants of *S. epidermidis* attachment to abiotic materials, at least *in vitro* in the absence of host proteins (Vacheethasanee *et al.*, 1998).

Subsequent to initial attachment, biofilm cells aggregate, eventually forming the typical mushroom-like towers, which are commonly observed under flow conditions *in vitro*. Factors influencing this aggregation step are manifold, but the most important appears to be the exopolysaccharide PIA mentioned above. Further factors important in this step are teichoic acids, proteinaceous factors, and possibly DNA.

While for some time, PIA was supposed to be absolutely critical for biofilm formation, several *S. epidermidis* strains have been isolated in the meantime, including from infection, that do not have PIA and appear to rely on proteinaceous factors for biofilm formation (Chokr *et al.*, 2006; Kogan *et al.*, 2006; Chokr *et al.*, 2007; Nilsdotter-Augustinsson *et al.*, 2007; Qin *et al.*, 2007; Rohde *et al.*, 2007). A critical proteinaceous biofilm factor appears to be AAP, the accumulation-associated protein of *S. epidermidis* (Hussain *et al.*, 1997; Rohde *et al.*, 2007). This protein forms fimbriae-like structures on the surface, thus possibly mediating cell-to-cell connections similar to the sugar-based PIA (Banner *et al.*, 2007). AAP may be identical to a protein, SSP, for which a similar phenotype was reported early (Veenstra *et al.*, 1996). Interestingly, AAP needs to be truncated by proteases, which may be host-derived, to become functional (Rohde *et al.*, 2005). Furthermore, it has been recently reported that *S. epidermidis* has the ability to switch to protein-dependent biofilms using AAP when PIA production is abolished (Hennig *et al.*, 2007). Another protein involved in *S. epidermidis* biofilm formation, including in strains without PIA production, is the biofilm-associated protein Bap (Bhp) (Tormo *et al.*, 2005a), originally found in bovine isolates of *S. aureus* where it is located on a mobile genetic element (Cucarella *et al.*, 2001; Cucarella *et al.*, 2004).

In the cases of several other factors, for which an influence on biofilm formation has been demonstrated, such as teichoic acids (Gross *et al.*, 2001), the effect on biofilm formation is probably not the main purpose. This is certainly true for DNA, whose role in biofilm formation has been recognized in several biofilm-forming bacteria (Spoering and Gilmore, 2006) including *S. aureus* (Rice *et al.*, 2007) and thus, it can be assumed that it has the same role in *S. epidermidis*. DNA as a main component of all organisms is released after cell lysis and may contribute to aggregation owing to its polyanionic character, similar to teichoic acids. However, there is no evidence to suggest that the release of DNA is a triggered process aimed to promote biofilm formation. In this regard, it is important to realize that PIA as a polycationic substance (owing to the free amino groups released by the deacylase IcaB (Vuong *et al.*, 2004b)) has a special role in making adhesive connections between the cell surface, which is anionic, and the anionic polymers such as teichoic acids and DNA (Otto, 2006). Thus, PIA may be called a true biofilm polymer, whereas teichoic acids and DNA do not by themselves promote biofilm formation. It appears to be important for the future of biofilm research to make a clear distinction between these two types of biofilm factors. Key contributors to the development of a biofilm can be recognized for example by regulatory effects that are aimed to control production in accordance with a biofilm lifestyle and are different from those factors that merely interact in a heterogeneous matrix with the genuine biofilm factors.

The detachment of cells or entire cell clusters from a biofilm for the purpose of dissemination is believed to be crucial for the spread of a biofilm-associated infection in the human body. While the mere cessation of the production of biofilm building material, such as PIA, may eventually lead to dissemination together with mechanical forces acting on the biofilm, there might be additional factors governing dissemination in a more specific and efficient manner. Some other bacteria produce enzymes that degrade PIA, which may promote a much more rapid biofilm dispersal, for example *Actinobacillus actinomycetemcomitans* (Kaplan *et al.*, 2003), but enzymes with such a function have not been discovered in *S. epidermidis* or other staphylococci. It is thus likely that the enzymatic degradation of PIA is not a factor contributing to biofilm dissemination in staphylococci. Recent research in our laboratory has shown that phenol-soluble modulins, amphipathic peptides, which also have a proinflammatory function, may contribute to biofilm detachment owing to their detergent-like properties (M. Otto, unpublished). Especially the phenol-soluble modulins of the β subtype appear to have a key role as biofilm surfactants and contribute to the formation of channels in the biofilm and at higher concentration, to biofilm detachment. Obviously, the production of these molecules needs to be under tight regulation for the efficient development of a biofilm and in fact we found that phenol-soluble modulins are strictly controlled by the *agr* quorum-sensing system, in clear contrast to the *ica* genes, which are not under *agr* control (Vuong *et al.*, 2003; Vuong *et al.*, 2004a). This regulatory influence likely has a main role in controlling biofilm expansion. Thus, the phenol soluble modulins appear to be functional analogues of the biofilm surfactant rhamnolipid that mediates biofilm detachment in *Pseudomonas aeruginosa* and is also under quorum-sensing control (Boles *et al.*, 2005).

MSCRAMMs and other surface proteins

For *S. epidermidis*, binding to host tissue is an important part of its commensal lifestyle on the skin and mucous membranes. In addition, the colonization of tissue surfaces is a crucial first step during the establishment of an infection. Therefore, similar to all staphylococci, *S. epidermidis* has a large number of surface proteins to bind to a variety of host matrix proteins with frequently redundant binding properties. These proteins have been termed MSCRAMMS (microbial surface components recognizing adhesive matrix molecules). Many surface proteins are linked to peptidoglycan by sortase and are easily recognizable by a common structure that contains an LPXTG motif at the C-terminus of the amino acid sequence (Mazmanian *et al.*, 1999). On this basis, there are ~12 sortase-linked surface proteins in the genome of *S. epidermidis* RP62A: the SdrG protein (or fibrinogen-binding protein

Fbe), which belongs to the serine/aspartate repeat family, and 11 proteins that have been named Ses (for *S. epidermidis* surface proteins) A through K, not all of which are present in all strains (Gill *et al.*, 2005). SesD is equivalent to Bhp (Bap) and SesF to Aap, whose role in biofilm formation has been discussed above. Interestingly, many of the *S. epidermidis* MSCRAMMS do not appear to be negatively regulated by the quorum-sensing regulator *agr* (Bowden *et al.*, 2005), a rule established in *S. aureus* and based on the idea that binding proteins are only needed during initial colonization (at low cell density) (Novick, 2003).

In addition to sortase-catalysed covalent linkage to peptidoglycan, surface proteins can be bound in a non-covalent form, such as the autolysin AtlE and other autolysins. In addition to the autolysins, non-covalently attached surface proteins in *S. epidermidis* comprise a large (~1MDa) fibronectin-binding protein (Ebh or Embp), the elastin-binding protein Ebp, 2 members of the serine/aspartate repeat family (SdrF and SdrH) and another fibrinogen-binding protein (Gill *et al.*, 2005).

Among the covalently attached surface proteins, probably most work has been performed on SdrG (Fbe), the fibrinogen-binding protein of *S. epidermidis*. SdrG is present in most *S. epidermidis* strains, has a size of 119 kDa, and is similar to the fibrinogen-binding proteins of *S. aureus* called clumping factors (ClfA, ClfB) (Nilsson *et al.*, 1998). Antibodies against SdrG can block adhesion to fibrinogen-coated surface and implanted catheters, demonstrating that this protein is a major factor mediating adhesion to fibrinogen in *S. epidermidis* (Pei *et al.*, 1999; Pei and Flock, 2001b). The presence of SdrG has significant implications for virulence, as demonstrated using an allelic replacement mutant in a rat model of intravascular catheter-associated infection (Vernachio *et al.*, 2006). SdrG binds to the beta chain of fibrinogen, like ClfB of *S. aureus*, but in contrast to ClfA of *S. aureus*, which binds to the gamma chain. In-depth biochemical studies of SdrG binding to fibrinogen revealed that SdrG binds to a peptide near the thrombin cleavage site in fibrinogen (Davis *et al.*, 2001). SdrG inhibits thrombin-induced clotting of fibrinogen by interfering with the release of fibrinopeptide B. As fibrinopeptide B functions as a chemotactic molecule, SdrG binding might reduce the influx of phagocytic neutrophils to the infection site. SdrG might therefore have two tasks to help bacterial survival in the host: it can serve as a binding molecule and reduces phagocytic elimination.

The Atl type of staphylococcal autolysins, the most important non-covalently attached surface proteins, consist of a glucosaminidase and an amidase part in addition to signal peptide and propeptide sequences (Otto, 2004b). There is increasing evidence to suggest that autolysins, in addition to their role in cell wall turnover, have an important second function as host matrix protein binding proteins. AtlE for example shows vitronectin binding (Heilmann *et al.*, 1997) and Aae, another recently discovered autolysin/adhesin of *S. epidermidis* binds to fibrinogen, fibronectin, and vitronectin (Heilmann *et al.*, 2003). Another example of a protein with a dual adhesive and enzymatic function is the collagen-binding lipase GehD of *S. epidermidis* (Bowden *et al.*, 2002).

Toxins and exoenzymes

The repertoire of toxins of *S. epidermidis* is very low compared with its much more aggressive relative *S. aureus*. The only characterized toxins are the phenol-soluble modulins (PSMs), a family of short peptides with haemolytic and proinflammatory function (Mehlin *et al.*, 1999; Liles *et al.*, 2001; Vuong *et al.*, 2004a). We have already mentioned the PSMs when discussing their role in biofilm detachment. These peptides are produced at substantial levels in *S. epidermidis* and may significantly contribute to its pathogenic potential. This is due to their strong capacity to cause activation, cytokine release, and at higher concentrations, lysis of human neutrophils, establishing a key role for these peptides in virulence and the interaction with innate host defence (R. Wang, M. Otto, unpublished results). The composition and function of PSMs in *S. epidermidis* has first been described by the group of S. Klebanoff, describing three peptides, PSMα, PSMβ, and PSMγ, the latter being identical to the previously described *S. epidermidis* δ-toxin (Fig. 9.2) (Mehlin *et al.*, 1999). (We suggest avoiding the term PSMγ for the δ-toxin, as this has caused much confusion.) In the meantime, we have found additional PSMs by reversed genetics, an approach that was necessary, as PSM genes are often smaller than the threshold level used for gene annotation in genome sequencing projects (Vuong *et al.*, 2004a; Yao *et al.*, 2005a). *S. epidermidis* contains four different PSMs of the shorter type (~ 20 amino acids), which we call the α-type. These are PSMα and PSMδ, which are encoded next to each other on the chromosome, PSMε, which is encoded without connection to another PSM gene, and the δ-toxin (*hld*), which is encoded within RNAIII, the gene coding for

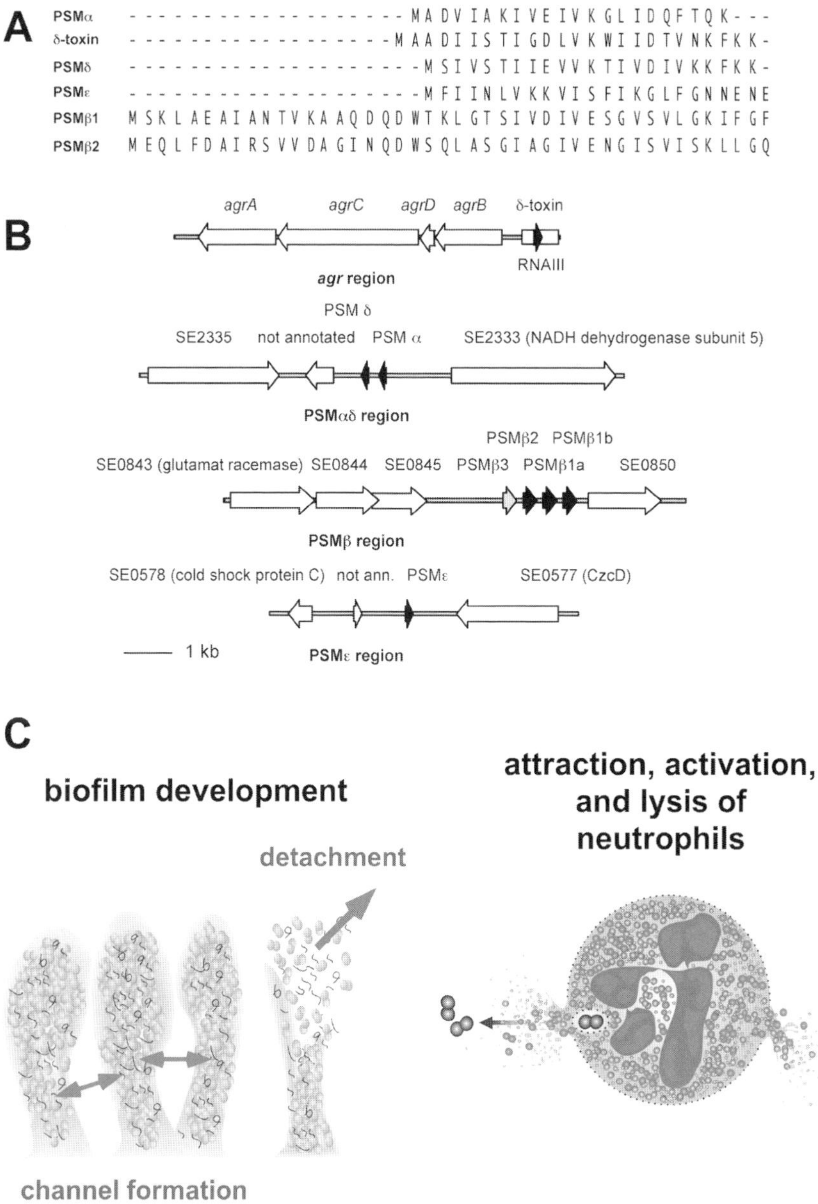

Figure 9.2 PSMs (Phenol-soluble modulins). (A) Amino acid sequences of the PSMs of *S. epidermidis*. (B) Genetic organization of PSM encoding loci in *S. epidermidis*. Gene numbers are from strain ATCC12228 (Zhang *et al.*, 2003). (C) Function of PSMs in biofilm development and inflammation. PSMs are believed to create channels and ultimately lead to cell cluster detachment owing to their detergent-like properties. In addition, they have considerable capacity to attract, activate and lyse human neutrophils and other immune cells (R. Wang and M. Otto, unpublished).

the regulatory RNA of the *agr* quorum-sensing system. The PSMβ peptides (β type) are larger (~40 amino acids) and encoded in an operon. This operon contains 3 or 4 genes, the difference arising from the duplication of the PSMβ1 gene in some strains (such as RP62A). The PSMβ3 gene product does not appear to be produced, as we have not detected the corresponding peptide in the culture filtrate of any *S. epidermidis* strain. In general, the proinflammatory activity appears to be more pronounced in the α type of PSM peptides, whereas the effect on biofilm formation is mostly executed by the β-type peptides.

Among the degradative exoenzymes, *S. epidermidis* contains a series of proteases with established or putative functions in virulence. It produces a glutamyl endopeptidase (SspA) (Dubin *et al.*, 2001), a cysteine protease (SspB) (Dubin *et al.*, 2004), and a homologue to the *S. aureus* aureolysin with additional elastase function (SepA) (Teufel and Gotz, 1993). The glutamyl endopeptidase is preferentially expressed in adherent culture, indicating a possible role in biofilm formation (Ohara-Nemoto *et al.*, 2002). The *agr*-controlled Zn^{2+}-metalloprotease SepA has been shown to degrade the human antimicrobial peptide dermcidin, and might thus play a key role in immune evasion processes on the human skin (Lai *et al.*, 2007). Like most proteases, SepA is under control of global regulatory systems, notably SarA, but also *agr* and *saeRS* (Lai *et al.*, 2007). These regulatory systems respond to the presence of antimicrobial peptides by a general upregulation of extracellular proteolytic activity. The *S. epidermidis* proteases may further contribute to virulence, functioning as maturation enzymes needed for the proteolytic removal of pro-peptides from other exported virulence factors, such as lipases, and by the degradation of host tissues during infection.

S. epidermidis has 5 putative genes coding for lipases (Gill *et al.*, 2005; Wei *et al.*, 2006). Two of these genes have been named GehC and GehD, (also known as Geh1 and Geh2). Lipase production is regulated by *agr* via regulation of the proteolytic maturation step (Vuong *et al.*, 2000a). The function of lipases during infection is mostly hypothetical, but may involve enabling the bacteria to persist in the fatty secretions on the human skin or by escaping phagocytosis.

Finally, a fatty-acid modifying enzyme (FAME) activity – first identified in *S. aureus* – has been found in 88% of *S. epidermidis* strains (Chamberlain and Brueggemann, 1997; Chamberlain, 1999). FAME inactivates bactericidal fatty acids by esterifying them to cholesterol. This putative extracellular enzyme may provide protection for *S. epidermidis* by inactivating these bactericidal lipids present on the skin. The molecular nature of FAME and the gene locus responsible for its expression are not known.

Capsule

S. epidermidis genome sequencing revealed the presence of the *cap* locus (Zhang *et al.*, 2003), which in *Bacillus anthracis* encodes production of the poly-γ-D-glutamate (PGA) capsule (Hanby and Rydon, 1946; Ashiuchi *et al.*, 2001; Ashiuchi and Misono, 2002), a key virulence factor of that important pathogen and bioterrorism agent. (Note that this *cap* locus is different from the polysaccharide *cap* loci of many bacteria including *S. aureus*.) In this polymer, D- (or sometimes L-) glutamate is linked by the γ-carboxy group to form a pseudopeptide chain. *B. anthracis* at the time was the only pathogen with a demonstrated function of PGA in virulence, which is to protect from phagocytosis. In addition, PGA is produced by a variety of non-pathogenic organisms, including halophilic bacteria, where it is believed to protect from high osmolarity (Oppermann-Sanio and Steinbuchel, 2002). In *S. epidermidis*, the *cap* locus was shown to represent a key virulence factor in catheter-associated infection and to protect from major components of innate host defence, including from neutrophil phagocytosis (similar to *B. anthracis*) and from human antimicrobial peptides (Kocianova *et al.*, 2005). In addition, presence of the *cap* locus facilitated *S. epidermidis* growth in high salt environment. The *cap* genes and production of PGA were found in all tested *S. epidermidis* strains and also in other related CNS species. In contrast to *B. anthracis*, but not uncommon compared to other bacteria, PGA in *S. epidermidis* consists of an about equal mixture of the D- and L-isomers of glutamate. Interestingly, the production of PGA in *S. epidermidis* is considerably lower than in *Bacillus* strains, by about 1 million times, and it is surprising how such a low amount of PGA capsule can have the observed significant influence on virulence, immune evasion properties, and especially resistance to high concentrations of salt (Kocianova *et al.*, 2005). A capsule is usually defined as a protective layer around a cell and is thus different from biofilm building material such as PIA that forms intercellular connections. Although in principle, the PGA polymer may contribute to the biofilm matrix owing to electrostatic interference with PIA, an influence of PGA on biofilm formation or the amount of PIA on the cell surface was not detected, most likely due to the minimal production levels (Kocianova *et al.*, 2005). However, it is interesting that PGA was produced in higher levels during biofilm formation, thus adding to the immune evasion properties of the *S. epidermidis* biofilm (Yao *et al.*, 2005a).

Iron acquisition

As almost all bacteria, *Staphylococcus* is dependent on iron acquisition for growth. Bacteria colonizing or infecting humans have invented two principal mechanisms to deal with the low availability of iron in human body fluids. The first involves high-affinity iron-binding molecules called siderophores and specific import systems, to which siderophore/iron complexes bind. *S. epidermidis* produces two siderophores called staphyloferrin A and staphyloferrin B, of 481 and 448 Da, respectively (Konetschny-Rapp *et al.*, 1990; Meiwes *et al.*, 1990; Drechsel *et al.*, 1993). The staphyloferrins have been linked to virulence in CNS, as they are produced more frequently by strains isolated from infection (Lindsay *et al.*, 1994). The second mechanism depends on direct binding of transferrin to a membrane-bound bacterial receptor. An iron-regulated ABC transporter of *S. epidermidis*, SitABC, is believed to be involved in iron uptake (Cockayne *et al.*, 1998).

Global regulators of virulence

Similar to *S. aureus*, several global regulators in *S. epidermidis* control the expression of virulence factors. Usually, these regulators have a profound impact on general bacterial physiology and the change in virulence factor expression forms just one aspect of the adaptation to changing environmental conditions, such as bacterial density, the triggering factor for quorum-sensing global regulators. Global regulatory systems that have been characterized in more detail in *S. epidermidis* comprise the quorum-sensing regulators *agr* and *luxS*, the alternative sigma factor *sigB*, and the accessory gene regulator *sarA*.

The accessory gene regulator (*agr*) of *S. epidermidis* system changes gene expression in response to cell density of the bacterial population, which is measured by the accumulation of an extracellular signal (Vuong *et al.*, 2000a). While the molecular details of the *agr* system have been investigated in *S. aureus* (Novick, 2003), the structure of the *S. epidermidis* signal was in fact the first to be determined (Otto *et al.*, 1998b). As shown for almost all staphylococcal Agr signal peptides, or AIPs (autoinducing peptides), the *S. epidermidis* AIP contains an internal thioester structure, linking the C-terminal carboxy group to the thiol group of a conserved cysteine residue (Otto, 2001). This posttranslational modification is required for the activity of all staphylococcal AIPs, with only the oxygen-ester structure of *S. intermedius* representing a slight variation to the common theme (Ji *et al.*, 2005). The staphylococcal AIPs show the interesting phenomenon of mutual inhibition between species or subgroups among one species and have therefore been suggested as lead substances for therapeutics that would specifically inhibit *agr*-dependent virulence factor expression (Ji *et al.*, 1997; Mayville *et al.*, 1999; Otto *et al.*, 1999; Otto *et al.*, 2001; Otto, 2004a).

The role of *S. epidermidis* agr has been investigated in much detail, including transcriptional profiling analysis and *in vitro* and *in vivo* influence on biofilm formation and biofilm-associated infections. The gene-expression data indicated that *agr* adapts bacterial physiology to stationary phase growth and controls a series of virulence factors, including degradative exoenzymes (Yao *et al.*, 2006). Remarkably, *agr* of *S. epidermidis* also regulates general and oxidative stress-response factors, including detoxifying enzymes of reactive oxygen species. These results obtained in *S. epidermidis* indicate that quorum-sensing regulation in staphylococci has important, previously unknown functions that contribute to protection from mechanisms of human innate host defence – and, therefore, to the pathogen's survival in the human host. An influence of *agr* on biofilm formation has first been recognized in *S. aureus*

(Vuong *et al.*, 2000b) and was subsequently investigated in detail in *S. epidermidis* (Vuong *et al.*, 2003; Vuong *et al.*, 2004c). Interestingly, deletion of *agr* leads to increased biofilm formation and an increased density of the *S. epidermidis* biofilm, which is likely due to the lack of PSM peptides of the β type (unpublished results). The positive influence of *agr* on virulence factors such as degradative exoenzymes and proinflammatory PSMs, and the negative influence on overall formation of a biofilm, are reflected by the increased *in vivo* colonization of catheters by an *agr* deletion strain, contrasting the increased infiltration into host tissue around the catheter by the corresponding wild-type strain (Vuong *et al.*, 2004c). Similar results were achieved with a deletion mutant of the *luxS* system, a more universal quorum-sensing system found in a variety of bacteria. Like with *agr*, there was increased colonization of catheters in an animal model in a *luxS* deletion mutant, an effect, which was however not due to PSM production, but increased expression of PIA in the *luxS* mutant (Xu *et al.*, 2006).

The DNA-binding protein SarA has been intensively investigated in *S. aureus* and in both *S. aureus* and *S. epidermidis*, it represents just one of a series of SarA paralogues with many different functions that we are just beginning to comprehend (Cheung and Zhang, 2002). Similar to *S. aureus*, SarA of *S. epidermidis* has a profound impact on the transcription of the *ica* gene locus, resulting in the almost compete absence of PIA production in a *sarA* mutant and abolished biofilm formation (Tormo *et al.*, 2005b). The alternative sigma factor SigB also has a considerable impact on PIA production, which is exerted by a negative influence on the *icaR* promoter controlling expression of the repressor of the *ica* system, IcaR (Conlon *et al.*, 2002; Knobloch *et al.*, 2004). Notably, the influence of SarA on PIA production appears to be independent from that of SigB and does not involve SarA-dependent regulation of SigB levels (Handke *et al.*, 2007). Except for the regulation of extracellular proteolytic activity by SarA of *S. epidermidis* (Fluckiger *et al.*, 1998; Lai *et al.*, 2007) and the differential lipase and protease production in a *sigB* mutant of *S. epidermidis* (Kies *et al.*, 2001), SigB and SarA in *S. epidermidis* have so far only been investigated with regard to their role in PIA production and are awaiting further characterization.

Resistance to antibiotics and innate host defence

Resistance to antimicrobials is one of the major reasons for the increasing spread of *S. epidermidis* infections. Resistance to antibiotics has been discussed above and most notably includes that to methicillin, which is now very prevalent among CNS and *S. epidermidis* (Raad *et al.*, 1998). Resistance to β-lactams is considerable (~ 80%) and there is also resistance to erythromycin and ciprofloxacin in a significant number of isolates, to list just a few examples (Arciola *et al.*, 2005). Notably, high-level resistance to vancomycin as an antibiotic of last resort in staphylococcal infections, such as described in a couple of *S. aureus* strains, has not been observed in *S. epidermidis* yet.

More recently, we have also made considerable progress in understanding how *S. epidermidis* is protected from mechanisms of host defence, most notably, innate host defence, which is crucial for defeating bacterial invaders within days after the infection. Biofilm formation has been recognized to contribute to this protection not only by functioning as a mechanical barrier, but in addition by establishing a physiological condition that is much less susceptible to attacks by antibacterial agents (Yao *et al.*, 2005a). Furthermore, there is upregulation in biofilms of molecules that have been characterized as key factors in providing resistance to innate host defence mechanisms, such as PGA. In addition, the main building substance of at least most *S. epidermidis* biofilms, PIA, has been found to provide protection form opsonic and non-opsonic phagocytosis, and from antimicrobial peptides (Vuong *et al.*, 2004b; Cerca *et al.*, 2006). Finally, the already mentioned proteases may degrade antimicrobial peptides.

In addition, we are beginning to gain insight into the regulatory and sensing mechanisms, by which *S. epidermidis* interacts with innate host defences and controls resistance mechanisms as a response to the release of antibacterial substances by the human body. In Gram-negative bacteria, a specific sensor is known to recognize the presence of antimicrobial peptides and upregulate specific mechanisms of antimicrobial peptide resistance. This response includes the alteration of the structure of lipid A, resulting in lesser binding of antimicrobial peptides to the bacterial surface. The sensor/transducer system is a classical bacterial two-component system named PhoP/PhoQ (Bader *et al.*, 2005). Until recently, a Gram-positive functional equivalent to this system in Gram-positive bacteria was not known. By determining the genome-wide gene regulatory response to human β-defensin 3 in *S. epidermidis*, we discovered an antimicrobial peptide sensor system that controls major specific resistance mechanisms of Gram-positive bacteria and is unrelated to the Gram-negative PhoP/PhoQ system (Li *et al.*, 2007). The controlled target loci comprise the *dlt* operon responsible for the D-alanylation of teichoic acids and the *mprF* gene, which incorporates positively charged phospholipids in the bacterial membrane (Peschel *et al.*, 1999; Peschel *et al.*, 2001). These two mechanisms lower the overall negative charge of the surface and thus limit the attraction of cationic antimicrobial peptides. Furthermore, the regulatory system controls the expression of the VraFG ABC transporter system, which is also likely involved in resistance to antimicrobial peptides (D.J. Cha, M. Otto, unpublished results). The regulatory system, which we named *aps* for antimicrobial peptide sensor, contains a classical two-component signal transducer and an unusual third protein, all of which are indispensable for signal transduction and antimicrobial peptide resistance. Furthermore, a very short, extracellular loop with a high density of negative charges in the sensor protein is responsible for antimicrobial peptide binding and the observed specificity for cationic antimicrobial peptides. Notably, this study showed that Gram-positive bacteria in general have developed an efficient and unique way of controlling resistance mechanisms to antimicrobial peptides, and represents another of the yet rare examples of a study performed in *S. epidermidis* with significance beyond the species.

Small colony variants and intracellular persistence

Internalization by endothelial cells and persistence inside pericatheter macrophages have been shown for *S. epidermidis* and might in part be responsible for the difficulty to clear CNS infections (Boelens *et al.*, 2000; Merkel and Scofield, 2001). Bacteria are believed to persist inside some cell types in a status of reduced metabolism and reduced production of extracellular virulence factors, similar to the status in so-called SCVs (small colony variants) (Proctor *et al.*, 2006). SCVs have been reported in *S. epidermidis* to cause bloodstream infections following pacemaker implantation (von Eiff *et al.*, 1999). Part of the increased capacity of *S. epidermidis* SCVs to cause biofilm-associated infection may be due to the increased expression of PIA in SCVs (Al Laham *et al.*, 2007). However, our knowledge of these processes and phenotypes, and to which extent they contribute to infection by *S. epidermidis*, is relatively limited.

Bacteriophages

In contrast to *S. aureus*, where bacteriophages are well characterized as structures significantly determining pathogenesis, and also an important genetic tool for molecular biology, we have a serious lack of knowledge about *S. epidermidis* phages. The first complete sequences of two *S. epidermidis* phages have only recently been determined (Daniel *et al.*, 2007). These two phages of the *Siphoviridae* family are closely related to phages of *S. aureus*. Further investigation of *S. epidermidis* phages will certainly have an important impact on our understanding of *S. epidermidis* epidemiology and pathogenesis.

Bacteriocins

Several *S. epidermidis* strains produce lantibiotics that kill other bacteria (Schnell *et al.*, 1988; Kaletta *et al.*, 1989; van de Kamp *et al.*, 1995; Israil *et al.*, 1996). Lantibiotics are bacteriocins that contain unusual amino acids with thioether bridges such as lanthionine and methyllanthionine (Jack and Jung, 2000). The genes responsible for the production of lantibiotics are usually clustered and are encoded on plasmids or on the chromosome. *S. epidermidis* strain Tü3298 produces a lantibiotic called epidermin (Schnell *et al.*, 1988). Structural genes for lantibiotics very similar to epidermin are found in several *S. aureus* chromosomes. Epidermin is active against staphylococci unless resistance genes are present, which are generally encoded in or near the biosynthetic cluster (Schnell *et al.*, 1992). Resistance factors include an ABC transporter system, which is found in many lantibiotic gene clusters and acts by removing the lantibiotic from the membrane by exporting it to the extracellular medium (Otto *et al.*, 1998a). Lantibiotics might act as weapons in bacterial competition, against staphylococci and also against other species. However, Gram-negative bacteria are not sensitive against this type of bacteriocins, because they are protected by the outer membrane (Stevens *et al.*, 1991).

Therapeutic intervention/vaccines

Target-directed drug discovery attempts are usually not directed against specific mechanisms of *S. epidermidis* virulence, as targets should be common to as many bacteria as possible. Thus, even for more specific putative therapeutics, targets are usually sought for in *S. aureus* as the more significant pathogen. Also, bacterial pathogenesis research that had paradigmatic implications for all staphylococci, or even Gram-positive pathogens, has only started in *S. epidermidis* a short time ago, most notably with the identification of the role of PIA, PGA, and most recently, the antimicrobial peptide sensor system (Mack *et al.*, 1996; Kocianova *et al.*, 2005; Li *et al.*, 2007).

The situation is somewhat different for vaccine development. A vaccine specific for *S. epidermidis* may have market potential for delivery in the hospital setting. Developing a vaccine against a commensal of humans may seem problematic, as one can expect immunity against *S. epidermidis* to be pre-established. However, such a vaccine may provide additional protection from *S. epidermidis* infection and similar approaches for some other commensal organisms, such as pneumococci, have proven effective. Broad systematic approaches to find the best antigens for vaccine development in *S. epidermidis* have been performed (Sellman *et al.*, 2005). Currently, aims to develop anti-*S. epidermidis* vaccines are mostly based on the surface location and ubiquity of antigens, and thus focused on polymers such as PIA and surface proteins like the fibrinogen-binding protein (McKenney *et al.*, 2000; Pei and Flock, 2001a). The frequency of PIA as a biofilm polymer in an increasing number of pathogens might render an anti-PIA vaccine more marketable. The efficacy of anti-PIA antibodies against PIA-producing *E. coli* has been demonstrated (Cerca *et al.*, 2007). In addition, PGA might represent an interesting target for vaccine development, as it is ubiquitous and has proven an effective antigen in vaccine development for *B. anthracis* (Leppla *et al.*, 2002; Kozel *et al.*, 2004). However, the role of antibody in preventing *S. epidermidis* disease is still speculative and may be complicated by the fact that *S. epidermidis* is part of the common human epithelial microflora, and we should be cautious after the failure of two large Phase III S. aureus vaccine trials.

Other coagulase-negative staphylococci

Besides *S. epidermidis*, relatively little is known about the other CNS, although some of them can cause similarly severe or even more serious human infections. The recent completion of *S. saprophyticus* and *S. haemolyticus* genomes however, has provided valuable insights into the physiology and pathogenesis of other CNS, for example, in the distribution of virulence factors among the species. Here, we will focus on virulence factors that are specifically found and characterized in the respective CNS species.

Staphylococcus saprophyticus – leading cause of uncomplicated urinary tract infection among Gram-positives

S. saprophyticus is considered the second most important causative agent of uncomplicated urinary tract infection (UTI) in young women, next to *E. coli* (Ronald, 2002). Occasionally, *S. saprophyticus* can also cause more severe infections in humans such as septicaemia, nephrolithiasis and endocarditis (Raz *et al.*, 2005). Comparative genomic analyses with strains of *S. epidermidis* and *S. aureus* revealed that *S. saprophyticus* share a core set of genes (approximately 67% of the genome) with the other two staphylococcal species (Kuroda *et al.*, 2005). In addition, approximately 24% of open reading frames (ORFs) and/or genes in *S. saprophyticus* genome are unique to the species. Of particular interest, ORFs classified under transport and regulation are most abundant in *S. saprophyticus* compared to the other staphylococci, which may be related to the rich metabolite contents in urine such as high salt and urea. For instance, there are more paralogues of transporters with a putative role in osmoprotection found in the genome of *S. saprophyticus*, compared to the common sodium ion/proton exchangers and proline, glycine/betaine transporters found in the other staphylococci. Although the genetic organization of the urease operon in the chromosome is the same in all three staphylococcal species, they appear to be regulated and/or activated differently. For example, the level of urease production was higher and reached a maximum much earlier during growth in *S. saprophyticus* compared to *S. aureus* and *S. epidermidis*. Urease production is important in *S. saprophyticus* for two reasons. Firstly, urease catalyses the hydrolysis of urea (prevalent in urine) to ammonia, an utilizable form of nitrogen for *S. saprophyticus*. Secondly, urease is an important virulence factor for S. *saprophyticus* in that it contributes to invasiveness by damaging bladder tissues (Gatermann *et al.*, 1989; Gatermann and Marre, 1989).

Similar to the other CNS, biofilm formation is probably also involved in urinary tract infections with *S. saprophyticus*. One of the factors involved in *S. saprophyticus* biofilm formation is the autolysin Aas, which mediates binding to fibronectin and promotes haemagglutination. Another surface-attached protein, Ssp (*S. saprophyticus* surface-associated protein), present in 98% of *S. saprophyticus* isolates, causes cell clumping and binding to uroepithelial cells (Gatermann *et al.*, 1992). Recent experimental data suggest that Ssp is a surface-associated lipase in *S. saprophyticus* (Sakinc *et al.*, 2005). In addition, two other surface-associated proteins with the LPXTG amino acid motif were found in *S. saprophyticus* (Kuroda *et al.*, 2005; Sakinc *et al.*, 2006). The UafA (uro-adherence factor A) adhesin is involved in haemagglutination and binding to eukaryotic cells from the urinary tract, while the serine-aspartate repeat protein, SdrI, binds to collagen. Interestingly, while the *agr* locus is well conserved among the staphylococcal species, *S. saprophyticus* lacks typical *agr*-controlled virulence factors. For instance, there is no ORF encoding δ-toxin within the RNAIII molecule of *S. saprophyticus* (Sakinc *et al.*, 2004). The presence of other putative virulence factors such as lipase, elastase and FAME activity has also been reported in *S. saprophyticus* (Lina *et al.*, 2000). Thus, it can be surmised that the prevalence of *S. saprophyticus* as an uropathogen, among the other staphylococci, may be attributed to its ability to adhere and grow rapidly in the urinary tract.

Staphylococcus haemolyticus

The human infections caused by *S. haemolyticus* are similar to those caused by other CNS species and include septicaemia, native valve endocarditis, peritonitis, infections of the bones and joints, and wound and urinary tract infections (Lina *et al.*, 2000). *S. haemolyticus* is the second most frequent CNS in human blood cultures, next to *S. epidermidis* (Ing *et al.*, 1997). In addition, decreased susceptibility to antimicrobial agents such as methicillin and glycopeptides (e.g. teicoplanin) has been reported early in *S. haemolyticus* (Froggatt *et al.*, 1989; Hiramatsu, 1998). A recent comparison between the genomes of *S. haemolyticus* and the other sequenced staphylococci (*S. aureus* and *S. epidermidis*) revealed that the three staphylococcal species share a core set of genes, whose sequences are well conserved and have similar genetic organization in the chromosome (Takeuchi *et al.*, 2005). Comparatively, *S. haemolyticus* may be better adapted for survival under nutrient scarce environments than the other two staphylococcal species, as denoted by the greater abundance of ORFs and/or genes in the genome dedicated for nutrient transports, amino acid and coenzyme biosynthesis. Similar to the mechanism of resistance in most MRSA and MRSE strains, resistance to methicillin in *S. haemolyticus* is mediated by *mecA*, located within the SCC*mec* element. In addition, resistance genes to an array of other antimicrobials are found as integrated plasmids (πSH1 and πSH2) in the genome and on three novel *S. haemolyticus* plasmids (pSHaeA, pSHaeB and pSHaeC). With regards to pathogenesis, only 2% of the genome contains virulence factors that are unique to *S. haemolyticus*, which includes at least three haemolysins and several hypothetical proteins. To date, the gonococcal growth inhibitor (GGI) peptide is one of the best characterized virulence factors identified in *S. haemolyticus* (Frenette *et al.*, 1984; Watson *et* *al.*, 1988). GGI is similar to the slush peptides of *S. lugdunensis* and PSMβ of *S. epidermidis* and is therefore likely to exhibit similar proinflammatory properties. Moreover, a majority of the putative virulence factors identified in the *S. haemolyticus* genome are similar to those of the other staphylococci, for example, PGA of *S. epidermidis* and capsular polysaccharide of *S. aureus* (Takeuchi *et al.*, 2005).

However, there are also unique features in the *S. haemolyticus* genome that distinguishes it from the other staphylococci (Takeuchi *et al.*, 2005). For instance, there was a greater abundance of insertion sequences in the *S. haemolyticus* chromosome (up to 4 fold more than *S. aureus* and 1.2 fold more than *S. epidermidis*), which may account for the high diversity among strains of this species. Additionally, the genetic diversity in *S. haemolyticus* is due to genes acquired horizontally within specific regions of the chromosome, *oriC* environ, from other staphylococci. For example, mannitol utilization is a metabolic marker that distinguishes *S. aureus* from the rest of the staphylococcal species. However, some strains of *S. haemolyticus* were also found to ferment mannitol. In the genome of the sequenced *S. haemolyticus* strain, a mannitol-specific phosphotransferase system (PTS), with high amino acid identity to that of *S. aureus* was found. Furthermore, in some strains of *S. haemolyticus*, part of the capsular polysaccharide operon (*cap*) share similar amino acid sequence to that of *S. aureus*, suggesting that such genes (i.e. mannitol-specific PTS and *cap* operon) were horizontally acquired. Recently, it was shown that frequent rearrangements of chromosomal loci within *oriC* environ in *S. haemolyticus* occurs and can lead to the loss of resistance to antimicrobial agents (Watanabe *et al.*, 2007). Taken together, it can be surmised that the diversity of *S. haemolyticus* strains may be associated with the acquisition of resistance factors and/or virulence factors from other staphylococci and possibly other genera.

Staphylococcus lugdunensis

Compared to other CNS, infections with the species *S. lugdunensis* can be more severe. *S. lugdunensis* can cause brain abscesses, sepsis, osteomyelitis and infective endocarditis (Vandenesch *et al.*, 1995). The species can bind to a variety of human matrix proteins such as fibrinogen and von Willebrand factor (vWf). The *S. lugdunensis* fibrinogen-binding protein, Fbl, is a member of the Sdr-family of proteins, with the typical serine-aspartate repeat regions. In addition, Fbl shares high amino acid sequence similarity to the clumping factor A (ClfA) of *S. aureus*, although Fbl exhibits 10 fold lesser affinity for fibrinogen than ClfA (Mitchell *et al.*, 2004; Nilsson *et al.*, 2004a). Similar to most staphylococcal cell surface-associated proteins, the vWf binding protein of *S. lugdunensis*, vWbl, has an N-terminal signal peptide and C-terminal LPXTG amino acids motif (Nilsson *et al.*, 2004b). Previously, protein A was identified to be the vWf-binding protein in *S. aureus* (Hartleib *et al.*, 2000). Interestingly, although protein A does not share any amino acid sequence homology to the vWbl of *S. lugdunensis*, both proteins bind to the same site in vWf. The *fbl* and *vwbl* genes appeared to be prevalent among clinical strains of *S. lugdunensis*.

Besides the production of adhesins, *S. lugdunensis* produces three similar peptides named slush (*Staphylococcus lugdunensis* synergistic haemolysin), which are similar to PSMβ of *S. epidermidis* and GGI of *S. haemolyticus* (Donvito *et al.*, 1997). In addition, several PSM-like peptides were found in *S. lugdunensis*. Interestingly, unlike the PSMs in *S. epidermidis*, the production of slush and other similar peptides in *S. lugdunensis* is not linked to the *agr* system and it is still unclear how the production of slush and other similar peptides is regulated in *S. lugdunensis*. Further, the production of FAME, esterase, lipase, and an extracellular slime matrix has been reported in *S. lugdunensis* (Lina *et al.*, 2000).

Staphylococcus warneri

S. warneri colonizes human skin in small populations and the species has been implicated occasionally in infections involving neonates (Center *et al.*, 2003; Cimiotti *et al.*, 2007). In rare cases, *S. warneri* can cause endocarditis in adults (Stollberger *et al.*, 2006). It can bind to human matrix proteins such as collagen and fibronectin and produces a variety of enzymes, including protease, lipase, esterase and urease (Lina *et al.*,

2000). In addition, the regulation and/or production of antimicrobial compounds have been the focus of a number of studies in strains of *S. warneri*. For instance, several *S. warneri* strains such as ISK-1 and RB4 produce a lacticin-481 type lantibiotic. In *S. warneri* ISK-1, the lantibiotic nukacin ISK-1 gene operon is located on a plasmid and contains at least six genes, *nukA*, *-M*, *-T*, *-F*, *-E* and -*G* and two ORFs, ORF1 and ORF7 (*nukH*) (Aso *et al.*, 2004; Nagao *et al.*, 2005). NukFEG (ABC transporter) and NukH (lantibiotic-binding protein) confer immunity of *S. warneri* ISK-1 to nukacin ISK-1. In *S. warneri* RB4, a nukacin ISK-1-like bacteriocin with a narrow spectrum of antibacterial activity, termed warnericin RB4, has been reported (Minamikawa *et al.*, 2005). An anti-*Legionella* peptide with high heat stability was also identified in *S. warneri* (Hechard *et al.*, 2005). Interestingly, unlike the *agr* system identified in the other staphylococci, the RNAIII molecule of *S. warneri* harbours two *hld* genes for δ-toxin that are similar, but non-identical (Tegmark *et al.*, 1998). The newly described species *S. pasteuri* is phenotypically similar to *S. warneri* (Chesneau *et al.*, 1993).

Staphylococcus simulans

S. simulans can cause infections in both animals and humans. Human infections caused by *S. simulans* include native valve endocarditis, urinary tract, wound, bone and joint infections, and septicaemia. Lipase, FAME, urease and bacteriocin production have been reported in *S. simulans* (Lina *et al.*, 2000; dos Santos Nascimento *et al.*, 2005). In addition, several strains of *S. simulans* harbour the *S. aureus bap* (biofilm-associated protein) orthologue gene, which may be important in device-related infections, particularly in CONS strains that are *ica* negative (Tormo *et al.*, 2005a). In the strain *S. simulans* biovar *staphylolyticus* ATCC1362, the genes encoding for lysostaphin production (*lss*) and immunity factor (*lif*) are encoded on a large plasmid (Heinrich *et al.*, 1987).

Staphylococcus capitis

S. capitis has been implicated in urinary tract infections, catheter-related bacteraemia, endocarditis, necrotizing enterocolitis in neonates and other infections (Lina *et al.*, 2000). The species can be further subdivided into two other subspecies, subsp. *ureolyticus* (produces urease) and subsp. *capitis* (no urease production). *S. capitis* can bind to a variety of human matrix proteins, including laminin, collagen and fibrinogen. Recently, the first collagen-binding Sdr protein was identified in *S. capitis* (Liu *et al.*, 2004). SdrX, a member of the Sdr family of MSCRAMM proteins, has the typical features of other staphylococcal surface-associated proteins, i.e. the serine-aspartate repeat regions and the LPXTG amino acids motif. Lipase and FAME activity have also been reported in the species (Lina *et al.*, 2000). In the strain *S. capitis* EPK1, a lysostaphin-like glycylglycine endopeptidase, ALE-1, is produced. The mode of action of ALE-1 is similar to that of lysostaphin, i.e. cleavage within the peptidoglycan interpeptide bridge (Sugai *et al.*, 1997a). An immunity gene, *epr*, found adjacent to the *ale-1* gene, encodes for a product which may act in a similar fashion to the lysostaphin immunity factor Lif of *S. simulans* (Sugai *et al.*, 1997b).

Staphylococcus hominis

S. hominis is a colonizer of human skin with low virulence. Occasionally, the species has been implicated in device-related infections in adults (Iyer *et al.*, 2005; Sunbul *et al.*, 2006) and sepsis in neonates (Chaves *et al.*, 2005). Binding to vitronectin, laminin, collagen and fibrinogen has been reported. In addition, S. esterase, lipase, urease and FAME activity have been demonstrated in *S. hominis* (Lina *et al.*, 2000).

Staphylococcus cohnii

S. cohnii subsp. *cohnii* is an exclusive colonizer of humans, whereas subsp. *urealyticum* can also be found in primates. The species is frequently isolated in hospitals and is rarely pathogenic (Szewczyk and Rozalska, 2000). However, more severe infections such as multiple brain abscesses have also been reported in *S. cohnii* (Yamashita *et al.*, 2005). Some *S. cohnii* strains harbour a number of plasmids encoding resistance factors to antimicrobial agents and the production of bacteriocins (Szewczyk *et al.*, 2004). A lantibiotic similar to epidermin has been reported in *S. cohnii* T strain (Furmanek *et al.*, 1999). Additionally, *S. cohnii* can bind to a variety of host matrix proteins including, vitronectin, laminin, fibronectin and collagen. Lipase and FAME activity have also been reported (Lina *et al.*, 2000).

Staphylococcus xylosus and *Staphylococcus carnosus*

S. xylosus and *S. carnosus* are used in sausage and fish fermentation. *S. carnosus* is frequently used as a heterologous host for the expression of virulence factors of other staphylococci and sometimes other genera for basic research and biotechnology projects (Gotz, 1990), and is being tested as a potential vaccine delivery system (Wernerus *et al.*, 2002). Although *S. xylosus* has been associated with human infections in very rare occasions (Mastroianni *et al.*, 1994; Siqueira and Lima, 2002), these species are usually considered non-pathogenic.

Staphylococcus saccharolyticus

S. saccharolyticus, formerly known as *Peptococcus saccharolyticus*, is part of the bacterial flora on human skin and is rarely pathogenic (Evans *et al.*, 1978). It is also the only anaerobic species within the genus *Staphylococcus*. To date, the species remains poorly described and has only been implicated in a few cases of anaerobic endocarditis and bacteraemia (Krishnan *et al.*, 1996; Steinbrueckner *et al.*, 2001).

Future

The recognition of *S. epidermidis* and many other CNS as important opportunistic and nosocomial pathogens has recently led to a substantially increased investigation of the molecular basis of virulence in these organisms. However, our knowledge is still lacking behind what we know about *S. aureus*. Important advances that have been made recently in *S. aureus* pathogenesis research will be transferred in the near future to CNS. These include genome sequencing of more than only one strain per species and genome-wide comparison of single nucleotide polymorphisms, sequencing of yet unknown CNS genomes, and in-depth evaluation of pathogenesis mechanisms including more elaborated animal models of infection. Furthermore, we will need to focus more on colonization models to be able to compare colonization factors with those needed for bacterial invasiveness. Finally, the elucidation of host factors that discriminate individuals susceptible for infection from mere colonizers has just started in *S. aureus* and we expect important findings to be achieved in that field in the case of CNS infections as well.

References

Al Laham, N., Rohde, H., Sander, G., Fischer, A., Hussain, M., Heilmann, C., Mack, D., Proctor, R., Peters, G., Becker, K., and von Eiff, C. (2007). Augmented expression of polysaccharide intercellular adhesin in a defined *Staphylococcus epidermidis* mutant with the small-colony-variant phenotype. J. Bacteriol. *189*, 4494–4501.

Arciola, C. R., Campoccia, D., Gamberini, S., Donati, M. E., Pirini, V., Visai, L., Speziale, P., and Montanaro, L. (2005). Antibiotic resistance in exopolysaccharide-forming *Staphylococcus epidermidis* clinical isolates from orthopaedic implant infections. Biomaterials *26*, 6530–6535.

Ashiuchi, M., and Misono, H. (2002). Biochemistry and molecular genetics of poly-gamma-glutamate synthesis. Appl. Microbiol. Biotechnol. *59*, 9–14.

Ashiuchi, M., Nawa, C., Kamei, T., Song, J. J., Hong, S. P., Sung, M. H., Soda, K., and Misono, H. (2001). Physiological and biochemical characteristics of poly gamma-glutamate synthetase complex of *Bacillus subtilis*. Eur. J. Biochem. *268*, 5321–5328.

Aso, Y., Sashihara, T., Nagao, J., Kanemasa, Y., Koga, H., Hashimoto, T., Higuchi, T., Adachi, A., Nomiyama, H., Ishizaki, A., *et al.* (2004). Characterization of a gene cluster of *Staphylococcus warneri* ISK-1 encoding the biosynthesis of and immunity to the lantibiotic, nukacin ISK-1. Biosci. Biotechnol. Biochem. *68*, 1663–1671.

Augustin, J., and Gotz, F. (1990). Transformation of *Staphylococcus epidermidis* and other staphylococcal species with plasmid DNA by electroporation. FEMS Microbiol. Lett. *54*, 203–207.

Bader, M. W., Sanowar, S., Daley, M. E., Schneider, A. R., Cho, U., Xu, W., Klevit, R. E., Le Moual, H., and Miller, S. I. (2005). Recognition of antimicrobial peptides by a bacterial sensor kinase. Cell *122*, 461–472.

Bae, T., Banger, A. K., Wallace, A., Glass, E. M., Aslund, F., Schneewind, O., and Missiakas, D. M. (2004). *Staphylococcus aureus* virulence genes identified by bursa aurealis mutagenesis and nematode killing. Proc. Natl Acad. Sci. USA *101*, 12312–12317.

Banner, M. A., Cunniffe, J. G., Macintosh, R. L., Foster, T. J., Rohde, H., Mack, D., Hoyes, E., Derrick, J., Upton, M., and Handley, P. S. (2007). Localized tufts of fibrils on *Staphylococcus epidermidis* NCTC 11047 are comprised of the accumulation-associated protein. J. Bacteriol. *189*, 2793–2804.

Boelens, J. J., Dankert, J., Murk, J. L., Weening, J. J., van der Poll, T., Dingemans, K. P., Koole, L., Laman, J. D., and Zaat, S. A. (2000). Biomaterial-associated persistence of *Staphylococcus epidermidis* in pericatheter macrophages. J. Infect. Dis. *181*, 1337–1349.

Boles, B. R., Thoendel, M., and Singh, P. K. (2005). Rhamnolipids mediate detachment of *Pseudomonas aeruginosa* from biofilms. Mol. Microbiol. *57*, 1210–1223.

Bowden, M. G., Chen, W., Singvall, J., Xu, Y., Peacock, S. J., Valtulina, V., Speziale, P., and Hook, M. (2005). Identification and preliminary characterization of cell-wall-anchored proteins of *Staphylococcus epidermidis*. Microbiol. *151*, 1453–1464.

Bowden, M. G., Visai, L., Longshaw, C. M., Holland, K. T., Speziale, P., and Hook, M. (2002). Is the GehD lipase from *Staphylococcus epidermidis* a collagen binding adhesin? J. Biol. Chem. *277*, 43017–43023.

Center, K. J., Reboli, A. C., Hubler, R., Rodgers, G. L., and Long, S. S. (2003). Decreased vancomycin susceptibility of coagulase-negative staphylococci in a neonatal intensive care unit: evidence of spread of *Staphylococcus warneri*. J. Clin. Microbiol. *41*, 4660–4665.

Cerca, N., Jefferson, K. K., Oliveira, R., Pier, G. B., and Azeredo, J. (2006). Comparative antibody-mediated phagocytosis of *Staphylococcus epidermidis* cells grown in a biofilm or in the planktonic state. Infect. Immun. *74*, 4849–4855.

Cerca, N., Maira-Litran, T., Jefferson, K. K., Grout, M., Goldmann, D. A., and Pier, G. B. (2007). Protection against *Escherichia coli* infection by antibody to the *Staphylococcus aureus* poly-N-acetylglucosamine surface polysaccharide. Proc. Natl Acad. Sci. USA *104*, 7528–7533.

Chamberlain, N. R. (1999). Identification and partial characterisation of an extracellular activator of fatty acid modifying enzyme (FAME) expression in *Staphylococcus epidermidis*. J. Med. Microbiol. *48*, 245–252.

Chamberlain, N. R., and Brueggemann, S. A. (1997). Characterisation and expression of fatty acid modifying enzyme produced by *Staphylococcus epidermidis*. J. Med. Microbiol. *46*, 693–697.

Chaves, F., Garcia-Alvarez, M., Sanz, F., Alba, C., and Otero, J. R. (2005). Nosocomial spread of a *Staphylococcus hominis* subsp. *novobiosepticus* strain causing sepsis in a neonatal intensive care unit. J. Clin. Microbiol. *43*, 4877–4879.

Chesneau, O., Morvan, A., Grimont, F., Labischinski, H., and el Solh, N. (1993). *Staphylococcus pasteuri* sp. nov., isolated from human, animal, and food specimens. Int. J. Syst. Bacteriol. *43*, 237–244.

Cheung, A. L., and Zhang, G. (2002). Global regulation of virulence determinants in *Staphylococcus aureus* by the SarA protein family. Front. Biosci. *7*, d1825–1842.

Chokr, A., Leterme, D., Watier, D., and Jabbouri, S. (2007). Neither the presence of *ica* locus, nor *in vitro*-biofilm formation ability is a crucial parameter for some *Staphylococcus epidermidis* strains to maintain an infection in a guinea pig tissue cage model. Microb. Pathog. *42*, 94–97.

Chokr, A., Watier, D., Eleaume, H., Pangon, B., Ghnassia, J. C., Mack, D., and Jabbouri, S. (2006). Correlation between biofilm formation and production of polysaccharide intercellular adhesin in clinical isolates of coagulase-negative staphylococci. Int. J. Med. Microbiol. *296*, 381–388.

Cimiotti, J. P., Haas, J. P., Della-Latta, P., Wu, F., Saiman, L., and Larson, E. L. (2007). Prevalence and clinical relevance of *Staphylococcus warneri* in the neonatal intensive care unit. Infect. Control Hosp. Epidemiol. *28*, 326–330.

Cockayne, A., Hill, P. J., Powell, N. B., Bishop, K., Sims, C., and Williams, P. (1998). Molecular cloning of a 32-kilodalton lipoprotein component of a novel iron-regulated *Staphylococcus epidermidis* ABC transporter. Infect. Immun. *66*, 3767–3774.

Conlon, K. M., Humphreys, H., and O'Gara, J. P. (2002). icaR encodes a transcriptional repressor involved in environmental regulation of ica operon expression and biofilm formation in *Staphylococcus epidermidis*. J. Bacteriol. *184*, 4400–4408.

Costerton, J. W., Stewart, P. S., and Greenberg, E. P. (1999). Bacterial biofilms: a common cause of persistent infections. Science *284*, 1318–1322.

Cramton, S. E., Gerke, C., Schnell, N. F., Nichols, W. W., and Gotz, F. (1999). The intercellular adhesion (*ica*) locus is present in *Staphylococcus aureus* and is required for biofilm formation. Infect. Immun. *67*, 5427–5433.

Cramton, S. E., Ulrich, M., Gotz, F., and Doring, G. (2001). Anaerobic conditions induce expression of polysaccharide intercellular adhesin in *Staphylococcus aureus* and *Staphylococcus epidermidis*. Infect. Immun. *69*, 4079–4085.

Cucarella, C., Solano, C., Valle, J., Amorena, B., Lasa, I., and Penades, J. R. (2001). Bap, a *Staphylococcus aureus* surface protein involved in biofilm formation. J. Bacteriol. *183*, 2888–2896.

Cucarella, C., Tormo, M. A., Ubeda, C., Trotonda, M. P., Monzon, M., Peris, C., Amorena, B., Lasa, I., and Penades, J. R. (2004). Role of biofilm-associated protein bap in the pathogenesis of bovine *Staphylococcus aureus*. Infect. Immun. *72*, 2177–2185.

Daniel, A., Bonnen, P. E., and Fischetti, V. A. (2007). First complete genome sequence of two *Staphylococcus epidermidis* bacteriophages. J. Bacteriol. *189*, 2086–2100.

Darby, C., Hsu, J. W., Ghori, N., and Falkow, S. (2002). *Caenorhabditis elegans*: plague bacteria biofilm blocks food intake. Nature *417*, 243–244.

Daum, R. S., Ito, T., Hiramatsu, K., Hussain, F., Mongkolrattanothai, K., Jamklang, M., and Boyle-Vavra, S. (2002). A novel methicillin-resistance cassette in community-acquired methicillin-resistant *Staphylococcus aureus* isolates of diverse genetic backgrounds. J. Infect. Dis. *186*, 1344–1347.

Davis, S. L., Gurusiddappa, S., McCrea, K. W., Perkins, S., and Hook, M. (2001). SdrG, a fibrinogen-binding bacterial adhesin of the microbial surface components recognizing adhesive matrix molecules subfamily from *Staphylococcus epidermidis*, targets the thrombin cleavage site in the Bbeta chain. J. Biol. Chem. *276*, 27799–27805.

Dominguez, M. A., Linares, J., Pulido, A., Perez, J. L., and de Lencastre, H. (1996). Molecular tracking of coagulase-negative staphylococcal isolates from catheter-related infections. Microb. Drug Resist. *2*, 423–429.

Donvito, B., Etienne, J., Denoroy, L., Greenland, T., Benito, Y., and Vandenesch, F. (1997). Synergistic hemolytic activity of *Staphylococcus lugdunensis* is mediated by three peptides encoded by a non-*agr* genetic locus. Infect. Immun. *65*, 95–100.

dos Santos Nascimento, J., Fagundes, P. C., de Paiva Brito, M. A., dos Santos, K. R., and do Carmo de Freire Bastos, M. (2005). Production of bacteriocins by coagulase-negative staphylococci involved in bovine mastitis. Vet. Microbiol. *106*, 61–71.

Drechsel, H., Freund, S., Nicholson, G., Haag, H., Jung, O., Zahner, H., and Jung, G. (1993). Purification and chemical characterization of staphyloferrin B, a hydrophilic siderophore from staphylococci. Biometals 6, 185–192.

Dubin, G., Chmiel, D., Mak, P., Rakwalska, M., Rzychon, M., and Dubin, A. (2001). Molecular cloning and biochemical characterisation of proteases from *Staphylococcus epidermidis*. Biol. Chem. *382*, 1575–1582.

Dubin, G., Stec-Niemczyk, J., Dylag, T., Silberring, J., Dubin, A., and Potempa, J. (2004). Characterisation of a highly specific, endogenous inhibitor of cysteine protease from *Staphylococcus epidermidis*, a new member of the staphostatin family. Biol. Chem. *385*, 543–546.

Evans, C. A., Mattern, K. L., and Hallam, S. L. (1978). Isolation and identification of *Peptococcus saccharolyticus* from human skin. J. Clin. Microbiol. *7*, 261–264.

Farrell, A. M., Foster, T. J., and Holland, K. T. (1993). Molecular analysis and expression of the lipase of *Staphylococcus epidermidis*. J. Gen. Microbiol. *139*, 267–277.

Fluckiger, U., Wolz, C., and Cheung, A. L. (1998). Characterization of a *sar* homolog of *Staphylococcus epidermidis*. Infect. Immun. 66, 2871–2878.

Frenette, M., Beaudet, R., Bisaillon, J. G., Sylvestre, M., and Portelance, V. (1984). Chemical and biological characterization of a gonococcal growth inhibitor produced by *Staphylococcus haemolyticus* isolated from urogenital flora. Infect. Immun. *46*, 340–345.

Froggatt, J. W., Johnston, J. L., Galetto, D. W., and Archer, G. L. (1989). Antimicrobial resistance in nosocomial isolates of *Staphylococcus haemolyticus*. Antimicrob. Agents Chemother. 33, 460–466.

Furmanek, B., Kaczorowski, T., Bugalski, R., Bielawski, K., Bohdanowicz, J., and Podhajska, A. J. (1999). Identification, characterization and purification of the lantibiotic staphylococcin T, a natural gallidermin variant. J. Appl. Microbiol. *87*, 856–866.

Galdbart, J. O., Morvan, A., Desplaces, N., and el Solh, N. (1999). Phenotypic and genomic variation among *Staphylococcus epidermidis* strains infecting joint prostheses. J. Clin. Microbiol. *37*, 1306–1312.

Gatermann, S., John, J., and Marre, R. (1989). *Staphylococcus saprophyticus* urease: characterization and contribution to uropathogenicity in unobstructed urinary tract infection of rats. Infect. Immun. *57*, 110–116.

Gatermann, S., Kreft, B., Marre, R., and Wanner, G. (1992). Identification and characterization of a surface-associated protein (Ssp) of *Staphylococcus saprophyticus*. Infect. Immun. *60*, 1055–1060.

Gatermann, S., and Marre, R. (1989). Cloning and expression of *Staphylococcus saprophyticus* urease gene sequences in *Staphylococcus carnosus* and contribution of the enzyme to virulence. Infect. Immun. *57*, 2998–3002.

Gerke, C., Kraft, A., Sussmuth, R., Schweitzer, O., and Gotz, F. (1998). Characterization of the N-acetylglucosaminyltransferase activity involved in the biosynthesis of the *Staphylococcus epidermidis* polysaccharide intercellular adhesin. J. Biol. Chem. *273*, 18586–18593.

Gill, S. R., Fouts, D. E., Archer, G. L., Mongodin, E. F., Deboy, R. T., Ravel, J., Paulsen, I. T., Kolonay, J. F., Brinkac, L., Beanan, M., *et al.* (2005). Insights on evolution of virulence and resistance from the complete genome analysis of an early methicillin-resistant *Staphylococcus aureus* strain and a biofilm-producing methicillin-resistant *Staphylococcus epidermidis* strain. J. Bacteriol. *187*, 2426–2438.

Gotz, F. (1990). *Staphylococcus carnosus*: a new host organism for gene cloning and protein production. Soc. Appl. Bacteriol. Symp. Ser. *19*, 49S-53S.

Gotz, F., Ahrne, S., and Lindberg, M. (1981). Plasmid transfer and genetic recombination by protoplast fusion in staphylococci. J. Bacteriol. *145*, 74–81.

Gross, M., Cramton, S. E., Gotz, F., and Peschel, A. (2001). Key role of teichoic acid net charge in *Staphylococcus aureus* colonization of artificial surfaces. Infect. Immun. *69*, 3423–3426.

Gu, J., Li, H., Li, M., Vuong, C., Otto, M., Wen, Y., and Gao, Q. (2005). Bacterial insertion sequence IS256 as a potential molecular marker to discriminate invasive strains from commensal strains of *Staphylococcus epidermidis*. J. Hosp. Infect. *61*, 342–348.

Hanby, W. E., and Rydon, H. N. (1946). The capsular substance of Bacillus anthracis. Biochemistry *40*, 297–307.

Handke, L. D., Slater, S. R., Conlon, K. M., O'Donnell, S. T., Olson, M. E., Bryant, K. A., Rupp, M. E., O'Gara, J. P., and Fey, P. D. (2007). SigmaB and SarA independently regulate polysaccharide intercellular adhesin production in *Staphylococcus epidermidis*. Can. J. Microbiol. 53, 82–91.

Hartleib, J., Kohler, N., Dickinson, R. B., Chhatwal, G. S., Sixma, J. J., Hartford, O. M., Foster, T. J., Peters, G., Kehrel, B. E., and Herrmann, M. (2000). Protein A is the von Willebrand factor binding protein on *Staphylococcus aureus*. Blood *96*, 2149–2156.

Hechard, Y., Ferraz, S., Bruneteau, E., Steinert, M., and Berjeaud, J. M. (2005). Isolation and characterization of a *Staphylococcus warneri* strain producing an anti-*Legionella* peptide. FEMS Microbiol. Lett. *252*, 19–23.

Heilmann, C., Gerke, C., Perdreau-Remington, F., and Gotz, F. (1996a). Characterization of Tn917 insertion mutants of *Staphylococcus epidermidis* affected in biofilm formation. Infect. Immun. *64*, 277–282.

Heilmann, C., Hussain, M., Peters, G., and Gotz, F. (1997). Evidence for autolysin-mediated primary attachment of *Staphylococcus epidermidis* to a polystyrene surface. Mol. Microbiol. *24*, 1013–1024.

Heilmann, C., Schweitzer, O., Gerke, C., Vanittanakom, N., Mack, D., and Gotz, F. (1996b). Molecular basis of intercellular adhesion in the biofilm-forming *Staphylococcus epidermidis*. Mol. Microbiol. *20*, 1083–1091.

Heilmann, C., Thumm, G., Chhatwal, G. S., Hartleib, J., Uekotter, A., and Peters, G. (2003). Identification and characterization of a novel autolysin (Aae) with adhesive properties from *Staphylococcus epidermidis*. Microbiol. *149*, 2769–2778.

Heinrich, P., Rosenstein, R., Bohmer, M., Sonner, P., and Gotz, F. (1987). The molecular organization of the lysostaphin gene and its sequences repeated in tandem. Mol. Gen. Genet. *209*, 563–569.

Hennig, S., Nyunt Wai, S., and Ziebuhr, W. (2007). Spontaneous switch to PIA-independent biofilm formation in an ica-positive *Staphylococcus epidermidis* isolate. Int. J. Med. Microbiol. *297*, 117–122.

Hiramatsu, K. (1998). Vancomycin resistance in staphylococci. Drug Resist. Updat. *1*, 135–150.

Hussain, M., Herrmann, M., von Eiff, C., Perdreau-Remington, F., and Peters, G. (1997). A 140-kilodalton extracellular protein is essential for the accumulation of *Staphylococcus epidermidis* strains on surfaces. Infect. Immun. *65*, 519–524.

Ing, M. B., Baddour, L. M., and Bayer, A. S. (1997). Bacteremia and infective endocarditis: pathogenesis, diagnosis, and complications., In The staphylococci in human diesease, K. B. Crossley, and G. L. Archer, eds. (New York: Churchill Livingstone), pp. 331–354.

Israil, A. M., Jack, R. W., Jung, G., and Sahl, H. G. (1996). Isolation of a new epidermin variant from two strains of *Staphylococcus epidermidis* – frequency of lantibiotic production in coagulase-negative staphylococci. Zentralbl. Bakteriol. *284*, 285–296.

Ito, T., Katayama, Y., Asada, K., Mori, N., Tsutsumimoto, K., Tiensasitorn, C., and Hiramatsu, K. (2001). Structural comparison of three types of staphylococcal cassette chromosome mec integrated in the chromosome in methicillin-resistant *Staphylococcus aureus*. Antimicrob. Agents Chemother. *45*, 1323–1336.

Iyer, M. N., Wirostko, W. J., Kim, S. H., and Simons, K. B. (2005). *Staphylococcus hominis* endophthalmitis associated with a capsular hypopyon. Am. J. Ophthalmol. *139*, 930–932.

Jack, R. W., and Jung, G. (2000). Lantibiotics and microcins: polypeptides with unusual chemical diversity. Curr. Opin. Chem. Biol. *4*, 310–317.

Ji, G., Beavis, R., and Novick, R. P. (1997). Bacterial interference caused by autoinducing peptide variants. Science *276*, 2027–2030.

Ji, G., Pei, W., Zhang, L., Qiu, R., Lin, J., Benito, Y., Lina, G., and Novick, R. P. (2005). *Staphylococcus intermedius* produces a functional *agr* autoinducing peptide containing a cyclic lactone. J. Bacteriol. *187*, 3139–3150.

Kaletta, C., Entian, K. D., Kellner, R., Jung, G., Reis, M., and Sahl, H. G. (1989). Pep5, a new lantibiotic: structural gene isolation and prepeptide sequence. Arch. Microbiol. *152*, 16–19.

Kaplan, J. B., Ragunath, C., Ramasubbu, N., and Fine, D. H. (2003). Detachment of *Actinobacillus actinomycetemcomitans* biofilm cells by an endogenous beta-hexosaminidase activity. J. Bacteriol. *185*, 4693–4698.

Katayama, Y., Ito, T., and Hiramatsu, K. (2000). A new class of genetic element, staphylococcus cassette chromosome *mec*, encodes methicillin resistance in *Staphylococcus aureus*. Antimicrob. Agents Chemother. *44*, 1549–1555.

Kies, S., Otto, M., Vuong, C., and Gotz, F. (2001). Identification of the *sigB* operon in *Staphylococcus epidermidis*: construction and characterization of a *sigB* deletion mutant. Infect. Immun. *69*, 7933–7936.

Kloos, W., and Schleifer, K. H. (1986). *Staphylococcus*, In Bergey's Manual of Systematic Bacteriology, S. PHA, M. S, S. ME, and H. JG, eds. (Baltimore: Williams & Wilkins).

Knobloch, J. K., Bartscht, K., Sabottke, A., Rohde, H., Feucht, H. H., and Mack, D. (2001). Biofilm formation by *Staphylococcus epidermidis* depends on functional RsbU, an activator of the *sigB* operon: differential activation mechanisms due to ethanol and salt stress. J. Bacteriol. *183*, 2624–2633.

Knobloch, J. K., Horstkotte, M. A., Rohde, H., Kaulfers, P. M., and Mack, D. (2002). Alcoholic ingredients in skin disinfectants increase biofilm expression of *Staphylococcus epidermidis*. J. Antimicrob. Chemother. *49*, 683–687.

Knobloch, J. K., Jager, S., Horstkotte, M. A., Rohde, H., and Mack, D. (2004). RsbU-dependent regulation of *Staphylococcus epidermidis* biofilm formation is mediated via the alternative sigma factor sigmaB by repression of the negative regulator gene *icaR*. Infect. Immun. *72*, 3838–3848.

Kocianova, S., Vuong, C., Yao, Y., Voyich, J. M., Fischer, E. R., DeLeo, F. R., and Otto, M. (2005). Key role of poly-gamma-DL-glutamic acid in immune evasion and virulence of *Staphylococcus epidermidis*. J. Clin. Invest. *115*, 688–694.

Kogan, G., Sadovskaya, I., Chaignon, P., Chokr, A., and Jabbouri, S. (2006). Biofilms of clinical strains of *Staphylococcus* that do not contain polysaccharide intercellular adhesin. FEMS Microbiol. Lett. *255*, 11–16.

Konetschny-Rapp, S., Jung, G., Meiwes, J., and Zahner, H. (1990). Staphyloferrin A: a structurally new siderophore from staphylococci. Eur J Biochem *191*, 65–74.

Kozel, T. R., Murphy, W. J., Brandt, S., Blazar, B. R., Lovchik, J. A., Thorkildson, P., Percival, A., and Lyons, C. R. (2004). mAbs to *Bacillus anthracis* capsular antigen for immunoprotection in anthrax and detection of antigenemia. Proc. Natl Acad. Sci. USA *101*, 5042–5047.

Kozitskaya, S., Cho, S. H., Dietrich, K., Marre, R., Naber, K., and Ziebuhr, W. (2004). The bacterial insertion sequence element IS256 occurs preferentially in nosocomial *Staphylococcus epidermidis* isolates: association with biofilm formation and resistance to aminoglycosides. Infect. Immun. *72*, 1210–1215.

Krishnan, S., Haglund, L., Ashfaq, A., Leist, P., and Roat, T. (1996). Prosthetic valve endocarditis due to *Staphylococcus saccharolyticus*. Clin. Infect. Dis. *22*, 722–723.

Kuroda, M., Yamashita, A., Hirakawa, H., Kumano, M., Morikawa, K., Higashide, M., Maruyama, A., Inose, Y., Matoba, K., Toh, H., *et al.* (2005). Whole genome sequence of *Staphylococcus saprophyticus* reveals the pathogenesis of uncomplicated urinary tract infection. Proc. Natl Acad. Sci. USA *102*, 13272–13277.

Lai, Y., Villaruz, A. E., Li, M., Cha, D. J., Sturdevant, D. E., and Otto, M. (2007). The human anionic antimicrobial peptide dermcidin induces proteolytic defence mechanisms in staphylococci. Mol. Microbiol. *63*, 497–506.

Leppla, S. H., Robbins, J. B., Schneerson, R., and Shiloach, J. (2002). Development of an improved vaccine for anthrax. J. Clin. Invest. *110*, 141–144.

Li, M., Lai, Y., Villaruz, A. E., Cha, D. J., Sturdevant, D. E., and Otto, M. (2007). Gram-positive three-component antimicrobial peptide-sensing system. Proc. Natl Acad. Sci. USA *104*, 9469–9474.

Liles, W. C., Thomsen, A. R., O'Mahony, D. S., and Klebanoff, S. J. (2001). Stimulation of human neutrophils and monocytes by staphylococcal phenol-soluble modulin. J. Leukoc. Biol. *70*, 96–102.

Lina, G., Etienne, J., and Vandenesch, F. (2000). Biology and pathogenicity of staphylococci other than *Staphylococcus aureus* and *Staphylococcus epidermidis*, In Gram-positive pathogens, V. A. Fischetti, R. P. Novick, F. J.J., D. A. Portnoy, and J. I. Rood, eds. (Washington, DC: ASM Press).

Lindsay, J. A., Riley, T. V., and Mee, B. J. (1994). Production of siderophore by coagulase-negative staphylococci and its relation to virulence. Eur. J. Clin. Microbiol. Infect. Dis. *13*, 1063–1066.

Liu, Y., Ames, B., Gorovits, E., Prater, B. D., Syribeys, P., Vernachio, J. H., and Patti, J. M. (2004). SdrX, a serine-aspartate repeat protein expressed by *Staphylococcus capitis* with collagen VI binding activity. Infect. Immun. *72*, 6237–6244.

Longshaw, C. M., Farrell, A. M., Wright, J. D., and Holland, K. T. (2000). Identification of a second lipase gene, gehD, in *Staphylococcus epidermidis*: comparison of sequence with those of other staphylococcal lipases. Microbiol. *146*, 1419–1427.

Ma, X. X., Ito, T., Tiensasitorn, C., Jamklang, M., Chongtrakool, P., Boyle-Vavra, S., Daum, R. S., and Hiramatsu, K. (2002). Novel type of staphylococcal cassette chromosome *mec* identified in community-acquired methicillin-resistant *Staphylococcus aureus* strains. Antimicrob. Agents Chemother. *46*, 1147–1152.

Mack, D., Fischer, W., Krokotsch, A., Leopold, K., Hartmann, R., Egge, H., and Laufs, R. (1996). The intercellular adhesin involved in biofilm accumulation of *Staphylococcus epidermidis* is a linear beta-1,6-linked glucosaminoglycan: purification and structural analysis. J. Bacteriol. *178*, 175–183.

Mack, D., Nedelmann, M., Krokotsch, A., Schwarzkopf, A., Heesemann, J., and Laufs, R. (1994). Characterization of transposon mutants of biofilm-producing *Staphylococcus epidermidis* impaired in the accumulative phase of biofilm production: genetic identification of a hexosamine-containing polysaccharide intercellular adhesin. Infect. Immun. *62*, 3244–3253.

Maiden, M. C., Bygraves, J. A., Feil, E., Morelli, G., Russell, J. E., Urwin, R., Zhang, Q., Zhou, J., Zurth, K., Caugant, D. A., *et al.* (1998). Multilocus sequence typing: a portable approach to the identification of clones within populations of pathogenic microorganisms. Proc. Natl Acad. Sci. USA 95, 3140–3145.

Massey, R. C., Horsburgh, M. J., Lina, G., Hook, M., and Recker, M. (2006). The evolution and maintenance of virulence in *Staphylococcus aureus*: a role for host-to-host transmission? Nat. Rev. Microbiol. *4*, 953–958.

Mastroianni, A., Coronado, O., Nanetti, A., and Chiodo, F. (1994). *Staphylococcus xylosus* isolated from a pancreatic pseudocyst in a patient infected with the human immunodeficiency virus. Clin. Infect. Dis. *19*, 1173–1174.

Mayville, P., Ji, G., Beavis, R., Yang, H., Goger, M., Novick, R. P., and Muir, T. W. (1999). Structure-activity analysis of synthetic autoinducing thiolactone peptides from *Staphylococcus aureus* responsible for virulence. Proc. Natl Acad. Sci. USA *96*, 1218–1223.

Mazmanian, S. K., Liu, G., Ton-That, H., and Schneewind, O. (1999). *Staphylococcus aureus* sortase, an enzyme that anchors surface proteins to the cell wall. Science *285*, 760–763.

McKenney, D., Pouliot, K., Wang, Y., Murthy, V., Ulrich, M., Doring, G., Lee, J. C., Goldmann, D. A., and Pier, G. B. (2000). Vaccine potential of poly-1–6 beta-D-N-succinylglucosamine, an immunoprotective surface polysaccharide of *Staphylococcus aureus* and *Staphylococcus epidermidis*. J. Biotechnol. *83*, 37–44.

McKevitt, A. I., Bjornson, G. L., Mauracher, C. A., and Scheifele, D. W. (1990). Amino acid sequence of a deltalike toxin from *Staphylococcus epidermidis*. Infect. Immun. *58*, 1473–1475.

Mehlin, C., Headley, C. M., and Klebanoff, S. J. (1999). An inflammatory polypeptide complex from *Staphylococcus epidermidis*: isolation and characterization. J. Exp. Med. *189*, 907–918.

Meiwes, J., Fiedler, H. P., Haag, H., Zahner, H., Konetschny-Rapp, S., and Jung, G. (1990). Isolation and characterization of staphyloferrin A, a compound with siderophore activity from *Staphylococcus hyicus* DSM 20459. FEMS Microbiol. Lett. *55*, 201–205.

Merkel, G. J., and Scofield, B. A. (2001). Interaction of *Staphylococcus epidermidis* with endothelial cells *in vitro*. Med. Microbiol. Immunol. *189*, 217–223.

Milisavljevic, V., Wu, F., Cimmotti, J., Haas, J., Della-Latta, P., Larson, E., and Saiman, L. (2005). Genetic relatedness of *Staphylococcus epidermidis* from infected infants and staff in the neonatal intensive care unit. Am. J. Infect. Control 33, 341–347.

Minamikawa, M., Kawai, Y., Inoue, N., and Yamazaki, K. (2005). Purification and characterization of Warnericin RB4, anti-Alicyclobacillus bacteriocin, produced by *Staphylococcus warneri* RB4. Curr. Microbiol. *51*, 22–26.

Miragaia, M., Couto, I., and de Lencastre, H. (2005). Genetic diversity among methicillin-resistant *Staphylococcus epidermidis* (MRSE). Microb. Drug Resist. *11*, 83–93.

Miragaia, M., Thomas, J. C., Couto, I., Enright, M. C., and de Lencastre, H. (2007). Inferring a population structure for *Staphylococcus epidermidis* from multilocus sequence typing data. J. Bacteriol. *189*, 2540–2552.

Mitchell, J., Tristan, A., and Foster, T. J. (2004). Characterization of the fibrinogen-binding surface protein Fbl of *Staphylococcus lugdunensis*. Microbiol. *150*, 3831–3841.

Nagao, J., Aso, Y., Sashihara, T., Shioya, K., Adachi, A., Nakayama, J., and Sonomoto, K. (2005). Localization and interaction of the biosynthetic proteins for the lantibiotic, Nukacin ISK-1. Biosci. Biotechnol. Biochem. *69*, 1341–1347.

Nedelmann, M., Sabottke, A., Laufs, R., and Mack, D. (1998). Generalized transduction for genetic linkage analysis and transfer of transposon insertions in different *Staphylococcus epidermidis* strains. Zentralbl. Bakteriol. *287*, 85–92.

Nilsdotter-Augustinsson, A., Koskela, A., Ohman, L., and Soderquist, B. (2007). Characterization of coagulase-negative staphylococci isolated from patients with infected hip prostheses: use of phenotypic and genotypic analyses, including tests for the presence of the ica operon. Eur. J. Clin. Microbiol. Infect. Dis. *26*, 255–265.

Nilsson, M., Bjerketorp, J., Guss, B., and Frykberg, L. (2004a). A fibrinogen-binding protein of *Staphylococcus lugdunensis*. FEMS Microbiol. Lett. *241*, 87–93.

Nilsson, M., Bjerketorp, J., Wiebensjo, A., Ljungh, A., Frykberg, L., and Guss, B. (2004b). A von Willebrand factor-binding protein from *Staphylococcus lugdunensis*. FEMS Microbiol. Lett. *234*, 155–161.

Nilsson, M., Frykberg, L., Flock, J. I., Pei, L., Lindberg, M., and Guss, B. (1998). A fibrinogen-binding protein of *Staphylococcus epidermidis*. Infect. Immun. *66*, 2666–2673.

Novick, R. P. (2003). Autoinduction and signal transduction in the regulation of staphylococcal virulence. Mol. Microbiol. *48*, 1429–1449.

Ohara-Nemoto, Y., Ikeda, Y., Kobayashi, M., Sasaki, M., Tajika, S., and Kimura, S. (2002). Characterization and molecular cloning of a glutamyl endopeptidase from *Staphylococcus epidermidis*. Microb. Pathog. *33*, 33–41.

Oppermann-Sanio, F. B., and Steinbuchel, A. (2002). Occurrence, functions and biosynthesis of polyamides in microorganisms and biotechnological production. Naturwissenschaften *89*, 11–22.

Otto, M. (2001). *Staphylococcus aureus* and *Staphylococcus epidermidis* peptide pheromones produced by the accessory gene regulator *agr* system. Peptides *22*, 1603–1608.

Otto, M. (2004a). Quorum-sensing control in Staphylococci – a target for antimicrobial drug therapy? FEMS Microbiol. Lett. *241*, 135–141.

Otto, M. (2004b). Virulence factors of the coagulase-negative staphylococci. Front. Biosci. *9*, 841–863.

Otto, M. (2006). Bacterial evasion of antimicrobial peptides by biofilm formation. Curr. Top. Microbiol. Immunol. *306*, 251–258.

Otto, M., Echner, H., Voelter, W., and Gotz, F. (2001). Pheromone cross-inhibition between *Staphylococcus aureus* and *Staphylococcus epidermidis*. Infect. Immun. *69*, 1957–1960.

Otto, M., Peschel, A., and Gotz, F. (1998a). Producer self-protection against the lantibiotic epidermin by the ABC transporter EpiFEG of *Staphylococcus epidermidis* Tu3298. FEMS Microbiol. Lett. *166*, 203–211.

Otto, M., Sussmuth, R., Jung, G., and Gotz, F. (1998b). Structure of the pheromone peptide of the *Staphylococcus epidermidis agr* system. FEBS Lett. *424*, 89–94.

Otto, M., Sussmuth, R., Vuong, C., Jung, G., and Gotz, F. (1999). Inhibition of virulence factor expression in *Staphylococcus aureus* by the *Staphylococcus epidermidis agr* pheromone and derivatives. FEBS Lett. *450*, 257–262.

Pei, L., and Flock, J. I. (2001a). Functional study of antibodies against a fibrogenin-binding protein in *Staphylococcus epidermidis* adherence to polyethylene catheters. J. Infect. Dis. *184*, 52–55.

Pei, L., and Flock, J. I. (2001b). Lack of *fbe*, the gene for a fibrinogen-binding protein from *Staphylococcus epidermidis*, reduces its adherence to fibrinogen coated surfaces. Microb. Pathog. *31*, 185–193.

Pei, L., Palma, M., Nilsson, M., Guss, B., and Flock, J. I. (1999). Functional studies of a fibrinogen binding protein from *Staphylococcus epidermidis*. Infect. Immun. *67*, 4525–4530.

Peschel, A., Jack, R. W., Otto, M., Collins, L. V., Staubitz, P., Nicholson, G., Kalbacher, H., Nieuwenhuizen, W. F., Jung, G., Tarkowski, A., *et al.* (2001). *Staphylococcus aureus* resistance to human defensins and evasion of neutrophil killing via the novel virulence factor MprF is based on modification of membrane lipids with l-lysine. J. Exp. Med. *193*, 1067–1076.

Peschel, A., Otto, M., Jack, R. W., Kalbacher, H., Jung, G., and Gotz, F. (1999). Inactivation of the dlt operon in *Staphylococcus aureus* confers sensitivity to defensins, protegrins, and other antimicrobial peptides. J. Biol. Chem. *274*, 8405–8410.

Proctor, R. A., von Eiff, C., Kahl, B. C., Becker, K., McNamara, P., Herrmann, M., and Peters, G. (2006). Small colony variants: a pathogenic form of bacteria that facilitates persistent and recurrent infections. Nat. Rev. Microbiol. *4*, 295–305.

Qin, Z., Yang, X., Yang, L., Jiang, J., Ou, Y., Molin, S., and Qu, D. (2007). Formation and properties of *in vitro* biofilms of *ica*-negative *Staphylococcus epidermidis* clinical isolates. J. Med. Microbiol. *56*, 83–93.

Raad, I., Alrahwan, A., and Rolston, K. (1998). *Staphylococcus epidermidis*: emerging resistance and need for alternative agents. Clin. Infect. Dis. *26*, 1182–1187.

Raz, R., Colodner, R., and Kunin, C. M. (2005). Who are you – *Staphylococcus saprophyticus*? Clin. Infect. Dis. *40*, 896–898.

Rice, K. C., Mann, E. E., Endres, J. L., Weiss, E. C., Cassat, J. E., Smeltzer, M. S., and Bayles, K. W. (2007). The *cidA* murein hydrolase regulator contributes to DNA release and biofilm development in *Staphylococcus aureus*. Proc. Natl Acad. Sci. USA *104*, 8113–8118.

Rohde, H., Burandt, E. C., Siemssen, N., Frommelt, L., Burdelski, C., Wurster, S., Scherpe, S., Davies, A. P., Harris, L. G., Horstkotte, M. A., *et al.* (2007). Polysaccharide intercellular adhesin or protein factors in biofilm accumulation of *Staphylococcus epidermidis* and *Staphylococcus aureus* isolated from prosthetic hip and knee joint infections. Biomaterials *28*, 1711–1720.

Rohde, H., Burdelski, C., Bartscht, K., Hussain, M., Buck, F., Horstkotte, M. A., Knobloch, J. K., Heilmann, C., Herrmann, M., and Mack, D. (2005). Induction of *Staphylococcus epidermidis* biofilm formation via proteolytic processing of the accumulation-associated protein by staphylococcal and host proteases. Mol. Microbiol. 55, 1883–1895.

Ronald, A. (2002). The etiology of urinary tract infection: traditional and emerging pathogens. Am. J. Med. *113 Suppl 1A*, 14S-19S.

Sakinc, T., Kleine, B., and Gatermann, S. G. (2006). SdrI, a serine-aspartate repeat protein identified in *Staphylococcus saprophyticus* strain 7108, is a collagen-binding protein. Infect. Immun. *74*, 4615–4623.

Sakinc, T., Kulczak, P., Henne, K., and Gatermann, S. G. (2004). Cloning of an *agr* homologue of *Staphylococcus saprophyticus*. FEMS Microbiol. Lett. *237*, 157–161.

Sakinc, T., Woznowski, M., Ebsen, M., and Gatermann, S. G. (2005). The surface-associated protein of *Staphylococcus saprophyticus* is a lipase. Infect. Immun. *73*, 6419–6428.

Schnell, N., Engelke, G., Augustin, J., Rosenstein, R., Ungermann, V., Gotz, F., and Entian, K. D. (1992). Analysis of genes involved in the biosynthesis of lantibiotic epidermin. Eur. J. Biochem. *204*, 57–68.

Schnell, N., Entian, K. D., Schneider, U., Gotz, F., Zahner, H., Kellner, R., and Jung, G. (1988). Prepeptide sequence of epidermin, a ribosomally synthesized antibiotic with four sulphide-rings. Nature *333*, 276–278.

Sellman, B. R., Howell, A. P., Kelly-Boyd, C., and Baker, S. M. (2005). Identification of immunogenic and serum binding proteins of *Staphylococcus epidermidis*. Infect. Immun. *73*, 6591–6600.

Siqueira, J. F., Jr., and Lima, K. C. (2002). *Staphylococcus epidermidis* and *Staphylococcus xylosus* in a secondary root canal infection with persistent symptoms: a case report. Aust. Endod. J. *28*, 61–63.

Spoering, A. L., and Gilmore, M. S. (2006). Quorum sensing and DNA release in bacterial biofilms. Curr. Opin. Microbiol. *9*, 133–137.

Steinbrueckner, B., Singh, S., Freney, J., Kuhnert, P., Pelz, K., and Aufenanger, J. (2001). Facing a mysterious hospital outbreak of bacteraemia due to *Staphylococcus saccharolyticus*. J. Hosp. Infect. *49*, 305–307.

Stevens, K. A., Sheldon, B. W., Klapes, N. A., and Klaenhammer, T. R. (1991). Nisin treatment for inactivation of *Salmonella* species and other gram-negative bacteria. Appl. Environ. Microbiol. *57*, 3613–3615.

Stollberger, C., Wechsler-Fordos, A., Geppert, F., Gulz, W., Brownstone, E., Nicolakis, M., and Finsterer, J. (2006). *Staphylococcus warneri* endocarditis after implantation of a lumbar disc prosthesis in an immunocompetent patient. J. Infect. 52, e15–18.

Sugai, M., Fujiwara, T., Akiyama, T., Ohara, M., Komatsuzawa, H., Inoue, S., and Suginaka, H. (1997a). Purification and molecular characterization of glycylglycine endopeptidase produced by *Staphylococcus capitis* EPK1. J. Bacteriol. *179*, 1193–1202.

Sugai, M., Fujiwara, T., Ohta, K., Komatsuzawa, H., Ohara, M., and Suginaka, H. (1997b). *epr*, which encodes glycylglycine endopeptidase resistance, is homologous to *femAB* and affects serine content of peptidoglycan cross bridges in *Staphylococcus capitis* and *Staphylococcus aureus*. J. Bacteriol. *179*, 4311–4318.

Sunbul, M., Demirag, M. K., Yilmaz, O., Yilmaz, H., Ozturk, R., and Leblebicioglu, H. (2006). Pacemaker lead endocarditis caused by *Staphylococcus hominis*. Pacing Clin. Electrophysiol. 29, 543–545.

Szewczyk, E. M., and Rozalska, M. (2000). *Staphylococcus cohnii* – resident of hospital environment: cell-surface features and resistance to antibiotics. Acta. Microbiol. Pol. *49*, 121–133.

Szewczyk, E. M., Rozalska, M., Cieslikowski, T., and Nowak, T. (2004). Plasmids of *Staphylococcus cohnii* isolated from the intensive-care unit. Folia Microbiol. (Praha) *49*, 123–131.

Takeuchi, F., Watanabe, S., Baba, T., Yuzawa, H., Ito, T., Morimoto, Y., Kuroda, M., Cui, L., Takahashi, M., Ankai, A., *et al.* (2005). Whole-genome sequencing of *Staphylococcus haemolyticus* uncovers the extreme plasticity of its genome and the evolution of human-colonizing staphylococcal species. J. Bacteriol. *187*, 7292–7308.

Tegmark, K., Morfeldt, E., and Arvidson, S. (1998). Regulation of *agr*-dependent virulence genes in *Staphylococcus aureus* by RNAIII from coagulase-negative staphylococci. J. Bacteriol. *180*, 3181–3186.

Teufel, P., and Gotz, F. (1993). Characterization of an extracellular metalloprotease with elastase activity from *Staphylococcus epidermidis*. J. Bacteriol. *175*, 4218–4224.

Thomas, J. C., Vargas, M. R., Miragaia, M., Peacock, S. J., Archer, G. L., and Enright, M. C. (2007). Improved multilocus sequence typing scheme for *Staphylococcus epidermidis*. J. Clin. Microbiol. *45*, 616–619.

Tormo, M. A., Knecht, E., Gotz, F., Lasa, I., and Penades, J. R. (2005a). Bap-dependent biofilm formation by pathogenic species of *Staphylococcus*: evidence of horizontal gene transfer? Microbiol. *151*, 2465–2475.

Tormo, M. A., Marti, M., Valle, J., Manna, A. C., Cheung, A. L., Lasa, I., and Penades, J. R. (2005b). SarA Is an Essential Positive Regulator of *Staphylococcus epidermidis* Biofilm Development. J. Bacteriol. *187*, 2348–2356.

Ueta, M., Iida, T., Sakamoto, M., Sotozono, C., Takahashi, J., Kojima, K., Okada, K., Chen, X., Kinoshita, S., and Honda, T. (2007). Polyclonality of *Staphylococcus epidermidis* residing on the healthy ocular surface. J. Med. Microbiol. *56*, 77–82.

Vacheethasanee, K., Temenoff, J. S., Higashi, J. M., Gary, A., Anderson, J. M., Bayston, R., and Marchant, R. E. (1998). Bacterial surface properties of clinically isolated *Staphylococcus epidermidis* strains determine adhesion on polyethylene. J. Biomed. Mater. Res. 42, 425–432.

van de Kamp, M., van den Hooven, H. W., Konings, R. N., Bierbaum, G., Sahl, H. G., Kuipers, O. P., Siezen, R. J., de Vos, W. M., Hilbers, C. W., and van de Ven, F. J. (1995). Elucidation of the primary structure of the lantibiotic epilancin K7 from *Staphylococcus epidermidis* K7. Cloning and characterisation of the epilancin-K7-encoding gene and NMR analysis of mature epilancin K7. Eur. J. Biochem. *230*, 587–600.

Vandenesch, F., Eykyn, S. J., Etienne, J., and Lemozy, J. (1995). Skin and post-surgical wound infections due to *Staphylococcus lugdunensis*. Clin. Microbiol. Infect. *1*, 73–74.

Veenstra, G. J., Cremers, F. F., van Dijk, H., and Fleer, A. (1996). Ultrastructural organization and regulation of a biomaterial adhesin of *Staphylococcus epidermidis*. J. Bacteriol. *178*, 537–541.

Venkatesh, M. P., Placencia, F., and Weisman, L. E. (2006). Coagulase-negative staphylococcal infections in the neonate and child: an update. Semin. Pediatr. Infect. Dis. *17*, 120–127.

Vernachio, J. H., Bayer, A. S., Ames, B., Bryant, D., Prater, B. D., Syribeys, P. J., Gorovits, E. L., and Patti, J. M. (2006). Human immunoglobulin G recognizing fibrinogen-binding surface proteins is protective against both *Staphylococcus aureus* and *Staphylococcus epidermidis* infections *in vivo*. Antimicrob. Agents Chemother. *50*, 511–518.

von Eiff, C., Vaudaux, P., Kahl, B. C., Lew, D., Emler, S., Schmidt, A., Peters, G., and Proctor, R. A. (1999). Bloodstream infections caused by small-colony variants of coagulase-negative staphylococci following pacemaker implantation. Clin. Infect. Dis. *29*, 932–934.

Vuong, C., Durr, M., Carmody, A. B., Peschel, A., Klebanoff, S. J., and Otto, M. (2004a). Regulated expression of pathogen-associated molecular pattern molecules in *Staphylococcus epidermidis*: quorum-sensing determines proinmmatory capacity and production of phenol-soluble modulins. Cell. Microbiol. *6*, 753–759.

Vuong, C., Gerke, C., Somerville, G. A., Fischer, E. R., and Otto, M. (2003). Quorum-sensing control of biofilm factors in *Staphylococcus epidermidis*. J. Infect. Dis. *188*, 706–718.

Vuong, C., Gotz, F., and Otto, M. (2000a). Construction and characterization of an *agr* deletion mutant of *Staphylococcus epidermidis*. Infect. Immun. *68*, 1048–1053.

Vuong, C., Kocianova, S., Voyich, J. M., Yao, Y., Fischer, E. R., DeLeo, F. R., and Otto, M. (2004b). A crucial role for exopolysaccharide modification in bacterial biofilm formation, immune evasion, and virulence. J. Biol. Chem. *279*, 54881–54886.

Vuong, C., Kocianova, S., Yao, Y., Carmody, A. B., and Otto, M. (2004c). Increased colonization of indwelling medical devices by quorum-sensing mutants of *Staphylococcus epidermidis in vivo*. J. Infect. Dis. *190*, 1498–1505.

Vuong, C., and Otto, M. (2002). *Staphylococcus epidermidis* infections. Microbes Infect. *4*, 481–489.

Vuong, C., Saenz, H. L., Gotz, F., and Otto, M. (2000b). Impact of the agr quorum-sensing system on adherence to polystyrene in *Staphylococcus aureus*. J. Infect. Dis. *182*, 1688–1693.

Vuong, C., Voyich, J. M., Fischer, E. R., Braughton, K. R., Whitney, A. R., DeLeo, F. R., and Otto, M. (2004d). Polysaccharide intercellular adhesin (PIA) protects *Staphylococcus epidermidis* against major components of the human innate immune system. Cell. Microbiol. *6*, 269–275.

Wang, X., Preston, J. F. I., and Romeo, T. (2004). The *pgaABCD* locus of *Escherichia coli* promotes the synthesis of a polysaccharide adhesin required for biofilm formation. J. Bacteriol. *186*, 2724–2734.

Wang, X. M., Noble, L., Kreiswirth, B. N., Eisner, W., McClements, W., Jansen, K. U., and Anderson, A. S. (2003). Evaluation of a multilocus sequence typing system for *Staphylococcus epidermidis*. J. Med. Microbiol. 52, 989–998.

Watanabe, S., Ito, T., Morimoto, Y., Takeuchi, F., and Hiramatsu, K. (2007). Precise excision and self-integration of a composite transposon as a model for spontaneous large-scale chromosome inversion/deletion of the *Staphylococcus haemolyticus* clinical strain JCSC1435. J. Bacteriol. *189*, 2921–2925.

Watson, D. C., Yaguchi, M., Bisaillon, J. G., Beaudet, R., and Morosoli, R. (1988). The amino acid sequence of a gonococcal growth inhibitor from *Staphylococcus haemolyticus*. Biochem. J. *252*, 87–93.

Wei, W., Cao, Z., Zhu, Y. L., Wang, X., Ding, G., Xu, H., Jia, P., Qu, D., Danchin, A., and Li, Y. (2006). Conserved genes in a path from commensalism to pathogenicity: comparative phylogenetic profiles of *Staphylococcus epidermidis* RP62A and ATCC12228. BMC Genomics *7*, 112.

Wernerus, H., Lehtio, J., Samuelson, P., and Stahl, S. (2002). Engineering of staphylococcal surfaces for biotechnological applications. J. Biotechnol. *96*, 67–78.

Williams, R. J., Henderson, B., Sharp, L. J., and Nair, S. P. (2002). Identification of a fibronectin-binding protein from *Staphylococcus epidermidis*. Infect. Immun. *70*, 6805–6810.

Wisplinghoff, H., Rosato, A. E., Enright, M. C., Noto, M., Craig, W., and Archer, G. L. (2003). Related clones containing SCCmec type IV predominate among clinically significant *Staphylococcus epidermidis* isolates. Antimicrob. Agents Chemother. *47*, 3574–3579.

Xu, L., Li, H., Vuong, C., Vadyvaloo, V., Wang, J., Yao, Y., Otto, M., and Gao, Q. (2006). Role of the *luxS* quorum-sensing system in biofilm formation and virulence of *Staphylococcus epidermidis*. Infect. Immun. *74*, 488–496.

Yamashita, S., Yonemura, K., Sugimoto, R., Tokunaga, M., and Uchino, M. (2005). *Staphylococcus cohnii* as a cause of multiple brain abscesses in Weber-Christian disease. J. Neurol. Sci. *238*, 97–100.

Yao, Y., Sturdevant, D. E., and Otto, M. (2005a). Genomewide analysis of gene expression in *Staphylococcus epidermidis* biofilms: insights into the pathophysiology of *S. epidermidis* biofilms and the role of phenol-soluble modulins in formation of biofilms. J. Infect. Dis. *191*, 289–298.

Yao, Y., Sturdevant, D. E., Villaruz, A., Xu, L., Gao, Q., and Otto, M. (2005b). Factors characterizing *Staphylococcus epidermidis* invasiveness determined by comparative genomics. Infect. Immun. *73*, 1856–1860.

Yao, Y., Vuong, C., Kocianova, S., Villaruz, A. E., Lai, Y., Sturdevant, D. E., and Otto, M. (2006).

Characterization of the *Staphylococcus epidermidis* Accessory-Gene Regulator Response: Quorum-Sensing Regulation of Resistance to Human Innate Host Defense. J. Infect. Dis. *193*, 841–848.

Zhang, Y. Q., Ren, S. X., Li, H. L., Wang, Y. X., Fu, G., Yang, J., Qin, Z. Q., Miao, Y. G., Wang, W. Y., Chen, R. S., *et al.* (2003). Genome-based analysis of virulence genes in a non-biofilm-forming *Staphylococcus epidermidis* strain (ATCC 12228). Mol. Microbiol. *49*, 1577–1593.

Ziebuhr, W., Krimmer, V., Rachid, S., Lossner, I., Gotz, F., and Hacker, J. (1999). A novel mechanism of phase variation of virulence in *Staphylococcus epidermidis*: evidence for control of the polysaccharide intercellular adhesin synthesis by alternating insertion and excision of the insertion sequence element IS256. Mol. Microbiol. *32*, 345–356.

Ziebuhr, W., Lossner, I., Rachid, S., Dietrich, K., Gotz, F., and Hacker, J. (2000). Modulation of the polysaccharide intercellular adhesin (PIA) expression in biofilm forming *Staphylococcus epidermidis*. Analysis of genetic mechanisms. Adv. Exp. Med. Biol. *485*, 151–157.

Staphylococci of Animals

10

J. Ross Fitzgerald and José R. Penadés

Abstract

Several staphylococcal species are notorious as human pathogens and are the focus of worldwide intensive research efforts. However, staphylococci are also associated with a large number of animal species and cause several infections of major economic importance. In addition, the zoonotic transmission of staphylococci to humans, especially those which are antibiotic-resistant, is a growing threat to public health. In this chapter, we will summarize selected aspects of the molecular pathogenesis of staphylococci which are pathogenic to animals and discuss how researchers are starting to investigate the host-adaptive evolution of staphylococci.

Introduction

Approximately 40 different species of staphylococci have been identified to date including many which are associated with animal hosts (Table 10.1). Research into the biology of animal staphylococci has traditionally lagged behind studies into human disease pathogenesis. However, several recent advances suggest that the near future will result in a much enhanced understanding of the evolution and pathogenesis of animal staphylococci. Here we will summarize what is known about several important animal-associated species and briefly discuss how current genome sequencing projects will improve the resources available to researchers in the field.

Staphylococcus aureus

Staphylococcus aureus colonizes a wide variety of members of the animal kingdom, including mammals, reptiles, and birds. In addition to the natural hosts identified, infection models for this clinically important bacterial species have

Table 10.1 Pathogenic staphylococcal species of animals

Staphylococcal species	Animal host species	Infection type
S. aureus	Ruminants (cows, sheep, and goats)	Mastitis,
	Horse	Skin infections
	Rabbit	Mastitis
	Poultry	'Bumble foot', chondronecrosis, septic arthritis
S. pseudintermedius	Dog	Commensal, pyoderma
S. delphini	Horse, mink, cow, dolphin, pigeon	Pyoderma
S. intermedius	Pigeon	
S. epidermidis	Ruminants (cow, sheep and goats)	Mastitis
S. hyicus	Pig	Exudative dermatitis
S. scuiri	Pig	Greasy pig syndrome
S. simulans	Ruminants (cow, sheep and goats)	Mastitis

been established in lower organisms, such as insects (*Drosophila melanogaster*) and worms (*Caenorhabditis elegans*) (Garcia-Lara *et al.*, 2005). Although usually a commensal, *S. aureus* can cause a variety of different animal infections including tick-associated pyaemia in lambs (Webster and Mitchell, 1989), staphylococcosis in rabbits (Hermans *et al.*, 2003), oedematous and necrotic dermatitis, septicaemia, abscesses and chondronecrosis in chickens (McNamee *et al.*, 1999), and pneumonia and osteomyelitis complex in turkeys (Alfonso and Barnes, 2006). However, the most economically important *S. aureus* infections of animals are intra-mammary infections (IMI) of lactating ruminants often leading to chronic mastitis (Miles *et al.*, 1992). Because this type of infection is very difficult to eradicate with antibiotic therapy, a premature culling of animals, involving substantial production losses, is often the only efficient strategy for control.

The ability of *S. aureus* to cause disease is due to a combination of virulence factors, including toxins, cell-surface associated adhesins, secreted exo-proteins and biofilm-related factors. The ability of *S. aureus* to cause infections in humans and animals raises the important question whether *S. aureus* strains are capable of causing infections of different host species or are largely adapted to infect a single host only. Population genetic studies have been employed to address this question.

Population structure of *S. aureus* from human and animal infections

Several studies have examined the diversity among natural populations of *S. aureus* infecting humans and animals (van Leeuwen *et al.*, 2005). Important insights have been obtained from studies using multilocus enzyme electrophoresis (MLEE) and more recently, Multilocus sequence typing (MLST). For example, in a classic population genetic study, MLEE was used to construct a framework of over 2000 isolates of *S. aureus* recovered from cases of human diseases and bovine and ovine mastitis. A total of 252 multilocus electrophoretic types (ET)s were identified representing 14 major lineages. The majority of isolates were assigned to 3 of the 14 major lineages and importantly, only 6 of 33 sublineages were shared between animals and humans. These data indicate that although some strains have the capacity to infect different hosts, the great majority of MLEE types are associated with a single host only. The bovine and ovine host-specialization phenotype is narrowly distributed throughout the dendrogram indicating that the ability to infect cows arose relatively few times in evolutionary history. More recently, a study using amplified fragment length polymorphism (AFLP) typing of animal and human isolates supported the previous evidence for host-specificity and added the suggestion that some strains may have adapted to specific tissues such as intramammary tissue rather than distinct hosts. Further, the authors reported that some virulence genes may be more commonly associated with strains infecting a particular host (van Leeuwen *et al.*, 2005). Of note, bovine mastitis is caused by a relatively small number of pathogenic clones with a broad geographic distribution (Fitzgerald *et al.*, 1997; Kapur *et al.*, 1995). Several studies have shown that the majority of bovine *S. aureus* strains are specific for cows and rarely cause infections of other hosts (Kapur *et al.*, 1995; Smith *et al.*, 2005; van Leeuwen *et al.*, 2005).

In spite of the common host-restriction of *S. aureus*, the development of specific animal models to reproduce the symptoms of human infections indicates that strains from humans have the capacity (under experimental conditions) to cross-infect animals such as rabbits (Wills *et al.*, 2005), rats (O'Riordan and Lee, 2004) and poultry (Rodgers *et al.*, 2006). Further, natural cross-infection between humans and domestic animals in the household has been reported (Simoons-Smit *et al.*, 2000). For example, a cat was implicated in an epidemic of methicillin-resistant *S. aureus* infections in a geriatric ward (Scott *et al.*, 1988). Further, dogs and cattle are potential reservoirs for the transmission of *S. aureus* strains to humans (Juhasz-Kaszanyitzky *et al.*, 2007).

Studies have shown that human and animal isolates of *S. aureus* often share some of the same virulence determinants. For example, bovine mammary isolates often harbour genes encoding superantigens, such as TSST-1 (Fitzgerald *et al.*, 2001a), exfoliative toxins (Endo *et al.*, 2003), and enterotoxins (Morandi *et al.*, 2007) which have all been shown to contribute to the pathogenesis of human infections, and these virulence factors were also made by *S. aureus* strains infecting horses (Holbrook *et al.*, 2003) and poultry (Smyth *et al.*, 2005). However, some determinants may be more commonly associated with host-specific lineages. For example, the *tst* gene, encoding the TSST-1 toxin, is found more often in the mastitis-associated *S. aureus* strains and the *cna* gene is not evenly distributed between the different AFLP clusters of the animal-related strains (van Leeuwen *et al.*, 2005). It is likely that, the combination of virulence factors produced by a pathogenic clone plays an important role in host- or tissue-tropism.

Intramammary infections of ruminants

S. aureus IMI of ruminants is usually initiated either by colonization of the teat canal with bacteria derived from the epidermis or by an influx of contaminated milk entering the gland during milking. Bacteria multiply in the milk and gain access to the upper part of the gland where they adhere to the ductular and alveolar epithelium. Adherent bacteria trigger macrophage activation and neutrophil migration from the blood into the milk resulting in an increase in the somatic cell count, (SCC), impairment of the host immune system, and epithelial cell damage. As a result, bacteria gain access to the basal subepithelial cell layers, bind fibrinogen and other host receptor proteins (Foster and Höök, 1998), and establish an infection which often becomes chronic. It is likely that *S. aureus* has the capacity to survive intracellularly which may facilitate immune and antibiotic avoidance (Dziewanowska *et al.*, 1999). The production of cytolytic toxins such as α-, β-, and a number of different leucotoxins may contribute to the severity of infection and high levels of expression may result in clinical mastitis (Bramley *et al.*, 1989; Cifrian *et al.*, 1996; Younis *et al.*, 2005; Barrio *et al.*, 2006).

S. aureus bovine host adaptation and insights from the sequencing project

In order to investigate the genetic basis for *S. aureus* adaptation to the bovine host, Herron and co-workers sequenced the complete genome of bovine strain RF122 (Herron *et al.*, 2002). Strain RF122 belongs to the common ET3 lineage previously identified by MLEE and ST151 as identified by MLST. RF122 produces TSST-1, SEC, and SEL, a common characteristic of bovine strains (Fitzgerald *et al.*, 2001a). Analysis of the genome sequence revealed evidence for the molecular adaptation of *S. aureus* to the intramammary milieu. In particular, high rates of amino acid substitution and premature truncations of predicted surface proteins were discovered. For example both ClfA and FnbpA which are important virulence factors of human staphylococcal diseases contain mutations leading to premature stop codons suggesting that these proteins are not essential for bovine intra-mammary pathogenesis. Furthermore, *sdrC* encoding another member of the MSCRAMM group of Gram-positive surface-associated factors is a pseudogene in strain RF122. In addition, Ebh is a very large fibronectin-binding protein that is highly conserved in sequenced *S. aureus* strains of human origin. In bovine strain RF122, the *ebh* gene contains several premature termination codons and several deletions such that only about 30% of the gene is present in comparison to human strains (Herron-Olson *et al*, unpublished data).

Remarkably, strain RF122 has the capacity to produce at least 30 different toxins including up to 10 different superantigen variants. SEC encoded by the pathogenicity island SaPIbov represents the bovine-specific allele, previously identified by Deringer *et al.* (1997). The authors demonstrated that SECbovine activated a unique profile of T-cells bearing specific Vbeta receptors which may be the result of host adaptation. RF122 contains several genes encoding toxins with leucotoxic properties including one LukM, lukF which is specific for bovine strains which has potent cytotoxic activity (Younis *et al.*, 2005; Barrio *et al.*, 2006). Interestingly, a feature of strain RF122 is the absence of genes encoding antibiotic resistance, consistent with the broad susceptibility of RF122 to most antibiotics except spectinomycin.

Other economically important *S. aureus* infections of animals

S. aureus can cause pyodermatitis, subcutaneous abscesses and mastitis in rabbits and is a major pathogen in rabbitries in various parts of the world (Hermans *et al.*, 2003). Two different strain types have been identified corresponding to high virulence (HV) strains which spread rapidly to most animals within a flock, or low virulence LV) which is generally restricted to sporadic infections of individual animals (Vancraeynest *et al.*, 2006). The high virulence strains belong to a single clone which has spread to different countries in Europe and an independent study in Spain indicated that some rabbit strains may be host-restricted (Segura *et al.*, 2007). The basis for the high virulence and host-specificity of rabbit strains remain to be explored.

S. aureus is one of the most important bacterial pathogens of broiler chickens causing infections such as septicaemia, osteomyelitis and chondronecrosis (Jordan and Pattison, 1996). A poultry biotype of *S. aureus* with unique phenotypic traits was identified in the 1970s (Devriese, 1984) and is not found among humans or other animals. Further, a single pulsed field gel electrophoresis type was shown to be responsible for epidemics of a broiler hatchery (Rodgers *et al.*, 1999). The evolutionary origin of this poultry-specific clone is unknown.

Methicillin-resistant *S. aureus* of animals

The first methicillin-resistant *S. aureus* (MRSA) strain of human origin was identified in the UK in 1961 (Jevons *et al.*, 1963) but the first major outbreak in the USA was not reported until 1968 (Barrett *et al.*, 1968). Subsequently it was shown that MRSA strains had acquired a gene (*mecA*) encoding an altered penicillin-binding protein which was responsible for methicillin resistance (Matsuhashi *et al.*, 1986). MRSA strains are now ubiquitous worldwide and are a significant healthcare problem (Appelbaum, 2006). A combination of lateral gene transfer of the *mecA* gene into distinct phylogenetic lines of *S. aureus*, and clonal dissemination, has been responsible for the emergence and spread of MRSA strains (Fitzgerald *et al.*, 2001b; Enright *et al.*, 2002). The first isolation of MRSA from a non-human source was from milk of a cow and several isolates of bovine MRSA have been identified since albeit with low frequency (Juhasz-Kaszanyitzky *et al.*, 2007; Lee, 2003; Moon *et al.*, 2007). Subsequently, MRSA has been found in a wide variety of domestic species including dogs (Guardabassi *et al.*, 2004), cats (Bender *et al.*, 2005), horses (Shimizu *et al.*, 1997) pigs (Huijsdens *et al.*, 2006), and chickens (Lee, 2003). Several reports have indicated an increase in the number of MRSA infections in dogs, cats and horses in recent years (Lee, 2003; Baptiste *et al.*, 2005; Loeffler *et al.*, 2005).The possibility that companion animals could act as the source of zoonotic infections of humans caused by MRSA is of interest. MRSA clones isolated from dogs and cats share similar genetic backgrounds to human nosocomial MRSA suggesting that the common strains spread from humans to animals rather than the emergence of new MRSA clones (Baptiste *et al.*, 2005). In contrast, equine strains appear to represent clones which are not represented among the common human hospital isolates indicating the possible emergence of new MRSA clones colonizing horses (Weese *et al.*, 2005; Cuny *et al.*, 2006). Further, it has been shown that pig farmers are at increased risk of nasal colonization by MRSA (Huijsdens *et al.*, 2006). Pig and pig farmer isolates also appear to be unrelated to hospital or typical human MRSA (Huijsdens *et al.*, 2006). Overall, the increasing frequency of isolation of MRSA from animals is a worrisome trend. Although vancomycin-resistant *S. aureus* has not yet been isolated from animal sources and is still not prevalent among human strains, Sung and Lindsay demonstrated that the bovine-specific lineage, ST151, was much more susceptible to the acquisition of vancomycin resistance genes from *Enterococcus* spp. than human clones (Sung and Lindsay, 2007). These data suggest that ST151 may represent a clone which has enhanced potential to become a risk to public health.

S. aureus clones have disseminated widely among animal species, causing a variety of infections. Although we have identified some host-specific and tissue-specific genes which contribute to disease pathogenesis in ruminants, it is likely that combinations of virulence factors and allelic variation of orthologous genes play an important role in host adaptation. Current genome sequencing projects in our laboratories will allow examination of the genetic basis for the host-specific pathogenesis of *S. aureus*.

Host adaptation and staphylococcal pathogenicity islands

Analysis of the sequences specific for bovine strains revealed that many are encoded by accessory genetic elements such as prophages or pathogenicity islands (PI)s (Fitzgerald *et al.*, 2001b; Herron *et al.*, 2002) suggesting the frequent exchange of genetic information among staphylococcal strains and less commonly with other bacterial species. In particular, staphylococcal pathogenicity islands (SaPIs) have been implicated in the pathogenesis and evolution of *S. aureus* including adaptation to new hosts (Novick and Subedi, 2007). As such, they are worthy of particular focus in this chapter. Pathogenicity islands been recognized as the repository of virulence genes in many organisms (Groisman and Ochman, 1996; Hochhut *et al.*, 2005). PIs are typically representative of large genomic regions (10–200 kb) with a G+C content distinct from the overall G+C content of the species-specific genome. PIs are frequently integrated adjacent to tRNA genes and are normally flanked by short direct repeats resembling attachment sites for phage integrases. Moreover, numerous PIs carry phage-related integrase genes, suggesting that the mechanism of integration and excision of PIs should correspond to that of phages (Hacker and Kaper, 2000). There are some reports of the genetic instability of PIs, suggesting integration, excision (deletion) or other rearrangements (Blum *et al.*, 1994; Hacker *et al.*, 1997). However, the majority of PIs described to date have degenerate integrase genes or recombination sites, and direct proof of integration and excision is lacking.

The situation in staphylococci is somewhat different. The first staphylococcal pathogenicity islands (SaPIs) identified were discovered because of their carriage of the genes for toxic shock syndrome toxin 1 (TSST-1) and other superantigens. They were soon found to be highly mobile, owing to their relationship with certain phages (Lindsay *et al.*, 1998). Not only are the SaPIs induced to excise and replicate by temperate phages, but they also induce the phage to produce special small capsids commensurate with their genome sizes, into which they are encapsidated (Ruzin *et al.*, 2001; Ubeda *et al.*, 2005), resulting in exceptionally high transfer frequencies, so that their transferability, like that of prophages, is clearly a strongly selected feature of their biology.

The SaPIs are inserted at specific chromosomal sites (att_S) and is always in the found in the same orientation. Usually, the functional ones are ~14–16 kb in size, although a highly degenerate SaPI, of only 3.14 kb, is present in 5 of the sequenced genomes. Currently, the complete sequence of 18 SaPIs are known, which form a highly coherent family with a conserved functional and genetic organization (Novick and Subedi, 2007). The SaPIs were initially identified as reservoirs of superantigen genes, especially tst (encoding TSST-1), and most, but not all of the SaPIs carry genes encoding superantigens or other products that specifically impact the host cell's phenotype.

Interestingly, and in contrast with many chromosomally encoded virulence factors, which are present in *S. aureus* isolates from different host origins, some SaPI-encoded genes have revealed a host-dependent association. For example, SaPIbov, the first PI identified in *S. aureus* from bovine strains, carries the genes encoding TSST-1, the host-specific allele of SEC, and a novel enterotoxin gene, SEL (Fig. 10.1) (Fitzgerald *et al.*, 2001a). It has been demonstrated that the recombinant SEL protein is expressed in *S. aureus*, and is mitogenic for bovine lymphocytes (Orwin *et al.*, 2003), indicating that SEL is a new superantigen.

Mobile genetic elements that encode and utilize site-specific integrases, are always located at an integrase-determined site. The known SaPIs carried five different integrase genes (*int*), which utilize five different sites (att_B) in the *S. aureus* genome, and therefore represent five different subclasses or families, defined by the combination int/att_B (Lindsay and Holden, 2004; Novick and Subedi, 2007). Given the broad diversity represented by the sequenced strains, it is likely these represent the majority of the SaPI sites in *S. aureus*.

The characterization of a second bovine SaPI (SaPIbov2) revealed the presence of an *int* (*sip*) gene identical to that found in SaPIbov1 and revealed a common int site adjacent to the GMP synthase gene (gmpS) (Ubeda *et al.*, 2003). Additionally, and in contrast to the integrase activity of the human strain-derived SaPI1,

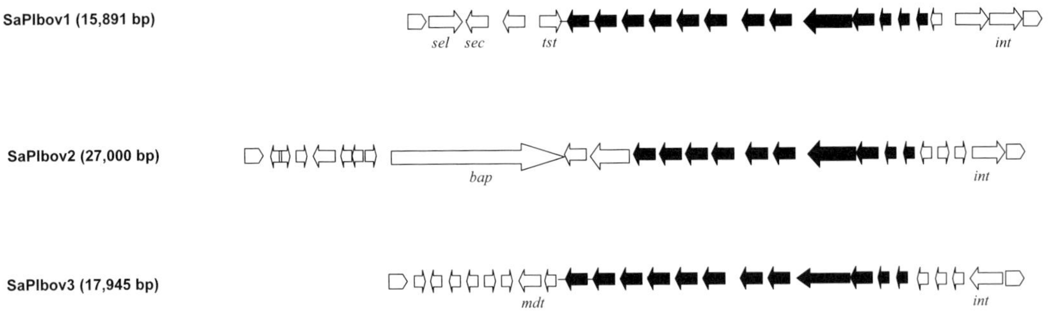

Figure 10.1 Bovine *S. aureus* pathogenicity islands. Conserved genes are indicated by black arrows, SaPI-specific genes by white arrows and direct repeats are indicated by hatched arrows. *sip*, staphylococcal integrase protein; *sel*, staphylococcal enterotoxin I; *sec*, staphylococcal enterotoxin C; *tst*, toxic shock syndrome toxin-1; *bap*, biofilm-associated protein; *mdt*, multidrug transporter.

the integrase encoded by both bovine islands can excise, form a circular element and integrate the islands site-specifically and RecA-independently into a chromosomal att site in an integrase-dependent manner (Ubeda *et al.*, 2003). This activity allows the SaPIbov islands to transfer using phages that do not induce replication but transfer the SaPIbov islands by a mechanism that probably involves the encapsidation of standard transducing fragments containing the intact island followed by Int-mediated excision, circularization and integration in the recipient strain (Maiques *et al.*, 2007).

The SaPIbov-int gene has only been found in bovine strains to date, whereas it was not possible to identify the gene in any isolate from human (Novick and Subedi, 2007) or rabbit origin (unpublished data), indicating the existence of a family of bovine-specific PIs integrated in the same attB site in the *S. aureus* chromosome. An additional SaPI with a unique *int* gene (SaPIbov3) contains genes encoding a predicted multidrug transporter (Fig. 10.1) in bovine strain RF122 at a site in the genome which often contains mobile genetic elements in human strains.

Finally, the third example of SaPI-mediated host-specificity is Bap (biofilm-associated protein). Bap is a surface protein involved in biofilm formation by *S. aureus* isolated from chronic mastitis infections (Cucarella *et al.*, 2001). The *bap* gene is carried by a putative composite transposon inserted into SaPIbov2, and encodes a 2276-amino-acid surface protein with a multidomain architecture characteristic of surface-associated proteins from Gram-positive bacteria. The N-terminal domain (819 amino acids) includes a signal sequence for extracellular secretion and is largely devoid of repeats. The region encompassing amino acid residues 820–2148 is composed of 13 identical repeats, each of which is 86 amino acids long. The C terminus of Bap contains a typical cell-wall attachment region comprising an LPXTG motif, a hydrophobic trans-membrane sequence, and positively charged cytoplasmic domain.

Production of Bap reduced infectivity during early infection by blocking adherence of MSCRAMMs to host proteins (Cucarella *et al.*, 2002). Later in infection, Bap contributed to bacterial persistence in the mammary gland most likely through the formation of biofilm (Ubeda *et al.*, 2003; Cucarella *et al.*, 2004).

The *bap* gene has not been found among *S. aureus* human isolates suggesting that specific host-dependent pathogenic factors evolved independently in human and ruminant strains (Cucarella *et al.*, 2001). In spite of this, the presence of *bap* is not restricted to *S. aureus*, and *bap* orthologues have been found in several coagulase-negative staphylococcal species, associated with mastitis including *S. epidermidis*, *S. chromogenes*, *S. xylosus*, *S. simulans*, and *S. hyicus* (Tormo *et al.*, 2005). Bap represents the paradigm for a group of surface proteins sharing structural and functional features, which have emerged as an important component of biofilm formation by diverse bacterial species (Lasa and Penades, 2006). In addition to staphylococci, members of this family have been described in *Enterococcus faecalis* (Esp; (Shankar *et al.*, 1999;

Toledo-Arana *et al.*, 2001)), *Burkholderia cepacia* (Bap; (Huber *et al.*, 2002)), *Pseudomonas putida* (mus20; (Espinosa-Urgel *et al.*, 2000)), and *Salmonella typhimurium* (Stm2689; (Latasa *et al.*, 2005)). All members of the Bap family are of high-molecular weight, have a signal sequence for secretion, a core domain of a variable number of repeats and confer the capacity to form a biofilm, leading to an important role in bacterial pathogenesis.

Although encoded by a mobile genetic element and only in bacteria causing mammary gland infections, Bap expression is controlled by *S. aureus* gene regulators, including SarA (Trotonda *et al.*, 2005), indicating integration into *S. aureus* global pathways. The presence of putative EF-hand motifs in the amino acid sequence of Bap prompted us to investigate the effect of calcium on the multicellular behaviour of Bap-expressing staphylococci (Arrizubieta *et al.*, 2004). We found that addition of millimolar amounts of calcium to the growth media inhibited intercellular adhesion and biofilm formation by Bap-positive *S. aureus* strain V329. Addition of manganese, but not addition of magnesium, also inhibited biofilm formation, whereas bacterial aggregation in liquid media was greatly enhanced by metal-chelating agents (Arrizubieta *et al.*, 2004). In contrast, calcium or chelating agents had virtually no effect on the aggregation of Bap-deficient strain M556. Site-directed mutagenesis of two of the putative EF-hand domains resulted in a mutant strain that was capable of forming a biofilm but whose biofilm was not inhibited by calcium. Our results indicated that Bap binds Ca^{2+} with low affinity and that Ca^{2+} binding renders the protein non-competent for biofilm formation and for intercellular adhesion (Arrizubieta *et al.*, 2004). The fact that calcium inhibition of Bap-mediated multi-cellular behaviour takes place *in vitro* at concentrations similar to those found in milk serum supports the possibility that this inhibition is relevant to the pathogenesis and/or epidemiology of the bacteria in the mastitis process, and suggest the existence of an additional mechanism of control of a virulence factor (in this case biofilm formation) by adaptation of protein structure to the characteristics present in the site of infection.

Staphylococcus intermedius group

Isolates traditionally identified as *S. intermedius* colonize the skin, hair coat and particularly mucocutaneous sites such as the nose, mouth and anus of dogs and other animals including horses, mink and pigeons (Allaker *et al.*, 1992a; Allaker *et al.*, 1992b).

Dogs with skin problems are commonly diagnosed with superficial pyoderma (also known as superficial bacterial folliculitis), and *S. intermedius* is regarded as the primary pathogen isolated from pyoderma cases. In fact, superficial bacterial folliculitis is the most frequent skin disease in dogs after flea allergy dermatitis (Gross *et al.*, 2005). Superficial bacterial pyoderma is a skin infection of the hair follicle (folliculitis) and the main clinical feature is the formation of follicular pustules which have an erythematous base and are filled with white or yellow purulent material. Canine cutaneous staphylococcal infections often occur as secondary infections to predisposing factors such as atopic dermatitis (AD), flea allergy dermatitis, demodicosis or hypothyroidism (Gross *et al.*, 2005). *S. intermedius* is also the primary pathogen in deep pyoderma which may develop from superficial pyoderma, impetigo and superficial spreading pyoderma.

Canine AD is characterized by a type I hypersensitivity reaction (predominantly IgE antibody associated) to environmental allergens such as house dust mites, plant pollens and moulds and the main clinical features are inflammation and pruritus of the skin (Hill and Olivry, 2001). *S. intermedius* attaches to epidermal cells of healthy dogs, but shows greater adherence to epidermal cells of atopic dogs (McEwan, 2000; Simou *et al.*, 2005b; McEwan *et al.*, 2006). It is possible that AD alters the availability of cutaneous receptors for staphylococci and facilitates bacterial adherence (Simou *et al.*, 2005b).

S. intermedius is not usually isolated from humans and seems to shows a host-specificity to canine corneocytes (Simou *et al.*, 2005a). However, transmission between humans and their pets can occur (Guardabassi *et al.*, 2004), and *S. intermedius* can be regarded as a true zoonotic pathogen (Tanner *et al.*, 2000). A common route of invasive infection in humans is through dog bite wounds and several instances of life-threatening infections of humans have been reported (Talan *et al.*, 2003; Pottumarthy *et al.*, 2004).

Population structure of the *S. intermedius* group

S. intermedius was first described by Hajek in 1976 who noted a remarkable variation in phenotypes of *S. intermedius* isolates from various hosts based on biotype differences such as mannitol fermentation and clumping factor production (Hajek, 1976). More recent studies have used genomic approaches such as ribotyping, PFGE and 16S-23S intergenic ribosomal DNA spacer polymorphism analysis (ITS-PCR) for population genetic studies of *S. intermedius* (Hesselbarth and Schwarz, 1995; Aarestrup, 2001; Bes *et al.*, 2002; Pinchbeck *et al.*, 2006).

Ribotyping studies which discriminate between strains based on the diversity of rRNA operons have shown that *S. intermedius* isolates from various host species (dog, pigeon, horse, mink) belong to three major clusters. Canine strains belonged to the same cluster whereas the other isolates varied widely in their ribotypes (Hesselbarth and Schwarz, 1995). These findings were supported by a study which analysed the ribotypes of *S. intermedius* isolates from various hosts of the canoidea superfamily (dog, skunk, weasel, racoon, red panda, bear) (Aarestrup, 2001). Aarestrup found that *S. intermedius* possesses remarkable genetic diversity but all isolates of the same host origin belonged to an identical ribotype cluster indicating host specificity or co-evolution between the host family and *S. intermedius* (Aarestrup, 2001). ITS-PCR, is a method which allows comparison of the length of the intergenic spacer region between the 16S and 23S rRNA genes of different isolates (Bes *et al.*, 2002). Bes *et al.* analysed 57 *S. intermedius* isolates from various hosts (human, camel, dog, pigeon, horse, mink) via ITS-PCR typing (Bes *et al.*, 2002) and found 12 ITS-PCR types among pigeon, horse and mink strains. However, the majority of human and canine strains belong to the same two ITS-PCR types, suggesting frequent transmission between humans and dogs (Bes *et al.*, 2002). Population genetic analyses of *S. intermedius* has been very limited due to the lack of a multilocus sequence typing system for *S. intermedius*. Recently, Sasaki *et al.* (Sasaki *et al.*, 2007) and our own group independently discovered the existence of three previously described species among strains phenotypically identified as *S. intermedius*, corresponding to *S. intermedius*, *S. pseudintermedius* and *S. delphini*, respectively. *S. pseudintermedius* and not *S. intermedius* was revealed to be the common canine pyoderma pathogen. Accordingly, we will refer to the common canine pyoderma pathogen as *S. pseudintermedius*.

Molecular pathogenesis of *S. pseudintermedius* canine pyoderma

Our understanding of the pathogenesis of *S. pseudintermedius* infections is very limited compared to *S. aureus*. However, we do know that *S. pseudintermedius* produces a wide array of virulence factors, many of which have structural and functional characteristics in common with *S. aureus*. For example *S. pseudintermedius* produces enzymes such as coagulase, proteases and thermonuclease and toxins including hemolysins, exfoliative toxins and enterotoxins. Surface proteins mediating fibrinogen- and fibronectin-binding and an immunoglobulin-binding protein similar to staphylococcal protein A have been identified but are yet to be characterized. In addition, *S. pseudintermedius* was recently shown to form biofilms (Futagawa-Saito *et al.*, 2004).

S. pseudintermedius produces α- and β-toxin and causes haemolysis of rabbit erythrocytes and hot-cold haemolysis of sheep erythrocytes (Hajek, 1976; Dziewanowska *et al.*, 1996; Futagawa-Saito *et al.*, 2004). Similarly to *S. aureus*, β-haemolysin has sphingomyelinase activity with a high affinity to sphingomyelin (Dziewanowska *et al.*, 1996). *S. pseudintermedius* also produces a bicomponent leukotoxin, Luk-I, which is encoded by two co-transcribed genes, *lukS* and *lukF* (Prevost *et al.*, 1995; Futagawa-Saito *et al.*, 2004). Luk-I has been shown to be leukotoxic on polymorphonuclear cells, but only slightly haemolytic on rabbit red blood cells (Prevost *et al.*, 1995). Futagawa-Saito *et al.* have examined the distribution of the *lukI* locus in canine and pigeon strains and found that all canine isolates were positive for *lukS* and *lukF* and were highly cytotoxic against rabbit leukocytes In contrast, pigeon strains contained *lukS*, but not *lukF* and demonstrated only low cytotoxicity (Futagawa-Saito *et al.*, 2004).

S. pseudintermedius produces an exfoliative toxin (SIET) which has been shown to have a rounding effect in cultured epithelial cells and an exfoliative effect in 1-day-old chickens, hamsters and dogs, but not in rats and mice (Terauchi *et al.*, 2003a; Terauchi *et al.*, 2003b). Dogs injected with purified SIET developed clinical signs such as erythema, exfoliation and crusting which are similar to the symptoms of canine pyoderma, and human staphylococcal scalded skin syndrome (Terauchi *et al.*, 2003b).

S. pseudintermedius has the capacity to produce several staphylococcal superantigens homologous to those made by *S. aureus* including SEA, SEB, SEC and SED and TSST-1 (Hendricks *et al.*, 2002). SEA and SEB have been shown to increase T cell blastogenesis in peripheral canine blood mononuclear cells *in vitro*, but a role in the pathogenesis of canine pyoderma remains to be defined (Hendricks *et al.*, 2002). In addition, a canine type C enterotoxin (SEC_{canine}) made by some *S. pseudintermedius* isolates has been shown to have emetic activity and an ability to activate T cell proliferation (Edwards *et al.*, 1997). A previous study identified 31 of a total of 247 canine *S. pseudintermedius* isolates which contained the sec_{canine} gene, whereas Futagawa-Saito *et al.* screened 106 pigeon and canine strains and found only a single sec_{canine}-positive isolate indicating that the enterotoxin may be encoded on a mobile genetic element (Becker *et al.*, 2001; Futagawa-Saito *et al.*, 2004). A second enterotoxin-related gene, *se-int*, was detected in all 44 canine strains tested (Futagawa-Saito *et al.*, 2004) but has not yet been characterized. The enterotoxigenic potential of *S. pseudintermedius* has not been fully clarified, but has been previously implicated in food-poisoning outbreaks in humans (Khambaty *et al.*, 1994).

Staphylococcal virulence is controlled by a number of global gene regulators including the accessory gene regulator (*agr*). The *agr* locus is widely distributed among staphylococcal species (Dufour *et al.*, 2002) including *S. pseudintermedius* (Sung *et al.*, 2006). The *S. pseudintermedius agr* locus consists of 3436 bp with five open reading frames (ORFs), *agrB*, *agrD*, *agrC*, *agrA*, and *hld*, encoding for the classic two-component regulatory system with AgrD being processed to AIP by AgrB (Sung *et al.*, 2006). The *S. pseudintermedius agr* system is functional and encodes an unusual AIP with a partially cyclic structure (Ji *et al.*, 2005). The expression of selected virulence factors was found to correlate with RNAIII levels during growth *in vitro*, consistent with a central role for *agr* in the regulation of *S. pseudintermedius* virulence (Sung *et al.*, 2006). So far, three different *agr* allelic types have been identified among *S. pseudintermedius* isolates, encoding for three different putative AIPs (Sung *et al.*, 2006).

Methicillin-resistant *S. pseudintermedius*

In recent years, an increasing number of isolates of methicillin-resistant *S. pseudintermedius* (MRSP) have been identified (Kania *et al.*, 2004; Waller, 2005; Morris *et al.*, 2006). In a recent study, Morris *et al.* have described that as many as 17% (57 out of 336) of *S. intermedius* isolates screened for antimicrobial susceptibility were methicillin resistant (Morris *et al.*, 2006). The comparison of *S. pseudintermedius* strains isolated from 13 dogs suffering from deep pyoderma and their owners by PFGE analysis revealed that six of the owners carried the same strain as their dogs, raising questions of interspecies transmission and the occurrence of resistance gene transfer between staphylococcal species (all strains tested were resistant to at least two antimicrobials, including penicillin, tetracycline and chloramphenicol) (Guardabassi *et al.*, 2004). In summary, these studies indicate the emergence of bacterial antimicrobial resistance in veterinary practice and the need for appropriate screening methods and selective antimicrobial use.

S. delphini

Our recent research in addition to others (Sasaki *et al.*, 2007), suggests that *S. delphini* may be commonly misidentified as *S. intermedius* and could therefore be much more clinically important than previously thought. To date, *S. delphini* has been isolated from skin infections of horses, cows, mink, pigeons and dolphins highlighting its broad host-range (Saski *et al.*, 2007).

Staphylococcus epidermidis

Staphylococcus epidermidis is an opportunistic pathogen whose ability to persist and multiply in a variety of environments results in a wide spectrum of diseases in both humans and animals. In humans, *S. epidermidis* infection is one of the most devastating complications of prosthetic joint surgery (Rupp and Archer, 1994). In contrast, in lactating animals, *S. epidermidis* has traditionally been considered a minor pathogen of bovine mastitis. However, recent studies have reported the increasing frequency of *S. epidermidis* intramammary infections of cows, sheep and goats. *S. epidermidis* infections are often milder than those produced by *S. aureus*, but can result in significant loss of milk production. For example, one study showed that *S. epidermidis* infections caused an 8.7% loss in milk production from a 305-day milk yield total (Timms and Schultz, 1987). The population genetics of *S. epidermidis* are not well understood. Although recently a MLST system was developed which has identified some genetic diversity between *S. epidermidis* strains (Thomas *et al.*, 2007), this study did not analyse the relationship between human and animal strains. Consequently, whether bovine strains of *S. epidermidis* represent host-adapted lineages or opportunistic environmental isolates is not yet established.

Although the ability of *S. epidermidis* to cause chronic mastitis is well recognized, very little is known about the virulence factors that contribute to pathogenesis. Interestingly, we have found the *bap* gene in a number of different staphylococcal species including *S. epidermidis*, *Staphylococcus chromogenes*, *Staphylococcus xylosus*, *Staphylococcus hyicus* and *Staphylococcus simulans*. The *bap* genes from coagulase negative staphylococci are involved in biofilm formation, similarly to *S. aureus* (Tormo *et al.*, 2005). As such, Bap may contribute to the intramammary pathogenesis of *S. epidermidis*.

Although we have found that SaPIbov2, the pathogenicity island carrying *bap*, can be transduced at a high frequency from *S. aureus* to CNS (Maiques *et al.*, 2007), SaPIbov2 does not encode Bap in *S. epidermidis*. In fact, novel mobile genetic elements appear to contain the *bap* gene in different staphylococcal species (unpublished results).

Staphylococcus hyicus

Staphylococcus hyicus is associated with the skin disease exudative epidermitis in pigs, also called Greasy Pig Disease. This organism has also been occasionally isolated from cows with mastitis (Birgersson *et al.*, 1992), horses with dermatitis (Devriese *et al.*, 1985), and chickens with exudative dermatitis or tenosynovitis (Kibenge *et al.*, 1983). In all cases, production of several exfoliative toxins have been involved in the pathogenesis of the disease. Using the skin disease exudative epidermitis in pigs as a model of infection, the exfoliative toxins responsible for the characteristic lesions associated with the disease have been identified, cloned and purified from strains in Japan and Denmark (Sato *et al.*, 1994; Sato *et al.*, 1999; Ahrens and Andresen, 2004). These toxins have been characterized as exo-proteins of approximately 30 kDa. The toxins isolated in Japan were designated SHETA and SHETB (Sato *et al.*, 2000), and the distinct toxins isolated in Denmark were designated ExhA, ExhB, and ExhC (Ahrens and Andresen, 2004). One additional toxin antigenically distinct from these was provisionally designated ExhD (Ahrens and Andresen, 2004).

Staphylococcus sciuri

Although *S. hyicus* is generally recognized as the causative agent of the skin disease exudative epidermitis in pigs, other species of *Staphylococcus* may also cause this disease. For example, a recent publication in which a highly pathogenic strain of *Staphylococcus sciuri* HBXX06 caused an acute form of exudative epidermitis in piglets in both natural and experimental conditions, suggested that the isolate of *S. sciuri* HBXX06 was a new emerging pathogen for exudative epidermitis in pigs (Chen *et al.*, 2007).

Members of the *S. sciuri* group are widely distributed in nature, and they can be isolated from a variety of animals and products of animal origin (Couto *et al.*, 1996) as well as from humans (Shittu *et al.*, 2004; Stepanovic *et al.*, 2003), but most of them are non-pathogenic to animals. However they are important human pathogens responsible for episodes of endocarditis, peritonitis, septic shock, urinary tract infection, pelvic inflammatory disease and wound infections (Chen *et al.*, 2007). Currently little information is known regarding the pathogenicity of *S. sciuri* in animals. It has been reported that *S. sciuri* can be associated with mastitis in ruminants such as goats (Poutrel, 1984) and cows (Rahman *et al.*, 2005).

Conclusions

This chapter provides an overview of what is known about the pathogenesis of staphylococcal infections of animals. While the host-specific tropism of several staphylococcal species has been well established, the evolutionary basis is poorly understood. Furthermore, recent evidence suggests that animals may represent reservoirs of antibiotic resistant isolates which could represent a threat to human health.

Future perspectives

At the time of writing the genome sequence of 10 strains of *S. aureus* of human origin had been determined, whereas the genome sequence of only a single strain of animal origin (bovine strain RF122) had been completed. Similarly, advances in understanding of the pathogenesis of staphylococcal diseases of animals have lagged behind studies of human infections. Currently however, genome sequencing projects of ovine, avian and rabbit-specific clones of *S. aureus* are underway in our laboratories. These data will represent an excellent resource for studies into the pathogenesis, evolution and horizontal gene transfer among animal *S. aureus* clones. We are also sequencing a strain of *S. pseudintermedius*, the common canine pyoderma pathogen, which will facilitate attempts to design an effective vaccine for the prevention of pyoderma and to explore the biology of a staphylococcal species which has co-evolved as a commensal with its canine host over many thousands of years. Overall, it is expected that the next few years will see an increase in our understanding of the pathogenesis and evolution of staphylococci infecting animals. It is hoped that these studies will ultimately lead to the design of novel therapeutics for the treatment or prevention of diseases of both economic and animal welfare importance.

Web resources

Department of Environment, Food and Rural Affairs, UK
http://www.defra.gov.uk/animalh/diseases/zoonoses/mrsaqa.htm
Questions and answers

BellaMoss foundation
http://www.thebellamossfoundation.com/
Charity supporting vets, pets and pet owners about MRSA

MRSA in animals
www.mrsainanimals.com
Collates information from the media, conferences, journals, etc.

Whole genome sequence of RF122, bovine *S. aureus*
http://www.ncbi.nlm.nih.gov/entrez/viewer.fcgi?db=nuccore&id=82655308

PubMed, Core nucleotide, AJ938182
www.ncbi.nlm.nih.gov/sites/entrez?db=PubMed

References

Aarestrup, F.M. (2001). Comparative ribotyping of *Staphylococcus intermedius* isolated from members of the Canoidea gives possible evidence for host-specificity and co-evolution of bacteria and hosts. Int. J. Syst. Evol. Microbiol. *51*, 1343–1347.

Ahrens, P., and Andresen, L.O. (2004). Cloning and sequence analysis of genes encoding *Staphylococcus hyicus* exfoliative toxin types A, B, C, and D. J. Bacteriol. *186*, 1833–1837.

Alfonso, M., and Barnes, H.J. (2006). Neonatal osteomyelitis associated with *Staphylococcus aureus* in turkey poults. Avian Dis. *50*, 148–151.

Allaker, R.P., Lloyd, D.H., and Bailey, R.M. (1992a). Population sizes and frequency of staphylococci at mucocutaneous sites on healthy dogs. Vet. Rec. *130*, 303–304.

Allaker, R.P., Lloyd, D.H., and Simpson, A.I. (1992b). Occurrence of *Staphylococcus intermedius* on the hair and skin of normal dogs. Res. Vet. Sci. *52*, 174–176.

Appelbaum, P.C. (2006). MRSA – the tip of the iceberg. Clin. Microbiol. Infect. *12 Suppl 2*, 3–10.

Arrizubieta, M.J., Toledo-Arana, A., Amorena, B., Penades, J.R., and Lasa, I. (2004). Calcium inhibits Bap-dependent multicellular behavior in *Staphylococcus aureus*. J. Bacteriol. *186*, 7490–7498.

Baptiste, K.E., Williams, K., Willams, N.J., Wattret, A., Clegg, P.D., Dawson, S., Corkill, J.E., O'Neill, T., and Hart, C.A. (2005). Methicillin-resistant staphylococci in companion animals. Emerg. Infect. Dis. *11*, 1942–1944.

Barrett, F.F., McGehee, R.F., Jr., and Finland, M. (1968). Methicillin-resistant *Staphylococcus aureus* at Boston City Hospital. Bacteriologic and epidemiologic observations. N. Engl. J. Med. *279*, 441–448.

Barrio, M.B., Rainard, P., and Prevost, G. (2006). LukM/LukF'-PV is the most active *Staphylococcus aureus* leukotoxin on bovine neutrophils. Microbes Infect. *8*, 2068–2074.

Becker, P., Hufnagle, W., Peters, G., and Herrmann, M. (2001). Detection of differential gene expression in biofilm-forming versus planktonic populations of *Staphylococcus aureus* using micro-representational-difference analysis. Appl. Environ. Microbiol. *67*, 2958–2965.

Bender, J.B., Torres, S.M., Gilbert, S.M., Olsen, K.E., and LeDell, K.H. (2005). Isolation of methicillin-resistant *Staphylococcus aureus* from a non-healing abscess in a cat. Vet. Rec. *157*, 388–389.

Bes, M., Saidi Slim, L., Becharnia, F., Meugnier, H., Vandenesch, F., Etienne, J., and Freney, J. (2002). Population diversity of *Staphylococcus intermedius* isolates from various host species: typing by 16S-23S intergenic ribosomal DNA spacer polymorphism analysis. J. Clin. Microbiol. *40*, 2275–2277.

Birgersson, A., Jonsson, P., and Holmberg, O. (1992). Species identification and some characteristics of coagulase-negative staphylococci isolated from bovine udders. Vet. Microbiol. *31*, 181–189.

Blum, G., Ott, M., Lischewski, A., Ritter, A., Imrich, H., Tschape, H., and Hacker, J. (1994). Excision of large DNA regions termed pathogenicity islands from tRNA-specific loci in the chromosome of an *Escherichia coli* wild-type pathogen. Infect. Immun. *62*, 606–614.

Bramley, A.J., Patel, A.H., O'Reilly, M., Foster, R., and Foster, T.J. (1989). Roles of alpha-toxin and beta-toxin in virulence of *Staphylococcus aureus* for the mouse mammary gland. Infec Immu 57, 2489–2494.

Chen, S., Wang, Y., Chen, F., Yang, H., Gan, M., and Zheng, S. J. (2007). A highly pathogenic strain of *Staphylococcus sciuri* caused fatal exudative epidermitis in piglets. PLoS ONE 2, e147.

Cifrian, E., Guidry, A.J., Bramley, A.J., Norcross, N.L., Bastida-Corcuera, F.D., and Marquardt, W.W. (1996). Effect of staphylococcal beta toxin on the cytotoxicity, proliferation and adherence of *Staphylococcus aureus* to bovine mammary epithelial cells. Vet. Microbiol. *48*, 187–198.

Couto, I., de Lencastre, H., Severina, E., Kloos, W., Webster, J.A., Hubner, R.J., Sanches, I.S., and Tomasz, A. (1996). Ubiquitous presence of a *mecA* homologue in natural isolates of *Staphylococcus sciuri*. Microb. Drug Resist. 2, 377–391.

Cucarella, C., Solano, C., Valle, J., Amorena, B., Lasa, I., and Penades, J.R. (2001). Bap, a *Staphylococcus aureus* surface protein involved in biofilm formation. J. Bacteriol. *183*, 2888–2896.

Cucarella, C., Tormo, M. A., Knecht, E., Amorena, B., Lasa, I., Foster, T.J., and Penades, J.R. (2002). Expression of the biofilm-associated protein interferes with host protein receptors of *Staphylococcus aureus* and alters the infective process. Infect. Immun. *70*, 3180–3186.

Cucarella, C., Tormo, M.A., Ubeda, C., Trotonda, M.P., Monzon, M., Peris, C., Amorena, B., Lasa, I., and Penades, J.R. (2004). Role of biofilm-associated protein bap in the pathogenesis of bovine *Staphylococcus aureus*. Infect. Immun. *72*, 2177–2185.

Cuny, C., Kuemmerle, J., Stanek, C., Willey, B., Strommenger, B., and Witte, W. (2006). Emergence of MRSA infections in horses in a veterinary hospital: strain characterisation and comparison with MRSA from humans. Euro. Surveill. *11*, 44–47.

Devriese, L.A. (1984). A simplified system for biotyping *Staphylococcus aureus* strains isolated from animal species. J. Appl. Bacteriol. *56*, 215–220.

Devriese, L.A., Nzuambe, D., and Godard, C. (1985). Identification and characteristics of staphylococci isolated from lesions and normal skin of horses. Vet. Microbiol. *10*, 269–277.

Dufour, P., Jarraud, S., Vandenesch, F., Greenland, T., Novick, R.P., Bes, M., Etienne, J., and Lina, G. (2002). High genetic variability of the *agr* locus in *Staphylococcus* species. J. Bacteriol. *184*, 1180–1186.

Dziewanowska, K., Edwards, V.M., Deringer, J.R., Bohach, G.A., and Guerra, D.J. (1996). Comparison of the beta-toxins from *Staphylococcus aureus* and *Staphylococcus intermedius*. Arch. Biochem. Biophys. *335*, 102–108.

Dziewanowska, K., Patti, J.M., Deobald, C.F., Bayles, K.W., Trumble, W.R., and Bohach, G.A. (1999). Fibronectin binding protein and host cell tyrosine kinase are required for internalization of *Staphylococcus aureus* by epithelial cells. Infect. Immun. *67*, 4673–4678.

Edwards, V. M., Deringer, J. R., Callantine, S. D., Deobald, C. F., Berger, P. H., Kapur, V., Stauffacher, C.V., and Bohach, G.A. (1997). Characterization of the canine type C enterotoxin produced by *Staphylococcus intermedius* pyoderma isolates. Infect. Immun. *65*, 2346–2352.

Endo, Y., Yamada, T., Matsunaga, K., Hayakawa, Y., Kaidoh, T., and Takeuchi, S. (2003). Phage conversion of exfoliative toxin A in *Staphylococcus aureus* isolated from cows with mastitis. Vet. Microbiol. *96*, 81–90.

Enright, M.C., Robinson, D.A., Randle, G., Feil, E.J., Grundmann, H., and Spratt, B.G. (2002). The evolutionary history of methicillin-resistant *Staphylococcus aureus* (MRSA). Proc. Natl. Acad. Sci. U S A *99*, 7687–7692.

Espinosa-Urgel, M., Salido, A., and Ramos, J.L. (2000). Genetic analysis of functions involved in adhesion of *Pseudomonas putida* to seeds. J. of Bacteriol. *182*, 2363–2369.

Fitzgerald, J.R., Meaney, W.J., Hartigan, P.J., Smyth, C.J., and Kapur, V. (1997). Fine-structure molecular epidemiological analysis of *Staphylococcus aureus* recovered from cows. Epidemiol. Infect. *119*, 261–269.

Fitzgerald, J.R., Monday, S.R., Foster, T.J., Bohach, G.A., Hartigan, P.J., Meaney, W.J., and Smyth, C.J. (2001a). Characterization of a putative pathogenicity island from bovine *Staphylococcus aureus* encoding multiple superantigens. J. Bacteriol. *183*, 63–70.

Fitzgerald, J.R., Sturdevant, D.E., Mackie, S.M., Gill, S.R., and Musser, J.M. (2001b). Evolutionary genomics of *Staphylococcus aureus*: insights into the origin of methicillin-resistant strains and the toxic shock syndrome epidemic. Proc. Natl. Acad. Sci. U S A *98*, 8821–8826.

Foster, T.J., and Höök, M. (1998). Surface protein adhesins of *Staphylococcus aureus*. Trends Microbiol. *6*, 484–488.

Futagawa-Saito, K., Sugiyama, T., Karube, S., Sakurai, N., Ba-Thein, W., and Fukuyasu, T. (2004). Prevalence and characterization of leukotoxin-producing *Staphylococcus intermedius* in Isolates from dogs and pigeons. J. Clin. Microbiol. 42, 5324–5326.

Garcia-Lara, J., Needham, A.J., and Foster, S.J. (2005). Invertebrates as animal models for *Staphylococcus aureus* pathogenesis: a window into host-pathogen interaction. FEMS Immunol. Med. Microbiol. *43*, 311–323.

Groisman, E.A., and Ochman, H. (1996). Pathogenicity islands: bacterial evolution in quantum leaps. Cell *87*, 791–794.

Gross, T.L., Ihrke, P.J., Walder, E.J., and Affolter, V.K. (2005). Skin diseases of the dog and the cat (Iowa: Blackwell)

Guardabassi, L., Schwarz, S., and Lloyd, D.H. (2004). Pet animals as reservoirs of antimicrobial-resistant bacteria. J. Antimicrob. Chemother. *54*, 321–332.

Hacker, J., Blum-Oehler, G., Muhldorfer, I., and Tschape, H. (1997). Pathogenicity islands of virulent bacteria: structure, function and impact on microbial evolution. Mol. Microbiol. *23*, 1089–1097.

Hacker, J., and Kaper, J.B. (2000). Pathogenicity islands and the evolution of microbes. Ann. Rev. Microbiol. *54*, 641–679.

Hajek, V. (1976). *Staphylococcus intermedius*, a new species isolated from animals. Int. J. Syst. Bacteriol. *26*, 401–408.

Hendricks, A., Schuberth, H.J., Schueler, K., and Lloyd, D.H. (2002). Frequency of superantigen-producing *Staphylococcus intermedius* isolates from canine pyoderma and proliferation-inducing potential of superantigens in dogs. Res. Vet. Sci. *73*, 273–277.

Hermans, K., Devriese, L.A., and Haesebrouck, F. (2003). Rabbit staphylococcosis: difficult solutions for serious problems. Vet Microbiol. *91*, 57–64.

Herron, L.L., Chakravarty, R., Dwan, C., Fitzgerald, J.R., Musser, J.M., Retzel, E., and Kapur, V. (2002). Genome sequence survey identifies unique sequences and key virulence genes with unusual rates of amino acid substitution in bovine *Staphylococcus aureus*. Infect. Immun. *70*, 3978–3981.

Hesselbarth, J., and Schwarz, S. (1995). Comparative ribotyping of *Staphylococcus intermedius* from dogs, pigeons, horses and mink. Vet. Microbiol. *45*, 11–17.

Hill, P.B., and Olivry, T. (2001). The ACVD task force on canine atopic dermatitis (V): biology and role of inflammatory cells in cutaneous allergic reactions. Vet. Immunol. Immunopathol. *81*, 187–198.

Hochhut, B., Dobrindt, U., and Hacker, J. (2005). Pathogenicity islands and their role in bacterial virulence and survival. Contrib. Microbiol. *12*, 234–254.

Holbrook, T.C., Munday, J.S., Brown, C.A., Glover, B., Schlievert, P.M., and Sanchez, S. (2003). Toxic shock syndrome in a horse with *Staphylococcus aureus* pneumonia. J Am. Vet. Med. Assoc. *222*, 620–623, 601–622.

Huber, B., Riedel, K., Kothe, M., Givskov, M., Molin, S., and Eberl, L. (2002). Genetic analysis of functions involved in the late stages of biofilm development in *Burkholderia cepacia* H111. Mol. Microbiol. *46*, 411–426.

Huijsdens, X.W., van Dijke, B.J., Spalburg, E., van Santen-Verheuvel, M.G., Heck, M.E., Pluister, G.N., Voss, A., Wannet, W.J., and de Neeling, A.J. (2006). Community-acquired MRSA and pig-farming. Ann. Clin. Microbiol. Antimicrob. *5*, 26.

Jevons, M.P., Coe, A.W., and Parker, M.T. (1963). Methicillin resistance in staphylococci. Lancet *1*, 904–907.

Ji, G., Pei, W., Zhang, L., Qiu, R., Lin, J., Benito, Y., Lina, G., and Novick, R.P. (2005). *Staphylococcus intermedius* produces a functional *agr* autoinducing peptide containing a cyclic lactone. J. Bacteriol. *187*, 3139–3150.

Jordan, F.T.W., and Pattison, M. (1996). Poultry Diseases, 4 edn (London: Saunders).

Juhasz-Kaszanyitzky, E., Janosi, S., Somogyi, P., Dan, A., van der Graaf-van Bloois, L., van Duijkeren, E., and Wagenaar, J.A. (2007). MRSA transmission between cows and humans. Emerg. Infect. Dis. *13*, 630–632.

Kania, S.A., Williamson, N.L., Frank, L.A., Wilkes, R.P., Jones, R.D., and Bemis, D.A. (2004). Methicillin resistance of staphylococci isolated from the skin of dogs with pyoderma. Am. J. Vet. Res. *65*, 1265–1268.

Kapur, V., Sischo, W.M., Greer, R.S., Whittam, T.S., and Musser, J.M. (1995). Molecular population genetic analysis of *Staphylococcus aureus* recovered from cows. J. Clin. Microbiol. *33*, 376–380.

Khambaty, F.M., Bennett, R.W., and Shah, D.B. (1994). Application of pulsed-field gel electrophoresis to the epidemiological characterization of *Staphylococcus intermedius* implicated in a food-related outbreak. Epidemiol. Infect. *113*, 75–81.

Kibenge, F.S., Rood, J.I., and Wilcox, G.E. (1983). Lysogeny and other characteristics of *Staphylococcus hyicus* isolated from chickens. Vet. Microbiol. *8*, 411–415.

Lasa, I., and Penades, J.R. (2006). Bap: a family of surface proteins involved in biofilm formation. Res. Microbiol *157*, 99–107.

Latasa, C., Roux, A., Toledo-Arana, A., Ghigo, J.M., Gamazo, C., Penades, J.R., and Lasa, I. (2005). BapA, a large secreted protein required for biofilm formation and host colonization of *Salmonella enterica* serovar Enteritidis. Mol. Microbiol. *58*, 1322–1339.

Lee, J.H. (2003). Methicillin (Oxacillin)-resistant *Staphylococcus aureus* strains isolated from major food animals and their potential transmission to humans. Appl. Environ. Microbiol. *69*, 6489–6494.

Lindsay, J.A., and Holden, M.T. (2004). *Staphylococcus aureus*: superbug, super genome? Trends Microbiol. *12*, 378–385.

Lindsay, J.A., Ruzin, A., Ross, H.F., Kurepina, N., and Novick, R.P. (1998). The gene for toxic shock toxin is carried by a family of mobile pathogenicity islands in *Staphylococcus aureus*. Mol. Microbiol. *29*, 527–543.

Loeffler, A., Boag, A. K., Sung, J., Lindsay, J.A., Guardabassi, L., Dalsgaard, A., Smith, H., Stevens, K.B., and Lloyd, D.H. (2005). Prevalence of methicillin-resistant *Staphylococcus aureus* among staff and pets in a small animal referral hospital in the UK. J. Antimicrob. Chemother. *56*, 692–697.

Maiques, E., Ubeda, C., Tormo, M. A., Ferrer, M.D., Lasa, I., Novick, R.P., and Penades, J.R. (2007). Role of staphylococcal phage and SaPI integrase in intra- and interspecies SaPI transfer. J. Bacteriol. 189, 5608–5616.

Matsuhashi, M., Song, M.D., Ishino, F., Wachi, M., Doi, M., Inoue, M., Ubukata, K., Yamashita, N., and Konno,M. (1986). Molecular cloning of the gene of a penicillin-binding protein supposed to cause high resistance to beta-lactam antibiotics in *Staphylococcus aureus*. J. Bacteriol *167*, 975–980.

McEwan, N.A. (2000). Adherence by *Staphylococcus intermedius* to canine keratinocytes in atopic dermatitis. Res. Vet. Sci. *68*, 279–283.

McEwan, N.A., Mellor, D., and Kalna, G. (2006). Adherence by *Staphylococcus intermedius* to canine corneocytes: a preliminary study comparing noninflamed and inflamed atopic canine skin. Vet. Dermatol. *17*, 151–154.

McNamee, P.T., McCullagh, J.J., Rodgers, J.D., Thorp, B.H., Ball, H.J., Connor, T.J., McConaghy, D., and Smyth, J.A. (1999). Development of an experimental model of bacterial chondronecrosis with osteomyelitis in broilers following exposure to *Staphylococcus aureus* by aerosol, and inoculation with chicken anaemia and infectious bursal disease viruses. Avian Pathol. *28*, 26–35.

Miles, H., Lesser, W., and Sears, P. (1992). The economic implications of bioengineered mastitis control. J. Dairy Sci. *75*, 596–605.

Moon, J.S., Lee, A.R., Kang, H.M., Lee, E.S., Kim, M.N., Paik, Y.H., Park, Y.H., Joo, Y.S., and Koo, H.C. (2007). Phenotypic and genetic antibiogram of methicillin-resistant staphylococci isolated from bovine mastitis in Korea. J Dairy Sci *90*, 1176–1185.

Morandi, S., Brasca, M., Lodi, R., Cremonesi, P., and Castiglioni, B. (2007). Detection of classical enterotoxins and identification of enterotoxin genes in *Staphylococcus aureus* from milk and dairy products. Vet. Microbiol. 124:66–72.

Morris, D.O., Rook, K. A., Shofer, F.S., and Rankin, S.C. (2006). Screening of *Staphylococcus aureus, Staphylococcus intermedius,* and Staphylococcus schleiferi isolates obtained from small companion animals for antimicrobial resistance: a retrospective review of 749 isolates (2003–04). Vet. Dermatol. *17*, 332–337.

Novick, R.P., and Subedi, A. (2007). The SaPIs: mobile pathogenicity islands of *Staphylococcus.* Chem Immunol Allergy *93*, 42–57.

O'Riordan, K., and Lee, J.C. (2004). *Staphylococcus aureus* capsular polysaccharides. Clin Microbiol Rev *17*, 218–234.

Orwin, P.M., Fitzgerald, J.R., Leung, D.Y., Gutierrez, J.A., Bohach, G.A., and Schlievert, P.M. (2003). Characterization of *Staphylococcus aureus* enterotoxin L. Infect. Immun. *71*, 2916–2919.

Pinchbeck, L.R., Cole, L.K., Hillier, A., Kowalski, J.J., Rajala-Schultz, P.J., Bannerman, T.L., and York, S. (2006). Genotypic relatedness of staphylococcal strains isolated from pustules and carriage sites in dogs with superficial bacterial folliculitis. Am. J. Vet. Res. *67*, 1337–1346.

Pottumarthy, S., Schapiro, J.M., Prentice, J.L., Houze, Y.B., Swanzy, S.R., Fang, F.C., and Cookson, B.T. (2004). Clinical isolates of *Staphylococcus intermedius* masquerading as methicillin-resistant *Staphylococcus aureus*. J. Clin. Microbiol. *42*, 5881–5884.

Poutrel, B. (1984). *Staphylococcus sciuri* subsp lentus associated with goat mastitis. Am. J. Vet. Res. *45*, 2084–2085.

Prevost, G., Bouakham, T., Piemont, Y., and Monteil, H. (1995). Characterisation of a synergohymenotropic toxin produced by *Staphylococcus intermedius*. FEBS Lett. *376*, 135–140.

Rahman, M.T., Kobayashi, N., Alam, M.M., and Ishino, M. (2005). Genetic analysis of *mecA* homologues in *Staphylococcus sciuri* strains derived from mastitis in dairy cattle. Microb. Drug Resist. *11*, 205–214.

Rodgers, J.D., McCullagh, J.J., McNamee, P.T., Smyth, J.A., and Ball, H.J. (1999). Comparison of *Staphylococcus aureus* recovered from personnel in a poultry hatchery and in broiler parent farms with those isolated from skeletal disease in broilers. Vet. Microbiol. *69*, 189–198.

Rodgers, J.D., McCullagh, J.J., McNamee, P.T., Smyth, J.A., and Ball, H.J. (2006). The anti-nuclease humoral immune response of broiler chickens exposed to *Staphylococcus aureus*, infectious bursal disease virus and chicken anaemia virus in an experimental model for bacterial chondronecrosis and osteomyelitis. Avian Pathol. *35*, 302–308.

Rupp, M.E., and Archer, G.L. (1994). Coagulase-negative staphylococci: pathogens associated with medical progress. Clin. Infect. Dis. *19*, 231–243; quiz 244–235.

Ruzin, A., Lindsay, J., and Novick, R.P. (2001). Molecular genetics of SaPI1 – a mobile pathogenicity island in *Staphylococcus aureus*. Mol. Microbiol. *41*, 365–377.

Sasaki, T., Kikuchi, K., Tanaka, Y., Takahashi, N., Kamata, S., and Hiramatsu, K. (2007). Reclassification of phenotypically-identified *Staphylococcus intermedius* strains. J. Clin. Microbiol.

Sato, H., Matsumori, Y., Tanabe, T., Saito, H., Shimizu, A., and Kawano, J. (1994). A new type of staphylococcal exfoliative toxin from a *Staphylococcus aureus* strain isolated from a horse with phlegmon. Infect. Immun. *62*, 3780–3785.

Sato, H., Watanabe, T., Higuchi, K., Teruya, K., Ohtake, A., Murata, Y., Saito, H., Aizawa, C., Danbara, H., and Maehara, N. (2000). Chromosomal and extrachromosomal synthesis of exfoliative toxin from *Staphylococcus hyicus*. J. Bacteriol. *182*, 4096–4100.

Sato, H., Watanabe, T., Murata, Y., Ohtake, A., Nakamura, M., Aizawa, C., Saito, H., and Maehara, N. (1999). New exfoliative toxin produced by a plasmid-carrying strain of *Staphylococcus hyicus*. Infect. Immun. *67*, 4014–4018.

Scott, G.M., Thomson, R., Malone-Lee, J., and Ridgway, G.L. (1988). Cross-infection between animals and man: possible feline transmission of *Staphylococcus aureus* infection in humans? J. Hosp. Infect. *12*, 29–34.

Segura, P., Martinez, J., Peris, B., Selva, L., Viana, D., Penades, J.R., and Corpa, J. M. (2007). Staphylococcal infections in rabbit does on two industrial farms. Vet. Rec. *160*, 869–872.

Shankar, V., Baghdayan, A.S., Huycke, M.M., Lindahl, G., and Gilmore, M.S. (1999). Infection-derived *Enterococcus faecalis* strains are enriched in *esp*, a gene encoding a novel surface protein. Infect. Immun. *67*, 193–200.

Shimizu, A., Kawano, J., Yamamoto, C., Kakutani, O., Anzai, T., and Kamada, M. (1997). Genetic analysis of equine methicillin-resistant *Staphylococcus aureus* by pulsed-field gel electrophoresis. J. Vet. Med. Sci. *59*, 935–937.

Shittu, A., Lin, J., Morrison, D., and Kolawole, D. (2004). Isolation and molecular characterization of multiresistant *Staphylococcus sciuri* and *Staphylococcus haemolyticus* associated with skin and soft-tissue infections. J. Med. Microbiol. *53*, 51–55.

Simoons-Smit, A.M., Savelkoul, P.H., Stoof, J., Starink, T.M., and Vandenbroucke-Grauls, C.M. (2000). Transmission of *Staphylococcus aureus* between humans and domestic animals in a household. Eur. J. Clin. Microbiol. Infect. Dis. *19*, 150–152.

Simou, C., Hill, P.B., Forsythe, P.J., and Thoday, K.L. (2005a). Species specificity in the adherence of staphylococci to canine and human corneocytes: a preliminary study. Vet. Dermatol. *16*, 156–161.

Simou, C., Thoday, K.L., Forsythe, P.J., and Hill, P.B. (2005b). Adherence of *Staphylococcus intermedius* to corneocytes of healthy and atopic dogs: effect of pyoderma, pruritus score, treatment and gender. Vet. Dermatol. *16*, 385–391.

Smith, E.M., Green, L.E., Medley, G.F., Bird, H.E., Fox, L.K., Schukken, Y.H., Kruze, J.V., Bradley, A.J., Zadoks, R.N., and Dowson, C.G. (2005). Multilocus sequence typing of intercontinental bovine *Staphylococcus aureus* isolates. J. Clin. Microbiol. *43*, 4737–4743.

Smyth, D.S., Hartigan, P.J., Meaney, W.J., Fitzgerald, J.R., Deobald, C.F., Bohach, G.A., and Smyth, C.J. (2005). Superantigen genes encoded by the egc cluster and SaPIbov are predominant among *Staphylococcus aureus* isolates from cows, goats, sheep, rabbits and poultry. J. Med. Microbiol. *54*, 401–411.

Stepanovic, S., Jezek, P., Vukovic, D., Dakic, I., and Petras, P. (2003). Isolation of members of the *Staphylococcus sciuri* group from urine and their relationship to urinary tract infections. J. Clin. Microbiol. *41*, 5262–5264.

Sung, J.M., Chantler, P.D., and Lloyd, D.H. (2006). Accessory gene regulator locus of *Staphylococcus intermedius*. Infect. Immun. *74*, 2947–2956.

Sung, J.M., and Lindsay, J.A. (2007). *Staphylococcus aureus* strains that are hypersusceptible to resistance gene transfer from enterococci. Antimicrob. Agents Chemother. *51*, 2189–2191.

Talan, D A., Abrahamian, F.M., Moran, G.J., Citron, D.M., Tan, J.O., and Goldstein, E. J. (2003). Clinical presentation and bacteriologic analysis of infected human bites in patients presenting to emergency departments. Clin. Infect. Dis. *37*, 1481–1489.

Tanner, M. A., Everett, C.L., and Youvan, D.C. (2000). Molecular phylogenetic evidence for noninvasive zoonotic transmission of *Staphylococcus intermedius* from a canine pet to a human. J. Clin. Microbiol. *38*, 1628–1631.

Terauchi, R., Sato, H., Endo, Y., Aizawa, C., and Maehara, N. (2003a). Cloning of the gene coding for *Staphylococcus intermedius* exfoliative toxin and its expression in Escherichia coli. Vet. Microbiol. *94*, 31–38.

Terauchi, R., Sato, H., Hasegawa, T., Yamaguchi, T., Aizawa, C., and Maehara, N. (2003b). Isolation of exfoliative toxin from *Staphylococcus intermedius* and its local toxicity in dogs. Vet. Microbiol. *94*, 19–29.

Thomas, J.C., Vargas, M.R., Miragaia,M., Peacock, S.J., Archer, G.L., and Enright, M.C. (2007). Improved multilocus sequence typing scheme for *Staphylococcus epidermidis*. J. Clin. Microbiol. *45*, 616–619.

Timms, L.L., and Schultz, L.H. (1987). Dynamics and significance of coagulase-negative staphylococcal intramammary infections. J. Dairy Sci. *70*, 2648–2657.

Toledo-Arana, A., Valle, J., Solano, C., Arrizubieta, M.J., Cucarella, C., Lamata, M., Amorena, B., Leiva, J., Penades, J.R., and Lasa, I. (2001). The enterococcal surface protein, Esp, is involved in *Enterococcus faecalis* biofilm formation. Appl. Environ. Microbiol. *67*, 4538–4545.

Tormo, M.A., Knecht, E., Gotz, F., Lasa, I., and Penades, J.R. (2005). Bap-dependent biofilm formation by pathogenic species of *Staphylococcus*: evidence of horizontal gene transfer? Microbiology *151*, 2465–2475.

Trotonda, M.P., Manna, A.C., Cheung, A.L., Lasa, I., and Penades, J.R. (2005). SarA positively controls Bap-dependent biofilm formation in *Staphylococcus aureus*. J. Bacteriol. *187*, 5790–5798.

Ubeda, C., Maiques, E., Knecht, E., Lasa, I., Novick, R.P., and Penades, J.R. (2005). Antibiotic-induced SOS response promotes horizontal dissemination of pathogenicity island-encoded virulence factors in staphylococci. Mol. Microbiol. *56*, 836–844.

Ubeda, C., Tormo, M. A., Cucarella, C., Trotonda, P., Foster, T.J., Lasa, I., and Penades, J.R. (2003). Sip, an integrase protein with excision, circularization and integration activities, defines a new family of mobile *Staphylococcus aureus* pathogenicity islands. Mol. Microbiol. *49*, 193–210.

van Leeuwen, W.B., Melles, D.C., Alaidan, A., Al-Ahdal, M., Boelens, H.A., Snijders, S.V., Wertheim, H., van Duijkeren, E., Peeters, J.K., van der Spek, P.J., *et al.* (2005). Host- and tissue-specific pathogenic traits of *Staphylococcus aureus*. J. Bacteriol. *187*, 4584–4591.

Vancraeynest, D., Haesebrouck, F., Deplano, A., Denis, O., Godard, C., Wildemauwe, C., and Hermans, K. (2006). International dissemination of a high virulence rabbit *Staphylococcus aureus* clone. J. Vet. Med. B. Infect. Dis. Vet. Public Health 53, 418–422.

Waller, A. (2005). The creation of a new monster: MRSA and MRSI – important emerging veterinary and zoonotic diseases. Vet. J. *169*, 315–316.

Webster, K.A., and Mitchell, G.B. (1989). Experimental production of tick pyaemia. Vet. Parasitol. *34*, 129–133.

Weese, J.S., Archambault, M., Willey, B.M., Hearn, P., Kreiswirth, B.N., Said-Salim, B., McGeer, A., Likhoshvay, Y., Prescott, J.F., and Low, D.E. (2005). Methicillin-resistant *Staphylococcus aureus* in horses and horse personnel, 2000–2002. Emerg. Infect. Dis. *11*, 430–435.

Wills, Q.F., Kerrigan, C., and Soothill, J.S. (2005). Experimental bacteriophage protection against *Staphylococcus aureus* abscesses in a rabbit model. Antimicrob. Agents Chemother. *49*, 1220–1221.

Younis, A., Krifucks, O., Fleminger, G., Heller, E.D., Gollop, N., Saran, A., and Leitner, G. (2005). *Staphylococcus aureus* leucocidin, a virulence factor in bovine mastitis. J. Dairy Res. *72*, 188–194.

Index

A

B

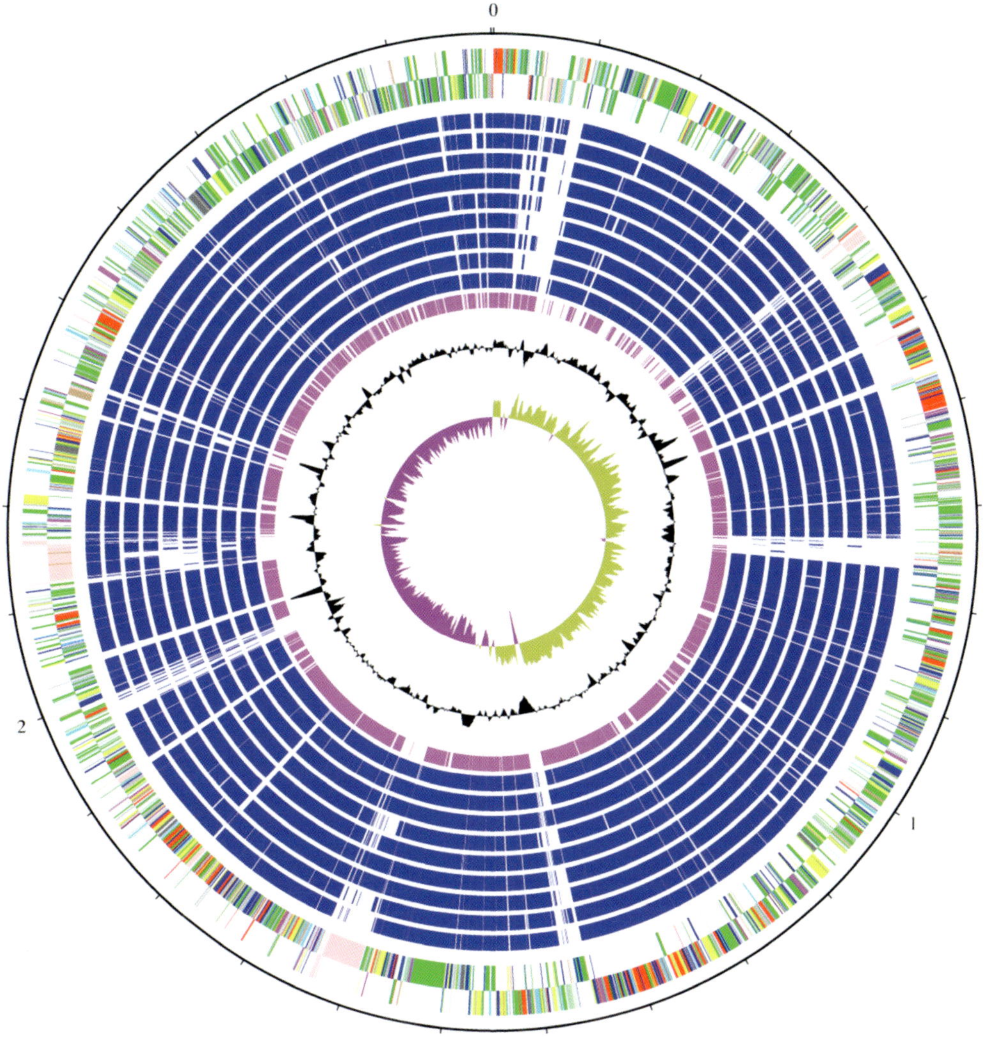

Figure 1.2 Schematic circular diagram of the *S. aureus* MRSA252 genome. Key for the circular diagram: scale (in Mb); annotated CDSs coloured according to predicted function represented on a pair of concentric circles, representing both coding strands; orthologue matches shared with other *S. aureus* strains, N315, Mu50, MW2, MSSA476, COL, RF122, USA300_FPR3757, NCTC8325 and JH9, blue; *Staphylococcus epidermidis* RP6a, purple; G + C% content plot; GC skew plot (>0% olive, <0% purple). Colour coding for CDS functions: dark blue; pathogenicity/adaptation, black; energy metabolism, red; information transfer, dark green; surface associated, cyan; degradation of large molecules, magenta; degradation of small molecules, yellow; central/intermediary metabolism, pale green; unknown, pale blue; regulators, orange; conserved hypothetical, brown; pseudogenes, pink; phage and IS elements, grey; miscellaneous.